CHEMISTRY OF PESTICIDES

CHEMISTRY OF PESTICIDES

By
N. N. MELNIKOV
University of Moscow

Edited by
FRANCES A. GUNTHER
and
JANE DAVIES GUNTHER
Riverside, California

Translated from the Russian by
RUTH L. BUSBEY

SPRINGER-VERLAG
NEW YORK · HEIDELBERG · BERLIN
1971

SB
951
M4

CHEMISTRY OF PESTICIDES by N. N. Melnikov is being published simultaneously as RESIDUE REVIEWS Volume 36 by Springer-Verlag New York Inc.

All rights reserved. No part of this book may be translated or reproduced in any form without written permission from Springer-Verlag.

© 1971 by Springer-Verlag New York Inc.
Library of Congress Catalog Card Number 62-18595.
Printed in the United States of America.

ISBN 0-387-90031-4 Springer-Verlag New York Heidelberg Berlin

Preface

This book was originally published as a Russian edition in mid-1968. It represented the first single-volume discussion of chemical pest control and of the detailed chemistry of all modern pesticide chemicals since D. E. H. Frear's classical book on this same subject, first published in 1942 (van Nostrand) with the latest (third) edition in 1955. Since 1955 many new pesticide chemicals have achieved commercial status, and many of the older ones have been supplanted. There is no up-to-date equivalent of this present volume in the world literature, with the exception of the encyclopedic and largely biologically oriented two-volume work "Chemie der Pflanzenschutz- und Schädlingsbekämpfungsmittel" (R. Wegler, ed.) published by Springer-Verlag in 1970. Professor Melnikov has updated the 1968 Russian edition, with emphasis on the primary Russian sources yet with excellent world-wide coverage of the latest chemicals to approach field stature in modern chemical pest control, for the present translation.

This edition of "Chemistry of Pesticides" therefore represents a thoroughly modern but concise summary of the basic principles of practical pest control by chemical means. The text is organized into sections on chemical types of compounds, rather than on categorization according to biological effectiveness, with emphasis on physical and chemical properties, including syntheses of both parent compounds and important intermediates, but recording toxicological properties as well. Professor Melnikov discusses all modern pest-control chemicals now in use around the world, including those developed or under development in the vast agricultural economy of the U.S.S.R. and associated countries.

The editors of "Residue Reviews" (Springer-Verlag) felt this volume should be made available to the western world in an English language edition. Therefore, we recommended its translation and publication not only as Volume 36 of "Residue Reviews", serving as a reference book for the specialist in pest control, but also as the present textbook edition for the beginner who needs an authoritative, comprehensive survey of those immensely diverse pesticide chemicals essential for adequate production of food and fiber and for continued protection of the public health.

In a few instances it was not possible to determine the exact sources involved in Professor Melnikov's general reference citations of the Russian literature in his original Russian language manuscript, even after lengthy correspondence.

Respectful and special appreciation is hereby accorded Mrs. Ruth L. Busbey, who as both chemist and expert linguist provided the present translation and first corrected the galley proofs. Any remaining errors in text and any deficiencies in the index are our responsibility. Because of intended use of this book both as a reference book and as a textbook, the index was prepared to be exhaustive; we shall welcome notification of any errors or of omissions.

Department of Entomology
University of California
Riverside, California
January 26, 1971

Francis A. Gunther
Jane Davies Gunther

Foreword

Chemical agents for the protection of plants from pests, diseases, and weeds and for the protection of animals from ectoparasites have attained great importance in agriculture and related fields. Their scale of use is continually increasing and the volume of scientific research is expanding in this area, which is important to the national economy. This is fully understandable because the use of various chemical preparations in agriculture makes it possible to save not less than a third of the crop.

Specialists in many different fields of science are engaged in the investigation of pesticides: chemists (synthetic, analytical, and physical chemists and chemical engineers), biochemists, entomologists, plant physiologists, phytopathologists, toxicologists, agronomists, physicians, specialists in agricultural machinery, and many others. Because the demand for contemporary pesticides is constantly increasing, every year new compounds appear for use in industry and agriculture. Their number already exceeds 900 and the number of formulations in which they are used in different countries is estimated at more than 100,000. Correlation of the vast amount of information that has accumulated is of great importance to specialists in many different fields. A systematic presentation of this information is especially important for students and graduate chemists who are specializing in the field of biologically active compounds.

The author has attempted to correlate and to characterize as briefly as possible the main classes of compounds and the individual substances that are used or are being investigated as pesticides. This treatise presents practically all of the principal materials that have received rather wide use in agriculture or industry as pesticides. Also included in this book are some promising new compounds that are presently being investigated under workink conditions.

This volume is arranged by classes of chemical compounds, and is the first experiment in describing the pesticidal properties of the compounds rather than their applied characteristics, as they have been described in almost all the well-known textbooks previously. The author hopes that this arrangement will promote a deeper study of the problem and arouse interest among chemists to set up new investigations on different classes of compounds that have received little study in this respect. Compounds whose chemical properties have been described in textbooks or monographs are considered only from the point of view of their use as pesticides. In each chapter the general characteristics of a given class of compounds are given and information is presented on the relationship of biological activity to structure.

A list of general references is provided the end of each chapter, which includes the principal monographs and review articles, mainly in the Russian language.

Naturally the construction of this book has resulted in some lack of uniformity in the description of the individual classes of compounds, but a detailed analysis of the properties of all the classes of compounds would result in an unjustifiable increase in the size of the book as a result of including material that is not yet of practical significance.

There may be individual deficiencies in the book; therefore, all indications of omissions and errors will be received by the author with appreciation.

University of Moscow
June 15, 1970

N. N. Melnikov

Table of Contents

Preface	v
Foreword	vii
I. Introduction	1
II. Pesticide formulations	12
III. Hydrocarbons	26
IV. Halogen derivatives of aliphatic hydrocarbons	34
V. Halogen derivatives of alicyclic hydrocarbons	42
VI. Halogen derivatives of aromatic hydrocarbons	67
VII. Nitro compounds	78
VIII. Amines and salts of quaternary ammonium bases	83
IX. Alcohols, phenols, and ethers	89
X. Aldehydes, ketones, and quinones	111
XI. Aliphatic carboxylic acids and their derivatives	118
XII. Alicyclic carboxylic acids and their derivatives	130
XIII. Aromatic carboxylic acids and their derivatives	141
XIV. Aryloxalkylcarboxylic acids and their derivatives	157
XV. Derivatives of carbonic acid	177
XVI. Derivatives of carbamic acid	183
XVII. Derivatives of thio- and dithiocarbamic acids	206
XVIII. Derivatives of urea and thiourea	225
XIX. Mercaptans, sulfides, and their derivatives	240
XX. Thiocyanates and isothiocyanates	252
XXI. Derivatives of sulfuric and sulfurous acids	262
XXII. Sulfonic acids and their derivatives	269
XXIII. Derivatives of hydrazine and azo compounds	278
XXIV. Organic mercury compounds	283
XXV. Organotin compounds	297
XXVI. Organophosphorus compounds	303
XXVII. Arsenic compounds	387
XXVIII. Heterocyclic compounds with one heteroatom in the ring	396
XXIX. Heterocyclic compounds with two heteroatoms in the ring	413
XXX. Heterocyclic compounds with three heteroatoms in the ring	427
XXXI. Inorganic pesticides	441
Subject Index	449

CHEMISTRY OF PESTICIDES

I. Introduction

Even in ancient Greece, about 2,000 years ago more than 500 different species of animals were known, and by the middle of the 19th century their number had reached several hundred thousand. At the present time the number of known species of animals and microorganisms exceeds two million. According to the calculations of Soviet entomologist N. N. Bogdanov-Kot'kov more than 68,000 different insects can inflict damage on man, domestic animals, plants, and a variety of materials. The number of different injurious microorganisms and plants is no less. Animals and microorganisms cause especially great damage to agriculture. It has been accurately determined that in agriculture not less than a third of the crop is lost because of pests and diseases, and if they were not systematically controlled, the loss would be even greater. Thus, according to the calculations of specialists, if the diseases and pests of agricultural crops are not systematically controlled, then at best not more than 37% of a normal harvest of potatoes can be gathered, not more than 22% of cabbage, 10% of appless, 9% of peaches, etc.

In world practice there are many examples of the mass destruction of different crops by diseases and plant pests. In 1845–1851 in Ireland as a result of a massive infection of potatoes by a phytophthora fungus, a large part of the harvest of this crop was lost, resulting in a famine and the death of about a million people. The misfortune was further multiplied by the superstitious population's refusal to use the phytophthora-infected potatoes for food. In 1930 in the United States, up to 30% of the wheat crop perished from stem rust, and in 1954 in Western Canada about three million tons of wheat were destroyed by stem rust.

Among the plant pests and causative agents of plant disease are representatives of various living organisms, including insects, mites, mollusks, nematodes (round worms), fungi, bacteria, viruses, parasitic plants, and the like. A large number of phytopathogenic microorganisms and insects infect potatoes, corn (maize), cereals, beans, fruit and vegetable crops, cotton, and flax.

Rodents cause tremendous damage to agriculture. One gray rat with its progeny in the course of a year destroys food products worth at least 440 dollars.

Substantial damage is caused by insect pests and microorganisms that destroy various goods such as wood, paper, products made of cotton, wool and silk, raw and cured rubber, plastics, leather, etc. The destruction of nonmetallic materials is especially rapid under the conditions of a humid, warm climate. Thus, wood and textiles of cotton and wool under tropical conditions are destroyed in one to two months. In the tropics, even optical glass may undergo deterioration.

Insect pests, mites, and other parasites also represent a great danger to animal husbandry, since many of them not only are vectors of infectious diseases but also damage animal hides and cause animals uneasiness, resulting in a diminished yield of milk and a smaller gain in weight by the animals. The loss from a parasite such as the warble fly in the USSR is estimated at about 550 million dollars a year.

The danger to human health caused by various insect pests and mites that are vectors of infectious diseases is difficult to evaluate, especially if we remember that comparatively recently such a disease as malaria infected more than 100 million persons throughout the world. In some regions of our planet this disease still takes millions of lives yearly. There is serious danger to man also from such diseases as encephalitis, typhus, relapsing fever, sleeping sickness, elephantiasis, and many others, the vectors of which are insects and arachnids. The danger is aggravated by the rapidity with which insects reproduce. Thus, a house fly at seven generations a season can produce progeny amounting to 3.5×10^{12} individuals, and a woolly apple aphid in twenty generations in the course of a season produces 35^{20} individuals. Only the high death rate of these insects from various causes prevents their accumulation in full measure.

Table I. *Consumption of fertilizers by different weeds*

Name of weed	Total (calc. as techn. fertilizer) (kg./ha.)	Consumption of nutritive substances (kg./ha.)		
		Nitrogen	P_2O_5	K_2O
Common thistle (field sow thistle)	739	138.1	31.0	116.9
Common cornflower	460	65.4	24.0	98.2
Polygonium	589	84.8	47.2	70.4
Willow herb	479	72.4	25.8	91.9
Coltsfoot	724	74.0	27.2	234.8
Thistle	600	67.0	28.7	159.7
Quackgrass	406	48.6	31.5	68.5
Common cyperus	1392	210.8	67.4	270.4

No less great are the agricultural losses from weeds, which deprive crop plants of the moisture and nutritive matter in the soil, shade the crop plants and hinder their normal growth, and contaminate harvested grain with seeds that are poisonous for man and animals. In some cases complete loss of the crop occurs as a result of the development of weeds. Thus, 1 kg. of weeds (calculated as dry matter) in the course of a season requires from 250 to 1,000 liters of water. The loss of nutritive substances from the soil with an average weed infestation (130–150 plants/m.²) of plantings is shown in Table I. World losses from pests, diseases, and weeds are estimated as 75 billion dollars a year. From what has been presented, it can be seen that the control of various harmful organisms has great importance for agriculture, industry, and public health.

One of the most convenient means for controlling harmful organisms is the chemical method using various compounds. Chemicals used to destroy any species of pests are called pesticides. Depending on the purpose for which they are used, pesticides are divided into the following basic groups:

Acaricides — for the control of mites or ticks.

Algicides — for the destruction of algae and other aquatic vegetation.

Antiseptics — for the protection of nonmetallic materials from damage by microorganisms.

Arboricides — for the destruction of undesirable arboreal and bushy vegetation.

Bactericides — for the control of bacteria and bacterial diseases of plants.

Fungicides — for the control of plant diseases and various fungi.

Herbicides — for the control of weeds.

Insecticides — for the control of harmful insects. Individual groups of insecticides also have more specific names such as *aphicides,* preparations for the control of aphids.

Limacides or molluskicides — for the control of various mollusks, including gastropods.

Nematocides — for the control of round worms (nematodes).

Zoocides — for the control of rodents. Materials of this group are often called rodenticides.

Pesticides include chemical compounds that stimulate or retard the growth of plants; they also include those that remove leaves *(defoliants)* or desiccate plants *(desiccants)* and are used for the purpose of mechanizing the work in harvesting cotton, soy beans, potatoes, and many other crops. The term pesticides also applies to compounds for repelling *(repellents),* attracting *(attractants),* and sterilizing (sexual *sterilants)* insects, although these last groups of compounds do not entirely fit the name and purpose of pesticides. Of the groups that have been listed, the compounds most widely used in agriculture and other areas are the herbicides, insecticides, and fungicides.

Insecticides are divided according to the nature of their penetration into the insect organism into the following principal subgroups: *Contact* insecticides, which kill insects by means of contact with any part of the body; *stomach* insecticides, which penetrate into the insect through the organs of its alimentary system and kill it as a result of the poison entering the gut; *systemic* insecticides, which are capable of moving through the vascular system of plants and poisoning insects that feed on the plants; and *fumigants,* which penetrate into the insect through its respiratory organs. It should be noted that the majority of compounds used can penetrate into the insect organism simultaneously by different routes. In this connection certain compounds are assigned to one subgroup or another on the basis of their main route of penetration into the insect. Thus, lindane shows contact, stomach, and fumigant action; however, it usually is assigned to the contact insecticides. Some insecticides act not as a result of penetration of a poison into the insect organism but as a result of purely physical causes — obstruction of the respiratory passages causing the insect to die of asphyxiation. Such compounds include mineral oils, silica gel (in finely dispersed condition), and some others.

Fungicides usually are divided into two main subgroups: *the fungicides used for control of diseases of growing plants* and *the seed disinfectants* that are used for the preplanting treatment of seeds to protect the seedlings from various diseases. The fungicides for green plants are divided in turn into *preparations with a prophylactic effect* that are used to protect plants from various infections, and *preparations with an eradicative effect* (curatives) that are used to cure the plants. As with the insecticides, there are among the fungicides compounds with contact and systemic action (the latter moving through the vascular system of the plants), but the latter group of substances is very small; the study and use of systemic fungicides is in the initial stage. The fungicides often are considered to include disinfectants for nonmetallic materials and bactericides which are not divided into separate groups. In many cases the same substances that are used for plant protection are employed as disinfectants and bactericides. It should be noted that some metabolic processes in the cells of bacteria and fungi have essential differences, so that many fungicides show weak bactericidal activity.

Herbicides are divided with respect to their action on plants into two principal subgroups: *nonselective* herbicides, acting on all species of plants, and *selective* herbicides, dangerous only for some species of plants and safe for others. This division is conditional, since the majority of the compounds, depending on the concentrations used and the rate of application per unit area treated, may be either nonselective or selective herbicides. With respect to the external indications of their effect on plants and the methods of use, all herbicides are divided into three subgroups:

1. *Contact herbicides.*
2. *Systemic herbicides.*
3. *Herbicides acting on the root system of plants or on germinating seeds.*

Contact herbicides include compounds injurious to plant foliage and stems that come in direct contact with them. When such compounds contact the foliage, the normal life processes of the plant are disturbed and it dies. It is necessary, however, to note that such herbicides injure only those parts of the plant with which they come in contact, and in some cases sprouting of new shoots and further development of the injured plant are observed. This is associated with the fact that this group of compounds is not able to move through the plant. The subgroup of systemic herbicides includes compounds that are capable of moving through the vascular system of plants. When such herbicides come in contact with the foliage and roots of a plant, they are quickly distributed through the whole plant causing its death. The use of systemic herbicides is especially valuable in controlling weeds with strong root systems and perennial weeds. The third subgroup consists of herbicides that are introduced into the soil to destroy seeds, germinating seeds, and the roots of weeds.

Defoliants and desiccants resemble the contact herbicides in the nature of their action; in many cases contact herbicides are used as defoliants and desiccants if they are safe for the seeds of the crop that is treated.

Both arboricides and algicides in most cases are assigned to the herbicides.

Plant growth regulators also find some use both to stimulate growth and to increase the yield of various crops, and to retard the sprouting of different root and tuber crops during long storage.

Economic efficiency of pesticide use. As already indicated, losses to the national economy from various pests are very large and the use of pesticides can sharply decrease the damage that is suffered. Expenditures for pesticides in agriculture (including expenditures for their application) pay for themselves several times over in the course of a single season. Thus, for example, by spending 8–10 cents/ha. for seed disinfectants for wheat the yield is increased by 150–250 kg./ha. or by 12–20 dollars.

The efficiency of pesticide use on different crops can be seen especially well by comparing the figures in Table II.

Table II. *Increase in yield of various agricultural crops as a result of control of plant pests in the USA in 1953 (information from a random survey of farmers and entomologists)*

Crop	Crop losses (%)		No. of times increase in yield as result of chemical treatment	Area treated with insecticides (%)
	Without chemical treatment	With chemical treatment		
Beans	66	13	2.6	60
Grapes	77	11	4.8	84
Cherries	78	8.5	4	87
Strawberries	34	13.5	1.3	41
Grain	13.5	12	1.1	9
Cabbage	82	12	4.9	92
Potatoes	72	15	3	90
Onions	64	9	2.5	69
Tomatoes	32	15	1.7	56
Apples and pears	93	15	12	95

The high economic efficiency achieved with the use of pesticides in agriculture and other branches of the economy has favored the rapid development of this branch of the chemical industry. Therefore, not only has there been an increase in the production of pesticide preparations, but also a continuous change and improvement in the assortment of pesticides available.

The production of new organic pesticides in the United States (in thousands of tons) amounts to:

1947	1952	1954	1955	1956	1957	1958	1960	1961	1963	1965
56.4	139.6	190.4	225.5	258.8	232.2	263.3	283.5	317.1	350.4	397.0

These data relate only to new organic compounds, the use of which yields the greatest economic effect. Furthermore, from 70–80,000 tons of sulfur preparations and 13–17,000 tons of copper sulfate and some other inorganic compounds are used in the United States every year.

The world assortment of pesticide formulations now is estimated at more than 100,000 items based on more than 900 chemical compounds belonging to the most varied classes of organic and inorganic compounds. In the Soviet Union the production of chemicals for plant protection and weed control has increased in the last five years by almost three times and the assortment of active constituents of formulations has increased by 22 items. In the plans for the development of the national economy, provision is made for the further increase in production of pesticides and the broadening of their assortment.

General requirements for pesticides. A large number of different chemical compounds are known that have high physiological activity toward many harmful organisms. For successful utilization of pesticides in agriculture, industry, and public health it is necessary that, in addition to their high biocidal activity for various pests, the preparations be sufficiently safe in both production and use for man, domestic animals, useful plants, and beneficial insects and microorganisms. Plants that have been treated with any pesticide must, after specified periods, contain only such residual amounts of the compound that complete safety is assured in their use as food not only for animals, but also for man.

This problem is of such great importance that now in many countries of the world standards have been set for the maximum content of pesticides in different food products used by man and domestic animals. In the Soviet Union special laboratories have been established which control the quality of agricultural products and their pesticide content and also the possible accumulation of pesticides in the environment (soil, water, etc.). In some countries the pesticide residues in food products are not regulated, but time intervals for the treatment of plants before harvest and standards for the dosages of the chemicals applied to plants have been legally established. The times for treating plants are so calculated that at harvest the preparation used should have completely or almost completely decomposed, and the food products should not contain amounts of pesticide hazardous to human health.

In speaking of the danger to man and animals of using pesticides, it also is necessary to note as an indispensable condition the absence of blastogenic effect.

There are different methods for evaluating the toxicity of chemicals for mammals. In the German Democratic Republic and the German Federal Republic, for example, all compounds are divided with respect to acute toxicity for animals into three groups, in which the LD_{50} (mean dose in mg./kg. live weight that causes 50% mortality of the experimental animals) is related not to the active ingredient but to the finished preparation:

1. Preparations whose median lethal dose (LD_{50}) is less than 100 mg./kg., live weight of animal.
2. Preparations whose LD_{50} is within the limits 100–300 mg./kg.
3. Preparations whose LD_{50} is higher than 300 mg./kg.

To the first group of pesticides belong the powerfully-acting substances that are issued to users only by special permit; to the second group belong the substances of medium toxicity; the substances included in the third group are of low toxicity.

In the Soviet Union it is assumed that compounds with LD_{50} less than 50 mg./kg. are powerful, those with LD_{50} from 50–200 mg./kg. are highly toxic, those with LD_{50} from 200–1,000 mg./kg. are of medium toxicity, and those with LD_{50} greater than 1,000 mg./kg. are of low toxicity. It must be noted, however, that this division is to a certain degree conditional, since the toxicity of pesticides for man and animals depends not only on the absolute value of the lethal dose of the preparation, but also on a number of other conditions. In determining the degree of toxicity of a preparation, it is necessary to give attention to its chronic toxicity, the possibility of accumulation in the organism of man and animals, the reversibility of the toxic effect, the route of entry into the organism, and a number of other factors, as well as the toxicity of the products of its metabolism (change of chemical compounds in different biological subjects and substrates) in the cells of plants and animals. The last property is important, because some pesticides when they are metabolized in animal organs (most frequently in the liver) are converted to substances that are more toxic than the starting compound. Because of this, in order to accurately determine the possible hazard of a pesticide, it is necessary not only to determine its LD_{50} but also to study its metabolism.

It is very important in determining an efficient method of using a pesticide to correctly evaluate the possible routes of its penetration into the human organism. Compounds of the first group which are capable of penetrating through the human skin (resorption) and through the respiratory passages are especially dangerous. Therefore, in studying the toxicity of preparations on experimental animals, the LD_{50} is determined not only for oral administration but also for dermal application, inhalation of the vapors, and other routes.

For an accurate determination of the hazard of a preparation to human health it is necessary to study its mechanism of action and to establish the main directions of reaction of the preparation with the most important enzyme systems of man and animals. Such data also are of great importance in developing preparations with selective toxicity, *i. e.*, toxic for some species of organisms and harmless for others.

Pesticide preparations used for the protection of plants from pests should be completely safe for the plants, and preparations used against parasites of animals should be safe for the animals. Furthermore, pesticides used for treating agricultural plants should not adversely affect the quality of the crop and should not change the flavor characteristics of the agricultural products.

The duration of retention of a preparation in the environment usually is called its persistence. The persistence of a preparation depends on both its physical and its chemical properties (volatility, hydrolytic stability, resistance to the action of soil microorganisms, and to oxidation by oxygen of the air, etc.).

Of the physical properties, volatility has the greatest importance; the less the volatility, the longer the pesticide may persist in the environment. The volatility can be characterized approximately by the magnitude of vaporizability of the substance at a given temperature. This can be calculated from the formula:

$$L = \frac{M \cdot p_T \cdot 1000}{760 \cdot B_T}$$

where L = the volatility of the substance at the given temperature (vaporizability), in g./m.^{3}; M = the molecular weight; p_T = the pressure of the saturated vapor at the given temperature, in mm. of Hg; and B_T = the volume of a mole at the same temperature, in liters.

The value of L characterizes the volatility of the substance in still air. Calculation of the volatility under dynamic conditions is rather complex, since not only the temperature and rate of movement of the air must be taken into account, but also the heat capacity of the substance and of all of the materials touching it, the shape of the surface, and so forth. The evaporation of drops of pesticides is very important, because under practical conditions most compounds are applied by the spraying of solutions, emulsions, or suspensions. The compounds are not employed in the pure form but in a complex mixture; therefore, calculation of the vaporizability of their drops presents certain difficulties.

Evaporation of drops of a pure compound in still air has been studied comparatively well and can be calculated from the Langmuir formula:

$$-\frac{dm}{dk} = 4\pi r D \frac{M p_{sat}}{RT}$$

where $\frac{dm}{dk}$ = the rate of evaporation from a whole drop (loss of mass of the drop per second), in g./sec.; r = the radius of the drop, in cm.; D = the coefficient of diffusion, in cm.^{2}/sec.; p_{sat} = the pressure of the saturated vapor of the substance evaporated, in mm. of Hg; R = the gas constant; T = the temperature, in °K; and M = the molecular weigth of the substance. From this equation it is possible to derive a formula for calculation of the time for complete evaporation of a drop of the substance (k, sec.):

$$k = \frac{d\,R\,T}{2\,d\,M\,p_{sat}} r^2$$

where d = the density of the substance in g./cm.^{3}. It should be noted that the rate of evaporation of drops in moving air is considerably greater and their lifetime is correspondingly shorter.

The chemical stability of the substance is no less important. The concept of chemical stability of pesticides includes the following factors:

1. Stability in storage under ordinary conditions (temperature and humidity) for the given locality.

2. Resistance to hydrolysis both in the pure form and on green foliage and in the soil.

3. Oxidizability by oxygen of the air in solutions and also on green foliage and in soil in the presence of soil microorganisms.

4. Resistance to sunlight, etc.

Depending on the purpose of a compound, it has different requirements with respect to persistence. In all cases, however, a constant requirement is the breakdown of the compound on plants by the time of harvest. For compounds introduced into the soil the persistence should be higher but should not exceed one season for herbicides and two seasons for agents to control soil-inhabiting pests. In this case the compound should not be absorbed by plants nor accumulate in fruits or in other parts of the plants that are utilized as food for man or domestic animals. If treatment of plants is carried out shortly before harvest, then the compound should be characterized by low stability and should break down completely before harvesting the crop.

The composition of conversion products of pesticides in the soil, plants, etc., also is very important. Preference should be given to compounds whose breakdown products are safe for man and animals.

Pesticides should be nonexplosive, nonflammable (with a high ignition temperature), convenient to handle, and not aggressive toward the usual metallic and nonmetallic materials employed in making containers and application equipment. In addition to these requirements, other highly important factors are economic considerations such as the availability of a source of raw materials for production, availability and cheapness of raw material and means of production, low rates of application of the compound per unit of area treated, low cost, etc.

Besides the general requirements, there are also particular requirements presented for individual groups of pesticides, for example for fumigants, compounds for control of parasites under domestic conditions, herbicides, seed disinfectants, and fungicides. These requirements are considered below.

The most important methods of pesticide use. Depending on the form of the preparations, different methods are used for applying them to plants, animals, and materials. In agriculture the widest use is made of dusting, spraying, and application of granular preparations to the soil; treatment with liquid preparations is rapidly replacing dusting. Small areas are most often treated with hand apparatus and large ones are treated with the aid of tractors or self-propelled sprayers and dusters, or more often by aerial application. In the ordinary spraying of plants the amount of liquid applied per hectare is from 200–2,000 liters, and in low-volume spraying it is from 5–50 liters. When low-volume spraying is used, the productivity of labor is sharply increased, and as a result a tendency is observed to use this more widely. For the control of parasites on animals, spraying under a pressure of several atmospheres, dipping, and more rarely dusting with solid preparations are used. Systemic preparations are introduced into the stomachs of animals by mixing them with the feed.

As disinfectants for nonmetallic materials, pesticides are applied by impregnation of the respective materials with solutions, suspensions, or emulsions, and also by surface application together with dyes.

To control rodents, poisoned baits that contain food products in addition to the pesticide are widely used. The bait method is also of some importance for the control of a limited number of insect pests, but because of the great amount of labor involved this method is being replaced by better ones. The bait method using sex attractants is recommended for predicting the development and estimating the distribution of individual species of insect pests. It is possible that in the future the sex attractants may be used for local eradication of pests.

General references

ANONYMOUS: Uspekhi v oblasti izucheniya pestitsidov [Advances in the field of pesticide research]. Foreign Literature Publishing House [USSR] (1962).
BROWN, A. W.: Insect control by chemicals. New York-London: Wiley (1951).
CRAMER, H. H.: Pflanzenschutz und Welternte. Leverkusen (1967).
FREAR, D.: Khimiya insektitsidov i fungitsidov [Chemistry of insecticides and fungicides]. Foreign Literature Publishing House [USSR] (1948). *
FURST, H.: Chemie und Pflanzenschutz. Berlin (1959).
HORSFALL, J. G.: Fungitsidy i ikh deistvie [Fungicides and their action]. Foreign Literature Publishing House [USSR] (1948). *
KABACHNIK, M. I.: Zhur. Vsesoyuz. Khim Obshchestva im. D. I. Mendeleeva 13, 242 (1968).
LAZAREV, N. V., ed. Vvedenie v geogigienu (Introduction to geohygiene). Publishing House "Nauka" [USSR] (1966).
MAIER-BODE, H.: Pflanzenschutzmittel-Rückstände. Insekticide. Verl. Eugen Ulmer Stuttgart (1965).
MED'VED, L. I.: Zhur. Vsesoyuz. Khim Obshchestva im. D. I. Mendeleeva 9, 561 (1964).
—, YA. S. KAGAN, and H. I. SPYNU: Zhur. Vsesoyuz. Khim. Obshchestva im. D. I. Mendeleeva 13, 263 (1968).
MEL'NIKOV, N. N.: Soobshchenie o nauchnykh rabotakh chlenov Vsesoyuz. Khim. Obshchestva im. D. I. Mendeleeva No. 2, 17 (1955).
— Khim. Nauka i Prom. 1, 160 (1966).
— Zhur. Vsesoyuz. Khim. Obshchestva im. D. I. Mendeleeva 5, 242 (1960).

---

* Exact edition unknown; cited here is date of presumed Russian translation from English. — R. L. B.

MEL'NIKOV, N. N.: Khim. v Sel'skom Khozyaistve No. 11, 22 (1965).
— Zhur. Vsesoyuz. Khim. Obshchestva im. D. I. Mendeleeva **13**, 248 (1968).
—, and YU. A. BASKAKOV: Khimiya gerbitsidov i regulyatorov rosta rastenii. [Chemistry of herbicides and plant growth regulators]. State Publishing House of Chemical Literature. [USSR] (1952).
PAVLOVSKII, E. N.: Rukovodstvo po parazitologii cheloveka [Handbook on human parasitology]. Publishing House of the Academy of Sciences USSR. (vol. I, 1946; vol. II, 1948).
PERKOW, W.: Die Insektizide. Heidelberg: A. Hüthig (1968).
POPOV, P. V.: Spravochnik po yadokhimikatam [Handbook on pesticides]. State Scientific and Technical Publishing House of Chemical Literature [USSR] (1956).
SEVULESCU, A.: Zashchita rastenii [Plant protection]. Bucharest: "Meridiane" (1963).
SHARVELLE, E. G.: The nature and uses of modern fungicides. Minneapolis: Burgess (1961).

II. Pesticide formulations

The successful use of pesticides to control harmful insects, mites, microorganisms, and weeds depends to a large degree on the formulation of the preparation and the conditions under which the chemical compound is brought into contact with the pests of plants, vectors of plant diseases, stored product pests, destroyers of nonmetallic materials, and undesirable plants. The different natures of the chemical compounds used and the variety of objects for which they are used makes it necessary to produce a large number of formulations suitable for employment under practical conditions.

Depending on the physicochemical properties of the active ingredient of the preparation, its purpose, and the means of application, the most efficient and economical formulation for its use under the given concrete conditions is selected.

The form selected for use also determines to an important extent the technology of production of an insecticidal, fungicidal, bactericidal, or herbicidal preparation. At the present time a tremendous number of different formulations is manufactured for use in industry, agriculture, and health protection. In the United States more than 1,200 formulations are manufactured that are based on DDT alone, and about 1,500 are based on other chlorinated hydrocarbons. The most important types of formulations are the following:

1. Powders (dusts) for dusting.

2. Granulated preparations for treatment of plants and for application to the soil.

3. Wettable powders, which yield a suspension of the preparation in water for spraying.

4. Solutions in water and organic solvents.

5. Emulsive concentrates, which on dilution with water form emulsions for spraying.

6. Aerosols and fumigants.

7. Other types of formulations (antiseptic and insecticidal soaps, antiseptic and insecticidal paints, lacquers, pastes, mastics, waxes, insecticidal pencils, insecticidal and bactericidal paper, various baits, etc.). These formulations, however, have a more limited distribution and are produced on a relatively small scale. The only exceptions are poisoned baits, which are the main form for control of rodents.

Powders (dusts)

Insecticidal, fungicidal, and herbicidal dusts and combination powders consist of a mechanical mixture of the active ingredient with an inert diluent, pulverized to a particle size of 3–30 μ. In the grinding process not only are the pesticide particles distributed among the particles of the diluent, but also they coat the particles of diluent. Consequently, grinding of the active ingredient together with the diluent gives a more efficient formulation than separate grinding of the ingredients with subsequent mixing. The finer the grinding of the active constituent in a powder or dust type formulation, the more effective it is for dusting. This is explained by the better adherence of more finely ground preparations on plants and insects and also by the greater uniformity of coverage. Usually from 10–40 kg. of dusts are used per hectare of area. For uniform distribution of a preparation over this area rather fine grinding is necessary.

Pesticides, depending on their properties, can be reduced to the powdered state more or less easily. Compounds with a well-formed crystalline structure and having rather high friability and low plasticity grind the best. For example, pure DDT, which melts not lower than 106° C., is easily ground. The technical grade product, however, which contains oily impurities, is more difficult to grind; it lumps on grinding and forms coarse aggregates that stick to the walls of the mill. The lower the melting point of compounds, the more difficult it is to grind them in ball mills. Grinding of solid organic pesticides in ball mills and similar mills as a rule is carried out in the presence of some inert diluent which hinders lumping of the preparation both in the grinding process and in storage. In this connection, the greater the sorbtive capacity of the diluent employed as the additive, the less of it need be used. Thus, for example, when technical grade DDT is ground with silica gel (with a bulk density of 0.14 g./cm.^{3}), it is sufficient to add 5–10% diluent, while not less than 25% ordinary kaolin is required. When a smaller amount of diluent is used, preparations are obtained that lump in storage. Such preparations lump especially readily when they are stored under tropical conditions.

The conditions for grinding depend also on the construction of the milling unit. Lumping of preparations occurs most often when they are ground in a ball mill. Apparently this is associated with the increase in temperature in the mill as a result of the impact of the balls and with partial melting of the compound being ground. Therefore, in grinding low-melting pesticides it is desirable to introduce a larger amount of diluent and to carry out the process at the lowest temperature possible.

In making dusts, the grinding of most organic compounds usually is carried out in ball mills with air separation. Sometimes the grinding is carried out in two stages: a concentrated powder is first prepared (on rollers, in a ball mill, or in another type of milling unit) which is then diluted with the necessary amount of diluent and grinding is completed in a ball mill. The best results in the preparation of dusts are obtained when air-jet mills are used, but the cost of the product in this case is increased as a result of the greater expenditure of electric power.

It has now been shown that when a low-melting organic compound is ground with a mineral diluent in a ball mill, a dust is obtained that differs slightly from a preparation obtained by impregnation of the diluent with a solution of the compound and subsequent removal of the solvent by distillation, because in the grinding process coating of the diluent with a thin layer of the pesticide takes place.

Some other methods have been described for the preparation of pesticide dusts. Of interest is the preparation of dusts by mixing the melted active constituent with the ground diluent. Several modifications of this method are known for preparing dusts of low-melting pesticides that have high plasticity. For example, dusts are easily produced in this way from a fused mixture of DDT with enriched benzene hexachloride, and also from a solution of DDT in mineral oil, etc. In some cases, in order to retard crystallization of the pesticide it is recommended that the melted preparation be added to the diluent that has been heated to 30°–40° C.

There also are other methods for the preparation of dusts from melted pesticides. For example, the melted pesticide is atomized in a chamber that is constructed on the model of apparatus for spray drying, into which the diluent is introduced in the suspended state. The drops of insecticide or another pesticide adhere to the particles of diluent and fall to the bottom of the chamber. By means of photomicrographs, the process can be controlled so that a completely homogeneous dust can be obtained. It is possible to prepare dusts by the condensation of vapors of the pesticide on the diluent, but this method can hardly be considered economically efficient.

When dusts are prepared from liquid organic compounds it is simple enough to mix the diluent with the active constituent. When the content of the active constituent in the dust is very small, the mixing is carried out in several steps. Usually two-stage mixing is sufficient. The active ingredient content in the dust is chosen so that with a dosage of 10–25 kg./ha. of the dust, the minimum amount of pesticide will be used. However, this amount will be sufficient to obtain good effects in controlling pests or diseases of plants and weeds. DDT dusts usually contain 5–10% of the technical grade compound, parathion 1%, and growth regulators 0.05–0.1%.

The usual diluents for the production of pesticide dusts are hydrophobic minerals of the talc and pyrophyllite type; more rarely chalk, gypsum, kaolin, kieselguhr, tripoli earth, silica gel, and various clays are used. The best diluents for most preparations used in dusting plants are pyrophyllite and talc, which have a lamellar structure and adhere well to the plant foliage. Furthermore, these minerals are quite inert and do not exert an adverse effect on most pesticides. Alkaline impurities in these minerals, however, may decrease the stability of dusts that contain esters of the phosphorus acids.

In choosing a diluent it is necessary to consider not only its physical properties but also its chemical properties, because some diluents can catalyze the decomposition of pesticides or affect their photochemical stability. Thus, for example, alkaline diluents are not recommended for use in the production of dusts from esters, halogen derivatives of hydrocarbons, and the like. Esters may be hydrolyzed also under the influence of compounds of an acid nature. When such diluents are used it is recommended that special additives be introduced to eliminate the harmful effect of the diluent. In some cases it is expedient to use a mixture of several diluents.

Use of hydrophilic minerals of the kaolin, clay, and bentonite type as diluents is not recommended when there is high humidity, since dusts with these diluents cake easily after even slight dampening. To increase the retention of the dusts on plants, hydrophobic compounds of the type of calcium stearate, mineral oils, and the like can be added to them. The adherence of powdered formulations and their retention on plants in many cases determine the duration of their action. These characteristics depend not only on the shape of the dust particles (which is very important), but also on a number of other factors including the chemical composition of the active ingredient and of special additives to the dusts. Hydrophobic compounds are retained better on hydrophobic surfaces, and hydrophilic ones on hydrophilic surfaces. Therefore, in selecting a formulation for any compound it is necessary to consider not only its method of use, but also the crops for which it is intended.

Granulated formulations

Instead of dusts for the control of plant pests, diseases, and weeds, frequent use is being made of granulated formulations, which often are more convenient and leave a smaller amount of undesirable contaminants on the plants. In this form herbicide compounds are even able to change somewhat their field of action.

Granulated formulations are widely used also for the control of soil-inhabiting pests, and also for poisoning plants through the root system in order to protect the young seedlings from damage by sucking pests.

The most important methods of preparing granulated formulations are impregnation of prepared granules or minerals of the type of perlite or vermiculite with liquid pesticides or solutions of them, and granulation of powder formulations on a suitable diluent with subsequent screening. Kaolin, bentonite, or similar minerals are most often used as diluents. For granulation, in addition to the diluent and the pesticide, various binding agents based on synthetic resins or other adhesive materials are added to the formulation. The process of preparing granulated formulations by the second method is complicated, because in addition to preparing the starting powder the steps of impregnation, granulation, drying, and screening must be carried out.

The size of the granules may vary, depending on the purpose for which the formulation is to be employed. The most widely used formulations have granule sizes from 0.2–1 mm. The strength of the granules also differs. In formulations for the treatment of plants, granules of relatively low strength are used, while for control of weeds in water reservoirs granules with high strength are used.

Granulated formulations in which a fertilizer (for example, superphosphate) is used as the carrier are employed on a limited scale. However, only a very small number of compounds are used in this way, because many pesticides are rapidly broken down when mixed with fertilizers.

Wettable powders

Powdered formulations that on dilution with water yield rather stable suspensions are called wettable powders. In most cases the use of pesticide suspensions is of greater interest than the use of dusts. Suspensions usually are more effective than dusts. Spraying plants and other surfaces almost always is more effective for the same dosage of pesticide than dusting. In treating residential premises to control mosquitoes, flies, and gnats and in disinfection, dusting is entirely unsuitable for completely understandable reasons. Spraying is gradually displacing dusting, and dusts are being replaced by various formulations for spraying.

The use of suspensions is especially convenient when dust-type preparations are poorly retained on surfaces; emulsions are easily absorbed and as a result considerable amounts of poisonous residues are retained on fruits. The particles of a suspension adhere well to surfaces, and they do not penetrate; so they can be easily washed off in case of necessity.

The requirements for wettable powders are as follows:

1. Stability in storage and absence of caking.
2. Rapid formation of a suspension and slow settling out of solid particles.
3. Good wettability of sprayed objects and easy spreading over their surface.
4. Retention on sprayed surfaces for a more or less prolonged time.

One of the principal conditions guaranteeing the effectiveness of a suspension in controlling plant pests, diseases, and weeds is a high degree of dispersion of the active ingredient of the preparation. Good wettable powders usually contain not less than 80% of particles with a size below 3 μ.

Wettable powders can be divided into three groups: preparations with a high content of the active ingredient (from 60–90%), with a medium content (from 30–60%), and with a low content (below 30%). The last are usually prepared from liquid or waxy compounds, since a high content of such compounds in a wettable powder yields a preparation that cakes easily.

Wettable powders differ both in their method of preparation and in the starting material used. Usually a diluent, a surface-active agent, and an auxiliary material enter into the composition of a wettable powder in addition to the active ingredient. Sometimes, in order to increase the retention of preparations on plants or other surfaces, special stickers also are added to them.

In Table III typical formulas are given for wettable powders containing 90% DDT.

Table III. *Examples of composition of DDT wettable powders*

Ingredients	Composition (%)			
	A	B	C	D
DDT	90	90	90	90
Diluent	7	6	7.5	6.5
Emulsifier	3	3	—	—
Auxiliary material	—	1	—	1
Film-forming material	—	—	2	2
Wetting agent	—	—	0.5	0.5

For producing formulations with a high active ingredient content, rather pure materials without oily impurities and a diluent with a low bulk density and relatively large sorbtive capacity are used in almost all cases. In particular, special kinds of silica gel with a bulk density not greater than 0.15 are recommended. Some other hydrophilic diluents also may be used: hydrated aluminium oxide, synthetic calcium silicate, etc. It is also possible to prepare wettable powders without a diluent if the active constituent of the preparation has a melting point high enough and grinds well.

With a poorer quality of diluent it is necessary to increase the content of surface-active agent, otherwise unstable suspensions are obtained. However, a poor quality diluent cannot always be compensated for by increasing the surface-active agent content of a wettable powder.

A great variety of detergents is used in the preparation of wettable powders, including sulfonates of the alkali metals (sodium dibutoxydiphenylsulfonate, sodium alkylbenzenesulfonates, especially sodium dodecylbenzenesulfonate, sodium alkanesulfonates, etc.), alkyl aryl ethers of polyethyleneglycol and polypropyleneglycol (compounds of the type of OP-7, OP-10), and many others. These compounds are used in making wettable powders and pastes both independently and in combination with other products, for example sulfite waste liquor.

As auxiliary materials the sodium salts of sulfo acids obtained by sulfonation of petroleum products and also the sodium salts of lignin sulfo acids are recommended. Of the film-forming agents, those most often used are carboxymethylcellulose, methylcellulose, polymers of unsaturated alcohols, gelatin, animal glues, caseinates, salts of resin acids, and the like.

High-concentration wettable powders are produced by grinding in a colloid mill a mixture of the active ingredient with the diluent and the other ingredients. Good results are obtained by grinding in air-jet mills of various construction. Grinding can be carried out also in ball mills, but coarser powders are obtained. Satisfactory results are obtained with ball mills by grinding the pesticide with the diluent in the presence of water; the paste is then dried and ground a second time.

Pesticides obtained by precipitation from aqueous solutions can be used for the production of wettable powders without milling. In this case all the materials necessary for preparation of the wettable powder are added to the precipitated and washed pesticide and then the mixture is dried in a spray drier. By this method wettable powders of such pesticides as copper oxychloride, ziram, cuprous oxide, and many others can be prepared. However, supplementary grinding is sometimes required. The composition of such wettable powders usually includes, in addition to the pesticide, 2–3% OP-7, 2–3% sulfite waste liquor, and up to 5% carboxymethylcellulose.

Wettable powders containing 70–80% active ingredient are the most widely used.

In Table IV formulas are given for some wettable powders.

Table IV. *Composition of some pesticide wettable powders*

Ingredients	Composition (weight %)			
	DDT	BHC	Ester-sulfonate	Dichlor-alurea
Pesticide	30–50	50–80	50–80	80
Kaolin	45.5–67.5	14–47	16.5–47.5	15–16
Sulfite waste liquor	1.5–2.5	1–2	1.5–2	—
OP-7	1–2	2–4	1–1.5	4–5

Pastes that contain a small amount of organic solvent and water along with auxiliary agents are close to wettable powders with respect to the suspensions obtained from them. The use of pastes, however, is less convenient, because their composition may change in storage as a result of evaporation of part of the water or organic solvent. According to data in the literature, stable pastes are obtained by using high molecular weight, water-soluble protective colloids that have a viscosity in 10% solution at 25° C. of not less than 10 centipoises. The most readily available materials with such a viscosity are methyl- and carboxymethylcellulose. Of the surface-active agents, in this case, alkyl aryl ethers of polyethyleneglycol with a molecular weight from 400 to 600 are used. Preparation of the paste can be carried out by emulsification in a liquid colloid mill at an elevated temperature.

For example, a DDT paste is manufactured with the following composition: technical grade DDT 50%, spindle oil 10%, sulfite waste liquor 15%, and water 25%. A paste containing 0.5% OP-7 in addition to the components mentioned has still better properties. To prepare this paste a mixture of the DDT and spindle oil is heated to 105°–110° C.

and stirred until a completely homogeneous solution is produced. Then an aqueous solution of sulfite waste liquor is added with good stirring and the coarse emulsion obtained is passed through a liquid colloid mill. After the product is cooled a soaplike mass is formed that emulsifies well in water.

Wettable powders with a compound action also have been proposed that contain several pesticides for different purposes at the same time. The use of such preparations in agriculture gives some economy of labor in treating plants.

Solutions of pesticides in water and organic solvents

Only compounds that are rather soluble in water can be used in the form of aqueous solutions. The main pesticides used in aqueous solutions are herbicides which are salts of organic acids with different bases, and some organophosphorus insecticides and fungicides.

Disinfection of seeds with aqueous solutions of formalin, organic mercury compounds, and also some other fungicides and bactericides is widely practiced. Such disinfection of seeds is, in many instances, more economical than dry disinfection, and the expenditure of disinfectant by this method is considerably less. It is especially efficient to disinfect scaly seeds, like those of oats, with aqueous solutions. The efficiency of disinfecting seeds with aqueous solutions of organic mercury compounds can be judged from the following data. For disinfecting wheat seed with Granosan about 15 g. Hg is applied per ton of seed, while with disinfection by an aqueous solution of ethylmercury phosphate the same result is obtained by applying about 3 g. Hg per ton. Comparatively, however, wet disinfection of seed is seldom used. More often semi–dry disinfection is used, where the treatment of the seed is carried out with a limited amount of a concentrated aqueous or oil solution of the disinfectant, and in a number of other cases suspensions also are used.

Aqueous solutions of compounds are also used to control some diseases and pests of growing plants (for example, brown spot of tomatoes, nematodes). Herbicides and some systemic insecticides that quickly penetrate into the plants and consequently are not washed off by rain are often applied in the form of aqueous solutions.

Various solutions of pesticides in organic solvents are widely used for so-called low-volume, finely dispersed spraying of plants. When plants are sprayed with these preparations, the productivity of labor is increased and the cost of treatment is lowered.

Solutions of pesticides in organic solvents also find use in controlling various insect pests — parasites of man and animals, for protecting non-metallic materials from damage by insects and microorganisms, and for disinfecting premises.

Also various compounds are used in the form of solutions in organic solvents to repel blood-sucking insects and ticks.

In choosing solvents for pesticides or mixtures of pesticides, it is necessary to take into account not only the solubility of the pesticides, but also other properties of the solutions and the solvents. In particular, if the solutions are intended for use on green plants it is necessary to consider the possibile phytocidal effect of the solvent and the solution; it is desirable to use solvents that do not cause burning of the plants. If solutions are prepared for use in enclosed premises, flammability and toxicity for man and domestic animals must be considered, and also the possibility of chemical reaction of the dissolved compounds with the solvent. If unsaturated compounds are used as solvents, it is possible that they may react with even such stable compounds as DDT, and the duration of efficacy of the preparations will be less, especially in light and at a high temperature.

Sometimes it is efficient to mix several solvents, for example when it is necessary to increase the solubility of some pesticide in an inexpensive solvent. For this purpose small amounts of a more expensive solvent in which the pesticide is considerably more soluble are added. Such solvents are called intermediate solvents. This means is often employed to increase the solubility of DDT and other organochlorines in mineral oils. However, additive solubility of the pesticide is not always observed in a mixture of two or more solvents.

Intermediate solvents for the preparation of concentrated solutions of pesticides in mineral oils are: cyclohexanone, methylcyclohexanones, mesityl oxide, tetrahydrofuran, thiophene, isophorone, methyl ethyl ketone, dimethylformamide, diethyl carbonate, alkyl acetates, xylenes, methyl- and polymethylnaphthalenes, and chlorobenzene.

Some of the solvents listed have a low ignition temperature which must be taken into consideration when using them in enclosed areas or in aerial application. For example, cyclohexanone and isophorone ignite already at 37.8° C.

Intermediate solvents for halogenated and organophosphorus insecticides are halogenated hydrocarbons of the aliphatic series, including carbon tetrachloride, methylene chloride, dichloroethane, trichloroethylene, and tetrachloroethylene. In domestic aerosol "bombs" Freons or mixtures of them with other compounds are used as solvents and propellents.

In practice, the most frequently used solvents for the preparation of pesticide solutions are the petroleum hydrocarbons: dearomatized kerosine, white spirit (turpentine substitute), mineral oils, and diesel fuel.

Some use is also made of solutions, which are "solid" or pasty at room temperature, of pesticides in solid organic compounds or mixtures of them. Such "solid" solutions include creams and ointments containing compounds that repel blood-sucking insects and ticks. Antiparasitic and bactericidal waxes are used for polishing floors and other smooth surfaces. Antiparasitic pencils also exist, which contain, in addition to an insecticide, wax, paraffin, and solid filler. To prepare the pencils the insecticide is melted together with the wax and paraffin, the melt is mixed with the solid filler and poured into molds. After solidification the pencils are ready for use.

With respect to the nature of their use and their properties, the pesticidal lacquers and paints prepared by introducing an insecticide, fungicide, or disinfectant into ordinary lacquers and paints must be related to the solutions. To increase the efficiency of the disinfectant or insecticide, substances that impart some porosity to the coating are often added to such paints. Silica gel and various silicates with a low bulk density are recommended as additives. The disinfectant and insecticidal properties of paints and lacquers become evident only after the formation of microscopic crystals of the active ingredient on the surface of the lacquer film on the painted object. If the active ingredient is covered by a film of lacquer, its contact with insects or microorganisms that get onto the surface of the film is prevented and the effect of its action is sharply decreased.

Emulsive concentrates

One of the convenient types of formulation of pesticides is emulsive concentrates that upon dilution with water give stable emulsions suitable for spraying plants and surfaces. In some instances this kind of pesticidal formulation has advantages over suspensions, since many liquid pesticides are suitable for the preparation of such concentrates without the use of solvents or diluents. Emulsions at equal concentrations of the active ingredient usually are more effective than the corresponding suspensions. In this respect they approach the solutions of pesticides in organic solvents, but the amount of solvent required is many times less.

Two kinds of pesticide emulsive concentrates are known:

1. Emulsive concentrates are prepared by dispersing an aqueous solution of the pesticide in a water-immiscible solvent by means of a colloid mill. These concentrates are more properly called concentrated emulsions.

By mechanical dispersion highly dispersed concentrated emulsions can be obtained that are stable in storage. Emulsions of this type are used in agriculture, but only pesticides that are stable to the action of water are suitable for their preparation. By this means oil emulsions of DDT, benzene hexachloride, anthracene oil, and some other compounds are produced. The concentrated emulsions are stabilized by the addition of sulfite waste liquor.

The process of producing concentrated emulsions of pesticides can be desribed as follows. First, an aqueous solution of sulfite waste liquor is mixed with a solution of the pesticide in oil or another organic solvent (to obtain sufficiently stable concentrated emulsions, the specific gravity of the solution of pesticide in the organic solvent should be about one). The coarse dispersion obtained is placed in a mechanical homogenizer. The liquid enters the rotor, which turns at high speed (about 6,000 rpm), and is thrown out to the periphery by centrifugal force; then it enters the space between the rotor and the stator. The initial drops are converted into threads that are further broken down into colloidal particles. At the present time many brands of mechanical homogenizers are well known that are suitable for the production of concentrated pesticide emulsions.

The homogenized concentrated emulsions withstand low temperatures and prolonged storage without substantial changes. Sometimes concentrated emulsions thicken in storage, but after stirring they return to their normal condition.

2. Emulsive concentrates of pesticides, known under the name of "miscible oils", usually are prepared under factory conditions and contain a considerable amount of pesticides. When the miscible oils are mixed with water they yield stable emulsions. Concentrates of this type consist of a pesticide, solvent, and emulsifier. For liquid pesticides the use of a solvent is not obligatory, although the use of organic solvents often permits decreasing the content of expensive emulsifiers in the concentrate.

Hydrocarbons, halogenated derivatives of hydrocarbons, esters, various petroleum products, creolin, coal-tar oils, and many other compounds may be used as solvents. Emulsifiers are calcium sulfonates, ethers of polyethylene- and polypropyleneglycols, monoesters of sorbitol and mannitol with the higher fatty acids, various soaps, salts of naphthenic acids, etc. Particularly good results are obtained by the use of mixtures of two or more emulsifiers, one of which is an ether of polyethyleneglycol and the other a calcium or ammonium alkylarylsulfonate prepared from the higher alcohols. A favorable property of calcium alkylbenzenesulfonates is their solubility in organic solvents, in contrast to the corresponding alkali metal salts, which are relatively poorly soluble in organic solvents and on standing quickly crystallize out from emulsive concentrates.

To prepare emulsive concentrates, the pesticide is dissolved in the selected solvent and then mixed with the emulsifier while being heated to 40°–80° C. After the concentrate has cooled it should be a homogeneous mass without precipitate. The concentrate is filtered to remove mechanical impurities.

For use in agriculture it is convenient to have concentrates that are not too viscous, but for highly toxic organophosphorus compounds it is less dangerous to use viscous concentrates, because if they get on the skin they are more difficultly absorbed and can easily be washed off with water.

To the emulsive concentrates belong the disinfectant and disinsectant preparations and the disinfectant soaps that contain antiseptics and pesticides, and also various disinfectant pastes for the protection of wood and other nonmetallic materials.

There is wide use of the so-called "inverted emulsions" that also are prepared by dilution of the appropriate concentrates with water. Enough water is added so that an emulsion of the water-in-oil type is obtained. These concentrates are used for low-volume spraying, since the presence of an oily film hinders evaporation of the drops.

To decrease evaporation of the drops in spraying by aircraft or by other means, special additives are introduced. An example of such additives is the English preparation "Lovo" (butylamine stearate).

Aerosols

One of the new forms in which pesticides are used in public health and agriculture is the aerosol. The development of this means of applying pesticides is receiving much attention. Three basic methods of producing insecticidal and fungicidal aerosols are well known:

1. Burning various compositions containing insecticides, fungicides, or bactericides that on combustion are sublimed and form smokes or clouds poisonous to insects, fungi, and bacteria. Vaporization of pesticides with the aid of special heating equipment, for example electric lamps, also is possible.

2. Spraying solutions of pesticides in readily volatile solvents, which upon evaporation leave the pesticide in the air in the form of a fine dispersion.

3. Atomization of solutions of pesticides by mechanical means with the use of special atomizing devices. Sometimes this method is combined with the second. Heated solutions of pesticides in organic solvents, mainly in petroleum products, are atomized. Upon atomization, part of the solvent evaporates leading to a shrinkage of the drops to a size close to that of aerosols. This method is well known in the literature under the name of low-volume finely dispersed spraying.

The simplest method of producing pesticide aerosols is burning special smoke pots, paper, and other combustible porous materials that have been impregnated with insecticides, fungicides, or bactericides. Besides the pesticide, the combustible smoke-forming compositions usually contain a combustible substance, filler, and an oxidizing agent. Nitrates, nitrites, chlorates, chromates, bichromates, persulfates, and mixtures of them are used as oxidizing agents. Fillers, which passivate combustion, include kaolin, infusorial earth, and the like. To maintain the necessary temperature sawdust, waste products from cellulose manufacture, carbon, bitumen, and other tars are added.

The compositions of some insecticidal smoke mixtures are given below:

1. Potassium chlorate 23–26%
   Crude anthracene 9–12%
   Ammonium chloride . . . 9–12%
   Dicyandiamide 4– 6%
   Benzene hexachloride . . . 50–52%
2. Potassium chlorate 19%
   Urotropine 8%
   Diatomite 9%
   Benzene hexachloride . . . 64%
3. Potassium chlorate 25.8%
   Urea 7.3%
   Thiourea 12%
   DDT 54.9%

Indications in the literature suggest that when such compositions are burned more than 80% of the pesticide is sublimed without decomposition and only 10–17% of it is converted to products that are nontoxic to insects.

Insecticide paper, which forms a smoke that is toxic to insects when it burns, is prepared by impregnating unsized paper with a solution of the insecticide. The paper is first impregnated with an aqueous solution of sodium or potassium nitrate and dried.

Insecticide smoke pots are effective in controlling bloodsucking diptera and stored products pests, and also in treating greenhouses. In the last case special cartridges with tetraethyl dithiopyrophosphate ("Bladafume") are recommended for practical use. It should be noted, however, that the use of compounds with such high toxicity to man in the form of a mist may present a serious danger. The use of smokes and mists obtained by burning smoke pots to control pests of plants under field conditions does not give a good protective effect.

Aerosols obtained by atomizing solutions of insecticides in volatile solvents usually are recommended for control of flies and other flying insects in enclosed premises. Such solutions are placed in metal aerosol cylinders equipped with an atomizing device. With the aid of carbon dioxide or a low-boiling solvent (Freon, methyl chloride, etc.) pressure is created in the cyclinder that facilitates good atomization of the preparation. After the solution has been sprayed into the air and the solvent has evaporated, the pesticide remains in the atmosphere in a finely dispersed condition. The size of the pesticide particles is associated with the size of the drops of the solution and depending on the pressure in the cylinder and the construction of the atomizing device it may vary within wide limits.

Compositions of various aerosol mixtures are given in Table V. The aerosols obtained by the first two methods usually are used to control insects that are very sensitive to the insecticide or to control microorganisms but not to control plant pests.

Table V. *Insecticide solutions for preparation of aerosols*

Ingredients	Composition (parts by wt.) [a]				
	I	II	III	IV	V
Principal insecticide	3	0.5	5	6.3	4
Supplementary insecticide	2	1	—	4.4	—
Freon-12	85	83.5	—	83.75	88
Acetone	—	—	35	4.4	—
Cyclohexanone	5	5	5	—	—
Methyl chloride	—	—	50	—	—
Kerosine	—	5	—	—	4
Liquid petrolatum (mineral oil)	—	—	—	1.1	—
Machine oil	5	—	5	0.05	—
Synergist	—	5	—	—	—
Insecticide stabilizer	—	—	—	—	0.5
Other auxiliary materials	—	—	—	—	3.5

[a] Pressure is created with the aid of carbon dioxide.

In agriculture aerosols are used that are produced by the mechanical method — by the dispersion of solutions of pesticides in organic solvents. The use of low-volume finely dispersed spraying is effective both in sheltered ground and for spraying large bodies of forest and fields with a small expenditure of the preparation.

The solvents used for low-volume finely dispersed spraying are petroleum products with a relatively high boiling point. To increase the solubility of the pesticides in them, intermediate solvents are added (xylene, polymethylnaphthalenes, etc.). In many designs for apparatus used in the mechanical dispersion of pesticide solutions, the exhaust gases of internal combustion engines are utilized. It must be kept in mind that with such dispersing apparatus some pesticide may be lost as a result of partial thermal decomposition. In this connection the higher the temperature to which the pesticide is heated, the greater the loss. Therefore, when this method of producing aerosols is used, strict observance of the temperature conditions is necessary.

Other types of pesticide formulations

The brief description that has been presented of the most important types of pesticide formulations used does not exhaust all the directions of utilization and production of individual kinds of formulations. Thus, for example, to control stored product pests the packages are treated with insecticides (lindane, pyrethrins, and some other materials). Cereals in such a package are practically undamaged by granary pests.

Extremely toxic systemic insecticides are placed in the soil in gelatine capsules prepared at the factory. This eliminates contact by workers with the toxic preparation, since the gelatine gradually breaks down in the soil and the insecticide enters the soil water and is absorbed by the plant roots.

The bait method is widely used for the control of rodents, and fumigants often are employed to eliminate pests of stored food products.

General references

ANONYMOUS: Aerozoli v sel'skom khozyaistve [Aerosols in agriculture]. Collected works. State Publishing House of Agricultural Literature [USSR] (1956).
BEZUGLYI, S. F.: Zhur. Vsesoyuz. Khim. Obshchestva im. D. I. Mendeleeva 5, 298 (1960).
—, and V. P. TROPIN: Zhur. Vsesoyuz. Khim. Obshchestva im. D. I. Mendeleeva 9, 546 (1965).
EICHLER, W.: Handbuch der Insektizidkunde. Berlin (1965).
KOULA, V. and M. DURASOVA: Aerozoli v zashchite rastenii [Aerosols in plant protection]. Foreign Literature Publishing House [USSR] (1957).
MEL'NIKOV, N. N., V. A. NABOKOV, and E. A. POKROVSKII: DDT svoistva i primenenie [DDT — Properties and use]. State Scientific and Technical Publishing House of Chemical Literature [USSR] (1954).
NAUCHNYI INSTITUT PO UDOBRENIYAM I INSEKTOFUNGITSIDAM IMENI YA. V. SAMOILOVA: Dusty, emul'sii i suspensii DDT i GKhTsG [Dusts, emulsions, and suspensions of DDT and BHC]. Coll. Works of Nauchnyi Institut po Udobreniyam i Insektofungitsidam im Ya. V. Samoilova. State Scientific and Technical Publishing House of Chemical Literature [USSR] (1955).
— Dusty, emul'sii i smachivayushchiesya porokhi organisheskikh insektofungitsidov [Dusts, emulsions and wettable powders of organic insectofungicides]. Coll. works of Nauchnyi Institut po Udobreniyam i Insektofungitsidam im. Ya. V. Samoilova. State Scientific and Technical Publishing House of Chemical Literature [USSR] (1959).
— Khimicheskie sredstva zashchity rastenii [Chemical agents for plant protection]. Coll. works of Nauchnyi Institut po Udobreniyam i Insektofungitsidam im. Ya. V. Samoilova. State Scientific and Technical Publishing House of Chemical Literature [USSR] (1961).
SHOGAM, S. M., E. I. FEN'KOVA, I. A. EFIMENKO, and T. B. EPSHTEIN: Zhur. Vsesoyuz. Khim. Obshchestva im. D. E. Mendeleeva 5, 312 (1960).
—, and V. I. ORLOV: Khim. Prom. No. 8, 26 (1956).

III. Hydrocarbons

General characteristics of pesticidal properties

The pesticidal activity of a large number of hydrocarbons of different series has now been studied. Only a few of them, however, are used for the control of harmful organisms.

The insecticidal properties of hydrocarbons of the aliphatic series from methane to hexacosane have been studied, including the hydrocarbons of isostructure, the olefins, and the acetylenes. It has been established that the insecticidal and acaricidal activity of the hydrocarbons increases with an increase in their molecular weight. For the paraffins and the isoparaffins it reaches a maximum at a molecular weight of 320–350, after which it remains constant in the compounds that have been studied. A similar relationship is observed also in the unsaturated and the alicyclic series, but the insecticidal properties of the alicyclic hydrocarbons are somewhat weaker than those of the unsaturated ones. The insecticidal activity of the unsaturated hydrocarbons increases when going from the olefins to the acetylenes and dienes. The paraffinic hydrocarbons with branched chains are more active as insecticides than those having normal structure with the same molecular weight.

Because of their comparatively high toxicity and low phytocidal action, the petroleum isoparaffins are recommended in the United States for spraying of fruit trees to control the San Jose scale and mites.

The fungicidal and herbicidal properties of the paraffinic hydrocarbons are insignificant. The activity is increased somewhat going to the unsaturated compounds. Ethylene has a powerful defoliating effect and also accelerates the ripening of tomatoes and some fruit crops. Plants placed in a chamber with a small amount of ethylene ($<1\%$) quickly lose their leaves. The defoliating effect of the homologs of ethylene and of acetylene is substantially weaker.

The pesticidal properties of the aromatic hydrocarbons have been studied in somewhat more detail, and some of them are used in agriculture and industry, for example: benzene, toluene, the xylenes, ethylbenzene, the trimethylbenzenes, isopropylbenzene, cymene, *sec.-* and *tert.-*butylbenzenes, pentamethyl- and hexamethylbenzenes, dihydronaphthalene, tetralin, naphthalene, acenaphthene, methyl- and polymethylnaphthalenes, diphenyl, diphenylmethane, stilbene, fluorene, anthracene, dihydroanthracene, phenanthrene, chrysene, retene, terphenyl, and some other polynuclear compounds.

The biocidal activity of the aromatic hydrocarbons is higher than that of the compounds of the aliphatic and alicyclic series. However, in comparison with the pesticides now in use, their physiological activity is small, although mixtures of hydrocarbons in the form of petroleum and coal-tar oils are used in pesticidal preparations.

The xylenes, homologs of naphthalene, tetralin, and other liquid aromatic hydrocarbons are employed as solvents for the chlorine- and phosphorus-containing insecticides and for some herbicides. Of the individual hydrocarbons, only naphthalene and diphenyl find limited use.

Naphthalene is employed in small amounts as an antimoth agent and as a fumigant for grain, but it is being replaced by other, more effective preparations.

Diphenyl, m. p. 70.5° C., d_4^{25} 1.156, although practically insoluble in water, is readily soluble in most organic solvents. Its vapor pressure at various temperatures is:

Temperature, °C.	70.6	117	187	229.4	254.9
Pressure, mm. of Hg.	1	10	100	400	760

One of the most important reactions for preparing diphenyl is the pyrolysis of benzene. It is obtained as by-product in various processes. The LD_{50} for rats is 3,280 mg./kg. However, prolonged breathing of fumes containing 0.005 mg. of diphenyl per liter of air is considered dangerous for man.

Diphenyl is capable of inhibiting the growth of fungi; hence it is employed to impregnate paper that is used to wrap citrus fruits. A fused mixture of diphenyl and paraffin (1:1) is recommended for treatment of the paper at a rate of 30 g. of diphenyl/kg. of paper and 100 g. of diphenyl/kg. of cardboard. In a number of countries the content of diphenyl in citrus fruits is fixed; it must not exceed 100 mg./kg.

Petroleum oils

The petroleum oils are the oldest pesticides and have been used for several decades to control plants pests and weeds. Crude petroleum was first used in 1778 to control plant pests, and the use of kerosine to control scale insects on oranges was started in 1865. Although petroleum oils in agriculture have been in use for such a long period, investigations in this field have continued up to the present.

Depending on their origin and conditions of formation, petroleums usually are divided into paraffinic, naphthenic, aromatic, and mixed; these designations are associated with the predominance in their composition of one or another class of hydrocarbons. In the USSR special petroleum products are not produced for control of plant pests, and in most cases oils manufactured for various industrial purposes are used. Petroleum products are used in agriculture as pesticides for the following principal purposes:

1. For summer spraying of fruit trees (summer oils).

2. To control overwintering stages of pests of fruit trees and ornamental plants (winter oils).

3. As solvents in various insecticide, fungicide, and herbicide emulsive concentrates and solutions for use as low-volume, finely dispersed sprays.

4. As solvents for insecticides and disinfectants to control domestic insect pests and to protect nonmetallic materials.

5. To control weeds in plantings of umbelliferous and some other crops.

The basic requirements for oils used to control plant pests are the following: (a) high toxicity for the plant pests, (b) safety for the treated plants, (c) low toxicity for man and animals, and (d) availability and low price. Oils used as solvents for pesticides also must dissolve the appropriate compounds in the necessary quantities.

Petroleum products employed in preparations that are used in enclosed premises or residences should not have an unpleasant odor nor contain substances toxic for man or animals. Kerosine fractions of petroleum completely freed of sulfur compounds and of aromatic and unsaturated hydrocarbons by treatment with concentrated sulfuric acid or oleum are most often employed for the production of preparations used in living quarters. Other methods may be recommended for the removal of sulfur compounds that are not eliminated by sulfuric acid. In some cases low-boiling fractions of petroleum and turpentine are used for making such preparations.

As has been indicated, one of the most important conditions set forth for petroleum products employed to control plant pests or weeds is selectivity of action and safety for beneficial plants. This problem has been subjected to careful study in recent decades by many Soviet and foreign investigators. As a result they have been able to establish that the phytocidal effects of mineral oils on most species of plants depend on the physical properties of the oil as well as its composition and the structure of the component hydrocarbons.

The most phytocidal fractions of mineral oils are the aromatic and unsaturated hydrocarbons that on oxidation by the oxygen of the air form various acids having high toxicity for plants. E. I. SVENTSITSKII has established that the paraffinic and naphthenic hydrocarbons are oxidized slowly by the oxygen of the air under practical conditions, and consequently their phytocidal effect is small. The herbicidal effect of petroleum products may appear only when they have a high content of aromatic compounds. In this case the compounds toxic to plants are obtained as a result of the joint oxidation of the aromatic hydrocarbons with compounds of other classes. In such a process, the most easily oxidized compounds are those having an absorption spectrum in the region close to the spectrum of sunlight on the earth's surface. It has been shown that the rate of oxidation of various aromatic compounds by the oxygen of the air has a direct relation to the light absorption spectra of these compounds. Oxidation takes place most rapidly with compounds that have an absorption maximum lying in the visible or the near ultraviolet part of the spectrum. Phytocidal products are also obtained by the reaction of hydrocarbons with ozone, which is always present in small amount in the atmosphere.

From what has been said it is clear that petroleum products containing more than 10% of aromatic and unsaturated hydrocarbons can be employed as herbicides, but only oils that have been freed of aromatic and unsaturated compounds can be used to control pests on plants.

Petroleum products with a high content of aromatic hydrocarbons can be used successfully as nonselective herbicides. In this case the content of aromatic hydrocarbons with b. p. 150°–300° C. should be not less than 70%. Petroleum oils with an aromatic hydrocarbon content of about 20% are used to control weeds in plantings of umbelliferous crops at rates of 325–890 liters/ha.

Oils for summer spraying of green plants should not contain substantial amounts of aromatic hydrocarbons, as determined by the residue after sulfonation with concentrated sulfuric acid. The unsulfonatable residue should be not less than 95% for summer oils. The viscosity by the Engler method should be within the limits 1.5°–1.8°. More viscous oils are dangerous for plants, since they may cause burning of the plants from purely physical causes. When even comparatively harmless oils fall on the foliage of plants they are quickly absorbed, and if their viscosity is high they may clog the vessels and disrupt the vitality of the plants.

Since acids have a phytocidal effect, the acidity of summer oils should not exceed 0.1 (mg. KOH/100 g. of oil). The emulsifier and other additives influence the phytocidal properties of the oils. For summer spraying of plants to control the San Jose scale and other similar pests, preparation no. 30 is used, the composition of which is given below:

Paraffinic distillate from select Surakhany petroleum	40%
Transformer oil	40%
Sulfite waste liquor	2%
OP-7	0.5%
Water	17.5%

Less severe requirements with respect to phytocidal effect are set forth for oils for spraying plants in the early spring period under leafless conditions. There have been observations, however, that systematic spraying of fruit trees with various oils lowers their resistance to frost, especially if the treatment is carried out in late autumn.

Spraying of plants before the buds open can be carried out even with very phytocidal oils, such as the green oil that is a still residue from products of petroleum cracking. It contains an appreciable amount of unsaturated and aromatic compounds. This oil is toxic for most pests of plants and stored products. Fruit and ornamental trees and shrubs are sprayed with emulsions of green oil to control scale insects, coccids, psyllids, aphids, mites, and some other plant pests.

The mechanism of action of oils on plant pests and their eggs is still not entirely clear; but it is supposed that the following factors influence the toxic effect:

(a) disruption of gaseous exchange (access of oxygen hindered);

(b) disruption of water balance of the insect and its eggs;

(c) disruption of the coverings (the cases that are especially important for the eggs);

(d) penetration of the oils into the organism of the insect or into the egg and disruption of enzymatic processes; and

(e) coagulation of the protoplasm and disruption of the tissue structure.

The most important factor apparently is the first, as confirmed by the high activity of those paraffinic and isoparaffinic oils that are resistant to oxidation and consequently can form stable films that hinder metabolism in the egg or in the body of the insect. In the main, the same requirements are set forth for petroleum oils used to dissolve pesticides as for oils employed independently, but the question of phytocidal effect in the former case has a somewhat different significance, since the oils are used in considerably lower concentrations with insecticides and fungicides. Consequently even some aromatic compounds can be used as solvents for pesticides.

It is interesting to note that when unsaturated compounds are present in the oils, the activity of halogen-containing insecticides and some organic fungicides is decreased. This is associated with partial breakdown of the fungicides by joint oxidation by the oxygen of the air. The halogen-containing insecticides enter into a telomerization reaction with the unsaturated compounds, and also with hydrocarbons that are components of turpentine.

Some petroleum products have a marked fungicidal effect. This is noticed particularly with petrolatum, which has been proposed for the protection of wood and for putties.

Coal-tar oils

Besides the petroleum oils, preparations based on various fractions of coal tar are used in agriculture and industry. Since many of the compounds that enter into the composition of coal tar not only have insecticidal and herbicidal effects but are also active fungicides, they are used to protect wood and some other materials from damage by microorganisms and also to control fungal and bacterial diseases of plants.

The distillate obtained by dry distillation of coal in the coking process is usually called coal tar. The composition of the tar depends on the operating conditions for the coking and on the type and characteristics of the coal that is coked. The crude coal tar is subject to fractional distillation into several fractions, from which individual compounds are further separated. Depending on the coal used and the coking process, coal-tar fractions can be obtained that differ in composition. Furthermore, depending on the direction of further treatment, the fractionation may be carried out in different temperature ranges.

Coal-tar fractions are characterized by the following indices: *light* oil distils at a temperature up to 210° C., *medium* or *carbolic* oil at 210° to 240° C., *heavy* or *creosote* oil at 240°–270° C., and finally *anthracene* oil at 270° C. and above. For the isolation of individual compounds from the various coal-tar fractions there are many possible chemical methods.

The main components of coal tar are the aromatic hydrocarbons, which are present in predominant amounts in all fractions. Besides the hydrocarbons, the coal-tar oils contain various groups of derivatives of an acid or basic nature. The acid fraction is represented mainly by aromatic hydroxy derivatives (phenols), and the basic fraction by heterocyclic nitrogen compounds of the pyridine, quinoline, and some other series. The acid and basic fractions of the coal-tar oils usually are separated by treatment with alkalies and acids, respectively. Furthermore, there are some neutral oxygen- and sulfur-containing compounds of the type of dibenzofuran and others in coal-tar oils.

Table VI. *Example of coal-tar composition*

Composition	Content (%)	Composition	Content (%)
Light oil		*Anthracene oil*	
Benzene and toluene (unpurified)	0.3	Phenanthrene	4.0
		Anthracene	1.1
Coumarone, indene, etc.	0.6	Carbazole and its analogs	2.3
Xylenes, cumene and its isomers	1.1	Polycyclic compounds	5.5
		Phenol and its homologs	2.2
Medium and heavy oils		Pyridine, quinoline, acridine and their homologs and analogs	2.3
Naphthalene	10.9		
Compounds close in boiling point to naphthalene	1.7	Other compounds	12.3
1-Methylnaphthalene	1.0	Pitch	44.7
2-Methylnaphthalene	1.5		100.0
Dimethylnaphthalenes	3.4		
Acenaphthene and its analogs	2.4		
Fluorene and its analogs	2.8		

The coal-tar obtained by high-temperature coking as a rule contains a large quantity of aromatic hydrocarbons and heterocyclic nitrogen bases, while with a lower-temperature coking process acids, paraffins, and naphthenes appear in it. As an example, the composition of one sample of coal tar is shown in Table VI. There also are compounds such as 2,3-benzopyrene in coal tar that have a carcinogenic effect; on continuous contact they are capable of causing the formation of cancerous tumors. Therefore, the use of coal-tar oils containing 2,3-benzopyrene for control of plant pests is not permitted.

To control plant pests, anthracene oil is most often applied as a 6–8% emulsion spray to plants while they are in the leafless state. Since anthracene oil severely burns plants (it can even be used as a nonselective herbicide), fruit trees and bushes can be sprayed with it only during the dormant period. Formulations of anthracene oil have not only insecticidal, but also fungicidal and bactericidal effects. It is used to control black rot of fruit trees, lichens, and some other plant diseases. In the Soviet Union wide use is made of a concentrated emulsion of anthracene oil, referred to as KEAM, which is produced by mechanical dispersion of anthracene oil in water in the presence of sulfite liquor. This preparation has the following composition:

Anthracene oil	56–60%
Water	36–39%
Sulfite liquor (dry basis)	4– 5%

KEAM is used to control overwintering stages of pests of fruit trees and shrubs. For this purpose preparations with additions of benzene hexachloride and other insecticides also are suggested.

Anthracene oil is used for wood preservation. As a disinfectant for the impregnation of cross ties anthracene oil with the following properties is used:

Density, not less than	1.08
Content of compounds insoluble in benzene, not more than	0.5%
Water, not more than	1.5%

Content of fractions distilling under:

210° C., not more than	5 vol. %
235° C.	25
355° C.	82
360° C.	not standardized
Content of sediment in oil at 40° C.	None

Depending on the climatic conditions, wood impregnated with coal-tar oils serves one-and-a-half to two times longer than untreated wood. For the protection of wood and other nonmetallic materials, solutions of various active fungicides in coal-tar oils and also other solutions of disinfectants in hydrogenation products of coal-tar oils are recommended. Preparations based on coal-tar oils (creolin) with or without added insecticides are used to control sheep scabies. Preparations based on oils obtained by dry distillation of peat have also been proposed for the control of plant pests. Samples of peat carbolineum have been studied using different fractions of peat tar: 200°–300° C., 240°–315° C., and 240°–300° C. An example of the composition of peat carbolineum follows:

Peat phenols	9.4%
Nitrogenous bases	4.6%
Neutral compounds	51.0%
Emulsifier	31.0%

Peat oil has less insecticidal and acaricidal activity than coal-tar oil.

When plants are treated with coal-tar oils, just as with petroleum oils, their resistance to frost is decreased; furthermore the toxicity of the coal-tar oils is greater than that of the petroleum oils.

The petroleum and coal-tar oils are gradually being displaced by synthetic organic preparations based on phenols and organophosphorus compounds for the control of plant pests.

General references

American Chemical Society: Agricultural applications of petroleum products. Adv. Chem. Series No. 7 (1952).
BEZUGLYI, S. F.: Dusty, emul'sii i suspensii DDT i GKhTsG [Dusts, emulsions and suspensions of DDT and BHC], p. 138. State Scientific and Technical Publishing House of Chemical Literature [USSR] (1955).
—, and V. S. LUKANINA: Khimicheskie sredstva zashchity rastenii [Chemical agents for plant protection], p. 103. Coll. works of Nauchnyi Institut po Udobreniyam i Insektofungitsidam im. Ya. V. Samoilova. State Scientific and Technical Publishing House of Chemical Literature [USSR] (1961).
—, and I. G. SARISHVILI: Dusty, emul'sii i smachivayushchiesya poroshki organicheskikh insektofungitsidov [Dusts, emulsions and wettable powders of organic insectofungicides], p. 78. Coll. works of Nauchnyi Institut po Udobreniyam i Insektofungitsidam im. Ya. V. Samoilova. State Scientific and Technical Publishing House of Chemical Literature [USSR] (1959).
DE ONG, R. E.: Chemistry and uses of pesticides, 2nd ed. New York: Reinhold (1956).
POKROVSKII, E. A.: Dusty, emul'sii i suspensii DDT i GKhTsG [Dusts, emulsions and suspensions of DDT and BHC], p. 174. State Scientific and Technical Publishing House of Chemical Literature [USSR] (1955).
SVENTSITSKII, E. I., E. A. POKROVSKII, and YA. A. MANDEL'BAUM: Organicheskie insektofungitsidy [Organic insectofungicides], p. 119. State Scientific and Technical Publishing House of Chemical Literature [USSR] (1955).

IV. Halogen derivatives of aliphatic hydrocarbons

General characteristics of pesticidal properties

The halogen derivatives of the aliphatic hydrocarbons are considerably more toxic to insects, microorganisms, and plants than the parent hydrocarbons. Furthermore, among the haloalkanes and alkenes there are interesting nematicides that have received practical use.

The toxicity of the haloalkanes increases in the order: chloroalkanes, bromoalkanes, iodoalkanes. An increase in molecular weight of the haloalkane also increases the toxicity of the compound. An exception is presented by the derivatives of methane, which exceed all the other compounds of this class in toxicity.

Table VII shows the toxicity of some monohaloalkanes to granary weevils. The haloalkanes of isostructure are less toxic than those of the same molecular weight with a normal structure of the carbon chain.

Table VII. *Toxicity of monohaloalkanes to granary weevils*

Radicals	LC_{50} (mg./l.) [a]			Radicals	LC_{50} (mg./l.) [a]		
	Alkyl				Alkyl		
	Chloride	Bromide	Iodide		Chloride	Bromide	Iodide
CH_3	166	3	2	C_4H_9	82	66	5
C_2H_5	1124	205	11	$(CH_3)_2CHCH_2$	200	130	–
C_3H_7	428	133	6	C_5H_{11}	73	35	4.6
$(CH_3)_2CH$	740	216	65	$(CH_3)_2CHCH_2CH_2$	93	45	–

[a] LC_{50} = mean concentration of compound causing death of 50% of the test organisms.

Accumulation of halogen atoms in the methane molecule leads to a decrease in toxicity of the compound, while accumulation of halogen atoms in derivatives of the homologs of methane increases the insecticidal effect somewhat. Isomeric compounds differ greatly in activity.

In Table VIII the toxicity of some polyhaloalkanes to granary weevils is given.

The insecticidal activity of the halogen derivatives of the alkenes is considerably higher than that of the alkane derivatives. This can be seen from the data on toxicity of some haloalkenes to the confused flour beetle in Table IX.

The biological activity of the haloalkanes and haloalkenes has a direct relation to their structure and reactivity. Toxicity to insects and nematodes runs parallel with reactivity. This observation shows that the action of the halogen derivatives of aliphatic hydrocarbons on insects and nematodes has a biochemical nature and is associated with the reaction of these compounds with their vital systems. A large number of halogen derivatives of aliphatic hydrocarbons have found practical use for the control of pests of stored products and for the fumigation of soil to control nematodes and soil-inhabiting insects and microorganisms. Many compounds are used not as individual substances but in a mixture with isomers, homologs, analogs, or some other compounds.

Table VIII. *Toxicity of polyhaloalkanes to granary weevils*

Compound	LC_{50} (mg./l.)	Compound	LC_{50} (mg./l.)
Methylene bromide	90	1,1,1-Trichloroethane	290
Methylene chloride	380	1,1,2-Trichloroethane	53
Chloroform	250	1,1,1,2-Tetrachloroethane	33
Carbon tetrachloride	275	1,1,2,2-Tetrachloroethane	15
1,2-Dichloroethane	99	Pentachloroethane	16
1,1-Dichloroethane	380		

Table IX. *Toxicity of haloalkanes and haloalkenes to the confused flour beetle*

Haloalkanes	LC_{50} (mg./l.)	Haloalkenes	LC_{50} (mg./l.)
Ethyl bromide	150	Allyl bromide	9
Butyl bromide	100	Methylallyl bromide	14
1,2-Dibromoethane	14	Methylallyl chloride	12
1,2-Dichloroethane	19	2,3-Dichloropropene-1	2.9
1,2-Dichloropropane	40	Hexachloropropene	1.1
1,3-Dichloropropane	10	Hexachlorobutadiene	4.0
1,4-Dichlorobutane	11		
Heptachloropropane	2.5		

The following compounds are used as fumigants to control pests of stored products and nematodes: methyl bromide, 1,2-dichloroethane, 1,2- and 1,3-dichloropropanes mixed with 1,2- and 1,3-dichloropropenes (DD Mixture), methylallyl chloride, 1-chloro-2,3-dibromopropane (Nemagon), etc. Good results have been obtained in the control of phylloxera of grape vines by hexachlorobutadiene, as suggested by L. M. KOGAN and YA. I. PRINTS. 1,1,1-Trichloro-2,3-dibromopropane, a mixture of trichloropropenes obtained by dehydrochlorination of 1,1,1,3-tetrachloropropane, and propargyl bromide also have satisfactory nematicidal and antiphylloxeral effects. Such compounds as methyl chloride, Freon-12, carbon tetrachloride, and polychloroethylenes have found use in mixtures.

Individual compounds

Methyl chloride is a gas, b.p. $-23.36°$ C., m.p. $-97.6°$ C., critical pressure 65.9 atm., and critical temperature 143° C. Latent heat of vaporization at b.p. is 102.45 Cal./kg., heat capacity of liquid product is 0.381 Cal./kg., vapor pressure at $-10°$ C. $= 1.784$, $0°$ C. $= 2.571$, $10°$ C. $= 3.622$, and $20°$ C. $= 4.985$ atm. At $0°$ C. 1 liter of water dissolves 3.4 liters of gaseous methyl chloride. The limits of flammability of mixtures with air are 8.25–18.2 vol. %. Methyl chloride is often used in aerosol bombs to control flies and other flying insects in enclosed premises. In the United States the maximum permissible concentration of methyl chloride vapor in air is 100 mg./m.^{3}; in the USSR it is 5 mg./m.^{3}. It is highly toxic to man and animals. At a concentration of 2 mg. of methyl chloride per liter of air white mice die after four hours exposure. It is a valuable starting material for the production of a large number of practically important organic compounds, including methylchlorosilanes, some herbicides (e. g., paraquat), etc.

The most important industrial methods for producing methyl chloride are:
1. chlorination of methane from natural gas
2. oxidative chlorination of methane
3. action of hydrogen chloride on methanol
4. reaction of phosphorus trichloride with methanol.

The last method usually is employed for the production of dimethyl phosphite, but methyl chloride is a valuable by-product.

The most important industrial method is the first one, in which chloroform and carbon tetrachloride are produced in addition to the methyl chloride. Because methyl chloride of high purity is required for all purposes, great attention is paid in its production by the first and second methods to purification by fractional distillation; it is obtained with a higher degree of purity by the third and fourth methods.

Carbon tetrachloride is a colorless liquid, b.p. 76.5° C., m.p. $-22.85°$ C., vapor pressure at $0°$ C. $= 33.4$, $10°$ C $= 56.33$, $20°$ C. $= 91.3$, $30°$ C. $= 141.1$ mm. of Hg. Latent heat of vaporization is 46.8 Cal./kg.; heat capacity of liquid product at $20°$ C. $= 0.205$ Cal./kg.; d_4^{20} 1.594; n_D^{20} 1.4601. At $25°$ C., 100 g. of water dissolves 0.08 g., and at the same temperature 100 g. of carbon tetrachloride dissolves 0.013 g. of water. It is not used alone, since it is considerably less toxic to noxious insects than dichloroethane. It is most often employed as a fire-preventive additive to easily flammable fumigants, for example, to carbon disulfide, dichloroethane, methyl formate, and others. It is rather toxic to man and animals. Even at a concentration of 2 mg./liter of air headaches appear in

30 minutes; the maximum permissible concentration of its vapors in air is 20 mg./m.³.

In the dry state it does not affect most metals at room temperature, but when it is heated in the presence of water it may cause corrosion. When it reacts with water at an elevated temperature, phosgene may be formed:

$$CCl_4 + H_2O \rightarrow COCl_2 + 2\,HCl$$

In industry carbon tetrachloride is produced by the chlorination of carbon disulfide or of the simplest hydrocarbons.

Methyl bromide is a colorless liquid, b.p. 3.6° C., m.p. −93.7° C., d_4^0 1.732. At room temperature it exists in vessels under some pressure (1,250 mm. of Hg at 20° C.). Its heat capacity (liquid) at 0° C. is 0.12 Cal./kg.; the latent heat of vaporization is 61.8 Cal./kg. At 17° C., 100 g. of water dissolves 1.83 g. of methyl bromide. A mixture of the vapors with air is explosive within the limits 13.5–14.5 vol. %. It is widely used to control pests of stored products and to fumigate the soil for the control of plant pests and weeds (dosage $\sim$ 25–60 g./m.³). It approaches HCN in effectiveness, but it is less dangerous to plants and seeds. In the United States 7,880 tons of this compound were produced in 1963.

Methyl bromide is highly poisonous for man and animals, and therefore it should be used with great caution; one should always wear a gas mask when working with it. The concentration in working areas should not exceed 1 mg./m.³ of air. With respect to its chemical properties methyl bromide is a characteristic representative of the monohaloalkanes. It readily enters into exchange reactions; its reactivity is higher than that of methyl chloride.

In connection with the use of methyl bromide for the fumigation of stored food products, its reaction with the components of grain has been studied with the aid of C-14 labeled material. These investigations have shown that at 20° C. and 760 mm. of Hg in a humid atmosphere it acts as a methylating agent. As a result of methylation, disruption of the normal vital processes of the grain occurs, and its germinating capacity is lowered. Apparently methyl bromide also reacts similarly with vitally important systems in insects, thereby killing them.

Methyl bromide is obtained in good yields by the following reactions:
1. From methanol and salts of HBr in the presence of sulfuric acid:

$$CH_3OH + NaBr + H_2SO_4 \rightarrow CH_3Br + H_2O + NaHSO_4$$

2. By the reaction of methanol with bromine and hydrogen sulfide:

$$2\,CH_3OH + Br_2 + H_2S \rightarrow 2\,CH_3Br + 2\,H_2O + S$$

3. By the reaction of methanol with bromine and sulfur dioxide:

$$2\,CH_3OH + Br_2 + SO_2 \rightarrow 2\,CH_3Br + H_2SO_4$$

All three methods can be carried out industrially by both batch and continuous processes.

1,2-Dichloroethane is a liquid, b.p. 83.7° C., m.p. −35.3° C., vapor pressure at 20° C. 78 mm. of Hg, d_4^{20} 1.2569, n_D^{20} 1.4443. At 20° C., 0.869 g. of dichloroethane dissolves in 100 g. of water. It is slowly hydrolyzed by water, giving off HCl. Caustic alkalies split out a molecule of HCl with the formation of vinyl chloride. Like most haloalkanes, dichloroethane readily enters into exchange reactions. The maximum permissible concentration of vapors in air for prolonged work is 10 mg./m.^{3}. It will burn. The ignition temperature is 14.4° C., the temperature of spontaneous combustion is 449° C. The dangerously explosive concentration of the vapors at 50° C. lies within the limits 5.7–16 vol. %. In the presence of 300 g./m.^{3} of carbon tetrachloride, dichloroethane does not ignite.

Dichloroethane is used to control pests of stored products, mainly for fumigation of grain, alone or mixed with chloropicrin or other compounds. The rate of application is from 200–600 g./m.^{3}. It also is employed to control soil-inhabiting pests at a dosage from 30–600 kg./ha. (1 kg./ha. = 0.89 lb./acre).

1,2-Dichloroethane is produced by direct chlorination of ethylene.

1,2-Dichloropropane is a colorless liquid, b.p. 95.4° C., m.p. −70° C., d_{20}^0 1.1595, vapor pressure at 19.6° C. 210 mm. of Hg, ignition temperature 20° C. At 20° C., 0.27 g. dissolves in 100 g. of water. Its chemical properties are similar to those of dichloroethane. It is obtained as a by-product in the production of glycerine by the chlorine method.

1,2-Dichloropropane is a component of DD-Mixture, which is used to control root nematodes by soil fumigation. It also enters into the composition of Dowfume EB-5, which contains 7.2% 1,2-dibromoethane, 29.5% 1,2-dichloropropane, and 63.6% carbon tetrachloride. The maximum permissible concentration of 1,2-dichloropropane in the air is 10 mg./m.^{3}.

Hexachloroethane is a white crystalline compound with a characteristic odor, m.p. in a sealed capillary 187° C. It is practically insoluble in water, but highly soluble in most organic solvents, and of relatively low toxicity to man and animals. It is used to control fly larvae in their breeding places, for treatment of manure, drain pits, etc. It is obtained by the chlorination of dichloroethane, ethane, and polychloroethanes with chlorine at an elevated temperature in the presence of catalysts:

$$C_2H_4Cl_2 + 4\,Cl_2 \rightarrow C_2Cl_6 + 4\,HCl$$

1,2-Dibromoethane is a colorless liquid, b.p. 131.5° C., m.p. 9.3° C., vapor pressure at 25° C. 11 mm. of Hg, d_{25}^{25} 2.172. At 30° C., 0.43 g. dissolves in 100 g. of water. It is highly soluble in most organic solvents. The LD$_{50}$ for white mice is 146 mg./kg. It is used as a fumigant for the control of soil-inhabiting pests, including nematodes, at a dosage of up to 100 kg./ha. Dowfume B-85 contains 83% 1,2-dibromoethane. It can be used in the form of emulsions and of granules on a carrier. It is produced by direct bromination of ethylene.

1-Chloro-2,3-dibromopropane (Nemagon) is a colorless or pale–yellow liquid, b.p. 196° C., d^{20} 2.08, n_D^{25} 1.5518, vapor pressure at 21° C. 0.8 mm.

of Hg. Its solubility in water is about 0.1%. The ignition temperature is about 80° C. It approaches dibromoethane in toxicity. Its LD_{50} is 170 to 260 mg./kg. It exceeds DD-Mixture by eight to ten times in nematicidal activity and is used as a soil fumigant at a dosage of from 20–100 kg./ha. It is used both in the form of an emulsion and in the form of granules with 10 and 25% active ingredient. The preparation Fumazone contains 47.6% of 1-chloro-2,3-dibromopropane, a petroleum solvent, and an emulsifier.

1-Chloro-2,3-dibromopropane is produced by bromination of allyl chloride.

1,3-Dichloropropene is used in agriculture as the technical product, usually consisting of a mixture of the *cis*- and *trans*-isomers with a small amount of 3,3-dichloropropene as an impurity. *cis*-1,3-Dichloropropene is a liquid, b.p. 104.2° C., d_4^{20} 1.224, n_D^{24} 1.4682; *trans*-1,3-dichloropropene has b.p. 112.1° C., d_4^{20} 1.217, n_D^{20} 1.4730. The technical grade product has d_4^{25} 1.210. At 20° C., 0.1 g. dissolves in 100 g. of water. Its ignition point is about 21° C. The LD_{50} for rats is 250–500 mg./kg.; the maximum permissible concentration in air is 5 mg./m.^{3}.

A technical mixture of dichloropropenes is used to control nematodes in the United States under the name of Telone. This mixture also enters into the composition of the preparation Dorlone (a mixture with dibromoethane). Good results are obtained in nematode control with the use of 200–250 kg./hg.

The 1,3-dichloropropenes are an important constituent of the Shell Company's DD-Mixture, which under technical conditions has the following composition:

cis-1,3-Dichloropropene	30–33%
trans-1,3-Dichloropropene	30–33%
1,2-Dichloropropane	30–35%
Trichloropropanes and more highly chlorinated products	up to 5%

Not less than 95% of the liquid should distill below 142° C., d_4^{20} 1.198, vapor pressure at 20° C. 31.3 mm. of Hg.

A mixture of 1,3-dichloropropenes is obtained as a by-product in the production of allyl chloride by chlorination of propene:

$$CH_3CH=CH_2 + Cl_2 \rightarrow ClCH_2CH=CH_2 + HCl$$
$$ClCH_2CH=CH_2 + Cl_2 \rightarrow ClCH_2CH=CHCl + HCl$$

Besides the 1,3-dichloropropenes some of the compounds isomeric with them are formed, and also products of more extensive chlorination. By substitutive chlorination of propene the isomeric dichloropropenes are obtained in the following proportions:

cis-1,3-Dichloropropene	42%
trans-1,3-Dichloropropene	45%
3,3-Dichloropropene	12%
2,3-Dichloropropene	Traces

The presence of halogens and a double bond determines the chemical properties of the dichloropropenes. Thus, they are capable of a large number of halogen addition and substitution reactions. The allyl chlorine is especially easily replaced.

On reaction with salts of HBr 1-bromo-3-chloropropene is obtained:

$$ClCH_2CH = CHCl + KBr \rightarrow BrCH_2CH = CHCl + KCl$$

which is used in the form of a technical mixture with b.p. 130°–180° C. as a nematicide known as CBP-55.

1-Chloro-2-methylpropene-2 (methallyl chloride) is a colorless liquid with a characteristic odor, b.p. 72.17° C., vapor pressure at 20° C. 102.3 mm. of Hg, d_4^{20} 0.9257, n_D^{20} 1.4276, solubility in water about 0.1%. It is used as an experimental fumigant in European countries and gives satisfactory results in the control of a complex of granary pests. A deficiency of this compound is its high flammability. Its chemical properties recall those of allyl chloride. It is obtained by substitutive chlorination of isobutylene. The reaction of isobutylene with chlorine proceeds very smoothly at relatively low temperatures. Thus, for example, when isobutylene is chlorinated at 0° C. a product of the following composition is formed:

tert.-Butyl chloride	0.7%
Methallyl chloride	83.4%
Isocrotyl chloride	2.5%
Isomeric dichlorides	13.4%

A mixture of dichloroisobutylenes under the name of Nemacure is manufactured by the Bayer company for the control of root nematodes. Depending on the temperature of the soil, Nemacure is employed at 350–550 kg./ha. mainly against the onion nematode. The principal component of this preparation is 1,3-dichloro-2-methylpropene (b.p. 132° C., d_0^{20} 1.1726).

Hexachlorobutadiene is a colorless liquid with a characteristic odor, b.p. 215° C., m.p. −22° C., d_4^{20} 1.6820, n_D^{20} 1.5542. About 20 mg. of it dissolves in 100 g. of water; it is highly soluble in organic solvents. Hexachlorobutadiene has been proposed for use as an agent to control phylloxera of grape vines. At 150–250 kg./ha., in a few months it practically completely frees the vines from phylloxera. The duration of effectiveness of the compound is about four years. At high dosages it can cause damage to grape vines. The LD_{50} is 200–250 mg./kg.

Hexachlorobutadiene is produced by thermal chlorination of polychlorobutanes at a temperature of about 400° C. either with or without catalysts.

trans-1,4-Dibromobutene is an interesting experimental nematicide and soil fungicide, m.p. 54° C., b.p. 205° C. It has a slight lachrymatory effect.

In recent years a number of other unsaturated compounds also have been suggested, which at present are being studied but still have not received very wide practical use. In particular, such compounds include propargyl bromide, 1,4-dichloro- and 1,4-dibromobutyne, and some others.

General references

AZINGER, F.: Parafinovye uglevodorody, khimiya i tekhnologiya [The paraffin hydrocarbons, their chemistry and technology]. State Scientific and Technical Publishing House of the Petroleum and Mineral-Fuel Industry [USSR] (1959).
— Khimiya i tekhnologiya monoolefinov [Chemistry and technology of the monoolefins]. State Scientific and Technical Publishing House of the Petroleum and Mineral-Fuel Industry [USSR] (1960).
CHEKALINA, F. I.: Khim. v Sel'skom Khozyaistve, No. 6, 36 (1965).
METCALF, R. L., Ed.: Adv. Pest Control Research, vol. III (1961).
KOGAN, L. M., Uspekhy Khim. **28**, 133 (1959).
POPOV, P. V.: Trudy Nauch. Inst. Udobreniyam i Insektofungitsidam im. Ya. V. Samoilova, No. 135, **167** (1939).
VOLODKOVICH, S. D.: Zhur. Vsesoyuz. Khim. Obshchestva im. D. I. Mendeleeva **9**, 512 (1964).

V. Halogen derivatives of alicyclic hydrocarbons

General characteristics of pesticidal properties

Among the halogen derivatives of the alicyclic hydrocarbons are many compounds that have practical importance. Some of them are produced on a large commercial scale.

With respect to volume of production, some individual representatives of this series are surpassed only by DDT.

In the series of mono- and dihalogen derivatives of alicyclic hydrocarbons the same relationships of insecticidal and fungicidal activity to composition and structure are observed as in the derivatives of aliphatic hydrocarbons. The bromine derivatives as a rule are more active than the corresponding chlorine derivatives. Accumulation of halogens in the molecule somewhat increases the activity of a compound, but only up to a certain limit. A systematic study of the insecticidal properties of a large number of different halogen derivatives of hydrocarbons has shown that compounds having a molecular weight lower than 480 and a melting point lower than 200° C. possess insecticidal properties.

Compounds with a higher molecular weight and a low solubility in lipoids are practically inactive as contact insecticides, a fact apparently associated with their slight penetration through the surface coverings of insects and their low rate of diffusion and transport in insect tissues.

Among the halogen derivatives of alicyclic hydrocarbons with molecular weight below 480 there are a number of valuable insectofungicides.

Of the monocyclic derivatives benzene hexachloride has been widely used. The halogen derivatives of polycyclic hydrocarbons are represented by chloroindan, heptachlor, aldrin, isodrin, dihydroheptachlor, and also dieldrin and endrin, which are oxidation products of aldrin and isodrin, respectively.

Benzene hexachloride (BHC, HCCH) and its analogs

Benzene hexachloride (1,2,3,4,5,6-hexachlorocyclohexane) is an important, active insecticide; 8 stereoisomers are known, but only one, the γ-isomer, is an active insecticide.

It has been established that the cyclohexane molecule has the chair form

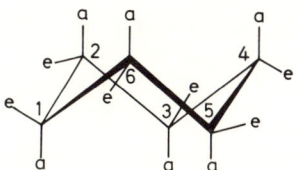

where atoms 1, 3, and 5 lie in one plane and atoms 2, 4, and 6 in another parallel plane. The bonds of the carbon atoms with the other atoms and groups may lie in two directions, perpendicular to the planes mentioned or directed toward the periphery of the molecule (usually these bonds lie at an angle of 109° 28'). The bonds of the first type are called axial (or polar) and are designated by the Latin letters a or (p); the bonds of the second type are called equatorial and are designated by the letter e. In Table X the position of the bonds of the chlorine atoms in the different isomers of BHC are given.

Table X. *Structure of isomers of benzene hexachloride*

Isomer	Orientation of Cl atoms	Isomer	Orientation of Cl atoms
α	aaeeee	ε	aeeaee
β	eeeeee	ζ	aaeaee
γ	aaaeee	η	aeaaee
δ	aeeeee	ϑ	aeaeee

In Table XI the melting points of the isomers of BHC and some related compounds are given. It should be noted that the compounds listed in Table XI, with the exception of the γ-isomer of BHC, are practically nontoxic for most insects and fungi. Only δ-BHC has an appreciable fungicidal effect.

To characterize their insecticidal properties, the comparative insecticidal activities of the individual isomers of BHC for various species of insects are given in Table XII, with the insecticidal activity of the γ-isomer taken as unity. In this same table are given the toxicities of the isomers of BHC for white rats. As seen, γ-BHC exceeds all the other isomers in insecticidal activity by 50–10,000 times. It is interesting to note that a mixture of 1,2,3,4,5,6-hexachloro-1-methyl-cyclohexanes obtained by photochemical chlorination of toluene also has high insecticidal properties for a number of insects.

The isomers of BHC differ not only in construction and crystalline structure, but also in solubility in various organic solvents. All the isomers are relatively poorly soluble in paraffinic and cycloparaffinic hydrocarbons, more soluble in unsaturated ones, and still more soluble in aromatic hydrocarbons. Their solubility in the chloro-derivatives of aromatic and aliphatic hydrocarbons is higher than in the hydrocarbons. The isomers of BHC are readily soluble in some alcohols, ketones, esters, and others. The solubilities of some of the isomers of BHC in organic solvents

Table XI. *Melting points of isomers of benzene hexachloride and analogs*

Compound	M. P. (°C.)	Compound	M. P. (°C.)
α-Benzene hexachloride	157.5–158.5	β-1,2,4,5-Tetrabromo-cyclohexane	218
β-Benzene hexachloride	309		
γ-Benzene hexachloride	112.8	β-1,2-Dibromo-4,5-dichlorocyclohexane	240
δ-Benzene hexachloride	138–139		
ε-Benzene hexachloride	218.5–219.3	α-Benzene hexabromide	221
ζ-Benzene hexachloride	88–89	β-Benzene hexabromide	253
η-Benzene hexachloride	89.8–90.5	γ-Benzene hexabromide	160
ϑ-Benzene hexachloride	124–125	β-1,2-Dibromo-3,4,5,6-tetrachlorocyclohexane	282
α-1,1,2,4,4,5-Benzene hexachloride	145	1,2,3,4-Tetrabromo-5,6-dichlorocyclohexane	203
β-1,2,4,5-Tetrachlorocyclohexane	228	1,2,4,5-Tetrabromo-1,4-dichlorocyclohexane	200
		α-Heptachlorocyclohexane	147

Table XII. *Comparative toxicity of isomers of benzene hexachloride*

Species of organism	Isomers of benzene hexachloride					
	α	β	γ	δ	ε	ζ
Grain weevil	1/900	1/5,000	1	1/5,500	—	—
Malaria mosquito	1/250	1/10,000	1	1/250	—	—
House fly	1/10,000	—	1	1/3,000	1/10,000	—
Leaf-eating coccinellid	1/520	1/220	1	1/55	—	1/410
Confused flour bettle	1/2,800	—	1	1/7.3	1/53	—
Thrips	1/1,000	1/10,000	1	1/10,000	—	1/10,000
Rats (LD_{50} in mg./kg.)	500	6,000	125	1,000	1,000	—

are given in Table XIII. The difference in solubility of the individual isomers in organic solvents is utilized in separating them and isolating pure γ-BHC. Usually methyl alcohol or acetone is used to extract the γ-isomer from the reaction mixture.

The γ-isomer of BHC is volatile (vapor pressure at 20° C., 9.4×10^{-6} mm. of Hg.) and to increase the duration of its effect polychlorodiphenyls, polychloronaphthalenes, or some other compounds often are added to it. The maximum permissible concentration of BHC in air is 0.1 mg./m.³, and that of the pure γ-isomer is 0.05 mg./m.³

To control plant pests BHC is utilized mainly in the form of 99–100% γ-isomer, which is known as "lindane". As an additive to seed disinfectants a preparation containing not <90% γ-BHC is often used. The technical grade product containing 10–14% γ-isomer is seldom used now.

Table XIII. *Solubility of isomers of benzene hexachloride*

Solvent	Solubility (g./100 g. solution)			
	α	β	γ	δ
Acetone	13.9	7.9	43.5	71.1
Benzene	9.9	1.12	28.9	41.1
Water	1×10^{-4}	0.5×10^{-4}	1×10^{-4}	—
Dichloroethane	7.9	0.6	28.9	27.3
Isooctane	12.7	6.9	35.7	58.5
Isopropyl alcohol	0.6	0.4	2.8	18.0
Xylene	8.5	3.3	24.7	42.1
Methyl alcohol	2.3	1.6	7.4	27.3
Chloroform	6.3	0.3	24.0	13.7
Ethyl alcohol	1.8	1.1	6.4	24.2

BHC differs comparatively little in its chemical properties from the other halogen derivatives of hydrocarbons that have halogen atoms on adjacent carbon atoms. Water acts slowly at room temperature on BHC and only on heating to 100° C. is an appreciable amount of HCl formed. Thus, when steam acts on BHC at about 102° C. for an hour about 0.13% HCl is split out from the γ-isomer. When γ-BHC is heated with water at 200° C. in sealed tubes 1,2,4-trichlorobenzene and its hydrolysis products are obtained.

The dehydrochlorination of BHC is accelerated by water in the light and in the presence of bases. It is assumed that the first product of the splitting out of HCl is pentachlorocyclohexene, which further goes to tetrachlorocyclohexadiene and trichlorobenzenes:

$$C_6H_6Cl_6 \xrightarrow[-HCl]{} C_6H_5Cl_5 \xrightarrow[-HCl]{} C_6H_4Cl_4 \xrightarrow[-HCl]{} C_6H_3Cl_3$$

Impurities of pentachlorocyclohexene and tetrachlorocyclohexadiene give BHC its characteristic odor. Very pure preparations of BHC do not have an odor and only on storage in the light and in the presence of moisture or bases do they acquire an unpleasant odor.

With caustic alkalies in alcohol solution and in aqueous solution on heating BHC splits out three molecules of HCl, going to trichlorobenzenes:

The main product of the reaction is 1,2,4-trichlorobenzene, which is obtained from the different isomers in yields of 75–95%. In addition to this isomer certain amounts of 1,2,3-trichlorobenzene (3–20%) and 1,3,5-trichlorobenzene (0–17.6%) also are formed. By taking advantage of the different reaction rates of caustic alkalies with the different isomers of BHC a kinetic method has been developed for the quantitative determination of the γ-isomer in a mixture. Splitting out of HCl also occurs upon reaction with lime, ammonia, and organic amines.

Breakdown of BHC into trichlorobenzenes and HCl proceeds at an elevated temperature in the presence of compounds capable of initiating such a decomposition (chlorine, iron, aluminum, their salts, and others). In the presence of these initiators breakdown of BHC takes place at 250° to 350° C. When chlorine is employed as the initiator of the reaction, part of the trichlorobenzene is chlorinated. With an excess of chlorine, hexachlorobenzene may be obtained in good yields. This reaction is utilized on an industrial scale for the preparation of tri-, tetra-, and hexachlorobenzenes from the nontoxic isomers of BHC.

By heating to 170° C. it is possible to isomerize individual isomers to the other isomers. In the presence of ferric chloride the reaction proceeds to the extent of not more than 50%, and the relative amount of γ-isomer formed in any one experiment did not exceed 4.4%.

BHC is reduced by zinc dust in acid medium to benzene and this reaction is utilized for its quantitative determination.

It reacts slowly with chlorine and it has been possible to chlorinate it to undecachlorocyclohexane. It reacts also with sulfides and thiocyanates of the alkali metals and some other compounds.

Three methods are known for the preparation of BHC:

1. Chlorination of cyclohexane:

$$C_6H_{12} + 6\,Cl_2 \rightarrow C_6H_6Cl_6 + 6\,HCl$$

2. Chlorination of cyclohexene:

$$C_6H_{10} + 5\,Cl_2 \rightarrow C_6H_6Cl_6 + 4\,HCl$$

3. Chlorination of benzene:

$$C_6H_6 + 3\,Cl_2 \rightarrow C_6H_6Cl_6$$

The first two methods have not found practical use and have purely theoretical significance, while the last reaction is widely used in industry. The chlorination of benzene is a chain reaction and proceeds through the stages of formation of dichlorocyclohexadiene and tetrachlorocyclohexene. Practically, the chlorination of benzene can be carried out in the presence of different initiators: organic peroxides, unsaturated compounds (ethylene, terpenes, etc.), and also with irradiation by actinic light or hard γ-radiation from radioactive cobalt (Co-60).

In industry the photochemical chlorination of benzene with irradiation of the reaction medium by ultraviolet light has received the widest use.

The chlorination is carried out either in excess benzene or in other organic solvents, most frequently methylene chloride. In the chlorination of benzene a mixture of isomers of BHC is obtained with the following content:

α-Isomer	53–70%
β-Isomer	3–14%
γ-Isomer	11–18%
δ-Isomer	6–10%
Other isomers	3–5%
Heptachlorocyclohexane	3–4%
Octachlorocyclohexane	0.5–1%
Oily compounds (composition undetermined)	0.5–3%

The optimum conditions for the preparation of BHC with the largest possible content of γ-isomer are the following:

1. Low temperature chlorination (from $-20°$ to $-30°$ C.). It is recommended that the process be carried out in organic solvents, since benzene crystallizes at a temperature below $+6°$ C. When the chlorination reaction is carried out in excess benzene it is recommended that the temperature be kept not higher than $24°$ C. Chlorination at a lower temperature considerably decreases the reaction rate and consequently the productivity of the equipment.

2. A high concentration of chlorine in the reaction medium also favors an increase in the γ-isomer content of the technical grade product.

3. The concentration of the chlorination products in the final reaction solution should be as low as possible.

The purity of the starting materials is very important because such impurities as oxygen of the air, iron compounds, and some other substances affect the process negatively. Traces of iron not only retard the main process, but they also promote the formation of side products of the reaction, products of the replacement of hydrogen by chlorine in the benzene molecule. After the chlorination is ended the product can be separated from the reaction solution by many methods: distilling off the solvent (or excess benzene) with steam, in vacuum, or under atmospheric pressure, fractional distillation of the solvent (or the excess benzene), and removal of part of the product by filtration with subsequent steam distillation of the rest of the solvent. In the last case it is almost exclusively the α- and β-isomers that are removed by the initial filtration, and after distillation of the rest of the solvent a technical product is obtained containing 30–40% γ-isomer. Moreover, drying of the final product is facilitated, since it has a low melting point and water is easily separated from the molten preparation. Furthermore, this process can be employed for the production of lindane (99.5–100% γ-isomer).

BHC also is produced on an industrial scale by chlorinating benzene in the presence of aqueous solutions of caustic alkalies or lime at a temperature of about 0° C. Usually the benzene, ice, and alkali solution are placed in the apparatus and liquid chlorine is added. During the reaction the temperature rises somewhat and sometimes reaches 30° C. This process is more difficult to regulate than that described, but the technical product obtained by this method contains up to 18% γ-benzene hexachloride.

Table XIV. *Compounds obtained from by-product isomers of benzene hexachloride*

Compound	Purpose
Trichlorobenzene	Solvent, liquid for transformers, intermediate for organic synthesis
1,2,4,5-Tetrachlorobenzene	Herbicide, intermediate for organic synthesis
Hexachlorobenzene	Seed disinfectant, intermediate for organic synthesis
2,5-Dichlorophenol	Intermediate for organic synthesis
2,4,5-Trichlorophenol	Disinfectant, intermediate for organic synthesis
Sodium 2,4,5-trichlorophenolate	Disinfectant
Copper 2,4,5-trichlorophenolate	Seed disinfectant for cotton, disinfectant for nonmetallic materials
2,4,5-Trichlorophenoxy acetic acid and its derivatives	Herbicides
2,4,5-Trichlorothiophenol	Intermediate for preparation of renacite
Pentachlorophenol	Disinfectant, desiccant, herbicide
Sodium pentachlorophenolate	Same
Pentachlorothiophenol	Intermediate for preparation of renacite
1,2,4,5-Tetrachloronitrobenzene	Sprout inhibitor for potatoes
Pentachloronitrobenzene	Soil fungicide, seed disinfectant
Hexachlorodihydroxydiphenyl-methane	Disinfectant
Tetrachlorobenzoquinone	Seed disinfectant, intermediate for organic synthesis
Hexachlorocyclohexenone	Herbicide
Trichlorodinitrobenzene	Selective fungicide
2-Methoxy-3,6-dichlorobenzoic acid	Herbicide
2-Methoxytrichlorobenzoic acid	Herbicide

The isolation of the pure γ-isomer from the technical product is an important problem. γ-BHC can be isolated by extraction with suitable solvents. To obtain the γ-isomer the crude product is treated at a definite temperature with the calculated amount of methanol (or other appropriate solvent), the undissolved α- and β-isomers are filtered off, and the filtrate is cooled. By this process technical γ-isomer is isolated after removal from the mother liquor and drying, and contains more than 90% of the principal compound. To obtain lindane the preparation is recrystallized and the γ-isomer is removed from it by extraction. Lindane can be obtained in this way in a >80% yield. The extraction process is affected not only by the temperature and the proportions of the components, but also by the size of

Halogen derivatives of alicyclic hydrocarbons

Diagram

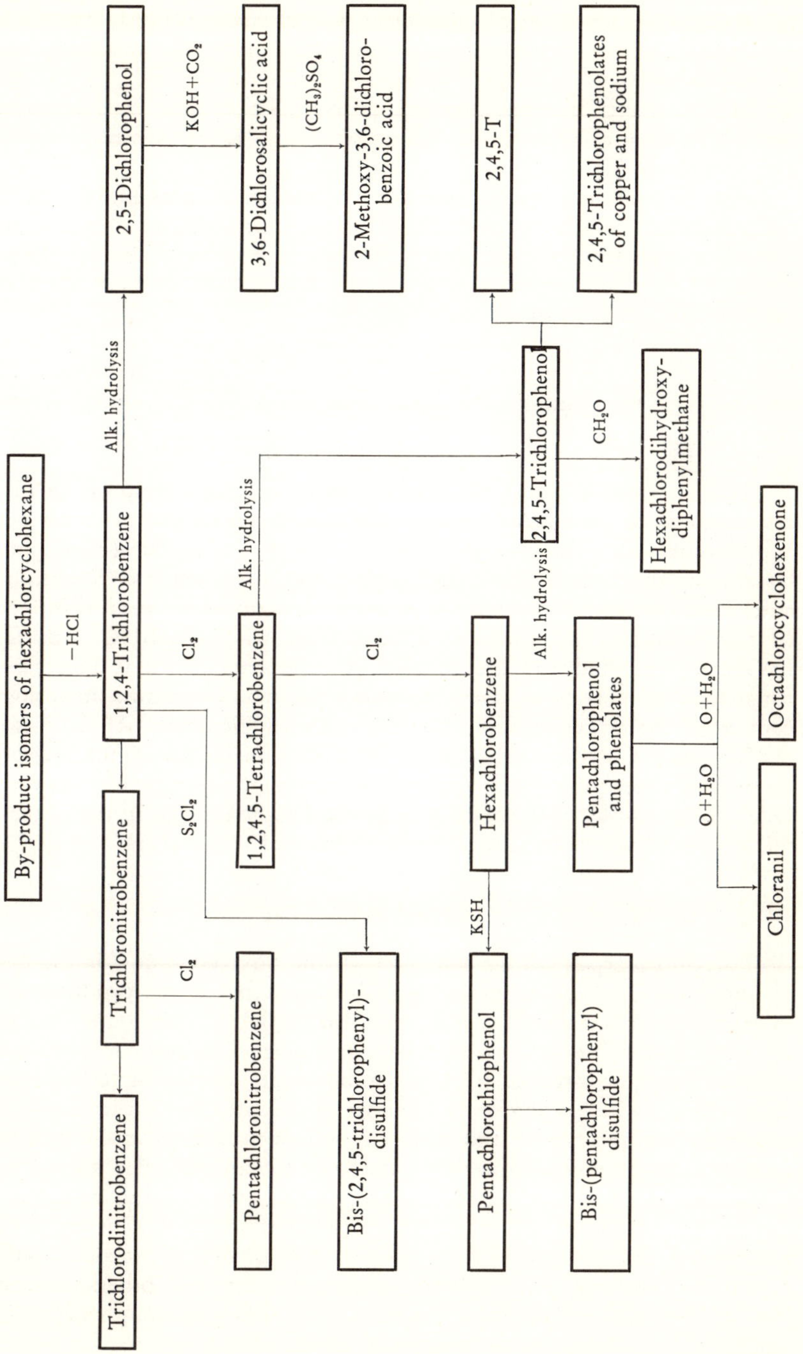

the crystals of the technical product as well as many other factors. γ-BHC is best extracted by repeated use of the mother liquor from the separation of the γ-isomer.

In all countries it is predominantly lindane that is manufactured because it is convenient to use, has no odor, and leaves less residue in food and forage products.

Also of practical interest are the by-product isomers known as "non-toxic isomers", which can be utilized for the production of valuable preparations in agriculture and industry. Some of these products are shown in Table XIV. Their production from the by-product isomers of BHC is presented in the diagram given on page 49.

BHC is used to control various harmful insects, plant pests, and animal parasites. However, its use in animal husbandry is decreasing because of the possibility of its accumulation in the tissues of animals, eventually entering the milk and meat. It is employed in various forms: dusts, wettable powders, emulsive concentrates, and smoke pots.

Technical grade BHC and lindane enter into the composition of seed disinfectants of compound action. Combination disinfectants with γ-BHC and ethylmercury chloride (Mercuran), hexachlorobenzene (Hexagamma), tetramethylthiuram disulfide, and copper trichlorophenolate (Phenthiuram) and many other fungicides are known. Usually such seed disinfectants contain, besides a fungicide and diluent, from 10–50% γ-BHC. δ-BHC also has been suggested for use in seed disinfectants, since it increases the fungicidal properties of some preparations, which permits lowering the content of the active ingredient of the main fungicide and of the γ-isomer.

Lindane is used together with DDT to control the Colorado potato beetle and other pests. Such a combined formulation undoubtly has advantages, for example, in the control of synanthropic insects. In the latter case, methoxychlor also enters into the composition of the mixture.

Polychloroterpenes

Polychloroterpenes were first used in the Soviet Union for the control of some parasites of animals. A preparation based on products of the chlorination of a pinene fraction from turpentine oleoresin and containing up to 55% chlorine was used as an insecticide. This preparation was unsuitable for the control of plant pests, however, because of its high phytocidal effect, which apparently is associated with the evolution of HCl under the action of light and water. A somewhat less phytocidal chlorination product of pinene containing up to 66–68% chlorine is known in the USSR as "chloropinene", and in other countries as "Strobane", but it is less insecticidal than other known chlorinated terpenes. The chlorination products of many terpenes have been studied including pinene, camphene, dipentene, fenchene, phellandrene, carene, thujone, and others, but only a few of them are of practical interest. 2,6,7-Trichlorocamphene, which is phytocidal, has appreciable insecticidal effect.

Toxaphene. The most well known terpene chlorination product is the polychlorocamphane ($C_{10}H_{10}Cl_8$) obtained by the chlorination of camphene to a chlorine content of 67–69%. Outside the Soviet Union this product is called "toxaphene". In the [Soviet] literature an analogous preparation is called chlorophene or polychlorocamphene.

To produce toxaphene and its analogs the chlorination of camphene is most often carried out with illumination of the reaction mixture by various light sources. It is possible to carry out the chlorination also in the presence of initiators of free radicals such, for example, as azo-bis(isobutyronitrile). At first, the chlorination of camphene proceeds very vigorously with the evolution of much heat, but as the chlorine is introduced the rate of chlorination sharply decreases, and it is necessary to carry it out at an elevated temperature. However, because of the relatively low thermal stability of toxaphene the reaction temperature should not be too high. To decrease the viscosity of the reaction medium the chlorination process very often is carried out in carbon tetrachloride, which is distilled off after the chlorination is ended.

Polychlorocamphene is a complex mixture of polychlorocamphenes and camphenes of different structure in which there also may be present some products of chlorination of tricyclene and related compounds:

In external appearance toxaphene is a waxy white substance, m.p. 65°–90° C., d^{20} 1.6. It is practically insoluble in water, but dissolves well in many organic solvents. When acted on by caustic alkalies it readily splits out part of the chlorine as HCl. A hexane solution of toxaphene on heating with pyridine and KOH in methyl alcohol gives a red color. A similar reaction is obtained also with some other chlorinated terpenes. For the analysis of toxaphene, IR spectroscopy (lines at 7.67 and 12.94 μ) or total chlorine determination is most often used. The LD_{50} for rats is 60 mg./kg. Toxaphene is used in the form of dusts or suspensions and also as emulsions or solutions for low-volume spraying.

An advantageous property of toxaphene is its low toxicity for bees. In the United States not more than 7 mg./kg. of toxaphene is permitted on fruits intended for food, and in the German Federal Republic, 0.4 mg./kg. Toxaphene also gives satisfactory results in combating rodents. For this purpose it can be employed in the form of food baits and with oils. In the latter case the entrance to the hole is treated with the preparation and the animal dies as a result of licking its contaminated fur.

Polychloropinene (chlorothene) is used in agriculture. It is obtained by photochemical or initiated chlorination of bornyl chloride to a chlorine content of 64–67%. For its production a pinene fraction of turpentine oleoresin generally is used which is hydrochlorinated by the action of HCl and the bornyl chloride obtained is further chlorinated with chlorine. The chlorination of bornyl chloride is carried out under almost the same conditions as the chlorination of camphene:

$$\text{pinene} + HCl \longrightarrow \text{bornyl chloride} \xrightarrow{Cl_2} C_{10}H_{10}Cl_8 + HCl$$

Polychloropinene is a viscous colorless oil, d^{20} 1.5–1.6. It is practically insoluble in water and highly soluble in organic solvents. Its chemical properties are similar to those of toxaphene. Its LD_{50} for experimental animals is 350–525 mg./kg. The maximum permissible concentration in air is 0.2 mg./m.^{3}.

Polychloropinene exceeds toxaphene in phytotoxicity. Unlike toxaphene it is toxic to bees. In the USSR polychloropinene is used to control the sugarbeet weevil in the form of emulsions or oil solutions for low-volume spraying. The application rate is 1–2 kg./ha.

Strobane. As already mentioned, Strobane is produced by the chlorination of pinene to a chlorine content of not less than 66%. This product is a thick, viscous mass, d^{20} 1.6, n_D^{20} 1.5790, vapor pressure at 20° C. 3×10^{-7} mm. of Hg. It is practically insoluble in water and highly soluble in organic solvents; e. g., the solubility in ethyl alcohol is 12–14%. The LD_{50} for white rats is 250 mg./kg. Strobane is used similarly to polychloropinene. It is approximately one and a half times weaker than toxaphene in insecticidal activity.

Practically all the polychloroterpenes that have been studied have some acaricidal effect and are very slightly active as fungicides and bactericides. Because of their low cost and simplicity of production, the polychloroterpenes are being used more and more in agriculture, including animal husbandry, and the scale of manufacture is continually expanding.

Polychlorocyclodienes

In agriculture and industry the polycyclic insecticides that are derivatives of bi-, tri-, and tetracyclic hydrocarbons have received wide distribution. These compounds in most cases are obtained by diene synthesis reactions using hexachlorocyclopentadiene as the diene.

Among these compounds we should name first: 1,2,3,4,7,7-hexachloro-5,6-bis(chloromethyl)bicyclo[2.2.1]heptene-2 (alodan) (I); 2,3,4,5,6,7,8,8-octachloro-4,7-*endo*methylene-2,3,3a,4,7,7a-hexahydroindene (chlordane)

(II); 1,4,5,6,7,8,8-heptachloro-4,7-*endo*methylene-3a,4,7,7a-tetrahydroindene (heptachlor) (III); 1,2,3,4,10,10-hexachloro-1,4-*endo*-5,8-*exo*-dimethylene-1,4,4a,5,8,8a-hexahydronaphthalene (aldrin) (IV) and its stereoisomer 1,2,3,4,10,10-hexachloro-1,4,5,8-di*endo*methylene-1,4,4a,5,8,8a-hexahydronaphthalene (isodrin) (V), and also the epoxides of the last two compounds, dieldrin (VI) and endrin (VII), respectively.

The principal intermediate for the synthesis of all the compounds enumerated and of some other similar compounds is hexachlorocyclopentadiene. Because of the great significance of this intermediate for the synthesis of insecticides, fungicides, and a large number of other materials important for national economy, the properties and methods of producing hexachlorocyclopentadiene are described below.

Hexachlorocyclopentadiene is a light yellow liquid with a sharp, unpleasant odor, b.p. 236°–238° C. at 760 mm. of Hg, m.p. ~7° C., d^{20} 1.70–1.71. It approaches phosgene in toxicity for man and animals. Two groups of methods are well known for the production of hexachlorocyclopentadiene: (a) from a cyclic raw material, and (b) from halogen derivatives of the aliphatic series.

Hexachlorocyclopentadiene was obtained for the first time in 1930 by the action of potassium hypochlorite on cyclopentadiene:

$$\text{C}_5\text{H}_6 + 6\text{Cl}_2 + 6\text{KOH} \rightarrow \text{C}_5\text{Cl}_6 + 6\text{KCl} + 6\text{H}_2\text{O}$$

Hexachlorocyclopentadiene is obtained by this reaction in 40–60% yield, calculated on the cyclopentadiene. In addition to the hexachlorocyclopentadiene, a considerable amount of side products is formed, predominantly chloro derivatives of dicyclopentadiene. Moreover, pentachlorocyclopentadienes are obtained that cannot be removed by simple fractional distillation. Because of this the hypochlorite method of producing hexachlorocyclopentadiene is not used on an industrial scale.

The best method of producing hexachlorocyclopentadiene from cyclic raw materials is direct chlorination of cyclopentadiene with chlorine, which usually is carried out in two stages. First cyclopentadiene is chlorinated at the lowest temperature possible and then the polychlorocyclopentanes that are formed are subjected to exhaustive chlorination at 350°–500° C. The octachlorocyclopentene obtained breaks down at an elevated temperature into chlorine and hexachlorocyclopentadiene:

$$\text{C}_5\text{H}_6 \xrightarrow{2\text{Cl}_2} \text{C}_5\text{H}_2\text{Cl}_4 \xrightarrow{\text{Cl}_2} \text{C}_5\text{Cl}_8 \xrightarrow{-\text{Cl}_2} \text{C}_5\text{Cl}_6$$

Cyclopentane from petroleum also may be used as the cyclic starting material. In this case it is initially chlorinated to polychlorocyclopentanes, which are further subjected to high-temperature chlorination.

The last two methods can be used for industrial production of hexachlorocyclopentadiene. However, if one considers the amount of cyclopentadiene necessary for the production of various other compounds, it is more expedient to make hexachlorocyclopentadiene from derivatives of the aliphatic hydrocarbons. The most readily available raw materials for the production of hexachlorocyclopentadiene are the pentanes or the amylenes or a mixture of these hydrocarbons, which are easily separable from appropriate petroleum products, and also piperylene [pentadiene].

Usually the chlorination of the pentanes or amylenes by chlorine is carried out with illumination or with the addition of substances capable of breaking down to give free radicals [for example, azo-bis(isobutyronitrile)]. The polychloropentanes obtained are subjected to high-temperature chlorination and cyclization. Chemically this process can be represented by the following reaction:

$$C_5H_{10} + xCl_2 \rightarrow C_5H_{12-(x+1)}Cl_{x+1} + (x-1)HCl$$

The chlorination of the pentanes or amylenes is carried on until a product with a density of 1.72–1.78 g./cm.^{3} is obtained, which corresponds to a content of 6–8 chlorine atoms in the polychloropentane molecule. These compounds are further subjected at an elevated temperature to complete chlorination and dehydrochlorination, as a result of which decachloropentane and octachloropentadiene-1,3 are formed:

$$CCl_3CCl_2CHClCHClCCl_3 \rightarrow CCl_3CCl = CClCCl = CCl_2 + 2\ HCl$$

The last compound undergoes cyclization by the Butlerov-El'tekov reaction and goes to octachlorocyclopentene. As a result of dehydrochlorination of the octachlorocyclopentene, hexachlorocyclopentadiene is obtained. The yield of hexachlorocyclopentadiene calculated on the starting hydrocarbons is 85–90%. Carbon tetrachloride, tetrachloroethylene, and hexachlorobutadiene are formed in certain amounts as by-products. The formation of these products takes place as a result of high-temperature chlorinolysis of both octachlorocyclopentene and polychloropentanes.

It is recommended that the chlorination of the polychloropentanes be carried out at 300°–450° C. The process can be conducted both in a hollow tube and in the presence of catalysts. To avoid the formation of hexachlorobenzene by cyclization it is best to carry it out at 400°–425° C.

Hexachlorocyclopentadiene is capable of diene synthesis reactions and allyl substitution. Its dimerization may serve as an example of a diene synthesis reaction which takes place upon prolonged heating; the exchange of halogens for various other atoms and groups is an example of allyl substitution reaction. When cuprous chloride in methanol acts on hexachlorocyclopentadiene, bis(pentachlorocyclopentadien-2,4-yl) is obtained in good yield; this compound is known as Pentac or Decachlor (m.p. 122° to 123° C., LD_{50} for rats 3,100 mg./kg.) and is used as a selective acaricide to control herbivorous mites on various crops.

Hexachlorocyclopentadiene has been proposed for the control of phylloxera of grape vines.

Chlordane (chloroindan, compound 1068) was the first insecticide of the cyclodiene series to be used in agriculture and industry. The technical grade product is a light yellow oil, d^{20} 1.59–1.63, n_D^{20} 1.57–1.58. It is practically insoluble in water but highly soluble in organic solvents; the aromatic hydrocarbons, their halogen derivatives, ketones, and esters are miscible with chlordane in all proportions.

Technical grade chlordane contains 60–75% of the *cis*- and *trans*-isomers of 2,3,4,5,6,7,8,8-octachloro-4,7-*endo*methylene-2,3,3a,4,7,7a-hexahydroindene and 25–40% of related compounds. The *cis*-isomer (α-chlordane) is a white crystalline substance, m.p. 106.5°–108° C.; the *trans*-isomer melts at 104.6°–106° C. Heptachlor, nonachlor (m.p. of one isomer 122°–123° C. and of the other 209°–211° C.), chlordene, and some other compounds also have been found in the technical grade product. Octachlor obtained as a result of substitutive chlorination of heptachlor may possibly be present in technical grade chlordane. This product has recently been found in technical grade heptachlor.

The structure of chlordane and the other compounds contained in the technical grade product are:

α – isomer   β – isomer
chlordane

cis – isomer   trans – isomer
nonachlor

chlordene   octachlor

Chlordane is obtained by the chlorination of chlordene at 50°–80° C. either in the presence of a solvent or without one. Chlordene is formed in practically quantitative yield when cyclopentadiene is condensed with hexachlorocyclopentadiene at 80°–90° C. Chlordene is a white crystalline substance, m.p. 154° C. (dec.). When the synthesis of chlordane is attempted by chlorination of dicyclopentadiene with chlorine, a complex mixture of chloro derivatives is formed that has a relatively weak insecticidal effect and a higher phytocidal effect as a result of gradual decomposition with the evolution of HCl.

Technical grade chlordane under the influence of caustic alkalies splits out part of the chlorine as HCl, but it is not possible to split out all eight chlorine atoms under ordinary conditions. When chlordane is acted on by caustic alkalies in alcohol solution or by zinc dust in an acid medium more than two chlorine atoms are split out. A small amount of HCl is given off when chlordane is acted on by water, and this must be taken into consideration when it is stored in a metal container. Moist chlordane strongly corrodes metals. To tie up the HCl evolved in storage, small amounts of epichlorohydrin are added to chlordane. A similar means is used also to stabilize chlorinated terpenes.

Chlordane gives several color reactions: when an alcoholic solution of it is heated with KOH and *p*-aminophenol a red-brown color is observed. Less than 1 mg. of chlordane in 1 ml. can be detected by this reaction. Upon heating chlordane with pyridine and ethylcellosolve in the presence of KOH a red color appears, and if the cellosolve is replaced by ethylene glycol a red-violet color is obtained. A violet color is formed when a methanol solution of chlordane is heated with diethanolamine.

The LD_{50} of chlordane for various experimental animals lies within the limits 100–500 mg./kg. The preparation has a high chronic toxicity and can cause poisoning by systematic action of small doses. In the United States not more than 0.3 mg./kg. of chlordane residue is permitted in food products that are in seasonal use, and the presence of chlordane is not permitted at all in products that are in continuous use. It is used as an agent to control various chewing pests and especially to protect nonmetallic materials from termites under tropical conditions. It is employed in the form of emulsions or solutions in organic solvents of petroleum origin.

Chlordane is gradually being displaced by the more effective heptachlor and other compounds of diene synthesis.

Heptachlor. The pure compound is a white crystalline substance with a weak camphor-like odor, m.p. 95°–96° C. It is practically insoluble in water, moderately soluble in ethyl alcohol (4.5 g. in 100 g. of alcohol), and highly soluble in kerosine and various aromatic hydrocarbons and halogen derivatives of hydrocarbons. The technical grade product is a waxy mass, m.p. 46°–74° C., that contains 65–72% heptachlor and 28–35% related compounds; besides heptachlor, chloroindan, nonachlor, and octachlor have been found in it.

Heptachlor is produced by substitutive chlorination of chlordene. This reaction proceeds at a low temperature in the presence of catalysts (infusorial earth, silica gel, and other similar materials):

The process of chlorinating chlordene is most often carried out at 0°–5° C. in carbon tetrachloride solution. After the chlorination is ended, the solvent is distilled off and the residue is used to prepare insecticidal preparations. To obtain a product that can be ground easily the technical grade compound is recrystallized from methanol. This recrystallization permits removal of most of the oily impurities, but it does not increase substantially the content of the principal compound in the product.

Heptachlor is obtained in good yield by the chlorination of chlordene with sulfuryl chloride, and also by the action of thionyl chloride on hydroxychlordene. This last compound is formed when chlordene reacts with selenium dioxide:

[chemical reaction scheme showing chlordene + SeO₂ + CH₃COOH → hydroxychlordene acetate + Se; then + HCl → hydroxychlordene + CH₃COCl; then + SOCl₂ → heptachlor + SO₂ + HCl]

This method is used only for the preparation of the very pure compound under laboratory conditions.

Heptachlor is more inert chemically than chlordane. It is resistant to the action of water and caustic alkalies. An interesting reaction of heptachlor is the splitting out of one chlorine atom under the action of silver nitrate in 80% acetic acid. This reaction is specific, since neither DDT nor toxaphene splits out halogen when heated with silver nitrate in acetic acid. An exception is octachlor, which under these conditions splits out two chlorine atoms.

When heptachlor is heated with diethanolamine and KOH in methanol, like chlordane it gives a colored product, which permits its determination in biological media. A similar reaction takes place in butylcellosolve on heating with KOH and ethanolamine.

An interesting reaction is the oxidation of heptachlor to the corresponding epoxide:

$$\text{[heptachlor]} + [O] \longrightarrow \text{[heptachlor epoxide]}$$

The reaction takes place readily under the influence of soil microorganisms, in animals (in the liver), and probably in insects.

Heptachlor epoxide is the main metabolite of heptachlor; it is more toxic than heptachlor to all species of organisms. It has been demonstrated that practically all the heptachlor in animal organisms and in the soil goes over to the epoxide. It is possible that the effect of heptachlor is based on the reaction of its epoxide with vitally important systems in animals and insects. Heptachlor lasts for a long time in the soil in the form of the epoxide. Thus, according to the observation of American investigators, when 25 kg./ha. of heptachlor is introduced into the soil, about 90% of the compound is decomposed in the first two years, and further decomposition takes place more slowly. After a five-year period about 0.4 mg./kg. of soil remains.

When heptachlor is oxidized with nitric acid 3,4,5,6,7,7-hexachlorobicyclo[2.2.1]heptene-4,5-dicarboxylic-1,2 acid (chlorendic acid) is formed in small yield.

The LD_{50} for different animals ranges within the limits 60–300 mg./kg., and chronically the toxicity of heptachlor is higher. In the Soviet Union and the United States the presence of heptachlor residues is not permitted in food and forage products; in the USA a heptachlor content of up to 0.1 mg./kg. is permitted only in some products of seasonal usage. The maximum permissible concentration in air is 0.1 mg./m.^{3}.

Heptachlor is used to control soil-inhabiting pests (wireworms, larvae of May and June beetles, gray sugarbeet weevil, etc.), and also as an insecticidal additive to seed disinfectants. In this case it has an advantage over lindane, since it is more stable in alkaline soils and considerably cheaper. Heptachlor is added to seed disinfectants in an amount of 10–20%. Like lindane, heptachlor not only protects the seedlings from damage by insects, but also stimulates the germination of the seeds.

Dihydroheptachlor is obtained by the action of anhydrous HCl on chlordene at elevated temperature and pressure. It is a white crystalline substance, m.p. 123° C. It is considerably less toxic than heptachlor for mammals, but its insecticidal properties have not yet been studied sufficiently.

Analogs of heptachlor and chlordane containing fluorine instead of two and four of the chlorine atoms also have been described, but the insecticidal properties of these compounds are weaker. These compounds have not yet received practical use.

Aldrin is a white crystalline substance, m.p. 104°–104.5° C., vapor pressure at 25° C. 6×10^{-6} mm. of Hg. The technical grade compound melts at 45°–60° C., and is practically insoluble in water. Its solubility in organic solvents is given in Table XV.

Table XV. *Solubility of aldrin in some organic solvents at 25° C.*

Solvent	Solubility (g./100 ml. of solvent)	Solvent	Solubility (g./100 ml. of solvent)
Acetone	109	Xylene	235
Amyl acetate	38	Methyl alcohol	5
Benzene	183	Methylcellosolve	15
n-Butyl alcohol	9	Methyl ethyl ketone	28
Dipentene	109	Pentane	3
Dichloroethane	302	Oleoresin turpentine	127
Isopropyl alcohol	4	Toluene	267
Dearomatized kerosine	28	Carbon tetrachloride	303
		Ethyl alcohol	5

The technical preparation is a brown mass containing 82% *endo-exo* isomer of 1,2,3,4,10,10-hexachloro-1,4,5,8-di*endo*methylene-1,4,4a,5,8,8a-hexahydronaphthalene, 12–13% analogs, and about 5% various other compounds, d^{20} 1.57–1.67. It contains not more than 0.1% water and not more than 0.1% free acid calculated as HCl.

The main method of producing aldrin is the reaction of hexachlorocyclopentadiene with an excess (2–3 moles) of bicyclo[2.2.1]heptadiene-2,5 at 100° C. Excess latter compound is necessary to exclude the possibility of the formation of dodecachlorohexacycloheptadecadiene, which is obtained from the reaction of two molecules of hexachlorocyclopentadiene with one molecule of bicyclo[2.2.1]heptadiene-2,5:

The yield of aldrin from this reaction is more than 80% calculated on the hexachloropentadiene. The excess bicyclo[2.2.1]heptadiene-2,5 can be practically completely regenerated and used for the next reaction; it is a colorless liquid with a characteristic odor, b.p. 90.3° C., d_4^{20} 0.9064, n_D^{20} 1.4702.

Bicyclo[2.2.1]heptadiene-2,5 is obtained by the reaction of cyclopentadiene with acetylene at 250°–360° C., and 4–20 atm. pressure. The reaction can be carried out either in a solvent (pentane, isopentane, etc.) or without a solvent with nitrogen as a diluent for the acetylene. The yield by this reaction is 30–60%. As by-products toluene, tricyclene, and some other compounds may be formed. It can also be obtained by the action of magnesium on 2,3-dichlorobicyclo[2.2.1]heptene-5 or of potassium *tert*-butylate on 2-chlorobicyclo[2.2.1]heptene-5:

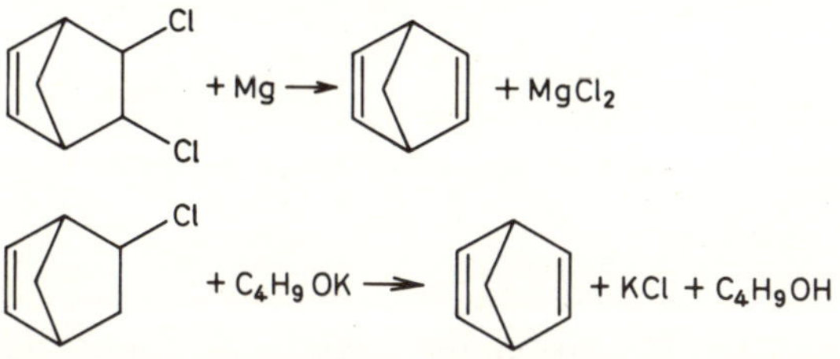

However, the yields of bicyclo[2.2.1]heptadiene-2,5 from the last two reactions are not good enough.

Aldrin is chemically stable. Water and caustic alkalies at room temperature do not break it down, although the technical grade product on boiling with water and alkalies gives off some HCl, associated with impurities in it that are hydrolyzed by water. Peroxides (H_2O_2, peracetic and other organic peracids) convert aldrin to dieldrin. The formation of dieldrin from aldrin occurs in the soil, plants, insects, and vertebrates. Aldrin is especially rapidly oxidized in the liver.

Zinc dust in acetic acid reduces aldrin:

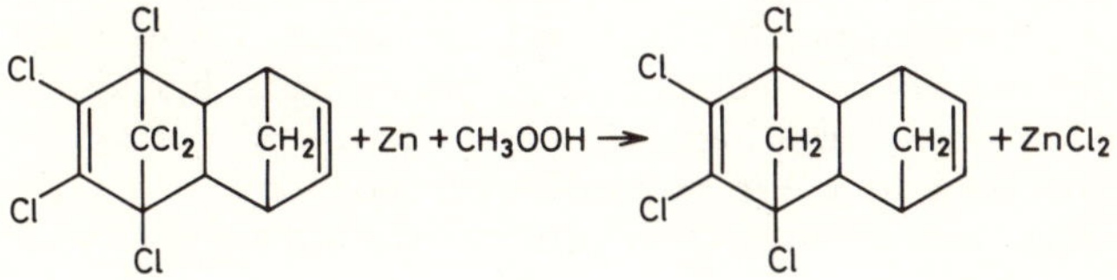

It readily reacts with phenylazide with the formation of aldrinphenyldihydrotriazole, and this compound when acted on by HCl splits out a molecule of nitrogen and goes to chloroaldrinaniline, which readily combines with diazo compounds to form brightly colored products:

This cycle of reactions is used for the colorimetric determination of aldrin in biological media. However, some terpenes also react with phenylazide, and therefore the determination of aldrin in plant materials must be preceded by careful removal of terpenes.

Aldrin adds a molecule of bromine, HBr, and the like at the double bond.

The LD_{50} for experimental animals is 40–50 mg./kg. With continuous introduction of aldrin its toxicity increases, although there is information that in these cases an equilibrium is attained. The amount of aldrin entering is equal to the amount of its conversion products excreted, as established by using C-14 labeled compound. In the United States the established tolerance for aldrin residues in products seasonally consumed is 0.1 mg./kg.; the maximum permissible concentration in air is 0.01 mg./m.[3]

The main purpose for which aldrin is used in agriculture is for the control of grasshoppers and soil-inhabiting insects. Aldrin is used as a soil insecticide not only as a component of poison baits, but also mixed with fertilizers. However, the high stability of aldrin in the soil has made it necessary to limit its use in mixtures with fertilizers. In England in 1965 the introduction of aldrin and dieldrin into the soil together with fertilizers was forbidden. The use of aldrin to control cotton pests gives good results, but when used it is necessary to take precautionary measures to protect the workers from inhaling it or coming into skin contact with it.

Up to the present time a large number of analogs of aldrin have been synthesized but no good insecticides have been found among them. It has been established that removal of the double bonds by the addition of different atoms and groups (for example, hydrogen, halogens, and the like) leads to a sharp decrease in the insecticidal activity of the compound. This activity is lowered also when some of the chlorine atoms are replaced by fluorine and other atoms or groups.

Dieldrin. Endo-exo-exo-1,2,3,4,10,10-hexachloro-1,4,5,8-diendomethylene-6,7-epoxy-1,4,4a,5,8,8a-hexahydronaphthalene is a white crystalline substance, m.p. 175°–176° C., vapor pressure at 25° C. 1.8×10^{-7} mm. of Hg. At 26° C., 5×10^{-6} g. of dieldrin dissolves in 100 g. of water; it is highly soluble in many organic solvents, but its solubility is considerably less than that of aldrin. It is obtained in about 90% yield by the oxidation of aldrin by organic peracids or H_2O_2 in acetic acid. The technical grade product is a light brown material with a solidification point above 95° C.; it contains about 85% of dieldrin.

Dieldrin approaches aldrin in chemical stability. Its reactions are determined to a considerable degree by the presence of the epoxy group. Thus, for example, with HBr it yields the corresponding bromohydrin. On heating with monoethanolamine for 12–16 hours at 140° C. it completely splits out the chlorine from the endomethylene bridge. The LD_{50} of dieldrin for various animals is within the limits 25–50 mg./kg.; the maximum permissible concentration in air is 0.01 mg./m.^{3}.

Dieldrin is used in many, very different forms, including seed disinfectants together with fungicides and aerosol bombs with pyrethrins and DDT. However, dieldrin is highly toxic for man and domestic animals. In the United States no residues of dieldrin are permitted in food and forage products. In the Soviet Union the use of aldrin and dieldrin is not allowed.

Isodrin is a white crystalline substance, m.p. 240°–242° C. It is obtained in 70–80% yield by the condensation of cyclopentadiene with 1,2,3,4,7,7-hexachlorobicyclo[2.2.1]heptadiene-2,5:

Hexachlorobicyclo[2.2.1]heptadiene-2,5 is a liquid, b.p. 125°–126° C. at 10 mm. of Hg, d_4^{20} 1.6606, n_D^{20} 1.5550, freezing point 0.1° C. It is produced by the condensation of hexachlorocyclopentadiene with acetylene at 5–18 atm. pressure and temperatures below 150° C. When the reaction is carried out at a pressure of 18 atm. the yield reaches 65%.

Isodrin is reminiscent of aldrin in its chemical properties. The LD_{50} for experimental animals is 7–15 mg./kg. In spite of the fact that isodrin has a strong insecticidal effect, it has still received practically no use in agriculture, but it is employed for the production of another insecticide, endrin.

Endrin is a white crystalline substance that decomposes on heating to 200° C., is insoluble in water and highly soluble in most organic solvents.

The technical grade product is brown and contains not less than 85% principal compound, d_4^{20} 1.77 g./cm.³ and vapor pressure at 25° C. 2×10^{-7} mm. of Hg.

Endrin is produced by the oxidation of isodrin with hydrogen peroxide in acetic acid at the lowest temperature possible. When the oxidation of isodrin is carried out at 100° C. the endrin rearranges to a ketone that has no insecticidal effect:

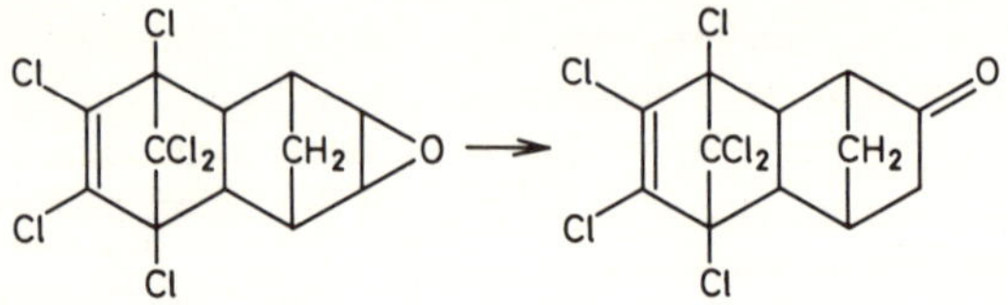

Such isomerization also occurs upon prolonged storage in the presence of compounds of an acid nature, and in connection with this special stabilizers, for example epichlorohydrin, are added to the technical grade compound.

Endrin exceeds many other compounds of this series in toxicity for man, domestic animals, and insects. The LD_{50} for rats is 17 mg./kg., and for rabbits it is 7–10 mg./kg.

Depending on the conditions and the formulation used, and also on the pest and the crop, endrin is employed at the rate of 60–600 g./ha. Endrin is now used to control the black-currant bud mite, against which all other compounds are ineffective, and also as a zoocide. However, in working with this compound it is necessary to use precautionary measures to avoid direct contact with it. An advantageous property of endrin is its low persistence, since in the light it is relatively rapidly isomerized to the non-toxic ketone.

In the Soviet Union the use of endrin is not permitted.

1,2,3,4,7,7 - Hexachloro - 5,6 - bis(chloromethyl)bicyclo [2.2.1] heptene-2 (alodan) is a white crystalline substance, m.p. 104°–106° C., insoluble in water but highly soluble in organic solvents. It is produced by the condensation of hexachlorocyclopentadiene with *cis*-1,4-dichlorobutene-2 at 150°–160° C. in an autoclave:

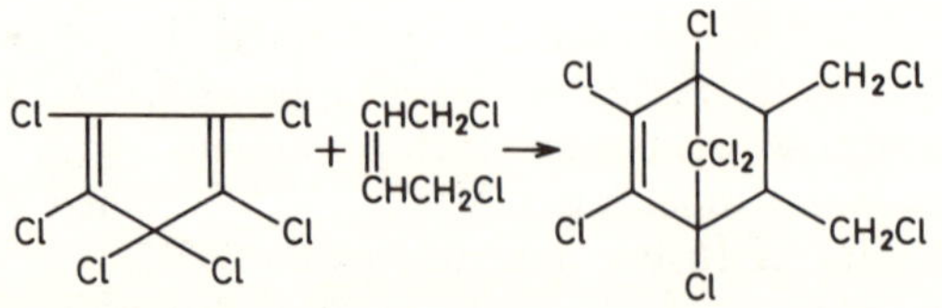

cis-1,2,3,4,7,7-Hexachloro-5,6-bis(chloromethyl)bicyclo[2.2.1]heptene-2 is a stable compound, but the presence of the chloromethyl groups makes it capable of substitution reactions. The chlorine of the side chains can be replaced by various atoms and groups. When the compound reacts with water and alkalies, part of the halogen splits out as HCl.

Alodan is distinguished by its very low toxicity for vertebrates. The LD_{50} for rats and rabbits is 15,000 mg./kg., and for dogs 8,000 mg./kg. Chronic poisoning takes place only when 3,200 mg./kg. of the compound is introduced. The maximum permissible concentration of alodan in air is 0.5 mg./m.^{3} It is used to control pests of stored products (in the treatment of granaries), and also to control parasites of animals. Because of its phytocidal effect this compound has not been used to treat plants.

A number of analogs of alodan have been synthesized in which insecticidal properties also have been found, but these compounds have still received practically no use.

Mirex. Related to a certain extent to the compounds prepared by the diene synthesis is dodecachloroheptacyclo[5.3.0.0^{2,6}.0^{4,10}.0^{5,9}]decane, which is obtained by the condensation of two molecules of hexachlorocyclopentadiene:

This compound has been proposed for the control of cotton pests and some species of ants; it is toxic for vertebrates.

Related to mirex is the compound Kepone, which has been proposed to control various pests that have chewing mouth parts. It is a solid substance, m.p. 350° C. (dec.), LD_{50} 126 mg./kg.

General references

BELIKOVA, N. A., L. G. VOL'FSON, K. V. KUZNETSOVA, N. N. MEL'NIKOV, A. I. PERSON, A. F. PLATE, and M. A. PRYANISHNIKOVA: Zhur. Priklad. Khim. **33**, 454 (1960).

BEZOBRAZOV, YU. N., A. B. MOLCHANOV, and K. A. GAR: Geksakhloran, ego svoistva, poluchenie i primenenie [Hexachloran, its properties, production and use]. State Scientific and Technical Publishing House of Chemical Literature [USSR] (1956).

BÜCHEL, K. H., A. G. GINSBERG, R. FISCHER: Chem. Ber. **99**, 405 (1966).

KOGAN, L. M.: Uspekhi Khim. **28**, 133–167 (1959).
MEL'NIKOV, N. N., and L. G. VOL'FSON: Khim. Prom. No. 10, 45 (1953).
—, S. D. VOLODKOVICH, L. G. VOL'FSON, and E. F. GRANIN: Zhur. Priklad. Khim. **34**, 2716 (1961).
—, S. D. VOLODKOVICH, L. G. VOL'FSON, and S. S. KUKALENKO: Reaktsii i metody issledovaniya organicheskikh soedinenii [Reactions and methods of investigating organic compounds], vol. II. State Scientific and Technical Publishing House of Chemical Literature [USSR] (1962).
POONAWALLA, N. H., and F. KARTE: J. Agr. Food Chem. **16**, 13 (1968).
TARASOVA, G. A., A. F. PLATE, N. N. MEL'NIKOV, L. G. VOL'FSON, and A. I. TISHCHENKO: Neftekhim. **1**, 65 (1961).
VOL'FSON, L. G.: Khim. Sredstva Zashchity Rastenii No. 2, 23 (1959).
— Zhur. Vsesoyuz. Obshchestva im. D. I. Mendeleeva **5**, 260 (1960).
—, and L. M. KOGAN: Khim. Prom. No. 10, 672 (1961).
—, N. N. MEL'NIKOV, A. F. PLATÉ, P. M. PEREL'MUTTER, S. D. VOLODKOVICH, M. A. PRYANISHNIKOVA, K. V. LEBEDEVA, and N. P. VOLOSHKEVICH: Khim. Prom. No. 10, 714 (1962).
— — —, G. A. TARASOVA, A. I. PERSON, and A. S. PLETNEVA: Trudy Nauch. Inst. Udobreniyam i Insektofungitsidam im. Ya. V. Samoilova No. 171, 52 (1962).
—, S. D. VOLODKOVICH, N. N. MEL'NIKOV, A. B. MOLCHANOV, and YU. N. SAPOZHKOV: Trudy Nauch. Inst. Udobreniyam i Insektofungotsidam im. Ya. V. Samoilova No. 158, 187 (1958).
VOLODKOVICH, S. D., S. S. KUKALENKO, and N. N. MEL'NIKOV: Trudy Nauch. Inst. Udobreniyam i Insektofungicidam im. Ya. V. Samoilova No. 158, 201 (1959).
—, L. G. VOL'FSON, L. M. KOGAN, N. N. MEL'NIKOV, and YU. N. SAPOZHKOV: Zhur. Priklad. Khim. **33**, 227 (1960).
— —, K. V. KUZNETSOVA, and N. N. MEL'NIKOV: Zhur. Priklad. Khim. **29**, 2837 (1959).

VI. Halogen derivatives of aromatic hydrocarbons

General characteristics of pesticidal properties

The pesticidal properties of many halogen derivatives of aromatic hydrocarbons have been studied, including the fluorine, chlorine, bromine, and iodine derivatives of benzene, toluene, xylenes, isopropylbenzene, and cymene and their homologs, the chlorine and bromine derivatives of naphthalene, acenaphthene, diphenyl, diphenylmethane, diphenylethanes and their homologs, fluorocene, anthracene, phenanthrene, pyrene, and others.

The insecticidal and acaricidal properties of the halogen derivatives of benzene depend on the nature of the halogen atoms, on the number of them in the benzene molecule, and on their position. The fluorobenzenes have relatively weak insecticidal properties, although their activity is higher than that of benzene itself. The insecticidal activity of the chlorobenzenes is somewhat higher and increases with an increase in the number of chlorine atoms in the molecule up to three; the most active compounds are the trichlorobenzenes; the insecticidal effect of hexachlorobenzene is considerably weaker. There is no very substantial difference in the activity of the three isomers of dichlorobenzene; the most effective, however, is *p*-dichlorobenzene. Bromobenzene and the dibromobenzenes are somewhat more active than the corresponding chlorine derivatives. With the accumulation of bromine atoms in the molecule of the benzene derivatives the activity of the compounds decreases sharply.

The halotoluenes containing halogen in the aromatic nucleus differ little in insecticidal activity from the corresponding benzene derivatives. The activity of the halobenzils is considerably higher; this apparently is associated with the greater capacity of the halogen in the side chain for reactions of nucleophilic substitution. Some halobenzils have been proposed for the control of nematodes, but because of their strong irritating effect on the mucous membranes they have received practically no use. According to patent data, pentachloroisopropylbenzene and some of its analogs have good insecticidal properties.

Compounds containing halogen in the side chain at a double bond are more active. Thus, for example, β-bromostyrene shows considerable insecticidal activity, and there also is information on nematicidal properties of 1-bromo-2-phenylethane.

The monochloro- and monobromonaphthalenes and also the halogen derivatives of other mononuclear hydrocarbons have relatively weak insecticidal properties. Exceptions are the derivatives of unsymmetrical diarylethanes to which DDT, its homologs, and its analogs belong.

Some use as a mothproofing agent has been made of *p*-dichlorobenzene, which also has been used in a mixture with other chlorine derivatives of benzene to control phylloxera and the sugarbeet weevil; more effective compounds are now used to control these pests.

The fungicidal activity of the halogen derivatives of benzene also changes with the nature, number, and position of the halogen atoms in the compound. The fungicidal activity of the chlorobenzenes increases from monochlorobenzene to hexachlorobenzene. The last compound is rather active as a selective fungicide and finds practical use for disinfection of grain seeds against various species of smut. The fungicidal properties of the di- and tribromobenzenes is somewhat higher than that of the corresponding chlorobenzenes, but hexabromobenzene is less active than hexachlorobenzene. This apparently is connected with its greater molecular weight and low rate of diffusion through the cell membrane of the fungus.

1-Chloronaphthalene also has fungicidal properties; it finds some use in a mixture with other antiseptics for the protection of wood from damage by microorganisms and insects.

The herbicidal properties of the chlorobenzenes increase as the number of halogens in the molecules rises. The most effective herbicides are the tri- and tetrachlorobenzenes, but when more than four halogen atoms are introduced into the benzene ring the herbicidal properties decrease. A mixture of trichlorobenzenes (Benachlor) is employed to control aquatic vegetation at an application rate of about 750 g./m.^{3} of water. 1,2,4,5-Tetrachlorobenzene at a dosage of 12–20 kg./ha. gives satisfactory results in control of wild oats and other grasses. It should be applied three to four months before planting time.

Hexachlorobenzene is a white crystalline substance, m.p. 226° C., d_4^{20} 2.049, vapor pressure at 20° C. 1.089×10^{-5} mm. of Hg, practically insoluble in water but highly soluble in benzene, chlorobenzene, alcohol, and other organic solvents. It is obtained in high yields by the chlorination of benzene, chlorobenzene, trichlorobenzene, and other halogen derivatives of benzene in the presence of catalysts for substitutive chlorination in either the liquid or gaseous phase. A good method for producing hexachlorobenzene is chlorination of the nontoxic isomers of BHC at 300°–400° C. Initially three molecules of HCl split out from the BHC, and the trichlorobenzene formed is chlorinated to hexachlorobenzene. Under normal working conditions the yield of hexachlorobenzene is more than 95%.

An interesting method of producing hexachlorobenzene from the by-product isomers of BHC is the oxidation of the latter with oxygen or air:

$$C_6H_6Cl_6 + 3O \rightarrow C_6Cl_6 + 3 H_2O$$

According to patent data the process should be carried out in the presence of catalysts used to produce chlorine from HCl by the Deacon method. By this process hexachlorobenzene can be obtained practically without an additional expenditure of chlorine. The yield of hexachlorobenzene varies within the theoretical limits of 80–95%. It is also formed in many thermal chlorination reactions of hydrocarbons and chlorine derivatives of hydrocarbons, including compounds of the aliphatic and aromatic series. Thus, for example, it is obtained in the thermal chlorination of polychloropentanes and polychlorohexanes at 450°–500° C. However, these methods have no industrial application because of the large expenditure of chlorine required and the small yield of hexachlorobenzene.

Hexachlorobenzene is an inert compound and does not react with many active substances. Neither water nor caustic alkalies react with it at room temperature. At 130°–200° C. it readily enters into reaction with caustic alkalies with the formation of pentachlorophenolates:

$$C_6Cl_6 + 2\,KOH \rightarrow C_6Cl_5OK + KCl + H_2O$$

The reaction with alkali metal hydrosulfides proceeds similarly with the formation of pentachlorothiophenol.

Hexachlorobenzene is moderately toxic to man and animals; the maximum permissible concentration in air is 0.9 mg./m.^{3} It is used to control smut fungi of grains as a component of seed disinfectants at dosages of 0.6–1 kg./ton of seed (active ingredient basis). The seed disinfectants used contain 20–40% hexachlorobenzene, 10–20% lindane, and a diluent. Seed disinfectants are marketed that contain a mercury fungicide in addition to the hexachlorobenzene. An example of such a disinfectant is Mercurihexane, which contains 1% ethylmercury chloride, 20% hexachlorobenzene, 20% γ-BHC (or heptachlor), and a diluent. The use of hexachlorobenzene permits cutting of the content of toxic ethylmercury chloride in half, which is very important for decreasing the total toxicity of the preparation.

1,1,1-Trichloro-2,2-bis(p-chlorophenyl)ethane (DDT) is one of the most important contemporary insecticides having received the widest use in agriculture and public health. DDT and its analogs occupy the first place in scale of production and use among the chemical agents for the control of harmful insects. In the United States alone in 1963 80,500 tons of DDT were produced. In spite of serious shortcomings of this compound, its production is continually increasing.

Pure 1,1,1-trichloro-2,2-bis(p-chlorophenyl)ethane is a white crystalline substance, m.p. 108.5°–109° C., b.p. 185° C. at 1 mm. of Hg (partial dec.), vapor pressure at 20° C. 1.9×10^{-7} mm. of Hg, solubility in water about 0.001 mg./l.; the solubility in some organic solvents is given in Table XVI. The best solvents for DDT are ketones, esters of the lower fatty acids, aromatic hydrocarbons, and halogen derivatives of hydrocarbons of the aliphatic and aromatic series. The higher the aromatic hydrocarbon content of petroleum products, the more DDT they will dissolve.

The technical grade product is a complex mixture of compounds in which the p,p'-isomer amounts to 75–76%. In Table XVII the composition of three samples of technical grade DDT is given. Of all the isomers of

Table XVI. *Solubility of DDT in some organic solvents at 27° C.*

Solvent	Solubility (g./100 g. of solvent)	Solvent	Solubility (g./100 g. of solvent)
Acetone	74	Methyl ethyl ketone	100
Acetophenone	65	Monomethylnaphthalene	51
Benzene	89	Pinene	16
Dimethylphthalate	29	Tetralin	63
Dioxane	89	Trichloroethylene	38
Dipentene	26	1,2,4-Trichlorobenzene	28
o-Dichlorobenzene	45	Tetrachloroethylene	23
Dichloroethane	47	Chlorobenzene	67
Isopropyl alcohol	14	Methylene chloride	66
m-Xylene	64	Cyclohexane	19
o-Xylene	66	Cyclohexanone	122
Cumene	43	Cymene	34
Methylnaphthalenes (mono- and di-)	56	Carbon tetrachloride	28
		Ethyl alcohol (95%)	2.2 [a]

[a] 24 °C.

Table XVII. *Composition of technical grade DDT*

Compound	Content (%) in samples with setting point		
	91.4° C.	91.2° C.	83.6° C.
1,1,1-Trichloro-2,2-bis-(p-chlorophenyl)ethane	72.7	72.9	70.5
1,1,1-Trichloro-2-(o-chlorophenyl)-2-(p-chlorophenyl)ethane	11.9	19.9	20.9
1,1,1-Trichloro-2,2-bis(o-chlorophenyl)ethane	0.011	—	—
1,1-Dichloro-2,2-bis(p-chlorophenyl)ethane	0.17	0.3	4.0
1,1-Dichloro-2-(o-chlorophenyl)-2-(p-chlorophenyl)ethane	0.044	—	—
p-Chlorophenyltrichloromethylcarbinol	—	0.2	—
p,p'-Dichlorodiphenyl sulfone	0.034	0.6	0.1
o-Chlorophenyltrichloromethylcarbinol p-chlorobenzenesulfonate	0.57	0.4	0.1
o-Chlorophenylchloroacetamide	—	—	0.007
p-Chlorophenylchloroacetamide	0.006	—	0.01
Sodium p-chlorobenzenesulfonate	—	0.02	—
Ammonium p-chlorobenzenesulfonate	0.005	—	—
Inorganic compounds	0.01	0.1	0.04
Unidentified compounds and losses	14.55	5.58	2.59

DDT only the p,p'-isomer has valuable insecticidal properties. The properties of the DDT isomers are given in Table XVIII.

Because technical grade DDT has a much lower melting point than pure p,p'-DDT, it grinds poorly in ball mills thereby hindering the preparation of wettable powders with a high content of the active ingredient.

For preparing such powders, DDT without oily impurities is required. This is achieved by producing DDT from specially purified chloral.

The principal method of producing DDT is the condensation of chlorobenzene with chloral:

$$2\ ClC_6H_5 + CCl_3CHO \rightarrow (ClC_6H_4)_2CHCCl_3 + H_2O$$

This reaction takes place in the presence of condensing agents such as concentrated sulfuric acid, oleum, chlorosulfonic acid, fluorosulfonic acid, hydrogen fluoride, anhydrous aluminium chloride, etc. In industry the most

Table XVIII. *Properties of DDT isomers*

Isomer	M. P. (°C.)	B. P. (° C. at mm. Hg)	Relative toxicity (house flies)
4,4'	108.5–109	185/1	1
2,2'	92.5–93	—	0.011
2,3'	—	181–189/0.1	0.015
2,4'	74–74.5	—	0.018
3,4'	—	169–170/0.3	0.9

frequently used method is condensation of chloral with chlorobenzene in the presence of concentrated sulfuric acid or weak oleum at a temperature not higher than 20° C., since at higher temperatures the amount of p-chlorobenzenesulfonic acid that is formed as a by-product increases sharply. Routes have now been developed for the use of this product for the production of acaricides, DDT synergists, and other compounds. It also is possible to regenerate chlorobenzene from p-chlorobenzenesulfonic acid in 95% yield by the action of dilute sulfuric acid at 200°–240° C.

p-Chlorobenzenesulfonic acid is produced in a minimal amount when the condensation is carried out in the presence of chlorosulfonic acid. However, because of the high cost of chlorosulfonic acid this method is almost never used in industry.

There are many variants of the preparation of DDT by the condensation of chloral with chlorobenzene in the presence of sulfuric acid. These variants differ with respect to details of their procedure.

The chloral necessary for the production of DDT is obtained by chlorination of ethyl alcohol or acetaldehyde. The chlorination of acetaldehyde proceeds through an enol form and can be represented by the general equation:

$$CH_3CHO + 3\ Cl_2 \rightarrow CCl_3CHO + 3\ HCl$$

The mechanism of chlorination of ethyl alcohol is more complex:

$$C_2H_5OH + Cl_2 \rightarrow [C_2H_5OCl] + HCl \rightarrow CH_3CHO + HCl$$
$$CH_3CHO + Cl_2 + C_2H_5OH \rightarrow CH_2ClCHClOC_2H_5 + H_2O$$
$$CH_3CHO + Cl_2 + 2\ C_2H_5OH \rightarrow CH_2ClCH(OC_2H_5)_2 + HCl + H_2O$$
$$CH_2ClCH(OC_2H_5)_2 + Cl_2 \rightarrow CHCl_2CH(OC_2H_5)_2 + HCl$$

$$CHCl_2CH(OC_2H_5)_2 + H_2O \rightarrow CHCl_2CH(OH)OC_2H_5 + C_2H_5OH$$
$$CHCl_2CH(OH)OC_2H_5 + Cl_2 \rightarrow CCl_3CH(OH)OC_2H_5 + HCl$$
$$CCl_3CH(OH)OC_2H_5 + H_2O \rightarrow CCl_3CHO + C_2H_5OH$$

The alcohol reacts also with the HCl that is formed and ethyl chloride is obtained as a by-product.

It can be seen from the equations presented that the chlorination of ethyl alcohol is best carried out in the presence of a small amount of water. It has been established experimentally that 14 kg. of water to 100 kg. of alcohol distillate is optimum. The process can be carried out as either a batch or a continuous process. The chlorination is carried out in the first stage at 50°–60° C., and then at 90° C.

The product obtained by the chlorination of ethyl alcohol contains chloral alcoholate and chloral hydrate, which on treatment of the reaction mixture with concentrated sulfuric acid go over to chloral:

$$CCl_3CH(OH)_2 + H_2SO_4 \rightarrow CCl_3CHO + H_2SO_4 \cdot H_2O$$
$$CCl_3CH(OH)OC_2H_5 + H_2SO_4 \rightarrow CCl_3CHO + C_2H_5OSO_3H + H_2O$$

It also is possible to obtain DDT by the condensation of chlorobenzene with pentachloroethane:

$$2\ C_6H_5Cl + CHCl_2CCl_3 \rightarrow (ClC_6H_4)_2CHCCl_3 + 2\ HCl$$

The DDT produced by this method is strongly contaminated with various by-products.

An interesting method for the synthesis of DDT and especially of its unsymmetrical analogs is the reaction of chlorobenzene with *p*-chlorophenyltrichloromethylcarbinol:

$$C_6H_5Cl + ClC_6H_4CH(OH)CCl_3 \rightarrow (ClC_6H_4)_2CHCCl_3 + H_2O$$

This reaction proceeds readily in the presence of sulfuric acid or oleum. The *p*-chlorophenyltrichloromethylcarbinol necessary for the synthesis of DDT is prepared from chloroform and *p*-chlorobenzaldehyde:

$$ClC_6H_4CHO + CHCl_3 \rightarrow ClC_6H_4CH(OH)CCl_3$$

For the synthesis of C-14 labeled DDT the following reactions are employed:

$$ClC_6H_4\ {}^{14}COCl + CH_2N_2 \rightarrow ClC_6H_4\ {}^{14}COCH_2Cl + N_2$$
$$ClC_6H_4\ {}^{14}COCH_2Cl + 2\ Cl_2 \rightarrow ClC_6H_4\ {}^{14}COCCl_3 + 2\ HCl$$
$$ClC_6H_4\ {}^{14}COCCl_3 + Al(OC_3H_7)_3 \rightarrow ClC_6H_4\ {}^{14}CH(OH)CCl_3$$
$$ClC_6H_4\ {}^{14}CH(OH)CCl_3 + C_6H_5Cl \rightarrow (ClC_6H_4)_2\ {}^{14}CHCCl_3 + H_2O$$

The chemical properties of DDT are determined by the presence in it of the aromatic rings and the trichloromethyl group. The pure *p,p'*-isomer of DDT is thermally stable. Its decomposition starts above 195° C. and proceeds according to the equation:

$$(ClC_6H_4)_2CHCCl_3 \rightarrow (ClC_6H_4)_2C = CCl_2 + HCl$$

Iron salt impurities (especially ferric chloride) sharply lower the decomposition temperature of DDT, and in the presence of 0.01% ferric chloride it is lowered to 120° C. This thermal decomposition of DDT in the presence of iron salts is retarded by the addition of special stabilizers. The simplest stabilizers are alkali or alkaline earth carbonates, which react with the ferric chloride and convert it to the DDT-insoluble carbonate or oxide. Good results are obtained by using chalk or magnesium carbonate.

When DDT decomposes under the influence of sunlight in alcohol solution the following reactions take place:

$$2\,(ClC_6H_4)_2CHCCl_3 + 2\,C_2H_5OH \rightarrow 2\,CH_3CHO +$$
$$+ (ClC_6H_4)_2CHCCl = CClCH(C_6H_4Cl)_2 + 4\,HCl$$

In the presence of the oxygen of the air the process goes farther and p,p'-dichlorobenzophenone is obtained:

$$(ClC_6H_4)_2CHCCl = CClCH(C_6H_4Cl)_2 \rightarrow$$
$$[(ClC_6H_4)_2C = C = C = C(C_6H_4Cl)_2] \rightarrow$$
$$CO(C_6H_4Cl)_2 + 2\,CO_2$$

Apparently similar processes also occur on the leaves of plants.

Pure DDT at room temperature does not affect most metals, but technical preparations, especially those containing water and salt solutions, cause more or less corrosion. Probably this is associated with the evolution of HCl as a result of hydrolysis of the DDT by water:

$$(ClC_6H_4)_2CHCCl_3 + 2\,H_2O \rightarrow (ClC_6H_4)_2CHCOOH + 3\,HCl$$

At room temperature this reaction proceeds slowly, but when an aqueous suspension of DDT is boiled the process is accelerated. Caustic alkalies, lime, barium hydroxide, and other alkaline agents increase the rate of hydrolysis of DDT. The first step of the reaction of DDT with alkalies is the splitting out of HCl and the formation of p,p'-dichlorodiphenylethylene, which further goes over at a higher temperature to p,p'-dichlorodiphenylacetic acid:

$$(ClC_6H_4)_2CHCCl_3 + KOH \rightarrow (ClC_6H_4)_2C = CCl_2 + KCl + H_2O$$

This reaction is employed for the quantitative determination of DDT and also of the p,p'-isomer of DDT in the technical grade product. Determination of p,p'-DDT is based on the different rate of splitting out of HCl by caustic alkalies from the isomers of DDT.

The breakdown of DDT in house flies that are resistant to DDT proceeds by way of splitting out HCl from its molecule under the influence of a special enzyme, DDT-dehydrochlorinase.

DDT is capable of entering into a telomerization reaction to form products of low toxicity for insects:

$$(ClC_6H_4)_2CHCCl_3 + xCH_2=CH_2 \rightarrow (ClC_6H_4)_2CHCCl_2(CH_2CH_2)_xCl$$

Because of this, when solutions of DDT in unsaturated hydrocarbons stand for a long time a decrease in toxicity of the preparation occurs. This reaction proceeds especially rapidly at an elevated temperature and in the light.

DDT is nitrated by concentrated nitric acid with the formation of dinitro and tetranitro products. This reaction is employed for colorimetric determination of small amounts of DDT, since tetranitro DDT gives a blue-colored product with sodium methylate.

The toxicity of DDT for mice, rats, rabbits, and guinea pigs lies within the limits 100–400 mg./kg. Lengthy experiments have been carried out in studying its chronic toxicity. It has been established that DDT is capable of accumulating in the fatty tissues of man and animals. When milk cows are fed products containing DDT considerable amounts of it appear in the milk. Taking into consideration the large scale use of DDT, such accumulation in food products might represent a hazard to the population, especially to children. Thus, the presence of DDT is not permitted in milk, butter, grain, meat, and other food products [in the Soviet Union] that are continuously consumed. In seasonal products in the United States a DDT content of only 1 mg./kg. is permitted, and in the Soviet Union also 1 mg./kg. The maximum permissible concentration of DDT in air is 0.1 mg./m.^{3} DDT is employed to control a great many different insects that are plant pests and vectors of infectious diseases. Now, however, its use is gradually being diminished and the question of complete cessation is being considered due to the accumulation of large amounts of DDT in nature as well as its systematic accumulation in man and animals, which may present a great hazard. DDT is formulated as dusts, concentrated emulsions, wettable powders, preparations for fine-droplet spraying, and aerosols.

A serious shortcoming of DDT is the absence of acaricidal effect on plant-infesting mites, necessitating application of it together with acaricides on a number of crops. Another disadvantage of DDT as an agricultural insecticide is its persistence.

In spite of numerous investigations, the mechanism of action of DDT on insects still has not been determined.

Extensive investigations have been carried out on the synthesis and on the biological activity of homologs and analogs of DDT. When chlorine in the aromatic radical of 1,1,1-trichloro-2,2-bis(*p*-chlorophenyl)ethane is replaced by hydrogen, bromine, iodine, hydroxyl, a higher hydrocarbon radical, amino, thiocyano, carboxyl, nitro, and cyano groups the insecticidal activity of the compound is substantially decreased. A lowering of the insecticidal activity also occurs when several methyl or alkoxy groups are introduced into the aromatic radicals. Replacement of chlorine by fluorine, methoxyl, methyl, or ethyl does not substantially change the insecticidal activity of the compound, but it lowers the toxicity for vertebrate animals and man. 1,1,1-Trichloro-2,2-bis(p-methoxyphenyl)ethane is one fortieth as toxic as DDT for mammals, while 1,1,1-trichloro-2,2-bis(p-ethoxyphenyl)ethane (m.p. 109° C.) is two-thirds as toxic.

Trichlorodinaphthylethane, trichlorodithienylethane, and others are weak insecticides. The unsymmetrical analogs also, as a rule, are considerably less active than DDT. Removal of the trichloromethyl group from the aromatic radicals leads to less active compounds, as does also the splitting out of HCl from compounds containing the trichloromethyl group. The corresponding dichlorodiarylethylenes are practically nontoxic for insects. The diaryltrichlorovinylmethanes also have weak insecticidal properties.

Replacement of one chlorine in the trichloromethyl group by hydrogen lowers the activity for insects by two to four times, and the toxicity for animals by five to 15 times. When a second atom of chlorine is replaced by hydrogen a fivefold decrease in toxicity for animals is observed in comparison with DDT and the insecticidal activity decreases five to 50 times. p,p'-Dichlorodiphenylethane has practically no insecticidal effect, but it is an acaricide. Replacement of chlorine by fluorine, alkyl, and other similar groups in most cases also leads to a substantial lowering of the insecticidal properties.

Of a large number of analogs and homologs of DDT that have been studied a few, which are described below, have found practical use.

Methoxychlor (1,1,1-trichloro-2,2-bis(p-methoxyphenyl)ethane). In addition to DDT, methoxychlor, which is considerably less toxic for man and animals, has been widely used. The LD_{50} is 6,000 mg./kg. Because of its low toxicity it has been used to control insects in places occupied by cattle. It is a white crystalline substance, m.p. 89° C. The technical grade compound melts at 70°–85° C., d^{25} 1.41, and is soluble in organic solvents, especially in ketones, aromatic hydrocarbons, and their halogen derivatives.

Methoxychlor is obtained in good yields by the condensation of chloral with anisole in the presence of sulfuric acid and other condensing agents:

$$2\ CH_3OC_6H_5 + \overset{CHO}{\underset{|}{CCl_3}} \rightarrow (CH_3OC_6H_4)_2CHCCl_3 + H_2O$$

In contrast to the manufacture of DDT, the use of oleum is not recommended in making methoxychlor, because in this process a very large amount of sulfonation products is obtained. The technical grade product contains not less than 88% of the p,p'-isomer and a small amount of the o,p'-isomer.

Methoxychlor is similar to DDT in its chemical properties, but its dehydrochlorination takes place considerably more slowly. Thus, while the rate constant of the reaction of DDT with KOH in alcohol at 40.19° C. is 0.186, that of the p,p'-isomer of methoxychlor at the same temperature is 0.00097.

Methoxychlor has limited use because not only does it cost more than DDT, but it also has little effectiveness toward a number of insects. It often is contained in various preparations based on DDT and lindane as a third component. Methoxychlor formulations are completely similar to those of DDT.

1,1-Dichloro-2,2-bis(p-chlorophenyl)ethane (DDD). In addition to methoxychlor, DDD which is also known as TDE (tetrachlorodiphenylethane) has found use in agriculture. It consists of white crystals, m.p. 112° C. The technical grade preparation has a setting point of about 86° C. and contains as the main impurity the o,p'-isomer. The o,p'-isomer of TDE has shown promising results in the medical treatment of malignant tumors of the adrenal glands. TDE is one of the metabolites of DDT. It is produced by the condensation of chlorobenzene with dichloroacetaldehyde, which must be quite pure and not contain chloral as an impurity:

$$2 \, ClC_6H_5 + CHCl_2CHO \rightarrow (ClC_6H_4)_2CHCHCl_2 + H_2O$$

TDE differs somewhat from DDT in its chemical properties. Thus, when it is heated with ferric chloride it splits out HCl and dichlorotolan:

$$(ClC_6H_4)_2CHCHCl_2 \rightarrow ClC_6H_4C \equiv CC_6H_4Cl + 2 \, HCl$$

It is low in toxicity to vertebrates. The LD_{50} for rats is 3,400 mg./kg.

1,1-Dichloro-2,2-bis(p-ethylphenyl)ethane (Perthane) has been proposed as a selective insecticide to control pests of stone fruits, lettuce, and spinach and also to control flies in animal husbandry. It is one-fifth to one-tenth as active as DDT for different insects, but its LD_{50} for mice is 6,600 mg./kg. As a result of this low toxicity to animals, up to 15 mg./kg. of Perthane is permitted in food products in the United States. It is a white crystalline substance, m.p. 56°–57° C., insoluble in water, highly soluble in organic solvents. It is produced by the condensation of dichloroacetaldehyde with ethylbenzene in the presence of sulfuric acid, and is used in the form of emulsions and suspensions, and also in aerosols.

1,1,1-Trichloro-2,2-bis(p-fluorophenyl)ethane (DFDT) is an interesting analog of DDT, b.p. 138°–140° C. at 0.2 mm. of Hg, m.p. 45° C., almost insoluble in water but more soluble than DDT in organic solvents. DFDT is produced by the condensation of chloral with fluorobenzene in the presence of sulfuric acid. In this reaction about 10% of the o,p'-isomer is obtained. It is similar to DDT in chemical properties. The LD_{50} for warm-blooded animals is 480 mg./kg. Its toxicity for insects is close to that of DDT. DFDT is less persistent than DDT, which in some cases is a great advantage. However, it is more expensive than DDT because of the high cost of fluorobenzene.

General references

IL'INSKAYA, N. B.: Mekhanizm deistviya DDT na nasemkomykh [Mechanism of action of DDT on insects]. Publishing House of Academy of Sciences USSR (1961).
MEL'NIKOV, N. N., V. A. NABOKOV, and E. A. POKROVSKII: DDT, ego svoistva i primenenie [DDT, its properties and use]. State Scientific and Technical Publishing House of Chemical Literature [USSR] (1954).
MÜLLER, P.: DDT. Basel: Geigy (1956).
OTT, D. E., and F. A. GUNTHER: Residue Reviews 10, 70 (1965).
VASHKOV, V. I., L. N. POGODINA, and N. A. SAZONOVA: DDT i ego primenenie [DDT and its use]. State Publishing House of Medical Literature [USSR] (1955).
WEST, T. E., and G. A. CAMPBELL: DDT and newer persistent insecticides. London· Chapman & Hall (1950).

VII. Nitro compounds

Aliphatic nitro compounds

The pesticidal activity of the nitro compounds of the aliphatic and aromatic series is higher than that of the hydrocarbons and it increases sharply in the halonitro compounds.

Table XIX gives the toxicities of the vapors of aliphatic nitro compounds for the confused flour beetle and the bedbug.

The bromonitro compounds are somewhat more toxic than the corresponding chloro derivatives.

The halonitro compounds of the aliphatic series have a wide spectrum of action, being not only insecticides, but also fungicides, nematicides, bactericides, and herbicides. Chloropicrin, dichloronitroethane, and chloronitropropane are used as agents to control various harmful organisms, and also soil sterilants against insects, nematodes, and weeds. Their use is somewhat limited by the large dosages and their relatively high cost. The biological activity of the halonitro compounds of the aliphatic series apparently is connected with their oxidizing ability. As is well known, dichloronitroethane and other halonitroalkanes containing halogen on the same carbon atom as the nitro group react readily with sulfhydryl compounds and derivatives of trivalent phosphorus.

The reaction of chloropicrin with sulfhydryl compounds and trialkylphosphines can be represented as follows:

$$2\,CCl_3NO_2 + 6\,RSNa \longrightarrow \begin{cases} 3\,RSSR + 2\,CO_2 + N_2 + 6\,NaCl \\ 3\,RSSR + 2\,CO + 2\,NO + 6\,NaCl \end{cases}$$

$$2\,CCl_3NO_2 + 3\,R_3P \longrightarrow \begin{cases} 3\,R_3PCl_2 + 2\,CO_2 + N_2 \\ 3\,R_3PCl_2 + 2\,CO + 2\,NO \end{cases}$$

The insecticidal and fungicidal activity increases on going to the unsaturated compounds. Compounds containing a nitro group on a carbon with a double bond are especially active. Examples of such compounds are nitroethylene and β-nitrostyrenes with different substituents in the aromatic radical. Some nitrostyrenes have been proposed for use as seed disinfectants and bactericides.

In Table XX are given the properties of the aliphatic chloronitro compounds that are of some practical interest.

Table XIX. *Toxicity of nitro compounds for the confused flour beetle and bedbug*

Formula	LC_{95} (mg./l.)		Formula	LC_{95} (mg./l.)	
	Confused flour beetle	Bedbug		Confused flour beetle	Bedbug
$CH_3CH_2NO_2$	14	32	CCl_3NO_2	4	5
$CH_3CH_2CH_2NO_2$	16	20	$CH_3CCl_2NO_2$	12	8
$CH_3CH_2CH_2CH_2NO_2$	10	32	$CH_3CH_2CCl_2NO_2$	11	—

Table XX. *Properties of aliphatic chloronitro compounds*

Name and formula	d_{20}^{20}	B. P. at 760 mm. of Hg (°C.)	Vap. press. at 25° C. (mm. Hg)	Sol. in H_2O at 20° C. (%)
Chloropicrin CCl_3NO_2	1.6579	112.4	18.3	0.16
1,1-Dichloro-1-nitroethane $CH_3CCl_2NO_2$	1.4271	124	16	0.25
1-Chloro-1-nitroethane $CH_3CHClNO_2$	1.2860	127.5	11.9	0.4
1-Chloro-1-nitropropane $C_2H_5CHClNO_2$	1.2090	139.5–143	5.8	0.5
1-Chloro-2-nitropropane $ClCH_2CHCH_3NO_2$	1.1980	174	25 [a]	0.8

[a] 81° C.

Most of the chloronitroalkanes have a lachrymatory effect and are toxic to animals and man. Thus, a concentration of 0.025 mg. of chloropicrin per liter of air causes lachrymation, and 0.1–0.4 mg./l. cannot be endured for one minute. With an increase in molecular weight the lachrymatory effect diminishes.

The best method of preparing the halonitro compounds is the chlorination of nitroalkanes with chlorine:

$$RCH_2NO_2 + Cl_2 \rightarrow RCHClNO_2 + HCl$$

Chloropicrin can also be obtained by oxidative chlorination of picric acid or other aromatic nitro compounds.

The nitrostyrenes are produced in good yields by the condensation of the appropriate aromatic aldehydes with nitromethane or its homologs:

$$ArCHO + CH_3NO_2 \rightarrow ArCH=CHNO_2 + H_2O$$

When nitrostyrenes are halogenated, dihaloarylnitroethanes are formed that also have high bactericidal and fungicidal activity:

$$ArCH=CHNO_2 + X_2 \rightarrow ArCHXCHXNO_2$$

The halonitrostyrenes, which are obtainable by the action of sodium carbonate on dihaloarylnitroethanes, are still more active:

$$ArCHXCHXNO_2 + Na_2CO_3 \rightarrow ArCX = CHNO_2 + NaCl + NaHCO_3$$

The chloronitrostyrene obtained by this method melts at 45° C., and the bromonitrostyrene at 65° C. The nitrostyrenes containing chloro, bromo, iodo, fluoro, hydroxy, nitro, and other groups in the benzene ring have been studied. All these compounds have high fungicidal, bactericidal, and insecticidal activity, but some of them strongly irritate the skin.

High insecticidal activity also is shown by the diarylnitroalkanes, among which 1,1-bis(p-chlorophenyl)nitrobutane (Bulan, m.p. 66.5° to 67.5° C.), 1,1-bis(p-chlorophenyl)-2-nitropropane (Prolan, m.p. 80.5° to 81.5° C.) and others, are especially outstanding.

Dilan. The compounds mentioned are produced in the United States as a technical mixture containing 26–27% 1,1-bis(p-chlorophenyl)-2-nitropropane, 53–54% 1,1-bis(p-chlorophenyl)-2-nitrobutane, and 20% related compounds under the name of Dilan for use in agriculture; this technical mixture is a waxy mass that melts completely at 23°–25° C., d_{15}^{15} 1.28.

Diarylnitroalkanes are produced by the following reactions:

$$ClC_6H_4CHO + RCH_2NO_2 \rightarrow ClC_6H_4CH(OH)CH(R)NO_2$$

$$ClC_6H_4CH(OH)CH(R)NO_2 + C_6H_5Cl \rightarrow (ClC_6H_4)_2CHCH(R)NO_2 + H_2O$$

The first of these reactions takes place in the presence of alkali metal alcoholates or alkalies. It has been suggested that the nitro compounds react in the guise of aciform salts. To avoid splitting out of water and formation of unsaturated compounds (homologs of nitrostyrene) the reaction is carried out at 0°–5° C. The second reaction proceeds well in the presence of sulfuric acid.

An advantageous property of Dilan is its lower toxicity to animals in comparison with DDT. Its LD_{50} for mice is 950 mg./kg. The chronic toxicity has been studied on rats into whose food a level of 650 mg./kg. was introduced for a period of a year.

The persistence of Dilan is less than that of DDT. When Dilan is heated in the light, especially in the presence of acids, it decomposes according to the reaction scheme of Neff:

$$2\,(4\text{-}ClC_6H_4)_2CHCH(NO_2)CH_3 \rightarrow 2\,(4\text{-}ClC_6H_4)_2CHC(O)CH_3 + H_2O + N_2O$$

In the presence of acids this reaction goes more easily and at a lower temperature. Oxidizing agents destroy Dilan, producing a mixture of the corresponding ketones and acids. For microdetermination of Dilan the reaction of the *aci*-form with ferric chloride is employed.

Dilan is used in the form of emulsions and wettable powders. Usually it is produced in the form of an 80% solution in xylene, from which 25% emulsive concentrates are prepared. It can be used in the form of dusts containing 1–2% active ingredient.

Aromatic nitro compounds

Compounds containing nitro groups in an aromatic ring also have pesticidal properties. Halonitro compounds such as halonitrobenzenes, halonitrotoluenes, nitrohalobenzils, and dinitrohalonaphthalenes are especially active. Such compounds as trichlorodinitrobenzene, tetrachloronitrobenzene, pentachloronitrobenzene, fluorodinitrobenzene, fluorobromodinitrobenzene, etc., have found practical use in agriculture and industry.

1,2,4-Trichloro-3,5-dinitrobenzene (Brassisan) (m.p. 102°–104° C.) is used as a selective fungicide to control club root of cabbage and other diseases. It is obtained by the nitration of 1,2,4-trichlorobenzene. A more active fungicide is 1,2,3-trichloro-4,5-dinitrobenzene, but it still has not received practical use.

Interesting antifungal compounds are the fluorodinitrobenzenes, which are recommended for the protection of nonmetallic aviation materials from damage by microorganisms. To preserve articles of polyvinyl acetate 0.25% 1-fluoro-3-bromo-4,6-dinitrobenzene is added, and to protect leather belts 1-fluoro-2,4-dinitrobenzene, 1-fluoro-3-chloro-, 1-fluoro-3-bromo-, and 1,3-difluoro-4,6-dinitrobenzenes have been proposed. All these compounds have given satisfactory results, also, as seed disinfectants for various crops.

The synthesis of 1-fluoro-2,4-dinitrobenzene can be carried out either by the nitration of fluorobenzene or by the reaction of 1-chloro-2,4-dinitrobenzene with potassium fluoride at an elevated temperature.

1,3-Dihalo-4,6-dinitrobenzenes are obtained in good yields by the nitration of 1,3-dihalobenzenes with concentrated nitric acid.

In the German Federal Republic 1,3,5-trichloro-2,4,6-trinitrobenzene (m.p. 196° C.) (Bulbosan) is used. Its synthesis is carried out by nitrating 1,3,5-trichlorobenzene with nitric acid in the presence of oleum.

A general disadvantage of all the aromatic nitro compounds enumerated is the explosion hazard, which substantially restricts the possibility of their widespread use. Moreover, these compounds have a very limited area of action. A compound with less explosion hazard is 1-chloro-2,4-dinitronaphthalene (a yellow crystalline substance, m.p. 148° C., solubility at 25° C. in 100 ml. of water 0.1 mg., of methanol 0.24 g., of acetone 4.6 g., of benzene 7.2 g.). The LD_{50} is 250 mg./kg. A 50% wettable powder of this compound is known as CRP-32.

1-Chloro-2,4-dinitronaphthalene is obtained by the action of thionyl chloride on 2,4-dinitronaphthol-1.

The high fungicidal activity of the halonitro compounds of the aromatic series is often associated with the reactivity of the halogen in these compounds. A study of the kinetics of the reaction of 1-fluoro-2,4-dinitrobenzene and other substances of this type with taurine, glycine, leucine, alanine, and aminobutyric acid has shown that the rate of reaction diminishes in the order in which these amino acids are listed.

2,3,5,6-Tetrachloronitrobenzene (Fusarex, Folsan, DB-905), which is produced by the nitration of 1,2,4,5-tetrachlorobenzene (m.p. 93° C., LD_{50} for rats 250 mg./kg.), has been used as a selective fungicide and a sprout inhibitor for potatoes.

Pentachloronitrobenzene (m.p. 146° C., LD_{50} 1,650 mg./kg., maximum permissible concentration in air 0.5 mg./m.[3]) is used to control various plant diseases by treatment of the soil. Its synthesis is carried out by nitrating pentachlorobenzene or chlorinating various chloronitrobenzenes that are formed as by-products in the production of *m*-chloronitrobenzene and 3,4-dichloronitrobenzene. The best chlorination takes place in the presence of an iron-iodine catalyst. An important condition for the chlorination is complete absence of moisture, since even a trace of water sharply decreases the rate of the reaction. To bind the water sometimes some chlorosulfonic acid is added to the reaction mixture.

Pentachloronitrobenzene is stable in storage and does not burn. With alkali alcoholates it gives an ether of pentachlorophenol:

$$C_6Cl_5NO_2 + CH_3OK \rightarrow C_6Cl_5OCH_3 + KNO_3$$

This reaction is used for the quantitative determination of pentachloronitrobenzene. It is used as a soil fungicide to control diseases of cotton, potatoes, tomatoes, and peppers at a rate of 5–40 kg./ha. Use of 20% dust also gives satisfactory results for seed disinfection against smut.

The insecticidal, fungicidal, and herbicidal properties of the nitro derivatives of diphenyl, phenanthrene, acenaphthene, fluorene, and a number of other aromatic hydrocarbons and their halogen derivatives have been studied, but they still have received practically no use.

General references

BASKAKOV, YU. A., and N. N. MEL'NIKOV: Khimicheskie sredstva zaderzhki prorastaniya klubne- i korneplodov pri dlitel'nom khranenie [Chemical agents to inhibit the sprouting of tuber and root crops in prolonged storage]. State Scientific and Technical Publishing House of Chemical Literature [USSR] (1958).

BYRDY, S., Z. ECKSTEIN, R. KOWALIK, and Z. PLENKIEWIC: Nitro compounds. Proc. Internat. Symposium, Warsaw (Sept. 18–20, 1963).

MEL'NIKOV, N. N.: Khim. Prom. No. 3 and 4 (1945).

—, and YU. A. BASKAKOV: Khimiya gerbitsidov i regulyatorov rosta rastenii [Chemistry of herbicides and plant growth regulators]. State Scientific and Technical Publishing House of Chemical Literature [USSR] (1962).

RICHARDSON, H.: J. Econ. Entomol. 35, 664 (1942).

— J. Econ. Entomol. 36, 110, 420 (1943).

VIII. Amines and salts of quaternary ammonium bases

One of the most important groups of physiologically active compounds, having great significance for the vitality of living organisms, is the amines. The pesticidal properties of various amines and their salts with organic and inorganic acids have been studied.

The maximum activity as a contact insecticide against larvae of the house fly is shown by di-n-octylamine; amines of the aliphatic series with a higher or lower molecular weight have less insecticidal activity. Triamylamine, di-(4-ethylnonyl)amine, and di-(3-methylamyl)amine are relatively toxic to lice.

Going from the amines of the aliphatic series to the aromatic ones the toxicity increases. Aniline, for example, is more toxic than hexylamine. Introduction into the aromatic nucleus of halogen atoms or some other substituents does not affect the insecticidal activity of the compound, while the nitro group increases the general toxicity of the compound both to insects and to animals. It has been reported that o-iodoaniline is toxic for caterpillars, and 2,5-dichloroaniline is also toxic for lice.

The diarylamines as a rule have a higher insecticidal activity. Diphenylamine was used as an agent to control pediculosis during the Second World War and gave completely satisfactory results as a 20–25% dust. The triarylamines also are relatively toxic to insects. In Table XXI the insecticidal activity of primary, secondary, and tertiary amines toward the pea aphid is shown.

Table XXI. *Toxicity of aromatic amines to the pea aphid*

Amine	LC_{90} (moles/l.)	Amine	LC_{90} (moles/l.)
Aniline	0.53–1	Benzylamine	0.19
Diphenylamine	0.03	Dibenzylamine	0.025
Triphenylamine	0.04	Tribenzylamine	0.07

Amines have not been used to control plant pests because of their phytotoxicity for most plants. Some aromatic amines are so phytocidal that they find use as herbicides. For the control of annual weeds the following amines have been recommended; 2,6-dinitro-4-trifluoromethyl-N,N-dipropylaniline (trifluralin), 2,6-dinitro-4-trifluoromethyl-N-butyl-N-ethylaniline (benefin), and 2,6-dinitro-4-methylsulfonyl-*N*,*N*-dipropylaniline (nitralin).

2,6-Dinitro-4-trifluoromethyl-N,N-dipropylaniline (trifluralin) is a crystalline substance, m.p. 48.5°–49° C., vapor pressure at 29.5° C. 1.99×10^{-4} mm. of Hg; solubility in water 40 p.p.m., acetone 40 g./100 ml., and xylene 58 g./100 ml. Its LD_{50} is 10,000 mg./kg. It is used to control weeds in cotton at dosages of 0.5 to 2 kg./ha. in soil. It is produced in the following way:

$$\underset{Cl}{\underset{|}{C_6H_4}}-CF_3 \xrightarrow{HNO_3} \underset{Cl}{\underset{|}{C_6H_3}}(CF_3)(NO_2) \xrightarrow{(C_3H_7)_2NH} \underset{N(C_3H_7)_2}{\underset{|}{C_6H_2}}(CF_3)(NO_2)_2$$

2,6-Dinitro-4-trifluoromethyl-N-butyl-N-ethylaniline (benefin) is a crystalline substance, m.p. 65°–66.5° C., vapor pressure 4×10^{-7} mm. of Hg at 25° C.; solubility in water 70 p.p.m., readily soluble in most organic solvents, but having a lower solubility in ethanol. Its LD_{50} is 10,000 mg./kg. Benefin is produced in a similar way to trifluralin.

2,6-Dinitro-4-methylsulfonyl-N,N-dipropylaniline (nitralin) is a crystalline substance, m.p. 150°–151° C., vapor pressure 1.5×10^{-6} mm. of Hg at 25° C.; solubility in water 0.6 p.p.m., acetone 36 g./100 ml., dimethylsulfoxide 33 g./100 ml. at 22° C. Its LD_{50} is 2,000 mg./kg. It is prepared in the following way:

$$\underset{Cl}{\underset{|}{C_6H_4}}-SCH_3 \xrightarrow{HNO_3} \underset{Cl}{\underset{|}{C_6H_2}}(SO_2CH_3)(NO_2)_2 \xrightarrow{(C_3H_7)_2NH} \underset{N(C_3H_7)_2}{\underset{|}{C_6H_2}}(SO_2CH_3)(NO_2)_2$$

Derivatives of 2-phenoxyethylamine, derivatives of benzylamine, and their salts also show herbicidal activity. Since the derivatives of phenoxyethylamine are more active toward dicotyledonous plants, it has been suggested that they are converted in the plants or in the soil to derivatives of phenoxyacetic acid.

The fungicidal and bactericidal activity of the amines is relatively low, but it increases going from the free amines to their salts with various organic and inorganic acids. The salts of amines containing aliphatic hydrocarbon radicals with not fewer than eight carbon atoms have significant microbiological activity.

The germicidal quaternary ammonium salts also have an insecticidal effect which appears when insects are fed baits treated with such compounds. Recently the quaternary ammonium salts have been proposed as systemic insecticides, but because of their phytocidal effect have received little use. The ammonium salts are used for disinfection in the home and in animal husbandry. Positive properties of the substituted ammonium salts are their high solubility in water (which permits their use in aqueous solutions), absence of odor, and comparatively low toxicity to animals and man. This last property permits their use not only as disinfectants and fungicides, but also as detergents.

Table XXII. *Bactericidal effect of tetraalkylammonium and trialkylbenzylammonium salts at 37° C.*

Compound	Bactericidal dilution	
	Staphylococcus aureus	*Escherichia typhi*
$[C_7H_{15}(CH_3)_3N^+]Br^-$	Not active	1:15
$[C_8H_{17}(CH_3)_3N^+]Br^-$	1:30	1:75
$[C_{12}H_{25}(CH_3)_3N^+]Br^-$	1:4,000	1:9,000
$[C_{14}H_{29}(CH_3)_3N^+]Br^-$	1:38,000	1:36,000
$[C_{16}H_{33}(CH_3)_3N^+]Br^-$	1:80,000	1:40,000
$[C_{18}H_{37}(CH_3)_3N^+]Br^-$	1:64,000	1:8,000
$[C_{16}H_{33}(C_4H_9)_3N^+]Br^-$	1:48,000	1:16,000
$[C_{12}H_{25}(CH_3)_2N^+CH_2C_6H_5]Br^-$	1:32,000	1:30,000
$[C_{15}H_{31}(CH_3)_2N^+CH_2C_6H_5]Br^-$	1:52,000	1:36,000
$[C_{16}H_{33}(CH_3)_2N^+CH_2C_6H_5]Br^-$	1:76,000	1:32,000
$[2\text{-}ClC_6H_4CH_2N^+(CH_3)_2CH_2C_6H_5]Br^-$	1:52,000	1:8,000

Tabelle XXIII. *Bactericidal effect of ammonium salts with ether groups in the cation at 20° C.*

Compound	Bactericidal dilution	
	Staphylococcus aureus	*Escherichia typhi*
$[C_4H_9C_6H_4OC_2H_4OC_2H_4(CH_3)_2N^+CH_2C_6H_5]Cl^-$	1:1,200	1:2,500
$[(CH_3)_3CCH_2C_6H_4OC_2H_4OC_2H_4(CH_3)_2N^+CH_2C_6H_5]Cl^-$	1:12,000	1:25,000
$[4\text{-}C_6H_5C_6H_4OC_2H_4OC_2H_4(CH_3)_2N^+CH_2C_6H_5]Cl$	1:6,000	1:2,000
$[4\text{-}C_6H_5CH_2C_6H_4OC_2H_4OC_2H_4(CH_3)_2N^+CH_2C_6H_5]Cl$	1:2,800	1:2,800

In Table XXII the bactericidal activities of the simplest tetraalkylammonium and trialkylbenzylammonium salts are given. Compounds containing radicals with ether groups also have high activity. The bactericidal activity of compounds of this type is shown in Table XXIII.

The microbiological activity of substituted ammonium salts is affected not only by the structure of the radicals entering into the composition of the cation, but also by the structure of the anion. Thus, the salts obtained by the reaction of esters of phosphoric, thio- and dithiophosphoric acids with dimethyloctadecylamine are several times as active as trimethyloctadecylammonium iodide and chloride. The bactericidal activities of these salts are given in Table XXIV.

Table XXIV. *Bactericidal effect of substituted ammonium salts with organophosphorus anions at 37° C.*

Compound		Bactericidal dilution	
		Staphylococcus aureus	*Escherichia coli*
$[C_{18}H_{37}N^+(CH_3)_3]$	$\begin{bmatrix} OP(S)OCH_2CH_2SC_2H_5 \\ \vert \\ OCH_3 \end{bmatrix}^-$	1 : 25,000	1 : 25,000
$[C_{18}H_{37}N^+(CH_3)_3]$	$\begin{bmatrix} OP(S)OC_6H_2Cl_3\text{-}2,4,5 \\ \vert \\ OC_2H_5 \end{bmatrix}^-$	1 : 12,000	
$[C_{18}H_{37}N^+(CH_3)_3]$	$\begin{bmatrix} OP(S)SC_2H_5 \\ \vert \\ OCH_3 \end{bmatrix}^-$	1 : 50,000	1 : 6,000
$[C_{18}H_{37}N^+(CH_3)_3]$	$\begin{bmatrix} OP(S)SCH_2CONHCH_3 \\ \vert \\ OCH_3 \end{bmatrix}^-$	1 : 25,000	1 : 12,000
$[C_{18}H_{37}N^+(CH_3)_3]_2$	$\begin{bmatrix} OP(S)SP(S)O \\ \vert \quad\quad \vert \\ H_5C_2O \quad OC_2H_5 \end{bmatrix}^{2-}$	1 : 50,000	1 : 400

The substituted ammonium salts are prepared by the reaction of tertiary amines with halogen derivatives at an elevated temperature. The reaction goes most rapidly with compounds containing labile halogen. The iodine derivatives react the most easily, and the chlorine derivatives with somewhat more difficulty. However, with such compounds as benzyl chloride, esters of monochloroacetic acid, chloromethyl alkyl ethers, and other effective alkylating agents the reaction takes place rapidly.

The substituted ammonium salts are colorless crystalline substances, highly soluble in water, but not in hydrophobic organic solvents of the hydrocarbon type. Some of them are so hygroscopic that they deliquesce on standing in the air. The solubility of the substituted ammonium salts in water decreases as the amine molecule becomes more complex: e. g., 1,3-dialkylbenzotriazolium salts are hardly soluble in water.

All the salts of quaternary ammonium bases form complexes with platinic chloride that are barely soluble in water. This reaction can be used for their identification and quantitative determination. At an elevated temperature the substituted ammonium salts decompose with the formation of the tertiary amine and an unsaturated compound:

$$[RCH_2CH_2\overset{+}{N}R_3']Cl^- \rightarrow RCH = CH_2 + R_3'\overset{+}{N}HCl^-$$

Many salts of quaternary ammonium bases are toxic and have a curare-like effect.

Mixtures of various salts of quaternary ammonium bases, which are marketed in the form of 10–50% aqueous solutions, are used as fungicides and disinfectants. An alkyldimethylethylammonium bromide consisting of a mixture of compounds containing alkyls with 9–18 carbon atoms is used for seed disinfection. Preparations based on alkyldimethylbenzylammonium chloride (where the alkyl is $C_{12}H_{25}$ to $C_{16}H_{33}$), which is prepared by the reaction of benzyl chloride with alkyldimethylamine, are used for disinfection.

Preparations are also known that have as the active ingredient a mixture of salts formed by the reaction of trimethylamine with alkylbenzyl chloride. These compounds can be synthesized by the following route:

$$C_6H_6 + RCH=CH_2 \rightarrow C_6H_5CH_2CH_2R$$
$$RCH_2CH_2C_6H_5 + CH_2O + HCl \rightarrow RCH_2CH_2C_6H_4CH_2Cl + H_2O$$
$$(CH_3)_3N + RCH_2CH_2C_6H_4CH_2Cl \rightarrow [RCH_2CH_2C_6H_4CH_2\overset{+}{N}(CH_3)_3]Cl^-$$

The aliphatic hydrocarbon radical in these compounds contains 9–15 carbon atoms.

Instead of benzyl chloride sometimes 3,4-dichlorobenzyl chloride is used, which is easily synthesized by the chlorination of benzyl chloride at a low temperature in the presence of catalysts for substitutive chlorination of the benzene nucleus. These compounds are distinguished from preparations that do not contain halogen in the benzene nucleus by their lower activity. Because of the additional step in their production, they are more expensive.

In spite of numerous investigations, the mechanism of action of the substituted ammonium salts still has not been established. There is also a question as to whether the action of these compounds is bactericidal or bacteriostatic. In the opinion of some investigators such compounds do not kill bacteria and fungi, but only inhibit their growth; however, there still is no definite proof for this point of view.

Of the free amines, only 2,6-dichloro-4-nitroaniline has been used as a fungicide; it is effective in controlling *Botrytis* and *Sclerotinia* on various crops.

2,6-Dichloro-4-nitroaniline (dichloran, dicloran) melts at 192° to 194° C., vapor pressure at 20° C. 1.2×10^{-6} mm. of Hg. Its LD_{50} for warm-blooded animals lies within the limits 1,500–4,000 mg./kg. It is used mainly in the form of a 50% wettable powder, but also can be used in dusts (4–8%). It is produced by direct chlorination of *p*-nitroaniline:

$$H_2N-\langle\rangle-NO_2 + 2Cl_2 \rightarrow O_2N-\langle\rangle(Cl)(Cl)-NH_2 + 2HCl$$

An interesting physiologically active amine derivative is chlorocholine chloride (CCC) which is a white crystalline, hygroscopic substance. It is highly soluble in water and hydrophilic organic solvents and poorly soluble in hydrophobic organic solvents. It can be prepared by several methods:

1. By the action of chlorides of phosphorus, thionyl chloride, and aromatic sulfonyl chlorides on choline hydrochloride:

$$HOCH_2CH_2N^+(CH_3)_2Cl^- + SOCl_2 \rightarrow ClCH_2CH_2N^+(CH_3)_3Cl^- + \\ + HCl + SO_2$$

2. By the reaction of an excess of dichloroethane with trimethylamine:

$$(CH_3)_3N + ClCH_2CH_2Cl \rightarrow ClCH_2CH_2N^+(CH_3)_3Cl^-$$

Chlorocholine chloride is used as an agent to stimulate growth of the root system and thickening of the stem of cereal grains, making it possible to prevent breakage in rainy weather and under other circumstances.

General references

BROWN, A.: Insect control by chemicals. New York-London: Wiley (1951).
CCC Symposium 14. 12. 1965 auf der Landwirtschaftlichen Versuchsstation Limburgerhof. BASF, Ludwigshafen (1966).
CLARK, N.: Chem. & Ind. 1960, 572.
GUSO, L.: Proc. S. Weed Control Conf., p. 121 (1966); U. S. Patent 3257190.
JUNG, J.: Naturwissenschaft **54**, 346 (1967).
MEL'NIKOV, N. N., and YU. A. BASKAKOV: Khimiya gerbitsidov i regulyatorov rosta rastenii [Chemistry of herbicides and plant growth regulators]. State Scientific and Technical Publishing House of Chemical Literature [USSR] (1962).
—, B. A. KHASKIN, G. N. PERSHIN, and S. N. MILOVANOVA: Khim. v Sel'skom Khozyaistve No. 7, 59 (1965).
—, and K. D. SHVETSOVA-SHILOVSKAYA: Khim. v. Sel'skom Khozyaistve No. 3, 60 (1965).
—, N. D. SUKHAREVA, and O. P. ARKHIPOVA: Zhur. Priklad. Khim. **20**, 643 (1947).
— — — Zhur. Priklad. Khim. **21**, 306 (1948).
— — — Zhur. Priklad. Khim. **22**, 1122 (1949).
— — — Trudy Tsentral. Dezinfekt. Nauch. Issledovatel. Inst. **5**, 88 (1949).
SCHIEFERSTEIN, R. H., and W. J. HUGHES: Proc. 8th Brit. Weed Control Conf., p. 377 (1966).
SOLOWAY, S. B., and K. D. ZWAHLER: French Patent 1453170.
TATTERSFIELD, F.: Ann. Applied Biol. **17**, 181 (1927).
VOROZHTSOV, N. N.: Osnovy sinteza poluproduktov i krasitelei [Principles of synthesis of intermediates and dyes]. State Scientific and Technical Publishing House of Chemical Literature [USSR] (1955).

IX. Alcohols, phenols, and ethers

Alcohols

The saturated aliphatic alcohols are relatively weak pesticides and still have not received practical use in this field. However, the activity of the alcohols is somewhat higher than that of the hydrocarbons. The physiological activity of the alcohols toward various species of organisms increases with increasing molecular weight, but after reaching a certain size it decreases. Thus, the maximum for fungicidal activity is reached at undecyl and dodecyl alcohols. The sprouting of potatoes is most effectively inhibited by nonyl and decyl alcohols.

The unsaturated and cyclic alcohols have stronger insecticidal, fungicidal, and bactericidal effects, and also stronger physiological activity toward plants. Allyl alcohol is used to control weeds by treatment of the soil with 0.2–1% aqueous solutions in greenhouses, tree nurseries, and tobacco plantations at dosages of 80–200 liters/ha. Butynediol is employed at a dosage of 4–5 kg./ha. for defoliation of cotton before it is picked.

Sex attractants for a number of insects also are found among the unsaturated alcohols. Thus, hexadecadien-10,12-ol (I) has been isolated from the glands of virgin female silkworm moths, and dextrorotatory cis-hexadecene-7-diol-1,10 10-acetate (II) from the glands of virgin female gypsy moths:

$$CH_3(CH_2)_2CH=CH-CH=CH(CH_2)_8CH_2OH \qquad (I)$$

$$CH_3(CH_2)_5\underset{\underset{\displaystyle OCOCH_3}{|}}{CH}CH_2CH=CH(CH_2)_5CH_2OH \qquad (II)$$

These compounds attract the males when 10^{-14} g. of material are used/trap.

cis-Octadecene-9-diol-1,12 12-acetate (b.p. 182° C. at 0.5 mm. of Hg), prepared from ricinoleic acid, also has been shown to be an active attractant for the gypsy moth. This compound, called gyplure for short, is prepared in the following way. Ricinoleic acid from castor oil, after purification, is reduced with lithium aluminum hydride to the corresponding alcohol, which is then acetylated with acetic anhydride. The acetate obtained is subjected to partial saponification with an alcoholic solution of sodium hydroxide. On saponification the acetic acid radical is split off first from the primary alcohol group and gyplure is formed. It should be noted that the compound used to attract insects must be of the highest purity, because contaminants weaken its effect.

$$CH_3(CH_2)_5CH(OH)CH_2CH=CH(CH_2)_7COOH \rightarrow$$
$$CH_3(CH_2)_5CHOHCH_2CH=CH(CH_2)_7CH_2OH \rightarrow$$
$$CH_3(CH_2)_5\underset{\underset{OCOCH_3}{|}}{CH}CH_2CH=CH(CH_2)_7CH_2OCOCH_3 \rightarrow$$
$$CH_3(CH_2)_5\underset{\underset{OCOCH_3}{|}}{CH}CH_2CH=CH(CH_2)_7CH_2OH$$

A large number of analogs of gyplure have been synthesized and studied, but they have a weaker attractant effect and the *trans*-isomer of gyplure is completely inactive.

Among the dihydric alcohols of the aliphatic series there are also compounds with an opposite effect. 2-Ethylhexanediol-1,3 — known as "Rutgers-612" — (b.p. 244° C., m.p. about −40° C., d_{20}^{20} 0.9422, n_D^{20} 1.4511) is used in various compositions to repel blood-sucking diptera. Its solubility in water is about 4.2%, vapor pressure at 20° C. is 0.01 mm. of Hg, LD_{50} for rabbits is 2.6 ml./kg.; poisoning may take place through the skin. Chronic poisoning has been observed with daily application of 2 ml./kg. for 90 days. 2-Ethylhexanediol-1,3 usually is used in a mixture with dimethyl phthalate and indalone. One part of 2-ethylhexanediol-1,3 is used to one part of indalone and three parts of dimethyl phthalate. This preparation on application to the skin of a man is effective for eight to ten hours, repelling all blood-sucking insects.

2-Ethylhexanediol-1,3, is produced by the reduction of the aldol of butyraldehyde:

$$CH_3CH_2CH_2CH(OH)CH(C_2H_5)CHO + 2H \rightarrow$$
$$\rightarrow CH_3CH_2CH_2CH(OH)CH(C_2H_5)CH_2OH$$

It is possible to obtain 2-ethylhexanediol-1,3 also by the Tishchenko condensation reaction of butyraldehyde in the presence of aluminum ethylate. In this procedure the formation of the aldol and its reduction are combined in one process:

$$3\ CH_3CH_2CH_2CHO \xrightarrow{Al(OC_2H_5)_3}$$
$$CH_3CH_2CH_2CH(OH)CH(C_2H_5)CH_2OCOC_3H_7$$
$$CH_3CH_2CH_2CH(OH)CH(C_2H_5)CH_2OCOC_3H_7 + KOH \rightarrow$$
$$\rightarrow C_3H_7COOK + CH_3CH_2CH_2CH(OH)CH(C_2H_5)CH_2OH$$

The yield of 2-ethylhexanediol-1,3 in this reaction is small because of the large number of side reactions.

When one or several hydrogen atoms in an alcohol molecule are replaced by halogen, the biological activity toward all species of organisms is increased. Thus, for example, ethylenechlorohydrin has several times stronger insecticidal and fungicidal effects than ethyl alcohol. Ethylenechlorohydrin can disrupt the dormancy of plants, and this effect is of practical interest in potato culture. By using ethylenechlorohydrin the planting of potatoes can be carried out as soon as the tubers are gathered, which is important in the southern regions where it is expedient to obtain two crops in one season.

Of the cyclic alcohols mention should be made of bicyclo[2.2.1]heptylcarbinol-2-(2-norcamphanemethanol), which has a systemic fungicidal effect (b.p. 83° C. at 5 mm. of Hg, d_{20}^{20} 1.005, solubility in water 0.1%, LD_{50} for rats 1,620 mg./kg.). For control of plant diseases it is recommended that the soil be irrigated with aqueous solutions of the compound. However, this compound has not yet been widely used.

1,2,3,4-Tetrachlorobicyclo[2.2.1]hepten-2-ol-5 and 1,2,3,4,7,7-hexachlorobicyclo[2.2.1]heptene-2-diol-5,6 have herbicidal properties.

The aromatic alcohols of the diphenylmethane series are quite interesting; among them are found active acaricides used in agriculture.

4,4'-Dichlorodiphenylmethylcarbinol (Dimite) is an active specific acaricide, m.p. 69.5°–70° C. It is approximately five times as volatile as DDT. The technical grade product contains as impurities the 2,2'- and 2,4'-isomers, and also small amounts of 2,2'-, 2,4'-, and 4,4'-dichlorobenzophenone. Dimite is produced by the reaction of 4,4'-dichlorobenzophenone with methylmagnesium bromide:

$$(ClC_6H_4)_2CO + CH_3MgBr \rightarrow (ClC_6H_4)_2C(CH_3)OMgBr \xrightarrow{+H_2O}$$
$$\rightarrow Mg(OH)Br + (ClC_6H_4)_2C(OH)CH_3$$

4,4'-Dichlorodiphenylmethylcarbinol is not very stable and when it is heated with acids it easily goes over to unsymmetrical 4,4'-dichlorodiphenylethylene:

$$(ClC_6H_4)_2C(OH)CH_3 \rightarrow (ClC_6H_4)_2C = CH_2 + H_2O$$

This reaction takes place in storage even at room temperature. The higher the temperature, the more rapidly the compound decomposes. At 45° C., 20% breaks down in six-and-a-half months, while at 195° C. 61% breaks down in 24 hours. Since the decomposition products are not toxic to mites, this reaction is very undesirable. When 4,4'-dichlorodiphenylmethylcarbinol is oxidized by various oxidizing agents it yields 4,4'-dichlorobenzophenone. Dimite approaches DDT in toxicity. It can be used in various types of formulations to destroy both adult mites and their eggs, but because of its relatively high toxicity its use is limited.

As a result of the discovery of the acaricidal properties of Dimite, a large number of its analogs and homologs have been studied. It has been found that 4,4'-dichlorobenzhydrol not only affects adult mites, but also is toxic to their eggs, while benzhydrol does not have such activity. Replacement of the methyl group in 4,4'-dichlorodiphenylmethylcarbinol by ethyl or cyclohexyl does not change the activity of the compound, while replacement by phenyl or benzyl leads to complete loss of acaricidal effect. The same result is observed when the hydroxyl is shifted from the tertiary carbon atom. Replacement of hydrogen atoms in the methyl group by halogen leads to an increased acaricidal effect; the maximum effect is achieved when all three hydrogen atoms are replaced by chlorine. 4,4'-Dichlorodiphenyltrichloromethylcarbinol is known in agriculture as Kelthane or dicofol.

High acaricidal activity is shown by 4,4'-dichlorodiphenylethoxymethylcarbinol (m.p. 58°–59° C., b.p. 155°–157° C., at 0.06 mm. of Hg), which is known as Ethoxynol, and also by 4,4'-dichlorodiphenylethynylcarbinol (m.p. 73°–74° C.).

Replacement of the methyl group of Dimite by carboethoxyl also leads to the formation of an active acaricide, chlorobenzilate:

$$(ClC_6H_4)_2C(OH)COOC_2H_5$$

Kelthane (chloroethanol, dicofol), m.p. 78.5°–79° C., is produced by the hydrolysis of 4,4'-dichlorodiphenyltrichloromethylchloromethane with dilute sulfuric acid:

$$(ClC_6H_4)_2CClCCl_3 + H_2O \rightarrow (ClC_6H_4)_2C(OH)CCl_3 + HCl$$

The 4,4'-dichlorodiphenyltrichloromethylchloromethane necessary for this synthesis is obtained by the chlorination of the dehydrochlorination product of DDT:

$$(ClC_6H_4)_2C=CCl_2 + Cl_2 \rightarrow (ClC_6H_4)_2CClCCl_3$$

Like most chlorinated pesticides, dicofol is unstable toward the action of caustic alkalies. Its LD_{50} for rats is 730 mg./kg. It has an appreciable mutagenic effect. The compound is an active acaricide and is used in the form of emulsions and suspensions against plant-feeding mites on various crops.

Pentachlorbenzyl alcohol (PCBA, blastin). Blastin, used for rice blast control in Japan as a 50% wettable powder, and a 4% dust, exists as colorless crystals, m.p. 198° C., solubility at 25° C. in water 0.2 p.p.m., in acetone 0.72%, in xylene 0.34%, in ethanol 0.56%. Its LD_{50} is 6,750 mg./kg. for mice.

Pentachlorbenzyl alcohol is produced by chlorination of toluene and hydrolysis of the product:

Pentachloromandelonitrile (Oryzon) has similar properties (m.p. 163° to 164° C.).

Phenols

The physiological activity of the phenols is considerably greater than that of the alcohols. They are more powerful insecticides, fungicides, bactericides, and herbicides. The pesticidal activity is increased when various substituents, for example a halogen, nitro group, thiocyano group, alkyl, and others are introduced into the aromatic radical.

The insecticidal activity is most strongly increased by introduction of a nitro group into the phenol molecule. The maximum activity is shown by the dinitrophenols; introduction of a third nitro group lowers the insecticidal effect of the compound. The alkyldinitrophenols are especially active. A similar relationship is observed also for the herbicidal and fungicidal activity. With an increase in the length of the alkyl chain of the substituent, the pesticidal activity of the dinitroalkylphenols increases to a certain limit, after which it falls. The maximum insecticidal activity for various insects may be different. Most often for the 2,4-dinitro-6-alkylphenols it lies between 2,4-dinitro-6-amyl- and 2,4-dinitro-6-octylphenols. The greatest herbicidal activity is shown by the substituted alkylphenols containing an alkyl with 4 and 5 carbon atoms. In Table XXV the constants of some phenols and nitrophenols and their insecticidal properties for aphids and the silkworm are given.

Table XXV. *Insecticidal activity of some nitrophenols*

Compound	M. P. (°C.)	B. P. (°C. at 0.1 mm. Hg)	LC_{50} for pea aphid (%)	LD_{50} for silkworm (μg./kg.)
Phenol	42–43	—	0.3	—
4-Nitrophenol	113.4	—	0.007	—
2,4-Dinitrophenol	114–115	—	0.001	—
o-Cresol	30.8	—	0.2	—
2,4-Dinitro-6-methylphenol	86.4	—	0.0005	—
2,4-Dinitro-6-ethylphenol	30	—	—	29
2,4-Dinitro-6-propylphenol	—	128	—	18
2,4-Dinitro-6-butylphenol	—	140	—	9
2,4-Dinitro-6-amylphenol	—	145	—	8
2,4-Dinitro-6-hexylphenol	—	162	—	4
2,4-Dinitro-6-heptylphenol	—	171	—	4
2,4-Dinitro-6-octylphenol	—	178	—	10

The insecticidal activity of other isomers of the dinitroalkylphenols is somewhat less, but the order of change in activity with change in molecular weight of the compound is preserved.

The fungicidal and bactericidal effects of the substituted phenols in most cases are subject to the same rules. The most active bactericides and fungicides are the haloalkylphenols, as can be seen in the example of the 4-chloro-2-alkylphenols, the fungicidal and bactericidal activities of which are given in Table XXVI.

With increasing molecular weight the toxicities of these compounds to mice decrease.

Introduction of a second lower alkyl radical into the molecule of chlorophenol increases the fungicidal and bactericidal properties of the compounds up to certain limits. When a hydrogen in the phenol molecule is replaced by an alicyclic or aromatic radical the activity of the compound also is increased. Thus, 2-hydroxydiphenyl and benzylphenol considerably exceed phenol in fungicidal and bactericidal effect.

Table XXVI. *Fungicidal and bactericidal properties of 4-chloro-2-alkylphenols*

Compound	Phenol coefficient [a] for		
	Bacillus typhi	Staphylococcus aureus	Streptococcus
Phenol	1	1	1
4-Chlorophenol	4.3	4.3	4.2
4-Chloro-2-methylphenol	12.5	12.5	11.7
4-Chloro-2-ethylphenol	28.6	34.4	27.5
4-Chloro-2-propylphenol	93.3	93.8	83.3
4-Chloro-2-butylphenol	141.0	257.9	160.0
4-Chloro-2-amylphenol	15.0	500	400
4-Chloro-2-hexylphenol	23.2	1250	500
4-Chloro-2-cyclohexylphenol	26.7	430	417
4-Chloro-2-heptylphenol	20.0	1500	667
4-Chloro-3-methyl-6-ethylphenol	64.3	50	60
4-Chloro-3-methyl-6-propylphenol	133	200	150
4-Chloro-3,5-dimethyl-6-ethylphenol	46.9	106	130
4-Chloro-2-benzylphenol	71.4	200	273
4-Chloro-5-methyl-2-benzylphenol	18.3	375	400
4-Chloro-2-phenylethylphenol	100	375	375
4-Bromo-2-propylphenol	62.5	62.5	—
4-Fluoro-2-propylphenol	21.0	—	—
4-Fluoro-2-butylphenol	66.0	60.0	—

[a] The phenol coefficient is a value that shows how many times more active the given compound is than phenol for the microorganism.

Replacement of hydrogen in the phenol molecule by fluorine or bromine has a less significant effect than the introduction of chlorine on the microbiological activity of the compound. The alkylfluorophenols practically do not differ from the corresponding alkylphenols in fungicidal effect.

The accumulation of chlorine atoms in the phenol molecule leads to an increase in fungicidal activity, but different isomers of di-, tri-, and tetrachlorophenols have different bactericidal and fungicidal activity. Halohydroxydiphenyls, hydroxydiphenylmethanes, and halonaphthols also are more active than the compounds that do not contain halogen atoms. However, the accumulation of more than three halogen atoms in the molecule of naphthols and hydroxydiphenyls leads to a decrease in activity.

The phenols severely burn plants, and consequently they cannot be used to control pests and diseases of plants during the vegetative period. They are used for that purpose in the early spring, before the buds are put out. Furthermore, some phenol derivatives are employed as seed disinfectants and as disinfectants for nonmetallic materials.

Because of their high phytotoxicity the phenols are used also as contact herbicides and desiccants. In addition to the phenols, their salts with amines, ammonia, and the alkali metals, and the water-insoluble copper phenolates are used as pesticides. Furthermore, a number of esters of the dinitrophenols are employed to control plant pests and diseases. The halodinitrophenols also have fungicidal, insecticidal, and herbicidal effects, but this group of phenols still has not received widespread use.

2-Hydroxydiphenyl has been widely used as an antiseptic for nonmetallic materials and to protect citrus fruits from mold when they are stored for a long time. It is a white crystalline substance, m.p. 57° C., b.p. 228° C., d_{25}^{25} 1.217. At 25° C., 0.07 g. of 2-hydroxydiphenyl dissolves in 100 g. of water. It is highly soluble in most organic solvents.

For the disinfection of materials, it is used in the form of water-soluble phenolates of the alkali metals, and in organic solvents. To preserve fruits they are wrapped in paper impregnated with hydroxydiphenyl. It also is employed in practical disinfection.

The technical grade product contains up to 15% 4-hydroxydiphenyl (m.p. 164° C.), small amounts of phenol and, if it is obtained in the production of phenol through benzenesulfonic acid, some thiophenol and diphenyl disulfide. Such a preparation is not suitable for preservation of fruits; it can be used only as a disinfectant for moist wood. For the treatment of wrapping materials for fruits only the highly purified compound, completely without odor is used.

2-Hydroxydiphenyl is obtained as a by-product in the production of phenol from chlorobenzene or sodium benzenesulfonate:

$$ClC_6H_5 + C_6H_5ONa \to C_6H_5C_6H_4OH + NaCl$$

It gives all the characteristic reactions of phenols. With NaOH at pH 12 to 13.5 sodium phenolate is formed, which crystallizes with four molecules of water. The sodium salt of 2-hydroxydiphenyl is employed as a water-soluble disinfectant for wood under the name of "Dowicide-A."

2-Hydroxydiphenyl is easily given off from the phenolate by the action of CO_2 and acids. Precipitation from alkaline aqueous solutions by acids is sometimes employed to free it from the 4-isomer, since the isomers precipitate from alkaline aqueous solutions at different pH. When it is chlorinated a mixture of 3-chloro-2-hydroxydiphenyl and 5-chloro-2-hydroxydiphenyl is obtained. These compounds and their alkali phenolates find use as disinfectants for wood; more often they are used together with other antiseptics.

When a second chlorine is introduced 3,5-dichloro-2-hydroxydiphenyl is obtained, which on further chlorination yields 3,5,4'-trichloro-2-hydroxydiphenyl:

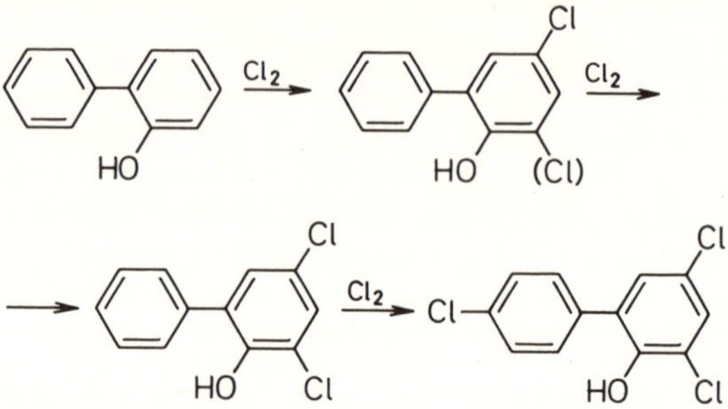

By direct nitration of 2-hydroxydiphenyl it is possible to obtain mono- and dinitro derivatives, depending on the conditions. When 2-hydroxydiphenyl is alkylated with alcohols or alkyl halides in the presence of appropriate catalysts, 2-hydroxy-5(3)-alkyldiphenyls are formed, which have powerful bactericidal and fungicidal effects on Gram-positive microorganisms.

2-Hydroxydiphenyl has comparatively low toxicity for mammals; the LD_{50} for white rats orally is 2,480 mg./kg. When animals were fed a diet containing 0.2% 2-hydroxydiphenyl for two years, no changes were observed in the animals. It and its homologs also have an insecticidal effect, but because of their high phytocidal activity they have received practically no use. The chlorinated 2-hydroxydiphenyls in combination with hexachlorobenzene are used in place of mercury compounds for seed disinfection.

Of the unsubstituted phenols, limited use for plant pest control is made of 2-naphthol (m.p. 122° C.), which is sometimes employed to impregnate trapping bands that are fastened around the trunks of fruit trees. The larvae of the codling moth hide under these bands and are killed. It also has a fungicidal effect. The fungicidal action of the naphthols increases when various substituents are introduced into the aromatic nucleus.

Nitrophenols

The nitrophenols were proposed for the control of harmful insects in the last century. From 1937 on the nitrophenols have been used also as selective contact herbicides.

For studies of the insecticidal, fungicidal, and herbicidal properties a large number of alkyl-, cycloalkyl-, and aryldinitrophenols have been synthesized, but practical use has been made of only a few compounds in this series, including *p*-nitrophenol, 2,4-dinitrophenol, 2,4-dinitro-6-methylphenol, 2,4-dinitro-6-*sec*-butylphenol, 2,4-dinitro-6-*sec*-amylphenol, and 2,4-dinitro-6-cyclohexylphenol and their derivatives. Furthermore, 2,4-

dinitro-6-*sec*-octylphenyl crotonate (dinocap, Karathane) can be used in agriculture for the control of both plant diseases and mites.

p-Nitrophenol, m.p. 113.4° C., has been proposed as an agent for preserving natural rubber and some other nonmetallic materials from destruction by microorganisms. It is poorly soluble in water, but more soluble in organic solvents. The LD_{50} for cats is 150–200 mg./kg. (cats are especially sensitive to different phenols).

p-Nitrophenol is produced by alkaline saponification of *p*-nitrochlorobenzene at an elevated temperature:

$$\underset{\text{Cl}}{\underset{|}{C_6H_4}}\text{-NO}_2 + \text{NaOH} \longrightarrow \underset{\text{OH}}{\underset{|}{C_6H_4}}\text{-NO}_2 + \text{NaCl}$$

It can be obtained also by direct nitration of phenol in a mixture with the *o*-isomer, which, because of its high volatility, is not used as a disinfectant. It is employed also as an intermediate for the production of parathion (thiofos), methyl parathion (metafos), and other insecticides.

2,4-Dinitrophenol, m.p. 114°–115° C., solubility in water about 0.5%. The alkali metal and ammonium phenolates are more soluble. 2,4-Dinitrophenol, especially the phenolates, are explosive and heating or shock may cause them to explode. The compound is also poisonous. Poisoning may occur if it penetrates into the organism by the respiratory route, the esophagus, or the skin; the LD_{50} is about 40 mg./kg. It is produced by alkaline hydrolysis of 2,4-dinitrochlorobenzene or by oxidative nitration of benzene in the presence of mercury salts:

$$C_6H_6 + O + 2\,HNO_3 \rightarrow HOC_6H_3(NO_2)_2 + 2\,H_2O$$

The insecticidal and herbicidal properties of 2,4-dinitrophenol are considerably weaker than those of the alkyldinitrophenols, and consequently it is scarcely used at all for the control of plant pests and weeds, but is employed in disinfectant compositions.

2,4-Dinitro-6-methylphenol (dinitro-o-cresol, DNOC, DINOC) is a yellow crystalline substance, m.p. 86.4° C., vapor pressure at 25° C. 5.2×10^{-5} mm. of Hg. The solubility of the compound in g./100 g. of solvent is: water 0.0128, acetone 100.6, benzene 37.15, methanol 7.33, and petroleum ether 0.31.

2,4-Dinitro-6-methylphenol with caustic alkalies, ammonia, and organic amines forms highly water-soluble salts, which in the dry state explode readily either from shock or detonation. Therefore, its salts are prepared either with inorganic diluents (sodium sulfate, ammonium sulfate, urea, etc.) or in aqueous solution.

Reducing agents convert it to amines (primarily one nitro group). Such reduction, in particular, occurs in warm-blooded animals. The product is

2-amino-4-nitro-6-methylphenol, a yellow substance, m.p. 173°–174° C. In the soil 2,4-dinitro-6-methylphenol is decomposed comparatively rapidly, probably both as a result of reduction and the splitting off of a nitro group with subsequent complete destruction of the molecule.

2,4-Dinitro-6-methylphenol stains the skin yellow. It belongs to those compounds toxic to man and animals. The minimum lethal dose for various experimental animals ranges within the limits 15–50 mg./kg., depending on the method of introduction into the organism and the species of animal. There are recorded poisoning cases of people by this compound, especially when it has been used in the form of oil solutions which are rapidly absorbed through the intact skin. Therefore, work with this compound in agriculture should be carried out with the strict observance of necessary precautionary measures. The maximum permissible concentration in air is 0.05 mg./m³.

2,4-Dinitro-6-methylphenol is produced by direct nitration of o-cresol with a nitrating mixture at a low temperature. In some cases the o-cresol is first sulfonated with sulfuric acid. For the sulfonation of o-cresol 70–93% sulfric acid is used:

$$\text{o-cresol} + H_2SO_4 \rightarrow \text{o-cresol-SO}_3H \xrightarrow{HNO_3} \text{2,4-dinitro-6-methylphenol}$$

It is obtained in small amounts also from the oxidative nitration of toluene in the presence of mercuric nitrate. Dinitro-m-cresol is the main product in this reaction.

2,4-Dinitro-6-methylphenol is used in agriculture to control plant pests and diseases, and for treatment of fruit trees before opening of the buds, either in the form of emulsions with oils (yellow oils) or more often in the form of aqueous solutions of its salts. The effectiveness of the compound increases when surface-active agents are present.

In the control of weeds it is used exclusively in the form of aqueous solutions of the salts. Good results are obtained in weed control in plantings of flax, grains, and some other crops at dosages of 3–6 kg./ha. calculated as the active ingredient.

2,4-Dinitro-6-sec-butylphenol (dinoseb). Preparations of 2,4-dinitro-6-methylphenol are being displaced more and more by preparations of its homologs, and primarily by 2,4-dinitro-6-sec-butylphenol, which has substantial advantages. It is less explosive, somewhat less toxic to man and domestic animals (LD_{50} 60 mg./kg.), and more effective in controlling plant pests, plant diseases, and weeds. Dinoseb surpasses 2,4-dinitro-6-methylphenol almost three times in insecticidal and herbicidal effect (because of the lower dosage, the cost of treatment per unit of area is substantially cheaper). Less of the compound remains on the plants, and the hazard in using it is decreased.

2,4-Dinitro-*sec*-butylphenol is a yellow crystalline substance, m.p. 38°–39° C. The technical grade product at room temperature is a light-brown liquid with crystals disseminated through it. Its solubility at 25° C. in water is 0.0734 wt.-%, in ethyl alcohol 23.40 wt.-%, and in diesel fuel 8.7 wt.-%.

Like other phenols, dinoseb with organic and inorganic bases forms salts that are highly soluble in water.

Several methods are known for preparing 2,4-dinitro-6-*sec*-butylphenol; the most important of them are the following:

1. Direct nitration of 2-*sec*-butylphenol with nitric acid in the presence of sulfuric acid. The 2-*sec*-butylphenol is first treated with 80–98% sulfuric acid and the sulfo mixture obtained is further added at 50°–100° C. to 50% nitric acid:

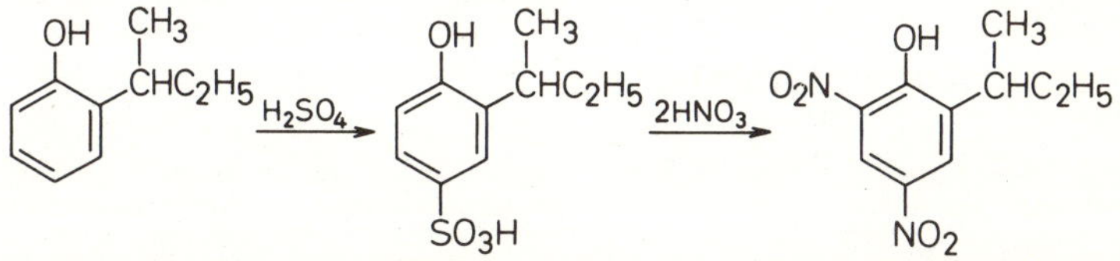

From 2–4 moles of nitric acid are used/mole of phenol, depending on the conditions. The 2-*sec*-butylphenol necessary for the production of dinoseb is obtained in good yields by the alkylation of phenol with butylene at 200°–250° C. in the presence of aluminum phenolate.

2. Alkylation of *p*-phenolsulfonic acid with butylene or butyl alcohol and subsequent nitration of the 2-butylphenol-4-sulfonic acid without separating it from the reaction mixture:

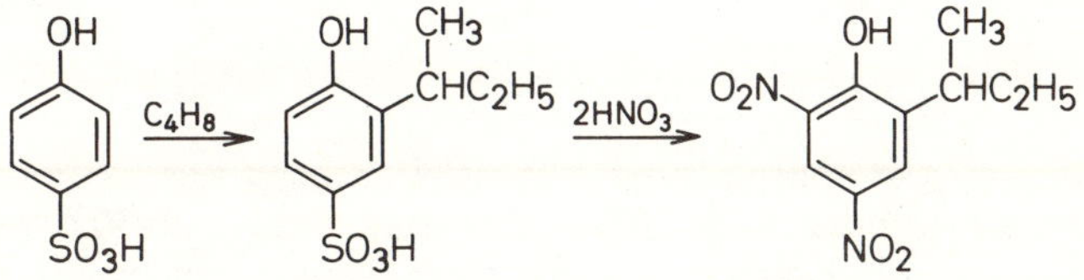

It is possible to obtain 2-*sec*-butylphenol by alkylation of *p*-bromophenol with subsequent removal of the bromine by reduction:

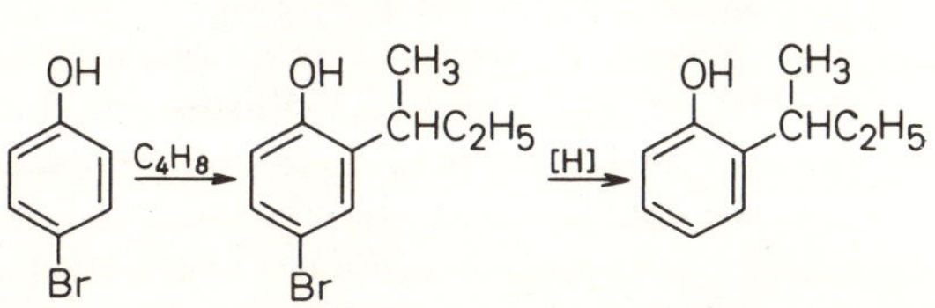

For the control of plant pests and weeds dinoseb is most often used in the form of aqueous solutions of the phenolates of ammonia and organic amines.

It is possible to use it in the form of emulsions and solutions in organic solvents, but these preparations are more dangerous. For the control of weeds it is employed at dosages of 1.5–3 kg./ha. calculated on the 100% compound.

For the control of plant pests, 2,4-dinitro-6-sec-butylphenyl acetate has begun to be used, which is somewhat less toxic than the phenol. Also 2,4-dinitro-6-sec-butylphenyl methacrylate (m.p. 67°–69° C., LD_{50} for rats 350 mg./kg.) has been used for the control of true powdery mildew on apple trees, along with some other esters of 2,4-dinitro-6-sec-butylphenol. These esters are obtained by the action of the anhydrides or acid chlorides of the appropriate acids on the phenolates. 2,4-Dinitro-6-sec-butylphenyl methacrylate has not only a fungicidal but also an acaricidal effect. A preparation based on it is known as binapacryl.

2,4-Dinitro-6-sec-amylphenol (b.p. 149° C. at 0.1 mm. of Hg) is similar to dinoseb in herbicidal effect.

On the basis of 2,4-dinitro-6-sec-butylphenol a very large number of different esters have been synthesized and tested for use as acaricides and fungicides.

2,4-Dinitro-6-sec-butylphenyl isopropyl carbonate (dinobuton, Acrex, Dessin, Dinofen). This is a non-systemic acaricide and fungicide (active against powdery mildews), pale yellow crystals, m.p. 61°–62° C. It is practically insoluble in water, but soluble in aliphatic hydrocarbons, alcohol, and aliphatic ketones. Its LD_{50} is 140 mg./kg. for rats, 2,540 mg./kg. for mice; its dermal toxicity to rats is greater than 5,000 mg./kg.

Dinobuton is produced by the condensation of an alkali salt of dinoseb with isopropylchloroformate:

$$O_2N-\underset{NO_2}{\underset{|}{C_6H_3}}-O-Na\ \underset{C_2H_5}{\overset{CH_3}{\underset{|}{CH}}} + (CH_3)_2CHOCOCl \longrightarrow O_2N-\underset{NO_2}{\underset{|}{C_6H_3}}-OCOOCH(CH_3)_2\ \underset{C_2H_5}{\overset{CH_3}{\underset{|}{CH}}} + NaCl$$

2,4-Dinitro-6-sec-butylphenyl isopropyl carbonate is used in the form of a wettable powder with 50% active ingredient and 30% emulsifiable concentrate. For greenhouse and field use against red spider mites and powdery mildew it is employed at 0.05%.

For weed control use has been made of 2,4-dinitro-6-tert-butylphenyl acetate (dinoterb-acetate) (m.p. 134°–135° C., LD_{50} 62 mg./kg.), 2,4-dinitro-3-methyl-6-tert-butylphenyl acetate (medinoterb-acetate) (m.p. 86°–87° C., LD_{50} 42 mg./kg.), and 2,4-dinitrophenyl-2,4-dinitro-6-sec-butylphenyl carbonate (tribonate) (m.p. 129°–131° C., LD_{50} 108 mg./kg.).

As a fungicide and acaricide for the control of powdery mildew, red spider mites, and other pests of plants use has been started of a mixture of isomers consisting substantially of methyl-2,4-dinitro-6-(1-ethylhexyl)-phenyl carbonate and methyl-2,4-dinitro-6-(1-propylpentyl)phenyl carbonate (dinocton-o) (LD_{50} 1,250 mg./kg.) and a mixture of isomers consisting substantially of methyl-2,6-dinitro-4-(1-ethylhexyl)phenyl carbonate and 2,6-dinitro-4-(1-propylpentyl)phenyl carbonate (dinocton-p) (LD_{50} 460 mg./kg.).

2,4-Dinitro-6-cyclohexylphenol (Dinex). Of the dinitrophenols, 2,4-dinitro-6-cyclohexylphenol and its cyclohexylamine salt have found wide use as insecticides. 2,4-Dinitro-6-cyclohexylphenol is a yellow crystalline substance, m.p. 106° C.; its solubility in water changes with the pH of the medium: at pH 1 at 25° C. its solubility is 1.8 mg./l., and at pH 6.5 it is 15 mg./l. With caustic alkalies and ammonia it forms salts that are highly soluble in water. The dicyclohexylamine salt of this phenol is practically insoluble in water (0.00515 g. of the salt dissolves in 100 g. of water at 25° C.) and difficultly soluble in most organic solvents. Because of its poor solubility in water this salt has little phytotoxicity and is used for the treatment of green plants.

2,4-Dinitro-6-cyclohexylphenol is completely similar to the compounds described in chemical properties. It is produced by the nitration of 2-cyclohexylphenol with a mixture of sulfuric and nitric acids at 60°–90° C. The 2-cyclohexylphenol can be prepared by the condensation of phenol with cyclohexene in the presence of aluminum phenolate or other catalysts at 200°–250° C.

2,4-Dinitro-6-cyclohexylphenol is poisonous for man and animals, but somewhat less so than DNOC and dinoseb. The LD_{50} for experimental animals is 50–125 mg./kg. It is used for the treatment of fruit and ornamental trees in the dormant state; the dicyclohexylamine salt may be used for the treatment of green plants to control mites, aphids, and some scale insects and thrips. It is used in the form of a wettable powder with 25–40% active ingredient.

2,4-Dinitro-6-sec-octylphenyl crotonate (dinocap, Karathane) is a brown liquid, which does not distill without decomposition at a pressure greater than 0.1 mm. of Hg, b.p. 138°–140° C. at 0.05 mm. of Hg. It is practically insoluble in water, but highly soluble in organic solvents.

To produce 2,4-dinitro-6-*sec*-octylphenyl crotonate, 2-*sec*-octylphenol is first nitrated and the dinitrophenol formed is further treated with crotonyl chloride in the presence of HCl acceptors, best of all organic tertiary amines in a hydrophobic solvent:

Dinocap is produced in the form of a 25% wettable powder and is employed to control powdery mildew on various crops (apple trees, pears, roses, cucurbits, etc.). The LD_{50} for rats is 980–1,100 mg./kg. Dinocap also has an acaricidal effect.

Other nitrophenols. For use as an insecticide, a fungicide for spring-fall eradicative spraying, and a herbicide, products of the nitration of coal-tar and shale phenols with b.p. up to 300° C. have been proposed. Nitration of these phenols can be carried out either with the preliminary preparation of the corresponding sulfonic acids or directly. The better results are obtained with preliminary sulfonation of the phenols and subsequent treatment of the sulfonic acid with sodium nitrate or ammonium nitrate. The nitration products are further neutralized with NaOH and used as a pesticide. Such a preparation based on shale phenols is called nitrofen or compound 125. The nitration process is carried out until a preparation is obtained that contains 1.5 nitro groups per molecule of phenol.

Nitrofen approaches preparations based on 2,4-dinitro-6-methylphenol in activity, and it is somewhat less toxic for man and domestic animals.

Considering that for the production of nitrofen cheap shale phenols are used, this is undoubtedly interesting. It is necessary only to carry out a strict standardization.

Halophenols

In agriculture and industry in addition to the nitrophenols use is made of the halophenols and their different derivatives.

2,4,5-Trichlorophenol is one of the important fungicides used both to control plant diseases and in the form of the sodium, copper, and zinc phenolates as a disinfectant for nonmetallic materials. It is a white crystalline substance, m.p. 66° C., practically insoluble in water, but highly soluble in organic solvents. It has the strong characteristic odor of phenols. It is toxic to warm-blooded animals; its LD_{50} is 150–250 mg./kg. depending on the route of administration and species of animal.

The most important method for producing 2,4,5-trichlorophenol is the alkaline hydrolysis of 1,2,4,5-tetrachlorobenzene; the reaction takes place in methyl alcohol at a temperature not lower than 160° C.:

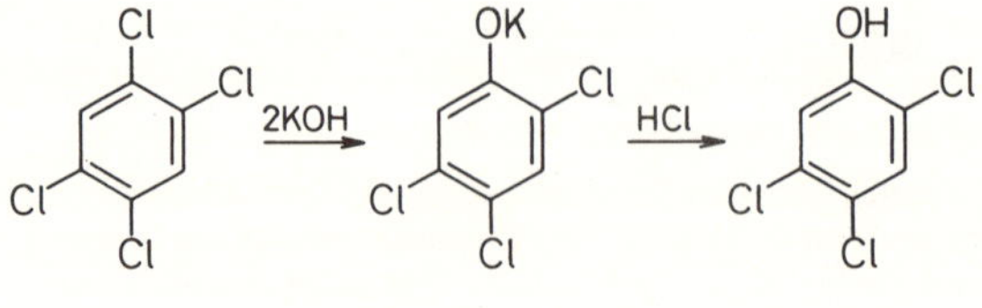

Some 2,4,5-trichloroanisole and also 2,3,6-trichlorophenol are formed as by-products. The latter is formed as a result of hydrolysis of 1,2,3,6-tetrachlorobenzene that usually is present in 1,2,4,5-tetrachlorobenzene.

2,4,5-Trichlorophenol, m.p. 61°–65° C. and synthesized from pure tetrachlorobenzene (m.p. not lower than 137° C.), is used for the production of copper and zinc trichlorophenolates. For the production of the herbicide 2,4,5-T purer 2,4,5-trichlorophenol (m.p. 65° C.) is required. Its purification is accomplished by fractional extraction with an aqueous solution of NaOH at 70° C. or by fractional precipitation from an aqueous solution of sodium 2,4,5-trichlorophenolate with mineral acids. For the extraction a 5% aqueous solution of NaOH is recommended, used in such amount that not more than 5% of the total trichlorophenols is taken into solution. Primarily 2,3,6-trichlorophenol dissolves, with a small amount of 2,4,5-trichlorophenol. To remove 2,4,5-trichloroanisole, the undissolved portion is treated with excess NaOH and the 2,4,5-trichloroanisole is steam-distilled off. The 2,4,5-trichlorophenol is isolated from the residue by acidification with a mineral acid.

The hydrolysis of 1,2,4,5-tetrachlorobenzene can be carried out without pressure, with ethylene glycol as the solvent, but in this process the regeneration of the solvent and purification of the wastewater is difficult. This last method of producing 2,4,5-trichlorophenol has still not found practical use in industry. The hydrolysis of 1,2,4,5-tetrachlorobenzene with NaOH can be carried out also in an aqueous medium at 250° C. under pressure.

2,4,5-Trichlorophenol is formed in practically quantitative yield when 2,5-dichlorophenol is chlorinated:

It can be obtained also by the chlorination of 3,4-dichlorophenol, but in a mixture with other isomers (47–56% 2,4,5-trichlorophenol, and 38–46% 2,3,4-trichlorophenol). The separation of this mixture is possible by fractional precipitation with mineral acids from aqueous alkali solutions.

Sodium trichlorophenolate is separated by neutralization of the 2,4,5-trichlorophenol with an equimolecular amount of NaOH and subsequent evaporation of the aqueous solution. Sodium 2,4,5-trichlorophenolate is a white crystalline, hygroscopic substance, insoluble in hydrophobic organic solvents. It is used as an antiseptic for natural rubber to protect it from mold. It also is an intermediate for the production of copper and zinc 2,4,5-trichlorophenolates, hexachlorophene, and 2,4,5-T.

For the synthesis of copper and zinc 2,4,5-trichlorophenolates it is possible to use the sodium 2,4,5-trichlorophenolate directly after the hydrolysis of 1,2,4,5-tetrachlorobenzene, but after removal of impurities such as 2,4,5-trichloroanisole. The solution should not contain excess NaOH.

The production of copper 2,4,5-trichlorophenolate takes place in an aqueous solution when a solution of copper sulfate (or other water-soluble copper salt) is mixed with a solution of sodium 2,4,5-trichlorophenolate:

Copper 2,4,5-trichlorophenolate settles out as a red-brown precipitate in quantitative yield. The precipitation of copper 2,4,5-trichlorophenolate can even be used for the determination of 2,4,5-trichlorophenol.

Copper 2,4,5-trichlorophenolate is used as a 20% preparation mixed with talc and kaolin for seed disinfection against gummosis at a dosage of 6–8 kg./ton of seed. It enters into the composition of the preparations Fenthiuram (10%) and Fenthiuram molybdate, which are used for disinfecting bean and cotton seeds. Besides copper 2,4,5-trichlorophenolate, Fenthiuram contains TMTD (40–50%), γ-benzene hexachloride or heptachlor (20%), and a diluent; Fenthiuram molybdate contains ammonium molybdate in addition to the compounds enumerated above.

The use of copper 2,4,5-trichlorphenolate for disinfection of cotton seed against gummosis permits practically complete elimination of cotton losses from this dangerous disease. Copper 2,4,5-trichlorophenolate was first proposed and developed in the Soviet Union. It is decomposed by mineral acids and alkalies to the free 2,4,5-trichlorophenol and the phenolate of the alkali metal, respectively.

Zinc 2,4,5-trichlorophenolate has similar properties. It finds use in the disinfection of cotton seed in the United States and some other countries, but its effect is considerably weaker than that of copper 2,4,5-trichlorophenolate.

2,4,5-Trichlorophenol is also an insecticide, but in this respect it is surpassed by other present-day compounds and has not been used for insect control.

Pentachlorophenol has an m.p. 190°–191° C., b.p. 310° C. at 760 mm. of Hg, vapor pressure at 20° C., 0.00017 mm. of Hg, solubility in 100 g. of water at 0° C. 0.5 mg., at 27° C. 1.8 mg., at 50° C. 3.5 mg. The solubility of pentachlorophenol in organic solvents depends greatly on the nature of the solvent. It is most soluble in methyl alcohol (57 g./100 g. of methyl alcohol at 20° C.), somewhat less soluble in ethyl alcohol, and barely soluble in alkanes. Its solubility in petroleum products increases with an increase in their content of aromatic compounds. When aromatic compounds are present at 25% content in petroleum products, the solubility of pentachlorophenol is 5%, and at a 40% aromatic content the solubility is 10%. The structure of the aromatic hydrocarbons in the petroleum product is very important. Pentachlorophenol is highly soluble in trichlorobenzene and in coal-tar and petroleum phenols. It has acid properties. Its salts with the alkali metals, ammonia, and organic bases are highly soluble in water and slightly soluble in hydrophobic organic solvents.

Copper pentachlorophenolate is a red-brown substance, insoluble in water and comparatively poorly soluble in most organic solvents. It can be used for quantitative determination of pentachlorophenol. The copper pentachlorophenolate is dissolved in 70% isopropyl alcohol, treated with potassium ferrocyanide, and the copper ferrocyanide formed is measured colorimetrically.

On oxidation in an aqueous medium pentachlorophenol gives chloranil:

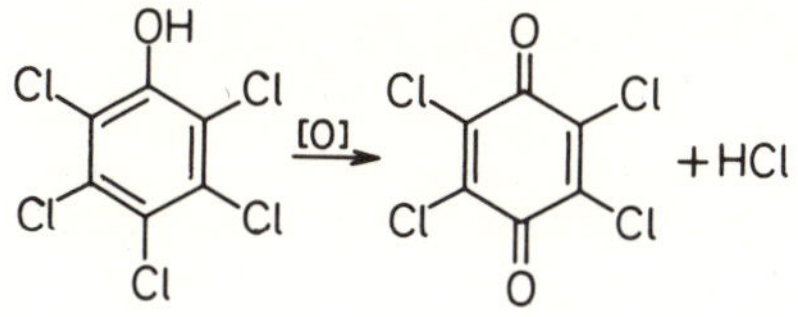

and on chlorination it gives derivatives of cyclohexene known as hexachlorophenols:

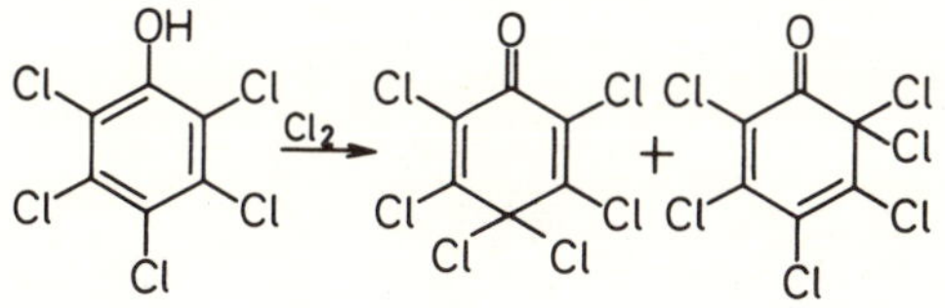

The hexachlorophenols are fungicidal and have been proposed for seed disinfection, but they are considerably less effective than other known seed disinfectants.

Technical grade pentachlorophenol produced in the United States contains as impurities up to 13% of other chlorophenols, of which isomeric tetrachlorophenols constitute the principal part. It is used as a disinfectant for nonmetallic materials, and as a herbicide and desiccant. For these purposes it is most often used in the form of solutions in mineral oils or as the highly water-soluble sodium pentachlorophenolate (100 g. of water dissolves 33 g. of the pentachlorophenolate). This compound's scale of use can be judged by the fact that in the United States from 13,000 to 17,000 tons of pentachlorophenol are produced annually.

Pentachlorophenol and its salts are toxic to man as well as animals, depending on the method of administration and the species of animal; the LD_{50} lies within the limits 36–210 mg./kg. Oil solutions of pentachlorophenol are absorbed through the skin and cause serious poisoning; the vapors and dust of pentachlorophenol and sodium pentachlorophenolate irritate the mucous membranes of the upper respiratory tract. Therefore, when pentachlorophenol is used it is necessary to protect the upper respiratory tract and to avoid contact between the compound and uncovered parts of the skin.

Pentachlorophenol is produced by two methods:

1. By direct chlorination of phenol, chlorophenols, and polychlorophenols with chlorine in the presence of catalysts (aluminum, antimony, their chlorides, and others) at an elevated temperature:

$$\text{C}_6\text{H}_5\text{OH} \xrightarrow{5\text{Cl}_2} \text{C}_6\text{Cl}_5\text{OH} + 5\text{HCl}$$

The chlorination process is carried out in two stages. First, the phenol is chlorinated at 30°–40° C. to 2,4,5-trichlorophenol; then the catalyst is introduced, and the chlorination is continued with gradual elevation of the temperature so that the reaction mixture is heated 5°–10° C. higher than the melting point. The pentachlorophenol formed contains a small amount of tetrachlorophenol as an impurity.

2. By alkaline hydrolysis of hexachlorobenzene in methanol, ethylene glycol, and other dihydric aliphatic alcohols, in water and mixtures of different solvents:

$$\text{C}_6\text{Cl}_6 \xrightarrow{\text{KOH}} \text{C}_6\text{Cl}_5\text{OK} \xrightarrow{\text{HCl}} \text{C}_6\text{Cl}_5\text{OH}$$

The hydrolysis of hexachlorobenzene in methanol takes place in an autoclave at 130°–170° C. and 2.25–2.5 moles of NaOH or KOH are used per mole of hexachlorobenzene. Depending on the temperature, the process is over in two to four hours. The hydrolysis of hexachlorobenzene in ethylene glycol is carried out under atmospheric pressure at 150°–160° C. A disadvantage of this method of making pentachlorophenol is the difficulty of complete regeneration of the expensive solvent and the necessity of purifying the wastewater. The yields of pentachlorophenol by both the first and the second method are 95–97% of theoretical.

Pentachlorophenol and its salts have not only a fungicidal and bactericidal effect, but an insecticidal effect as well. When these compounds are used, wood can be protected not only from destruction by microorganisms, but also from insects. Because of its high phytotoxicity pentachlorophenol is not used for the control of plant pests.

Pentachloroanisole and other ethers of pentachlorophenol have a stronger insecticidal effect. Pentachlorophenol and its phenolates are used as desiccants for various crops and as nonselective herbicides both alone and mixed with other materials.

2,2'-Dihydroxy-5,5'-dichlorodiphenylmethane (G-4) is a disinfectant for cotton goods. It is known in American and Canadian literature as G-4, m.p. 177°–178° C. The technical grade product melts at 164° C., is soluble in water at 25° C. about 0.003%, and has a vapor pressure 1×10^{-10} mm. of Hg. With caustic alkalies this compound forms highly water-soluble phenolates. For the impregnation of textile goods its monosodium derivative is used which is offered for sale in the form of a 40% aqueous solution (Preparation G-4-40). Its LD_{50} for experimental animals lies within the limits of 1,000–2,000 mg./kg.; the chronic toxicity also is slight.

It is produced by the condensation of p-chlorophenol with formalin in water in the presence of sulfuric acid:

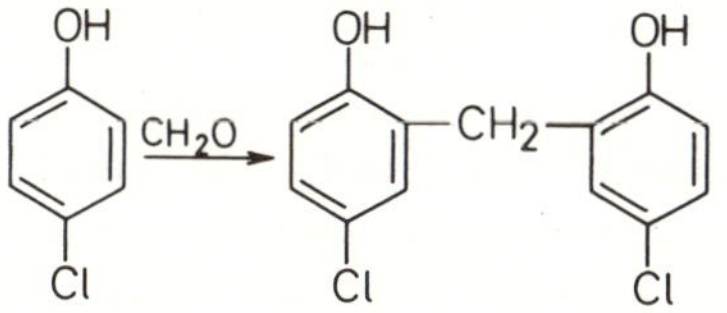

Another disinfectant, hexachlorophene (2,2'-dihydroxy-3,4,6,3',4',6'-hexachlorodiphenylmethane, m.p. 164°–165° C.) is also produced in a similar way:

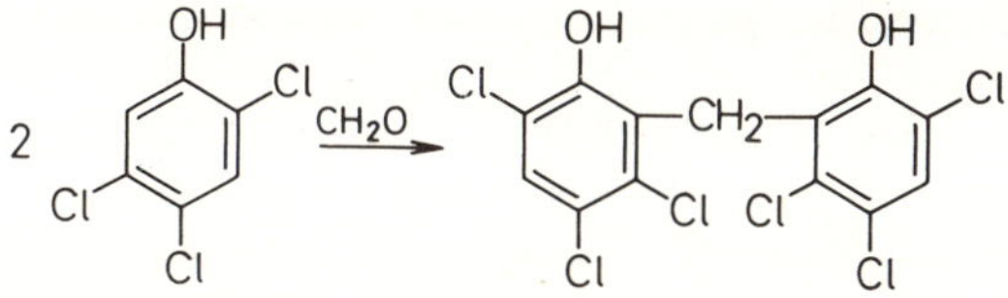

It is used for disinfection of the hands and in various other cases.

Ethers

Many aliphatic ethers have a strong narcotic effect, which is evident not only with respect to vertebrate animals, but also with insects. However, the toxicity of the aliphatic ethers for insects is relatively low. It increases when halogens and other atoms and groups are introduced into the molecule of the aliphatic ethers. The position of the substituent has a great effect on the toxicity of the ethers. Thus, for example, 2,2'-dichlorodiethyl ether (chlorex) is toxic for wireworms and is used as a soil fumigant, while its 1,2-isomer is practically nontoxic for this pest. Dichlorodimethyl ether has a low toxicity. This probably is connected with the low stability of these compounds in water and in other reagents. When fluorine is introduced into the molecule of an ether, toxic compounds also are formed; therefore, some fluorine-containing alcohols and ethers are used for controlling rodents (Glyfluor). 2-Fluoroethylformal [CH$_2$(OCH$_2$CH$_2$F)$_2$] is insecticidal, and 2,2'-bis(4-fluoroethoxyphenyl)propane has been used to control rodents under the name of difluoran.

Octachlorodipropyl ether, CCl$_3$CHClCH$_2$OCH$_2$CHClCCl$_3$, has been proposed as a synergist for pyrethroids.

The toxicity of the ethers increases sharply going from compounds of the aliphatic series to those of the aromatic series. Especially toxic are ethers containing various functional groups or halogens in the aromatic radical. Thus, for example, 2,4-dinitroanisole has such a powerful insecticidal effect that for several years it was used by the United States Army to control pediculosis. The chloroanisoles are also powerful insecticides, and their activity increases with the accumulation of chlorine atoms in the molecule.

Substituted ethers of the benzyl phenyl series containing halogen in the *p*-position to the ether group have insecticidal and acaricidal effects. The diaryl ethers in many cases show a fungicidal effect. This refers to ethers containing hydroxyl groups in one or both aromatic radicals. Introduction of halogen into a compound where there already is a hydroxyl strengthens the fungicidal effect. Many such compounds have been proposed for use as disinfectants for nonmetallic materials.

Some aryl ethers of ethylene glycol have systemic fungicidal properties, for example, 4-chloro-3,5-dimethylphenoxyethanol (m.p. 44°–45° C.). As a repellent for flies, a butoxypropylene glycol with molecular weight 400–800 has been proposed.

Alkyl aryl ethers of polyethylene and polypropylene glycols, (for example, the auxiliary materials OP-7 and OP-10) are used as emulsifiers for pesticidal preparations. OP-7, a detergent, is a product of the condensation of 1 mole of octylphenol with 7 moles of ethylene oxide:

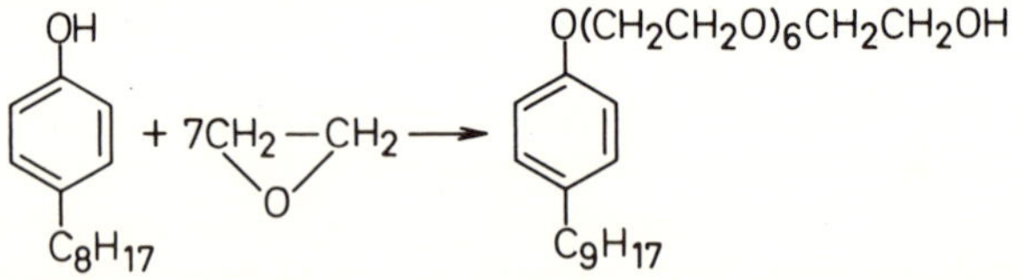

A number of alkylaromatic ethers have a high physiological activity for plants and some of them have received practical use. Thus, 1-naphthyldimethyl ether (b.p. 134° C. at 11 mm. of Hg) inhibits sprouting of potatoes in prolonged storage; 2,4-dichlorophenoxyethanol and its benzoate (sesin, m.p. 66° C.) and also 2,4,5-trichlorophenoxyethyl α,α-dichloropropionate (erbon, b.p. 161°–164° C. at 4 mm. of Hg) are used to control weeds.

3,4-Methylenedioxy-6-propylbenzyl *n*-butyldiethyleneglycol ether is a synergist for pyrethroids; it is known under the name of piperonyl butoxide (b.p. 180° C. at 1 mm. of Hg, d^{25} 1.06); its LD$_{50}$ for rats is 5,200 mg./kg. It is synthesized by the reaction of chloromethyldihydrosafrole with the sodium alcoholate of butylcarbitol:

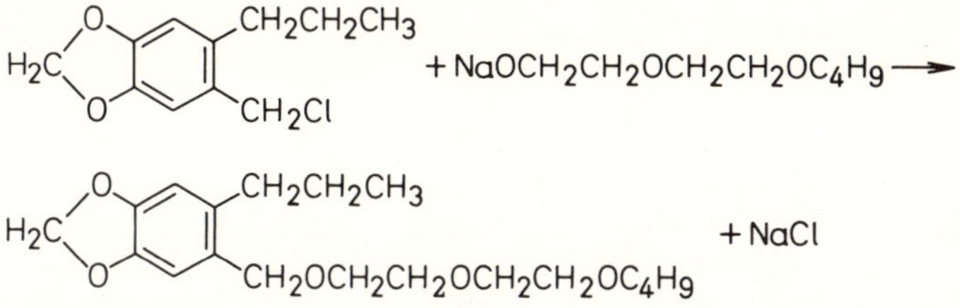

[Structure: methylenedioxybenzene with CH₂CH₂CH₃ and CH₂Cl substituents] + NaOCH₂CH₂OCH₂CH₂OC₄H₉ ⟶

[Structure: methylenedioxybenzene with CH₂CH₂CH₃ and CH₂OCH₂CH₂OCH₂CH₂OC₄H₉ substituents] + NaCl

The necessary chloromethyldihydrosafrole is prepared by direct chloromethylation of dihydrosafrole with formalin and hydrogen chloride:

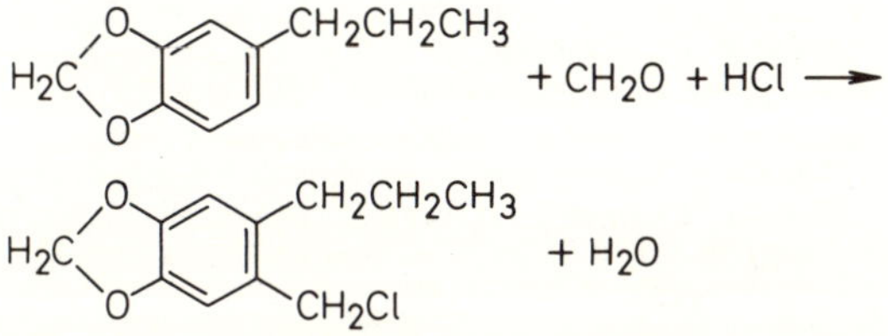

[Structure: dihydrosafrole with CH₂CH₂CH₃] + CH₂O + HCl ⟶

[Structure: chloromethyldihydrosafrole with CH₂CH₂CH₃ and CH₂Cl] + H₂O

A large number of analogs and homologs of piperonyl butoxide have been synthesized which are considerably weaker in activity.

2,4-Dichlorophenyl-4'-nitrophenyl ether, known as TOK E-25, has been proposed as a herbicide for the control of weeds in plantings of peas, carrots, cabbage, and rice. It is a pale-yellow crystalline substance, m.p. 70°–71° C.; at 22° C. about 1 mg. dissolves in one liter of water. It is highly soluble in most organic solvents; vapor pressure at 40° C. 8×10^{-6} mm. of Hg. Resistant to hydrolysis, it is converted by reducing agents to 2,4-dichlorophenoxyaniline. Its LD_{50} for rats is about 3,000 mg./kg. It is produced when sodium 2,4-dichlorophenolate is heated with p-nitrochlorobenzene:

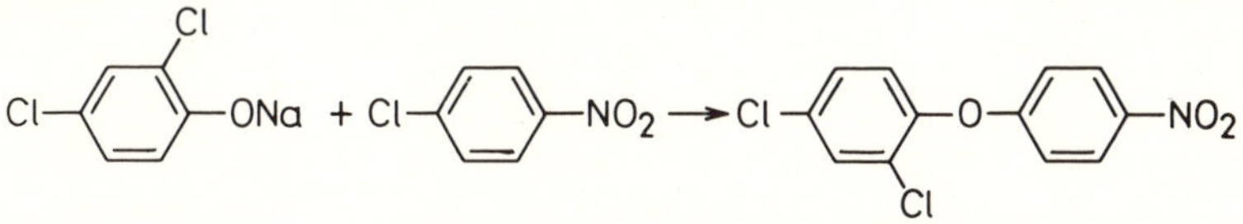

[Structure: 2,4-dichlorophenol sodium salt] + [Structure: p-nitrochlorobenzene] ⟶ [Structure: 2,4-dichlorophenyl-4'-nitrophenyl ether]

The reaction usually is carried out in an aqueous or alcoholic medium at an elevated temperature and slight pressure in more than 90% yield. This compound is used in the form of an aqueous emulsion prepared from an emulsive concentrate containing 25–40% active ingredient.

For the control of weeds use has been started also of 4-nitrophenyl-2,4,6-trichlorophenyl ether (MO Granules) (m.p. 107° C., LD_{50} 10,800 mg./kg.) and 4-nitrophenyl-2′-nitro-4′-trifluoromethylphenyl ether (C-6989) (m.p. 91.6°–92.2° C., solubility in water 2 p.p.m., LD_{50} 10,000 mg./kg.).

The dimethyl ether of 2,5-dichlorohydroquinone is used for the control of some soil fungi.

General references

BAZHIN, V. F., A. I. KULIKOV, I. P. KURILINA, I. M. POLYAKOV, and N. A. SHIPINOV: Coll. vol. "Khimiya i tekhnologiya goryuchikh slantsev i produktov ikh pererabotki" [Chemistry and technology of oil shales and products of their processing], No. 9, p. 276. State Scientific and Technical Publishing House of Chemical Literature [USSR] (1960).

BERGMAN, E. D., and A. KALUSZNER: J. Org. Chem. **23**, 1306 (1958).

BORECKI, Z., CZERWINSKA, Z. ECKSTEIN, and R. KOWALIK: Chemiczne grzybobójcze, Warsaw (1965).

EBNER, L., D. GREEN, and P. PANDE: Proc. 9th Brit. Weed Control Conf. **2**, 1026 (1968).

KULIKOV, A. E., I. P. KURILINA, I. M. POLYAKOV, N. A. NEPINOV, E. E. FEOFILOV, M. F. KOROLEVSKAYA, A. E. PETROVA, and G. N. GARNOVSKAYA: Coll. vol. "Khimiya i tekhnologiya goryuchikh slantsev i produktov ikh pererabotki" [Chemistry and technology of oil shales and products of their processing], No. 8, p. 152. State Scientific and Technical Publishing House of Chemical Literature [USSR] (1960).

LAPTEV, N. G.: In "Reaktsii i metody issledovaniya organicheskikh soedinenii" [Reactions and methods for the investigation of organic compounds], vol. 7. State Scientific and Technical Publishing House of Chemical Literature [USSR] (1958).

MEL'NIKOV, N. N.: Khim. Prom. No. 6, 181 (1950).

—, A. V. SKALOZUBOVA, E. I. ANDREEV, and A. S. DESHEVAYA: Trudy Nauch. Inst. Udobreniyam i Insektofungitsidam im. Ya. V. Samoilova No. 154, 238 (1955).

—, I. L. VLADIMIROVA, and S. N. IVANOVA: In "Khimicheskie sredstva predokhraneniya nemetallicheskikh materialov ot razrusheniya mikroorganizmami" [Chemical agents for the protection of nonmetallic materials from destruction by microorganisms]. Publishing House of the Laboratory of Scientific and Technical Investigations of the Nauch. Inst. Udobreniyam i Insektofungitsidam im. Ya. V. Samoilova (1959).

PIANKA, M.: J. Sci. Food Agr. **17**, 47 (1966).

— J. Sci. Food Agr. **18**, 355, 447 (1967).

— Chem. Ind. **1967**, 1625.

—, and P. J. SWEET: J. Sci. Food Agr. **19**, 667, 672, 676 (1968).

—, and J. D. EDWARDS: J. Sci. Food Agr. **19**, 60 (1968).

SKIBINSKAYA, M. B.: Poluchenie pentakhlorfenola iz geksakhlorbenzola [Preparation of pentachlorophenol from hexachlorobenzene]. Avtoreferat Dissertatsii, Publishing House of Nauch. Inst. Udobreniyam i Insektofungithsidam im. Ya. V. Samoilova (1954).

TROITSKII, S. V.: Prom. Org. Khim. **7**, 240 (1940).

X. Aldehydes, ketones, and quinones

Aldehydes

Aldehydes are more toxic than alcohols for insects, bacteria, fungi, and higher plants. Thus, formaldehyde, having powerful stomach action on insects, is a bactericide and fungicide. It is employed to control flies and has been used for several decades as a disinfectant for grain seeds. It gives especially good results in the disinfection of glumaceous crops, the seeds of which are treated with a concentration of 0.1–0.13% of the compound. Paraformaldehyde has considerably less fungicidal activity; therefore, in agriculture when formalin solutions are stored, measures are taken to retard the polymerization of the formaldehyde to paraldehyde. The simplest example of retardation of the polymerization is the neutralization of formaldehyde with alkalies, and the addition of small amounts of methyl alcohol.

With increasing molecular weight of the aldehydes, the contact insecticidal effect increases up to a certain point, after which a decrease in activity occurs. For adult granary weevils the maximum activity is reached at enanthic aldehyde.

Acetaldehyde has not as yet been used in agriculture, but its polymer metaldehyde is widely employed to control gastropods (snails). It is most often used in the form of baits or sprays on protein-rich food wastes (for example, on bran). The active-ingredient content in these preparations is 2–5%.

Metaldehyde is produced by the polymerization of acetaldehyde in the presence of mineral acids. It is a solid, easily subliming when heated to 110°–120° C., m.p. 246° C. (sealed capillary), LD_{50} 600–1,000 mg./kg. Since metaldehyde does not give the usual aldehyde reactions, a cyclic structure is assigned to it:

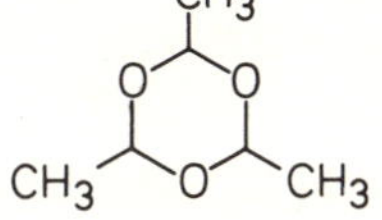

The unsaturated aldehydes have stronger insecticidal, fungicidal, and herbicidal effects. Acrolein and crotonic aldehyde considerably surpass the corresponding saturated compounds in the strength of their action. Acrolein (b.p. 52.5° C., LD_{50} 45 mg./kg.) is used as an agent to prevent water reservoirs from becoming overgrown and for combating slime formation in the paper industry. A serious disadvantage is its lachrymatory effect.

The insecticidal and fungicidal activity of the aromatic aldehydes in a number of cases is higher than that of the compounds of the aliphatic series with the same number of carbon atoms. The aldehydes containing halogen atoms and hydroxyl in the aromatic nucleus are especially active as fungicides.

Some aldehydes have high activity as plant growth regulators, and this apparently is associated with their conversion to the corresponding acids as a result of oxidation. Aldehydes like 3-indolylacetaldehyde, 1-naphthylacetaldehyde, and others that form the corresponding acids on oxidation are physiologically active. 2-Methoxy-5-acetylbenzaldehyde (m.p. 144° C.) has high physiological activity; it retards the growth of tomatoes, peppers, corn, and other plants at a concentration of 0.005%. Various chloro- and polychlorobenzaldehydes also are active.

Ketones

The pesticidal activity of the ketones is weaker than that of the corresponding aldehydes; they also are safer for plants, so that it is possible to use them as solvents in the production of pesticidal formulations.

Some use for the control of undesirable vegetation has been made of hexachloroacetone (b.p. 204° C., LD_{50} 1,200 mg./kg., maximum permissible concentration in air 0.5 mg./m.^{3}), which is added to petroleum oils for nonselective eradication of vegetation on uncultivable areas. It is marketed for this purpose in the form of 20–40% solutions in petroleum oils and applied at the rate of 450–1,300 liters/ha. of 2% solution in aromatic petroleums. It is produced by the direct exhaustive chlorination of acetone:

$$CH_3COCH_3 \xrightarrow{6Cl_2} CCl_3COCCl_3 + 6\,HCl$$

The technical grade product contains 85% hexachloroacetone, 4% pentachloracetone, and a small amount of high-boiling compounds.

Hexafluoroacetone trihydrate (GC 7787) has been presented primarily for the control of annual and perennial weeds and woody plants at doses of 10–20 kg./ha. It is liquid at 20° C. (b.p. 105° C.) and soluble in water. Its LD_{50} is 190 mg./kg.

Octachlorocyclohexenone (m.p. 106°–108° C.) is used under the name of octone. It is produced by the chlorination of pentachlorophenol in acetic acid:

It is recommended for use in the form of solutions in petroleum oils to control weeds in plantings of beans and garden beets at dosages of 4 to 10 kg./ha. For the preparation of an aqueous emulsion, an emulsifier is added to the solution of octachlorocyclohexenone in oil. Most often the compound is used before the seedlings appear or before planting.

4,6-Diisopropyl-1,1-dimethyl-5-propionylindan plus 4,6-diisopropyl-1,1-dimethyl-7-propionylindan (D 263) and 6-isopropyl-1,1,4-trimethyl-5-propionylindan plus 6-isopropyl-1,1,4-trimethyl-7-propionylindan (D 497) are two herbicides which are mixtures of related isomers; they are effective in controlling germinating grass seedlings at doses of 1–3 kg./ha. Both are being developed for the pre-emergence control of annual grass weeds in rice, cotton, and other crops.

D 263 is a liquid, b.p. 147°–154° C. at 3 mm. of Hg, solubility in water 1 p.p.m., highly soluble in most organic solvents. Its LD_{50} is 4,550 mg./kg. D 497 boils at 130°–135° C. at 2 mm. of Hg, LD_{50} 4,000 mg./kg.

3-m-Hexyl-5-(3,4-methylenedioxyphenyl)-2-cyclohexenone is used as a synergist for the pyrethrins under the name of piperonyl cyclonene. The technical grade product usually contains 56–58% of the ketone, 24–25% of its 6-carboethoxy derivative, and up to 20% related compounds. It is produced by the condensation of methyl hexyl ketone with 3,4-methylenedioxybenzaldehyde:

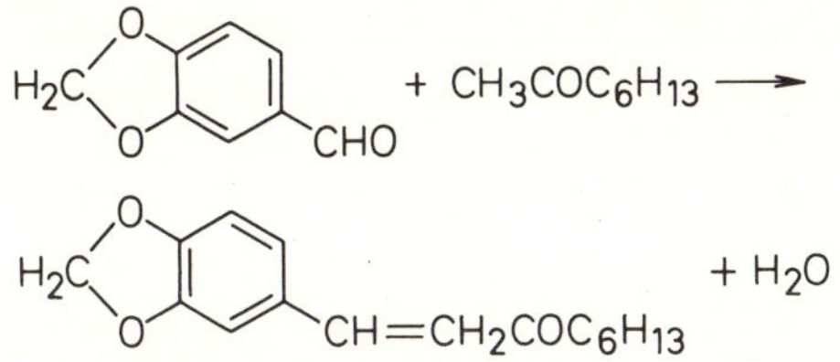

The hexyl 3,4-methylenedioxystyryl ketone formed in this reaction reacts with acetoacetic ester to give piperonyl cyclonene:

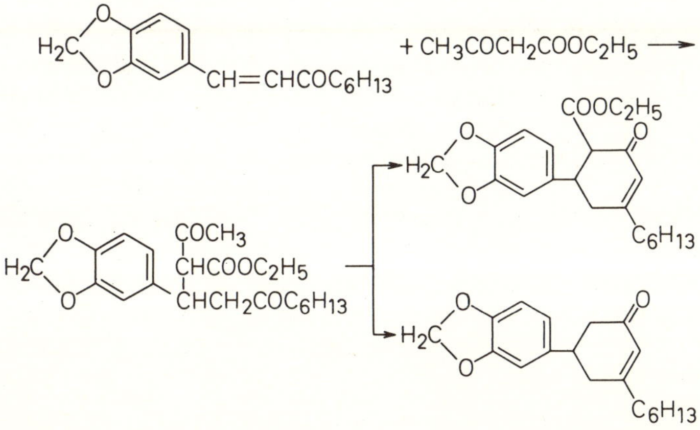

Product (I) is a light, viscous oil that does not distill without decomposition in vacuum; product (II) is a white crystalline substance, m.p. 59° C. The LD_{50} of this mixture is about 4,500 mg./kg.

In addition to these ketones, various cyclic diketones have been studied; of these the alkyl- and arylindandiones-1,3 are of practical interest. These compounds are produced by the condensation of esters of phthalic acid with ketones in the presence of alkali alcoholates as condensing agents:

[phthalate ester] + CH₃COR' → [2-acylindandione-1,3] —CR' + 2ROH

The simplest representative of the diketone series is 2-pivalylindandione-1,3 (m.p. 108.5°–110.5° C.; 1.8 mg. of the compound dissolves in 100 g. of water at 25° C.) used under the names of Pindone, Pival, Pivalin, and others for rodent control in the form of poisoned baits containing 0.025% active ingredient. The LD_{50} for rats is 50 mg./kg.; its chronic toxicity is greater. It is produced by the condensation of dimethyl or diethyl phthalate with pinacolin:

[diethyl phthalate] + CH₃COC(CH₃)₃ → [2-pivalylindandione] —CC(CH₃)₃

Pinacolin in turn is prepared in good yield by dehydration of pinacone hydrate:

$$(CH_3)_2C(OH)C(OH)(CH_3)_2 \rightarrow CH_3COC(CH_3)_3 + H_2O$$

2-Isovalerialylindandione-1,3 (m.p. 67°–68° C.) has similar properties, but it is not as high in zooicidal activity as pivalylindandione. In addition to its toxicity to rodents, it also has insecticidal properties although it has not been widely used for insect control.

In the Soviet Union the compound ratindan is used to control rodents; its active ingredient is 2-diphenacylindandione-1,3, produced by the condensation of diphenylmethyl methyl ketone with esters of phthalic acid (m.p. 145°–147° C.):

[diethyl phthalate] + CH₃COCH(C₆H₅)₂ → [2-diphenacylindandione] —CCH(C₆H₅)₂

It is used in food baits containing 1–3% active ingredient. Ratindan is low in toxicity for sheep and other domestic animals, but it has appreciable toxicity to rodents (LD_{50} 40–60 mg./kg.).

All the acyl derivatives of indandione-1,3 are anticoagulants and when they are systematically introduced into the animal organism, there is continuous bleeding from any wound as a result of poor coagulability of the blood.

Quinones

The quinones have considerably higher pesticidal activity than the aldehydes and ketones; the quinones are especially active as fungicides.

The benzoquinones show comparatively low fungicidal effect, but when halogens and hydrocarbon radicals are introduced into the ring the activity is strongly increased. Thus, p-benzoquinone is less active than chlorobenzoquinone and toluquinone, and likewise naphthoquinone-1,4 is less active than 2-chloronaphthoquinone-1,4 and 2,3-dichloronaphthoquinone-1,4. However, further accumulation of chlorine atoms in the molecule of naphthoquinone-1,4 lowers the fungicidal properties and 2,3,5,6,7,8-hexachloronaphthoquinone-1,4 is completely without fungicidal effect.

For disinfection of bean and cotton seeds some use has been made of tetrachlorobenzoquinone (chloranil, m.p. 290° C. in sealed capillary, LD_{50} for rats 4,000 mg./kg), which is employed in the form of the 96–98% compound, sometimes with DDT (3%). It is produced by oxidative chlorination of phenol or chlorophenols, and also by oxidative hydrolysis of pentachlorophenol:

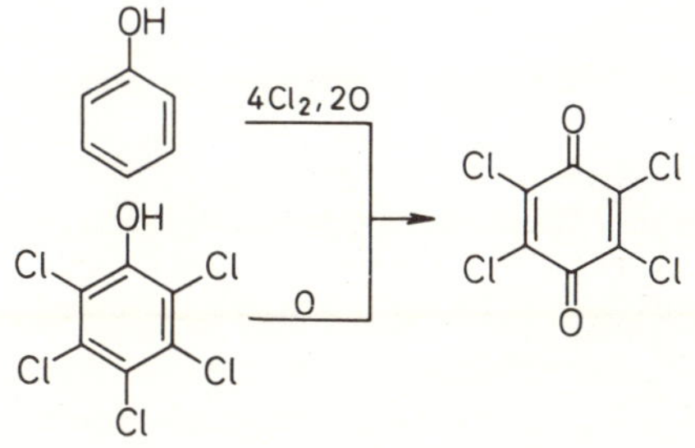

Wider use is made of 2,3-dichloronaphthoquinone-1,4, also known as Phygon and dichlone. This compound is a yellow crystalline substance, m.p. 193° C., LD_{50} for rats 1,300 mg./kg. At a high temperature and humidity it causes irritation of the skin. It is used for disinfecting the seeds of various crops, for spraying green plants to control powdery-mildew fungi, and also to control algae and lichens. Usually it is marketed either in the form of a 50% wettable powder containing, besides the naphthoquinone, a diluent, a surface-active agent and auxiliary materials, or in the form of 1–4% dust for seed disinfection. In an acid medium the compound is stable, but in alkali it is rapidly hydrolyzed and resinified, and it therefore is incompatible with alkalies. It can be produced by several methods, the most important of which are the following:

1. Chlorination of 1-naphthol in sulfuric acid in the presence of iron salts.
2. Chlorination of naphthionic acid.
3. Chlorination of naphthoquinone-1,4 or 1-nitronaphthalene in aqueous, weakly alkaline medium:

$$\text{1-naphthol} \xrightarrow{2Cl_2+2O} $$
$$\text{naphthionic acid} \xrightarrow{4Cl_2+7O} \text{2,3-dichloronaphthoquinone-1,4}$$
$$\text{naphthoquinone-1,4} \xrightarrow{Cl_2} $$

The second method is used most often for producing the compound with a m.p. not lower than 188° C.

2,3-Dichloronaphthoquinone-1,4 is poorly soluble in water. Therefore, regardless of the method of preparation, it precipitates in an aqueous medium, and after filtration and drying it is used for the production of appropriate formulations.

To prepare 2,3-dichloronaphthoquinone-1,4 from naphthionic acid (method 2) the acid is dissolved in 50% sulfuric acid, a small amount of ferric sulfate is added, and the mixture is chlorinated at a low temperature. When the chlorination is ended the precipitate is filtered off, washed with water, and dried.

Because of its high phytocidal activity, 2,3-dichloronaphthoquinone-1,4 is used at relatively low concentrations, from 0.05 to 0.1%.

Recently the benzoylhydrazone of quinone oxime (Cerenox) has been proposed as a seed disinfectant for sugar beets; this compound is a yellow powder decomposing at 207° C. (the technical grade product decomposes at 195° C.); LD_{50} for rats and mice is 100 mg./kg. It is obtained by the reaction of freshly prepared p-nitrosophenol with an aqueous solution of benzoylhydrazine in an acid medium:

$$HO\text{-}C_6H_4\text{-}NO + H_2NNHCOC_6H_5 \rightarrow C_6H_5CONHN=C_6H_4=NOH$$

The solid product that has separated in this process is filtered off, washed with water, dried, and used without purification for the preparation of formulations.

For dry seed disinfection a 10% preparation is used, and for wet disinfection 20 and 50% wettable powders are used. About 3 kg. of the 10% preparation is applied per ton of seed. Cerenox also is produced in the form of a complex seed disinfectant containing organic mercury compounds (1–2%) in addition to the benzoylhydrazone of quinone oxime.

In conclusion, let us note that glucochloralose, which is obtained by the reaction of chloral with glucose, has been used as a bird repellent:

$$CCl_3CHO + C_6H_{12}O_6 \rightarrow CCl_3CH\begin{array}{c}O-CH\\ \|\\ O-C-CHOHCHOHCHOHCH_2OH\end{array} + H_2O$$

It exists as an α-form, m.p. 187° C., and a β-form, m.p. 227°–230° C., causes narcosis in birds, and is used as a bait containing 0.5% active ingredient.

General references

BORECKI, Z., E. CZERWINSKA, Z. ECKSTEIN, and R. KOWALIK: Chemiczne grzybobojcze. Warsaw (1965).
FREAR, D.: Khimiya insektitsidov i fungitsidov [Chemistry of insecticides and fungicides]. Foreign Literature Publishing House [USSR] (1948) *.
METCALF, R. L.: Organic insecticides. New York: Interscience (1955).
PETERSEN, S., W. GAUSS, and E. URBSCHAT: Angew. Chem. 67, 217—231 (1965).
VANAG, G., ed.: Tsiklicheskie beta-ketony [Cyclic beta-ketones]. Publishing House Latvian Academy of Sciences. Riga (1961).
ZBIROVSKY, M., and J. MYSKA: Insekticidy, fungicidy, rodenticidy. Prague: Nakl. Ceskosl. Akad. Ved. (1957).

---

* Exact edition unknown; cited here is date of presumed Russian translation from English. — R.L.B.

XI. Aliphatic carboxylic acids and their derivatives

General characteristics of pesticidal properties

The saturated aliphatic monobasic and dibasic carboxylic acids have a relatively low pesticidal activity and a practical use has not been found for them. For the pea aphid the insecticidal activity of the acids and their salts increases with increasing molecular weight and reaches a maximum in lauric acid, after which it falls. The unsaturated acids are more active than the corresponding saturated analogs.

Calcium propionate [$(CH_3CH_2COO)_2Ca$], which is used in the baking and cheese industries to preserve the quality of bread products and cheese, has a fungistatic and bacteriostatic effect.

The salts of the higher fatty acids containing 16–18 carbon atoms have been used as surface-active agents. The unsaturated acids, for example, undecylenic acid (b.p. 275° C. at 700 mm. of Hg, m.p. 24.5° C.) and some of its derivatives also show herbicidal properties, but they have not received practical use.

The pesticidal activity of the acids rises sharply when hydrogen atoms in the alkyl radical are replaced by halogens. Thus, even the monohaloacetic acids have considerable pesticidal activity and some of them are used in agriculture and public health.

Table XXVII. *Toxicity to animals of esters of fluoroalkylcarboxylic acids*

$F(CH_2)_nCOOR$	LD_{50} (mg./kg.)	$F(CH_2)_nCOOR$	LD_{50} (mg./kg.)
FCH_2COOR [a]	15	$F(CH_2)_7COOR$	9
$F(CH_2)_2COOR$	200	$F(CH_2)_9COOR$	10
$F(CH_2)_3COOR$ [a]	—	$F(CH_2)_{10}COOR$	100
$F(CH_2)_4COOR$	160	$F(CH_2)_{11}COOR$	20
$F(CH_2)_5COOR$	4		

[a] Methyl esters; the other acids were tested as the ethyl esters.

In the fluoroalkylacetic acid series a regular change in toxicity to warm-blooded animals is observed in relationship to the number of CH_2 groups separating the fluorine from the carbonyl carbon. Compounds with an uneven number of CH_2 groups are more toxic. Table XXVII shows the toxicity to animals of some fluoroalkylcarboxylic acids in propylene glycol.

This alternation in activity in the series of fluoroalkylcarboxylic acids is explained by the fact that the compounds with an even number of CH_2 groups upon β-oxidation go to fluoroformic acid and further to carbon dioxide, while the acids with an uneven number of CH_2 groups give the very toxic fluoroacetic acid:

$$FCH_2CH_2CH_2COOH \rightarrow FCH_2CH=CHCOOH$$
$$\rightarrow FCH_2COOH + CO_2 + H_2O$$

The toxicity of the latter acids is associated with a series of metabolic processes, the most important being the formation of fluorocitric acid from fluoroacetic acid. The salts and some amides of fluoroacetic acid are used as very toxic zoocides.

A simple method for making fluoroacetic acid is the reaction of esters of monochloroacetic acid with potassium fluoride, which takes place at 200° to 220° C. under pressure:

$$ClCH_2COOCH_3 + KF \rightarrow FCH_2COOCH_3 + KCl$$
$$FCH_2COOCH_3 + H_2O \rightarrow FCH_2COOH + CH_3OH$$

It is interesting to note that with sodium fluoride it is not possible to obtain the monofluoroacetate.

Fluoroacetic acid is also produced by the reaction of carbon monoxide, formaldehyde, and hydrogen fluoride under pressure:

$$HF + CO + CH_2O \rightarrow FCH_2COOH$$

For rodent control sodium fluoroacetate and barium fluoroacetate, white crystalline substances highly soluble in water, have been used. These salts are used in the form of aqueous solutions or food baits. The LD_{50} is from 0.22 to 4 mg./kg. In view of the high toxicity of these compounds, only specially trained persons are permitted to work with them.

The amide and anilide of fluoroacetic acid (m.p. 108° C. and 75° to 76° C., respectively), which are less toxic to vertebrates and safer to use, also have been proposed as zoocides. The LD_{50} of fluoroacetamide for warm-blooded animals is 4–5 mg./kg., and that of fluoroacetanilide is 10–12 mg./kg.

The salts and amides of fluoroacetic acid have a systemic insecticidal effect and have been proposed for the control of various species of aphids. For some species of aphids and caterpillars fluoroacetanilide is toxic in concentrations of 25 mg./liter. However, because of their high toxicity to animals the derivatives of fluoroacetic acid have very limited use.

The chloroalkylcarboxylic acids are less toxic to vertebrate animals than the fluoroalkylcarboxylic acids, but their phytocidal activity is much greater. As a result of this the series of chlorosubstituted acids and their derivatives are of interest as herbicides. In particular, the mono-, di-, and trichloroacetic acids, the di- and trichloropropionic acids, and their various derivatives have herbicidal properties. Furthermore, dichloroisobutyric acid shows gametocidal properties (plant sterilant). The physiological activity of the halogen-substituted acids is influenced substantially by the position of the halogen with respect to the carbonyl group. The α-halogen-substituted acids have the greatest activity. Because of their high cost bromine- and iodine-substituted acids have not been widely used in agriculture.

Monochloroacetic acid, m.p. 63° C., is highly soluble in water and many organic solvents, and belongs to the strong acids which burn plants. It is used in agriculture in the form of the sodium salt, which is obtained by neutralization of the acid with Na_2CO_3 or NaOH at a low temperature, since at an elevated temperature sodium monochloroacetate easily goes over to glycolic acid. Sodium monochloroacetate is used as a herbicide and defoliant at a dosage of 5–20 kg./ha.

Monochloroacetic acid and its sodium salt are used for the production of such important herbicides as 2,4-D, MCPA, 2,4,5-T, and others. It is now produced by the following methods:

1. Chlorination of acetic acid in the presence of catalysts (iodine, sulfur, phosphorus trichloride, and combinations of them):

$$CH_3COOH + Cl_2 \rightarrow ClCH_2COOH + HCl$$

2. Hydration and hydrolysis of trichloroethylene:

$$CHCl = CCl_2 + H_2SO_4 \rightarrow ClCH_2CCl_2OSO_3H$$
$$ClCH_2CCl_2OSO_2H + H_2O \rightarrow H_2SO_4 + ClCH_2CCl_2OH$$
$$ClCH_2CCl_2OH \rightarrow ClCH_2COCl + HCl$$
$$ClCH_2COCl + H_2O \rightarrow ClCH_2COOH + HCl$$

The first method is the better one and its various modifications are used in industry in many countries. Small amounts of dichloro- and trichloro-acetic acids are obtained as by-products in the chlorination of acetic acid. Under more severe conditions this mixture by chlorination is converted to trichloroacetic acid and goes into the production of herbicidal preparations. Pure monochloroacetic acid is most often isolated by crystallization of the distilled reaction mixture, from which the main quantity of monochloroacetic acid precipitates on cooling. Meanwhile, the mixture of more highly chlorinated products and part of the acetic acid remains in the liquid condition and is removed by centrifugation.

The use of the esters and amides of monochloroacetic acid as herbicides also has been proposed. The esters of monochloroacetic acid also have nematicidal activity.

Trichloroacetic acid is used in agriculture to control monocotyledonous weeds; m.p. 57.3° C., b.p. 195° C. at 760 mm. of Hg. It is used for weed control almost exclusively in the form of salts with the alkali metals or amines. The trichloroacetates of the alkali metals are used mainly to control monocotyledonous weeds in plantings of sugar beets, alfalfa, sugarcane, and other crops at relatively large dosages of the compound (12–60 kg./ha.). The compound is used either before the seedlings appear or before planting of the crops, since such large dosages may damage the crop plants. The LD_{50} of sodium trichloroacetate is 3,300–5,000 mg./kg.

Trichloroacetic acid is produced either by direct catalytic chlorination of acetic, monochloroacetic, and dichloroacetic acid and mixtures of them:

$$CH_3COOH + 3\,Cl_2 \rightarrow CCl_3COOH + 3\,HCl$$

or by oxidation of chloral with nitric acid:

$$CCl_3CHO + O \rightarrow CCl_3COOH$$

The latter method is somewhat simpler and gives good yields of trichloroacetic acid. The salts of this acid are obtained by neutralization with the appropriate bases at the lowest possible temperature, since raising the temperature leads to decomposition of the acid with the formation of chloroform:

$$CCl_3COOH \rightarrow CHCl_3 + CO_2$$

α,α-Dichloropropionic acid. The best known of the aliphatic carboxylic acids is α,α-dichloropropionic acid (b.p. 193°–197° C. at 760 mm. of Hg), which is used for the control of monocotyledonous weeds in the form of the highly water-soluble sodium salt (m.p. 174°–176° C., dec.), LD_{50} is 6,000 to 8,000 mg./kg. It is marketed in various countries in the form of the 83–85% compound as dalapon.

The dichloropropionates of the alkali metals react with water to form pyruvic acid, which is inactive as a herbicide; because of this their aqueous solutions should not be kept for long periods of time.

Dalapon is capable of moving through the plant; it, therefore, shows good activity in the control of many monocotyledonous plants against which other herbicides are not very active. It is used at dosages of 12–40 kg./ha.

Dichloropropionic acid is produced by direct chlorination of propionic acid in the presence of catalysts:

$$CH_3CH_2COOH + 2\,Cl_2 \rightarrow CH_3CCl_2COOH + 2\,HCl$$

This reaction proceeds slowly at an elevated temperature and goes better in the light. The same materials are used as catalysts as for the chlorination of acetic acid. The technical product contains a slight amount of α,β-dichloropropionic acid and α-monochloropropionic acid. To prepare the sodium salt the concentrated acid is treated with 40% NaOH solution with cooling, the salt that crystallizes out is filtered off and dried, and the mother liquor is used as the solvent for the preparation of a new batch of the salt. By this method rather pure salt is obtained with minimal losses. When dilute solutions are neutralized and then evaporated, decomposition of a considerable part of the compound takes place and a strongly contaminated product is obtained.

The 2,4,5-trichlorophenoxyethyl ester of dichlorpropionic acid, known as erbon (m.p. 49°–50° C., b.p. 161°–164° C. at 4 mm. of Hg, LD_{50}

700–1,100 mg./kg.) is used as a nonselective herbicide. The dosage of the compound for nonselective eradication of plants is 14–20 kg./ha.

In addition to dalapon, the sodium salt of α,α,β-trichloropropionic acid, which is used in agriculture at dosages of 4–12 kg./ha., gives satisfactory results in the control of many monocotyledonous annual weeds. Trichloropropionic acid is produced by the chlorination of acrylonitrile with subsequent hydrolysis of the trichloropropionitrile in 50% sulfuric acid:

$$CH_2 = CHCN + 2\ Cl_2 \rightarrow ClCH_2CCl_2CN + HCl$$
$$ClCH_2CCl_2CN + 2\ H_2O + H_2SO_4 \rightarrow ClCH_2CCl_2COOH + NH_4HSO_4$$

Under conditions similar to those for the preparation of dalapon, trichloropropionic acid is converted to the sodium salt.

The sodium salt of dichloroisobutyric acid has been proposed as a plant sterilant (causing male sterility). Its synthesis is:

$$CH_2 = C(CH_3)COOCH_3 + Cl_2 \rightarrow ClCH_2CCl(CH_3)COOCH_3$$
$$ClCH_2CCl(CH_3)COOCH_3 + NaOH \rightarrow$$
$$ClCH_2CCl(CH_3)COONa + CH_3OH$$

The sodium salt melts at 170°–175° C. (dec.); the free acid boils at 130° to 134° C. at 35 mm. of Hg.; the LD_{50} for rats is about 8,000 mg./kg.

Among the halogen-substituted acids we should also point out 2,3,4,5,5-pentachloropentadienoic acid (m.p. 124°–125° C.), which is obtained by the action of KOH on hexachlorocyclopentenone:

$$\text{(hexachlorocyclopentenone)} + 2KOH \longrightarrow Cl_2C=CCl-CCl=CClCOOK + KCl + H_2O$$

This acid has been proposed for use as a desiccant and herbicide, but its use is still in the investigational stage.

The sodium salt of cis-β-chloroacrylic acid (m.p. 162° C., solubility in water 40%) is a defoliant for cotton and other crops. At dosages of 1–2.5 kg./ha. it completely and satisfactorily removes cotton foliage. This compound (PREP-Defoliant) has moderate toxicity to animals; the LD_{50} for rats is 320 mg./kg. and the toxicity for goldfish is 5,000 mg./liter for 48-hour exposure.

Esters of aliphatic carboxylic acids

Esterification of an acid often increases the insecticidal activity. A number of esters are used to control various harmful organisms.

Table XXVIII gives the insecticidal activity by fumigation of formic and acetic acid esters against the granary weevil. It can be seen from the data that in this series a general relationship is observed between the change in activity and the molecular weight. For the formates the activity decreases with increasing molecular weight, but for the acetates the reverse relationship is noted. Methyl formate is employed as a fumigant, but because of its high flammability its use is very limited.

The esters of the unsaturated acids, like methyl acrylate or methyl crotonate, for example, have somewhat higher insecticidal and fungicidal activity. The esters of the dibasic acids have been used in agriculture and public health, and some of them are satisfactory and cheap repellents for

Table XXVIII. *Toxicity of alkyl formates and alkyl acetates to adults of the granary weevil*

Alkyl group	Mean lethal concentration for adult granary weevils (mg./l.)	
	Formate	Acetate
CH_3	15	84
C_2H_5	35	56
C_3H_7	28	45
$CH(CH_3)_2$	34	90

harmful insects and ticks. Thus, dibutyl succinate (Tabatrex) is used to repel flies from cattle and also to repel cockroaches and ants. It is a colorless liquid, b.p. 108° C. at 4 mm. of Hg, m.p. −29° C., d_4^0 0.9963; its flash point in an open dish is 135° C. It is practically insoluble in water, but highly soluble in organic solvents with an LD_{50} for rats of about 8,000 mg./kg.

Dibutyl succinate usually is used as a 20% emulsion concentrate in petroleum products with an emulsifier. It is produced by direct esterification of succinic acid with butyl alcohol in the presence of ion-exchange resins or sulfonated coal as catalyst:

$$\begin{array}{l} CH_2COOH \\ | \\ CH_2COOH \end{array} + 2\,C_4H_9OH \rightarrow \begin{array}{l} CH_2COOC_4H_9 \\ | \\ CH_2COOC_4H_9 \end{array} + 2\,H_2O$$

The water formed in the esterification is distilled off as an azeotropic mixture with the excess butyl alcohol or with a hydrophobic organic solvent (benzene, toluene, etc.). The succinic acid needed for the synthesis of dibutyl succinate is formed as a by-product in the production of adipic acid.

Other esters of succinic acid also have the property of repelling insects, but they have no as yet been used.

Satisfactory repellent action for ixodid and other ticks is shown by dibutyl adipate (b.p. 183° C. at 14 mm. of Hg, d_4^{20} 0.9652), used under the name ZPS; its LD_{50} is 12,000 mg./kg. It is produced in a manner similar to dibutyl succinate:

$$\begin{array}{l} CH_2CH_2COOH \\ | \\ CH_2CH_2COOH \end{array} + 2\,C_4H_9OH \rightarrow \begin{array}{l} CH_2CH_2COOC_4H_9 \\ | \\ CH_2CH_2COOC_4H_9 \end{array} + 2\,H_2O$$

For experimental purposes this compound is put out in the form of an aerosol (15% ester and 85% aerosol solvents) and in the form of emulsive concentrates (90% active ingredient) for the impregnation of clothing.

The esters of the dibasic unsaturated acids, for example esters of fumaric, chlorofumaric, bromo- and dibromofumaric acids, are powerful fungicides, especially for fungi that cause rust. However, these compounds have not yet seen practical use.

Amides and imides of aliphatic carboxylic acids

The unsubstituted amides of alkylcarboxylic acids have relatively weak pesticidal effects, but many arylamides of even the simplest alkylcarboxylic acids are phytocidal and can be used as herbicides. The halogen-substituted anilides are especially active.

3,4-Dichloropropionanilide (Propanide, Stam F-34, Surcopur) is a white crystalline substance, m.p. 91°–92° C. (technical grade product melts at 88°–91° C.). It is slightly soluble in water, but dissolves in most organic solvents. It is an active herbicide for the control of miliary and other weeds in rice plantations. It is marketed in the form of 25 and 34–36% emulsive concentrates and is used at dosages of 2–4 kg./ha. (active ingredient) to control weeds. In addition to the active ingredient, the concentrates contain solvent and emulsifier. The LD_{50} for rats is about 1,300 mg./kg.; for rabbits it is 500 mg./kg.

3,4-Dichloropropionanilide is produced by the reaction of propionic acid with 3,4-dichloroaniline:

$$CH_3CH_2COOH + H_2N-\text{C}_6H_3(Cl)-Cl \rightarrow CH_3CH_2CONH-\text{C}_6H_3(Cl)-Cl + H_2O$$

The 3,4-dichloroaniline necessary for the synthesis is obtained by reduction of 3,4-dichloronitrobenzene:

$$O_2N-\text{C}_6H_3(Cl)-Cl + 3H_2 \rightarrow Cl-\text{C}_6H_3(Cl)-NH_2 + 2H_2O$$

The latter compound can be obtained by chlorination of *p*-chloronitrobenzene in the presence of iron and iodine:

$$O_2N-\text{C}_6H_4-Cl + Cl_2 \rightarrow O_2N-\text{C}_6H_3(Cl)-Cl + HCl$$

or by nitration of o-dichlorobenzene:

$$C_6H_4Cl_2 + HNO_3 \rightarrow C_6H_3Cl_2NO_2 + C_6H_3Cl_2NO_2 + H_2O$$

o-Dichlorobenzene on nitration yields two products, of which 3,4-dichloronitrobenzene predominates. The yield in this reaction, however, does not exceed 75%. The second isomer can be converted to pentachloronitrobenzene by further chlorination.

A second anilide, which has been used in controlling weeds in plantings of carrots and strawberries, as well as some other crops, is Karsil, the 3,4-dichloroanilide of α-methylvaleric acid (m.p. 101°–102° C., $LD_{5}0$ about 10,000 mg./kg.). Still better results are obtained in plantings of carrots, onions, and strawberries with use of the compound Solan, the 3-chloro-4-methylanilide of α-methylvaleric acid (m.p. 85°–86° C., LD_{50} 10,000 mg./kg.), which is marketed in the form of an emulsive concentrate containing 46–47% active ingredient. This preparation at dosages of 3–5 kg./ha. decreases the weediness of plantings by 70–100%.

Among the anilides of the unsaturated acids mention should be made of dicryl, the 3,4-dichloroanilide of methacrylic acid (m.p. 127°–128° C., LD_{50} 3,100 mg./kg.), which is recommended for controlling weeds in plantings of cotton, carrots, cauliflower, parsley, and some other crops at dosages of 2–5 kg./ha. At a dosage of 3 kg./ha. this compound eradicates annual weeds well and does not harm the crops enumerated above.

Diallylchloroacetamide, b.p. 92° C. at 2 mm. of Hg is about 2% soluble in water and soluble also in organic solvents. In the United States it has been used to control weeds in corn, barley, soy, flax, sugar beets, and some other crops. At a dosage of 3–4.5 kg./ha. it suppresses the growth of a number of grasses, including foxtail, Kentucky blue grass, chess, wild oat, and others. The LD_{50} is 700 mg./kg. It is known as Randox. It has the serious disadvantage that it irritates the skin, especially during prolonged work with it. For presprouting treatment of grain plantings some small use is made of a mixture of diallychloroacetamide with trichlorobenzyl chloride (Randox-T, a liquid mixture of isomers of $Cl_3C_6H_2CH_2Cl$ with b.p. 93° to 98° C. at 1 mm. of Hg, LD_{50} 3,000 mg./kg., obtained by direct chlorination of benzyl chloride). This mixture has the same disadvantage as diallylchloroacetamide.

Chloralmonochloroacetamide, obtainable by the condensation of chloral with chloroacetamide, has begun be used for the control of weeds in sugar beet plantings:

$$ClCH_2CONH_2 + CCl_3CHO \rightarrow ClCH_2(CONHCH(OH)CCl_3)$$

Table XXVIII a. *Several herbicidal amides of aliphatic acids*

Common names and chemical names	Structural formula	M.P. (°C.)	Sol. in H_2O	LD_{50} (mg./kg.)	Crops, types of activity
Propachlor 2-chloro-N-isopropyl-N-phenylacetamide	$(CH_3)_2HC-N-COCH_2Cl$ with phenyl	liquid	0.07%	800	Soil, vegetable crops
CP 31 675 2-chloro-N-(2-methyl-6-t-butyl-phenyl)acetamide	$NHCOCH_2Cl$, $C(CH_3)_3$, CH_3 on phenyl	115	300 p.p.m.	1,780	Soil, vegetable crops
CP 50 144 2-chloro-N-(2,6-diethyl)phenyl-N-methoxymethylacetamide	$CH_3OCH_2-N-COCH_2Cl$, C_2H_5, C_2H_5 on phenyl	liquid	148 p.p.m.	1,200	Soil, vegetable crops
CP 52 223 2-chloro-N-(2,6-dimethyl)phenyl-N-isopropoxymethylacetamide	$C_3H_7OCH_2-N-COCH_2Cl$, CH_3, CH_3 on phenyl	liquid	59 p.p.m.	1,775	Sugar beets, soil
Monalide N-(4-chlorophenyl)-2,2-dimethyl-valeroamide	$NHCO-C(CH_3)(CH_2CH_2CH_3)-CH_3$ with 4-Cl phenyl	87–88	22.8 p.p.m.	4,000	Contact, soil

Table XXXVIII a (continued).

Common names and chemical names	Structural formula	M.P. (°C.)	Sol. in H_2O	LD_{50} (mg./kg.)	Crops, types of activity
TO-2 (CMPT) 5-chloro-4-methyl-2-propionamido-thiazole	(structure)	158–160	180 p.p.m.	2,080	Wheat, contact
Cypromid N-(3,4-dichlorophenyl)cyclopropane-carboxamide	(structure)	131	—	215	Maize, contact
DMSA, B 995 Succinic acid 2,2-dimethylhydrazide	$CH_2CONH(CH_3)_2$ | CH_2COOH	154–156	—	8,400	Plant growth regulant

It is used in herbicidal preparations as a mixture with sodium trichloroacetate. Table XXVIIIa gives the properties of several herbicidal amides of aliphatic acids.

The anilides of chloromaleic, dichloromaleic, and mucochloric acids show fungicidal properties. The ester amides of the dibasic acids have been proposed as repellents. A large number of other amides have been described that have some type of pesticidal effect, but they have not as yet been practically utilized.

In addition to the amides, studies have been made of the pesticidal properties of a large number of different nitriles that have high insecticidal and acaricidal activity, and some of them are herbicides. To characterize the insecticidal properties of the aliphatic nitriles, their toxicities to adults of the granary weevil and the confused flour beetle are shown in Table XXIX.

Table XXIX. *Toxicity of aliphatic nitriles for beetles and weevils*

Formula	LD_{100} (mg./l.)		Formula	LD_{100} (mg./l.)	
	Granary weevil	Confused flour beetle		Granary weevil	Confused flour beetle
$CH_2=CHCN$	0.8	1.2	CH_3CCl_2CN	2.4	6.0
CH_3CN	27.4	27.4	$ClCH_2CCl_2CN$	0.7	2.1
$ClCH_2CN$	0.6	1.2	$CH_3CH_2CH_2CN$	7.9	8.0
Cl_2CHCN	0.7	3.4	$ClCH_2CH_2CH_2CN$	4.4	3.3
CCl_3CN	1.4	4.3	$CH_3CHClCH_2CN$	6.9	6.4
CH_3CH_2CN	11.7	27.4	$CH_3CH_2CCl_2CN$	11.7	11.7
$CH_3CHClCN$	1.6	7.5	$CH_3CHClCCl_2CN$	2.0	2.0
$ClCH_2CH_2CN$	1.7	2.3	$ClCH_2CH_2CCl_2CN$	2.4	6.0
			$ClCH_2CHClCCl_2CN$	1.4	2.2

Acrylonitrile and trichloroacetonitrile show high activity as fumigants. A disadvantage of acrylonitrile is its high flammability, because of which it is not used in the Soviet Union.

Trichloroacetonitrile (m.p. 42° C., b.p. 85° C.) is sold as Tritox in the USA as a fumigant for grain and for vacant premises at dosages of 2–37 g./m.[3] It has a lachrymatory effect and is highly poisonous. Sometimes a mixture of trichloroacetonitrile and methyl bromide is used for fumigation.

Trichloroacetonitrile is produced by direct chlorination of dry acetonitrile at an elevated temperature:

$$CH_3CN + 3\ Cl_2 \rightarrow CCl_3CN + 3\ HCl$$

In the moist state it corrodes metals. To prevent the formation of HCl by the action of moisture and corrosion, up to 2% of anhydrous Na_2CO_3 is added to it. It also shows a herbicidal effect, but α,α,β-trichlorpropionitrile, which has been tested on various crops at dosages of 8–25 kg./ha., is a stronger herbicide.

The amides of various acids of the following general formula have been widely studied:

$$R-N=\underset{\underset{R'}{|}}{C}-NR''_2$$

and also N-(4-chloro-2-methylphenyl)-N',N'-dimethylformamidine:

[Structure: 4-chloro-2-methylphenyl group with -N=CHN(CH$_3$)$_2$]

which has a good acaricidal effect. It is a crystalline substance, m.p. 35° C., b.p. 162°–165° C. at 14 mm. of Hg, LD$_{50}$ for rats 340 mg./kg. It is marketed in the form of an emulsive concentrate and is used in 0.05–0.08% active ingredient.

N-(3-Methylcarbamidophenyl)-N',N'-dimethylformamidine (Fundal) also has been proposed as a specific acaricide:

[Structure: phenyl ring with -N=CHN(CH$_3$)$_2$ and OC(=O)NHCH$_3$ substituents]

General references

BASKAKOV, YU. A.: Zhur. Vsesoyuz. Khim. Obshchestva im. D. I. Mendeleeva 9, 486 (1964).
CRAFTS, A. S.: The Chemistry and mode of action of herbicides. New York-London: Interscience (1961).
DITTRICH, V.: J. Econ. Entomol. 59, 889 (1966).
— VI Internat. Pflanzenschutz-Kongreß Abstr. Vienna (1967).
FREAR, D.: Khimiya insektitsidov i fungitsidov [Chemistry of insecticides and fungicides]. Foreign Literature Publishing House [USSR] (1948) *.
KURT, H.: Chemische Unkrautbekämpfung. Jena: Fischer Verlag (1963).
MEL'NIKOV, N. N., and YU. A. BASKAKOV: Khimiya gerbitsidov i regulyatorov rosta rastenii [Chemistry of herbicides and plant growth regulators]. State Scientific and Technical Publishing House of Chemical Literature [USSR] (1962).
PIANKA, M.: J. Sci. Food Agr. 17, 47 (1966).
— J. Sci. Food Agr. 18, 889 (1967).
SAUNDERS, B.: Some aspects of the chemistry and toxic action of organic compounds containing phosphorus and fluorine. London and New York: Cambridge Univ. Press (1957).
ZBIROVSKY, M., J. MYSKA, and J. ZEMANEK: Herbicidy. Prague (1960).

---

* Exact edition unknown; cited here is date of presumed Russian translation from English. — R.L.B.

XII. Alicyclic carboxylic acids and their derivatives

General characteristics of pesticidal properties

The free alicyclic acids have low pesticidal activity. In agriculture only the salts of the naphthenic acids, which are employed as surface-active agents, are used. The water-insoluble copper naphthenate, which has a powerful fungicidal effect, is used to control plant diseases and to disinfect nonmetallic materials.

Copper naphthenate is prepared from sodium naphthenate and water-soluble copper salts. It is a viscous, oily substance with a greasy consistency; it is practically insoluble in water, but rather soluble in petroleum products and various organic solvents. Aqueous emulsions of its solutions in organic solvents are used at a concentration of 0.1–0.2% to treat plants for control of fungus diseases. Copper naphthenate is strongly phytocidal, and great care must be used in treating plants with it. It has practically the same toxicity for vertebrates as the inorganic salts of copper.

Various other derivatives of the alicyclic acids have practical importance; among them are found active insecticides, herbicides, plant growth stimulators, and repellents. Such groups of natural compounds as the pyrethrins, gibberellins, and their synthetic analogs belong to the alicyclic acid derivatives.

The pyrethrins and their synthetic analogs

Use of the pyrethrins, in the form of flowers of the Dalmation daisy (pyrethrum) ground to a fine powder, to control harmful insects has been known for many centuries, but their structure was only established in the 1950's. The active insecticidal constituents of pyrethrum flowers consist of four compounds: cinerin I (I), cinerin II (II), pyrethrin I (III), and pyrethrin II (IV). The first and third of these compounds are esters of chrysanthemum monocarboxylic acid with the alcohols cinerolone and pyrethrolone; the second and fourth are esters of the same alcohols with chrysanthemum dicarboxylic acid.

$$CH_3C(CH_3)=CHCH-CHCO-[C(CH_3)_2]-O-[cyclopentenone]-CH_2CH=CHCH_3 \quad (I)$$

$$CH_3O\overset{O}{\underset{\|}{C}}-\overset{CH_3}{\underset{|}{C}}=CHCH\underset{H_3C\diagdown C\diagup CH_3}{-}CH\overset{O}{\underset{\|}{C}}O-\underset{=O}{\overset{CH_3}{\bigcirc}}-CH_2CH=CHCH_3 \quad \text{(II)}$$

$$\overset{CH_3}{\underset{|}{CH_3C}}=CHCH\underset{H_3C\diagdown C\diagup CH_3}{-}CH\overset{O}{\underset{\|}{C}}O-\underset{=O}{\overset{CH_3}{\bigcirc}}-CH_2CH=CHCH=CH_2 \quad \text{(III)}$$

$$CH_3O\overset{O}{\underset{\|}{C}}-\overset{CH_3}{\underset{|}{C}}=CHCH\underset{H_3C\diagdown C\diagup CH_3}{-}CH\overset{O}{\underset{\|}{C}}O-\underset{=O}{\overset{CH_3}{\bigcirc}}-CH_2CH=CHCH=CH_2 \quad \text{(IV)}$$

Table XXX. *Properties of synthetic analogs of the pyrethrins*

Compound	Boiling point (° C. at mm. Hg)	n_D (t °C.)	d (t °C.)	LD_{50} for rats (mg./kg.)
Pyrethrin I	136–137 / 0.005	1.5071 (20)	—	960
Cinerin I	100–101.5 / 0.001	1.5018 (20)	—	1,050
Barthrin	158–168 / 0.01	—	—	950
Dimethrin	147–150 / 0.01	—	0.986	40,000
Allethrin	127–130 / 0.002	1.5070 (20)	1.005 (25)	680
Cyclethrin	100 / 0.02	1.5170 (30)	—	900
Furethrin	187–188 / 0.4	1.5202 (25)	—	700
Neopynamin [a]	185–190 / 0.1	—	1.108 (20)	20,000

[a] M.P. 65°–80° C.

All four compounds have insecticidal properties, but the esters of chrysanthemum monocarboxylic acid are the more effective.

The pyrethrins and cinerins are liquids with a slight odor, are readily hydrolyzed by alkalies, and are rapidly inactivated in the air, apparently as a result of oxidation and hydrolysis. The pyrethrins, which are obtained by extraction of the pyrethrum flowers, are most often used in the food industry and in households to control various harmful insects. They are used in a mixture with synergists, which, as a rule, are not themselves insecticidal. However when added in very small amounts (1:1 and more) to the pyrethrins or their analogs they increase this effect on insects. The mechanism of synergism has still not been investigated sufficiently.

Since the synthesis of the pyrethrins and cinerins themselves is complicated, various synthetic analogs have been developed and proposed for use; these are the so-called "synthetic pyrethroids," use of which under practical conditions costs 10–20% less than the natural compounds extracted from pyrethrum flowers. The following compounds of this group have received some practical use in controlling harmful insects (mainly flies): allethrin (V), furethrin (VI), cyclethrin (VII), barthrin (VIII), dimethrin (IX), and Neopynamin (X); their properties are shown in Table XXX.

$$\text{CH}_3\underset{\underset{\underset{H_3C}{\diagdown}\underset{CH_3}{\diagup}}{C}}{\overset{\overset{CH_3}{|}}{C}}=\text{CHCH}-\overset{\overset{O}{\|}}{\text{CHCOCH}_2}\text{—}\underset{\underset{CH_3}{}}{\overset{\overset{CH_3}{}}{\bigcirc}} \qquad (IX)$$

$$\text{CH}_3\underset{\underset{\underset{H_3C}{\diagdown}\underset{CH_3}{\diagup}}{C}}{\overset{\overset{CH_3}{|}}{C}}=\text{CHCH}-\overset{\overset{O}{\|}}{\text{CHCHCOCH}_2}\text{N}\underset{\underset{O}{\|}}{\overset{\overset{O}{\|}}{\bigcirc}} \qquad (X)$$

The synthetic pyrethroids usually are produced by the reaction of the acid chloride of chrysanthemum monocarboxylic acid with the appropriate alcohol in the presence of tertiary amines, most often pyridine or triethylamine:

$$(CH_3)_2C = CHCH - CHCOCl + ROH \rightarrow$$
$$\diagdown\!\diagup$$
$$C(CH_3)_2$$
$$(CH_3)_2C = CHCH - CHCOOR + HCl$$
$$\diagdown\!\diagup$$
$$C(CH_3)_2$$

The synthesis of the cyclic alcohols of formula (XI) is accomplished by the following general route:

$$\underset{\underset{ONa}{|}}{CH_3C} = CHCOOC_2H_5 + RCl \rightarrow \underset{\underset{O}{\|}}{CH_3C} - \underset{\underset{R}{|}}{CHCOOC_2H_5} + NaCl$$

$$\underset{\underset{O}{\|}}{CH_3C} - \underset{\underset{R}{|}}{CHCOOC_2H_5} + NaOH \rightarrow \underset{\underset{O}{\|}}{CH_3C} - \underset{\underset{R}{|}}{CHCOONa} + C_2H_5OH$$

$$\underset{\underset{O}{\|}}{CH_3C} - \underset{\underset{R}{|}}{CHCOONa} + H_2SO_4 \rightarrow \underset{\underset{O}{\|}}{CH_3CCH_2R} + CO_2 + NaHSO_4$$

$$\underset{\underset{O}{\|}}{CH_3CCH_2R} + CO(OC_2H_5)_2 \xrightarrow{CH_3ONa} \underset{\underset{O}{\|}}{RCH_2CCH_2COOC_2H_5} + C_2H_5OH$$

$$\underset{\underset{O}{\|}}{RCH_2CCH_2COOC_2H_5} + KOH \rightarrow \underset{\underset{O}{\|}}{RCH_2CCH_2COOK} + C_2H_5OH$$

$$RCH_2CCH_2COOK + CO_2 + H_2O \rightarrow RCH_2CCH_2COOH + KHCO_3$$
$$\underset{O}{\|} \qquad\qquad\qquad\qquad\qquad \underset{O}{\|}$$

$$RCH_2CCH_2COOH + CH_3COCHO \rightarrow \left[CH_3\underset{\underset{O}{\|}}{\overset{\overset{OH}{|}}{C}}H - \underset{\underset{O}{\|}}{\overset{\overset{COOH}{|}}{C}}HCCH_2R \right] \rightarrow$$
$$\underset{O}{\|}$$

$$CH_3\underset{\underset{O}{\|}}{\overset{\overset{OH}{|}}{C}}HCH_2\underset{\underset{O}{\|}}{C}CH_2R + CO_2$$

$$CH_3\underset{\underset{O}{\|}}{\overset{\overset{OH}{|}}{C}}HCH_2\underset{\underset{O}{\|}}{C}CH_2R + NaOH \rightarrow \underset{\text{HO}}{\overset{\text{CH}_3}{\diagup\!\!\diagdown R}}\!\!\!-\!\!\text{O} + H_2O \qquad\qquad XI$$

The first step is the synthesis of the alkylacetoacetic ester by reaction of the sodium derivative of acetoacetic ester with the appropriate alkyl halide. This reaction is most often carried out in absolute alcohol with more or less prolonged heating. Depending on the alkyl used, the yield of alkylacetoacetic ester is 40–85% of theoretical. To obtain the ketone the alkylacetoacetic ester is treated with 5% NaOH solution and after the saponification 50% sulfuric acid is added to the reaction mixture. The ketone obtained is converted to the γ-alkylacetoacetic ester by the action of diethyl carbonate in the presence of sodium methylate. This reaction goes still better in the presence of sodium hydride. The ester formed is saponified with KOH and converted by the action of carbon dioxide to the free γ-alkylacetoacetic acid, which without isolation is condensed with pyruvic aldehyde. After decarboxylation the diketone formed is cyclized at pH 10 and the cyclopentenolone is separated and purified by distillation in vacuum.

For the synthesis of analogs of the pyrethrins, instead of the alkylmethylcyclopentenolones, substituted benzyl alcohols are used for making the compounds barthrin and dimethrin.

The route to the chrysanthemumic acid is no less complicated. The main intermediate for the synthesis of chrysanthemum monocarboxylic acid is 2,5-dimethylhexadiene-2,4 (m.p. 13.9° C., b.p. 134°–134.5° C.), which can be obtained by two methods.

The first method starts with 1-chloro-2-methylpropene-2, a product of the thermal chlorination of isobutylene at a relatively low temperature. 1-Chloro-2-methylpropene-2 with magnesium in absolute alcohol gives a 61—86% yield of 2,5-dimethyl-hexadiene-1,5:

$$2\ ClCH_2C(CH_3) = CH_2 + Mg \rightarrow$$
$$CH_2 = C(CH_3)CH_2CH_2C(CH_3) = CH_2 + MgCl_2$$

After purification by distillation, this compound is isomerized to 2,5-dimethylhexadiene-2,4 by passage over active alumina at 250°–275° C.:

$$CH_2=C(CH_3)CH_2CH_2C(CH_3)=CH_2 \xrightarrow{Al_2O_3} (CH_3)_2C=CHCH=C(CH_3)_2$$

In the second method 2,5-dimethylhexadiene-2,4 is obtained by condensing acetone with acetylene in the presence of solid KOH by the Favorskii method; the 2,5-dimethylhexynediol-2,5 that is formed, after isolation and purification, is hydrogenated with hydrogen at 70°–90° C. in the presence of a nickel catalyst under 6–8 atm. pressure. Then the 2,5-dimethylhexanediol-2,5 is dehydrated on alumina at an elevated temperature and the 2,5-dimethylhexadiene-2,4 is isolated:

$$2\ CH_3COCH_3 + CH\equiv CH \xrightarrow{KOH} (CH_3)_2C(OH)C\equiv CC(OH)(CH_3)_2$$

$$(CH_3)_2C(OH)C\equiv CC(OH)(CH_3)_2 + 2\ H_2 \rightarrow$$
$$(CH_3)_2C(OH)CH_2CH_2C(OH)(CH_3)_2$$

$$(CH_3)_2C(OH)CH_2CH_2C(OH)(CH_3)_2 \xrightarrow{Al_2O_3}$$
$$(CH_3)_2C=CHCH=C(CH_3)_2 + 2\ H_2O$$

Both of the methods described are used in the United States for the industrial synthesis of 2,5-dimethylhexadiene-2,4.

In the synthesis of chrysanthemumic acid the greatest difficulty is presented by the reaction of 2,5-dimethylhexadiene-2,4 with diazoacetic ester, which is highly explosive. This reaction takes place at 125°–130° C.:

$$(CH_3)_2C=CHCH=C(CH_3)_2 + N_2CHCOOC_2H_5 \rightarrow$$
$$N_2 + (CH_3)_2C=CHCH-CHCOOC_2H_5$$
$$\diagdown\ \diagup$$
$$C(CH_3)_2$$

This process usually is carried out in a special building by remote control. The reaction proceeds in an excess of the hydrocarbon, which is distilled off after the process is over, and the ethyl esters of the isomeric chrysanthemumic acids are distilled in vacuum (10 mm. of Hg.). To obtain the free acid, the ester is saponified with alcoholic alkali and the acid is separated by acidification with sulfuric acid. The chrysanthemumic acid is extracted from the aqueous solution with heptane.

To isolate the pure *cis*- and *trans*-isomers, the acid produced is crystallized from ethyl acetate. The physiologically inactive *cis*-acid melts at 115°–116° C., while the natural *trans*-acid melts at 54° C.

Another method of synthesizing the chrysanthemum monocarboxylic acid has been proposed, the starting materials for which are 5-methylhexen-4-one-2 and ethyl bromoacetate. By the condensation of 5-methylhexen-4-one-2 with ethyl bromoacetate in the presence of sodamide, the ester of 2-acetyl-4-methylpenten-3-carboxylic-1 acid (b.p. 68°–72° C. at 0.02 mm.

of Hg) is formed; this compound is converted by the action of methylmagnesium bromide to the corresponding tertiary alcohol:

$$(CH_3)_2C=CHCH_2COCH_3 + BrCH_2COOC_2H_5 \rightarrow$$
$$(CH_3)_2C=CHCHCH_2COOC_2H_5$$
$$\quad\quad\quad\quad\quad\quad |$$
$$\quad\quad\quad\quad\quad\quad COCH_3$$

$$(CH_3)_2C=CHCHCH_2COOC_2H_5 + CH_3MgBr \rightarrow$$
$$\quad\quad\quad\quad |$$
$$\quad\quad\quad\quad COCH_3$$
$$(CH_3)_2C=CHCHCH_2COOC_2H_5$$
$$\quad\quad\quad\quad |$$
$$\quad\quad\quad\quad HOC(CH_3)_2$$

When the ester obtained is hydrolyzed it gives a lactone (b.p. 70°–71° C. at 0.12 mm. of Hg, m.p. 57°–58° C.), which on reaction with HCl forms the chlorine-substituted acid (b.p. 89°–118° C. at 0.6 mm. of Hg):

$$(CH_3)_2C=CHCHCH_2COOC_2H_5 \dashrightarrow$$
$$\quad\quad\quad\quad |$$
$$\quad\quad\quad\quad HOC(CH_3)_2$$

[lactone structure] $\xrightarrow{HCl}$

$$(CH_3)_2C=CHCHCH_2COOH$$
$$\quad\quad\quad\quad\quad |$$
$$\quad\quad\quad\quad\quad (CH_3)_2CCl$$

By the action of alcoholates or other condensing agents on this acid, it is converted to the chrysanthemum monocarboxylic acid:

$$(CH_3)_2C=CHCHCH_2COOH \quad\quad (CH_3)_2C=CHCH-CHCOOH$$
$$\quad\quad\quad\quad | \quad\quad\quad\quad\quad \rightarrow \quad\quad\quad\quad\quad\quad\quad \diagdown \diagup$$
$$\quad\quad\quad\quad (CH_3)_2CCl \quad\quad\quad\quad\quad\quad\quad\quad\quad\quad\quad C(CH_3)_2$$

This second method, although it is no less complicated and involves several steps, avoids the explosive hazard present when working with diazoacetic ester. To obtain the acid chloride, the mixture of the isomeric acids is treated with thionyl chloride.

The pyrethroids mixed with synergists are most often used in the form of aerosols to control house flies, and also in the food industry to control stored-product pests. The pyrethroids are used also in spraying formulations, in dusts, and other forms. Recently they have been combined with more persistent insecticides.

The gibberellins

A second group of natural derivatives of alicyclic acids are the gibberellins, which are isolated from products of the activity of fungi of the genus *Fusarium*. The gibberellins were first discovered by Japanese investigators in products of the vital activities of the phytopathogenic gibberella fungi, which cause "bad shoot disease" of rice. Up to the present time fifteen compounds have been isolated that have growth-stimulating activity. The structural formulas of five of them are given below:

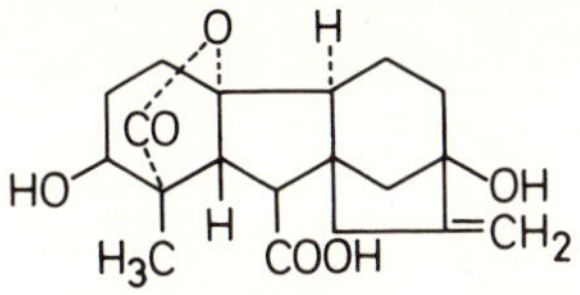

G_1

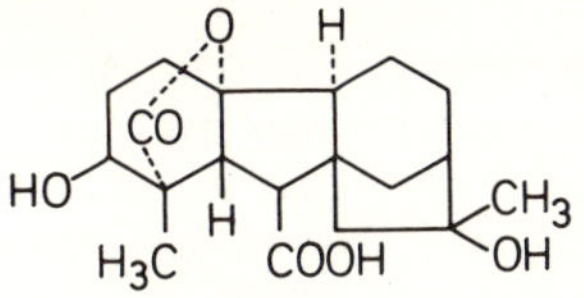

G_2

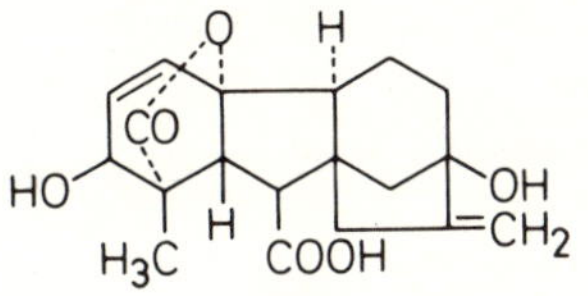

G_3
(gibberellic acid)

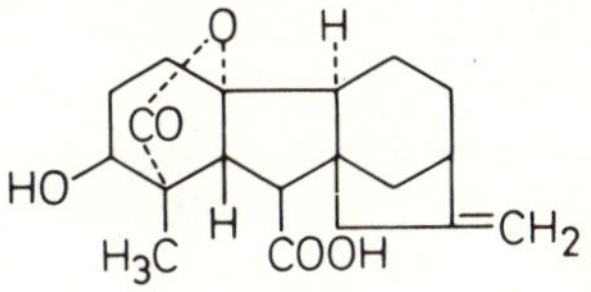

G_4

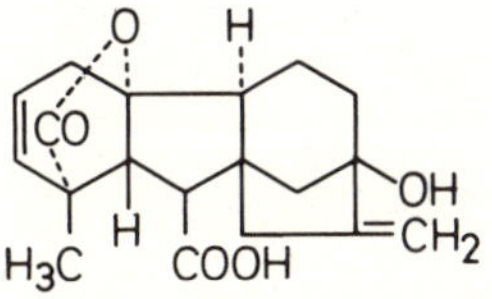

G_5

Table XXXI. *Properties of the gibberellins* [a]

Com- pound	Molecular formula	Free acid		Methyl ester	
		M. P. (°C.)	α_D^{25} (degrees)	M. P. (°C.)	α_D^{25} (degrees)
G_1	$C_{19}H_{24}O_6$	255–258	+36	209–210	+75
G_2	$C_{19}H_{26}O_6$	235–237	+12	234–235	+46
G_3	$C_{19}H_{22}O_6$	233–235	+92	190–192	+28
G_4	$C_{19}H_{24}O_5$	214–215	− 3	176	0
G_5	$C_{19}H_{22}O_5$	260–261	−77	190–191	−75

[a] Some gibberellins are polymorphic and the different modifications differ in melting point.

Table XXXI shows the properties of these gibberellins and also of their methyl esters. G_3, known as "gibberellic acid," has the greatest activity as a plant growth stimulator.

The action of the gibberellins on plants is characterized by an acceleration of growth, which occurs upon repeated application of a solution of the compound at a concentration from 10–200 mg./liter (depending on the species of plant and the conditions of use) to the top of the plant. When the gibberellins act on plants, elongation and division of the cells are noted, leading to lengthening of the stems and growth of the plants. Growth is most rapid when there is an optimum auxin content in the plant. The effect of the gibberellins is different on short-day and long-day plants. The gibberellins are used in growing ornamental plants and grapes. When individual varieties of grapes are treated with gibberellins the harvest is considerably increased. The gibberellins usually are produced microbiologically from different *Fusarium* strains.

A number of analogs of the gibberellins have been synthesized, and the physiological activity of some of them surpasses the natural compounds. Their synthesis, however, is complicated and is still not of practical importance. Even slight changes in the molecule of the gibberellins substantially affect the activity of the compounds.

The gibberellins are relatively unstable and easily converted to other compounds by the action of various reagents. Thus, gibberellic acid on heating to 50° C. in weakly acid solution goes over to allogibberoic and gibberellinoic acids. A diagram of the conversions of gibberellic acid under the influence of various reagents is presented below:

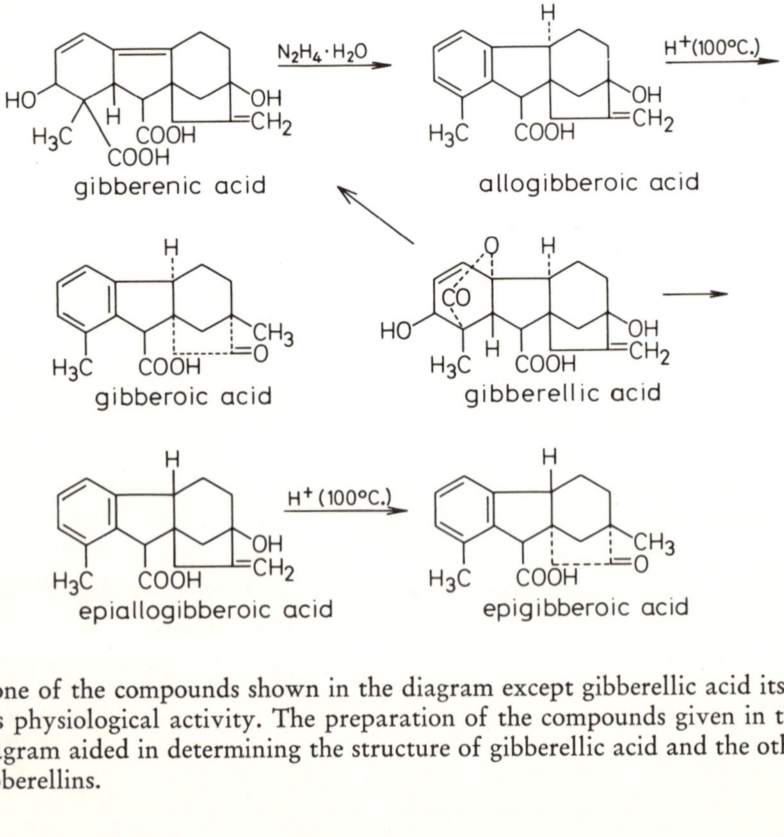

None of the compounds shown in the diagram except gibberellic acid itself has physiological activity. The preparation of the compounds given in this diagram aided in determining the structure of gibberellic acid and the other gibberellins.

Other alicyclic acids and their derivatives

A large number of other alicyclic acids and their derivatives have been studied. Most of them, however, have relatively low physiological activity and have not been practically utilized.

Of the derivatives of alicyclic acids that have been used mention should be made of the dimethyl ester of *cis*-bicyclo[2.2.1]heptene-2-dicarboxylic-5,6 acid, which is known as dimethyl carbate or Dimelone. It is a colorless liquid, b.p. 114°–115° C. at 3 mm. of Hg, d^{24} 1.1637, n_D^{25} 1.4829. At 35° C., 1.32 g. of dimethyl carbate dissolves in 100 ml. of water. Its toxicity for rats (LD_{50}) is more than 1,000 mg./kg. The compound does not irritate the skin nor the respiratory passages.

Dimethyl carbate is used mainly in the form of mixtures with other compounds as a repellent for blood-sucking diptera (it repels mosquitoes of the genus *Aedes* especially well). Thus, for example, the preparation DID contains 75% dimethyl phthalate, 20% indalone, and 5% dimethyl carbate, and the preparation M-1616 consists of 60% dimethyl phthalate, 20% dimethyl carbate, and 20% indalone. The preparation 622 contains 20% dimethyl carbate. To repel insects the preparation is applied to the uncovered parts of the skin or to the clothing. The preparation DID when applied to the skin repels insects for 5–5.5 hours (usually about 1 g. of the preparation is used).

Dimethyl carbate is produced by the condensation of cyclopentadiene with dimethyl maleate at an elevated temperature:

$$\text{cyclopentadiene} + \begin{array}{c} \text{CHCOOCH}_3 \\ \| \\ \text{CHCOOCH}_3 \end{array} \longrightarrow \text{bicyclic product with two COOCH}_3 \text{ groups}$$

It is possible to prepare it also by esterification with methyl alcohol of the anhydride of bicyclo[2.2.1]heptene-2-dicarboxylic-5,6 acid.

Some use is made as a synergist for the pyrethrins and especially for allethrin of the 2-ethylhexylimide of bicyclo[2.2.1]heptene-2-dicarboxylic acid, which is obtained by the action of 2-ethylhexylamine on the anhydride of bicyclo[2.2.1]heptene-2-dicarboxylic acid. The imide is a colorless liquid, b.p. 158° C. at 2 mm. of Hg, d^{18} 1.05, that is miscible with petroleum oils, and dissolves various insecticides well. Its LD_{50} for rats by dermal application is 2,800 mg./kg., and that for rabbits is 470 mg./kg. The use of anilides of cyclopropanecarboxylic acid as herbicides has been proposed.

The systematic study of the properties of cyclohexanecarboxylic acid derivatives has shown that compounds of this type are capable of attracting insects. In particular, such properties are exhibited by the *sec*-butyl ester of 6-methylcyclohexene-3-carboxylic acid, known as siglure or ENT-21486 (b.p. 113°–114° C. at 15 mm. of Hg). The *trans*-isomer is the more active. Still more effective is the product of condensation of HCl with siglure, the *sec*-butyl ester of 4-(or 5-)chloro-2-methylcyclohexanecarboxylic acid (medlure), which is a good attractant for the male Mediterranean fruit fly (b.p.

92°–94° C. at 0.2 mm. of Hg). Attractant properties are shown also by the *tert*-butyl ester of this latter acid, trimedlure (b.p. 107°–113° C. at 0.6 mm. of Hg).

General references

British Patent 933,193; German Federal Republic Patent 1,149,713; US Patent 3,077,596; Chemisches Zentralblatt **136**, 31–1269 (1965).
CHAILAKHYAN, M. KH.: Gibberelliny rastenii [Plant gibberellins]. Publishing House of Academy of Sciences USSR (1961).
FREAR, D.: Khimiya insektitsidov i fungitsidov [Chemistry of insecticides and fungicides]. Foreign Literature Publishing House [USSR] (1948).
Gibberelliny i ikh deistvie na rastenii [The gibberellins and their action on plants]. Publishing House of the Academy of Sciences USSR (1963).
HORSFALL, J. G.: Fungtsidy i ikh deistvie [Fungicides and their action]. Foreign Literature Publishing House [USSR] (1948).
KUCHEROV, V. F., and N. YA. GRIGOR'EVA: Uspekhi Khim. **35**, 2044 (1966).
LEBEDEVA, K. V.: Zhur. Vsesoyuz. Khim. Obshchestva im. D. I. Mendeleeva **13**, 272–282 (1968).
MEL'NIKOV, N. N., and K. D. SHVETSOVA-SHILOVSKAYA: Khim. Prom. No. 3, 50 (1955).

XIII. Aromatic carboxylic acids and their derivatives

General characteristics of pesticidal properties

The insecticidal activity of the free acids (benzoic, phenylacetic, and naphthoic), their homologs, halo and nitro derivatives, and salts with the alkali metals and ammonia is low. Esters of a number of aromatic acids have appreciable acaricidal activity. Thus, the benzyl ester of benzoic acid is powerful acaricide for some species of mites. Introduction of halogen atoms into the benzoic acid and benzyl alcohol groups increases the biological activity of the compounds. In this connection, the toxicity to the adult mites is reinforced in compounds containing chlorine in the *para*-position of the benzyl radical, and the toxicity to the eggs is strengthened in compounds containing chlorine in the *para*-position of the benzyl radical. The acaricidal activity is increased when nitro, amino, and hydroxy groups are introduced, but in the first and second cases the toxicity to mammals also is increased. The aliphatic esters of salicyclic, anthranilic, and anisic acids are moderately toxic to body lice and codling moths.

Of the esters of aromatic carboxylic acids, only ethyl 4,4′dichlorobenzilate (b.p. 141°–142° C. at 0.6 mm. of Hg, d_4^{20} 1.281, n_D^{20} 1.5727, LD_{50} 729 mg./kg. against mice), known as chlorobenzilate, has been used as a selective acaricide. At 0.02% it gives satisfactory results in the control of many plant-feeding mites and also of the tracheal mite, a parasite of bees. Chlorobenzilate is produced by the following method:

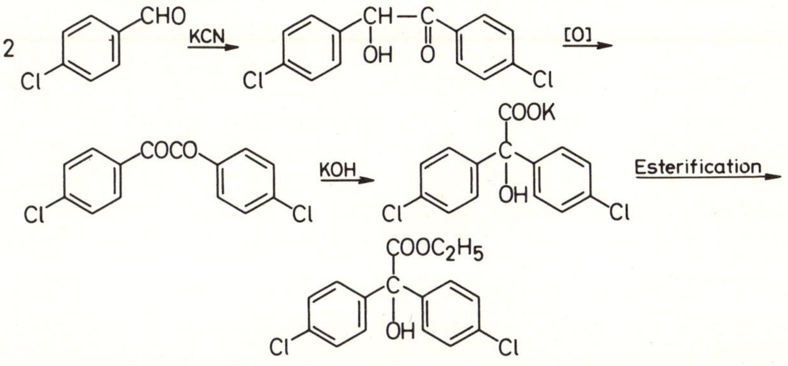

Chlorobenzilate is easily saponified by alkalies with the formation of salts of dichlorobenzilic acid. The free acid is unstable and is rapidly decarboxylated. When dichlorobenzilic acid is nitrated, nitro compounds are obtained that give colored products with sodium methylate.

In agriculture chlorobenzilate is used in the form of a 25% emulsifiable concentrate in xylene, a 25% wettable powder, a 3% dust, and a smoke-forming composition for treatment of greenhouses.

The 2-fluoroethyl esters of 4-phenylphenylacetic and some other acids are used as acaricides. These esters show a powerful and prolonged acaricidal effect, but they are very toxic to warm-blooded animals (LD_{50} about 1 mg./kg.).

The esters of phthalic acid repel blood-sucking insects. Dimethyl phthalate (b.p. 282°–285° C. at 760 mm. of Hg, LD_{50} 8 g./kg.) has the strongest repellent effect. With the increasing molecular weight of the dialkyl phthalate the repellent effect decreases, but the duration of the effect increases. In connection with this, dibutyl phthalate (b.p. 340° to 345° C. at 760 mm. of Hg, LD_{50} 20 g./kg.) is used usually for the impregnation of clothing, and its effect lasts for many days.

The repellent effect of the esters of isophthalic and terephthalic acids is considerably weaker. Introduction of halogen and other substituents into the aromatic nucleus of phthalic acid does not increase the repellent activity.

The dialkylamides of the substituted benzoic acids have the strongest repellent effect on blood-sucking insects and ticks. The dialkylamides of *m*-toluic and *m*-chlorobenzoic acids are especially active. The corresponding *o*- and *p*-isomers are less active. The diethylamide of *m*-toluic acid is known as DETA or deet, a universal, long-acting repellent.

Table XXXII shows the comparative repellent properties of the diethylamides of several aromatic acids.

The nitriles are more powerful insecticides as compared with the acids, but the activity of the compounds that have been studied is not great enough for practical use in agriculture, industry, or domestic life.

The aromatic carboxylic acids affect plant cells considerably more than the aliphatic acids, and many of them are fungicides, herbicides, and plant growth regulators. The first representative of this series, benzoic acid, has a significant fungicidal effect, which increases when halogens, a nitro group, or hydroxyl are introduced into the molecule. The fungicidal and bactericidal effects are increased most when hydrogens in the aromatic nucleus are replaced simultaneously by halogen and hydroxyl. All the halohydroxybenzoic acids that have been made surpass in fungicidal and bactericidal effect not only benzoic acid, but also the corresponding halobenzoic acids.

The hydroxybenzoic acids, and especially their esters and amides, show high fungicidal activity. The esters of *p*-hydroxybenzoic acid are used as antiseptics in lipsticks and creams, and the anilide of salicyclic acid is used to combat brown-spot of tomatoes (leaf mold) and as a disinfectant for nonmetallic materials. Introduction into the hydroxybenzoic acid molecule of a second hydroxyl group does not substantially change the fungicidal properties of the compound, but the toxicity to mammals is decreased somewhat. Introduction of lower hydrocarbon radicals increases the microbiological activity of the compounds.

The physiological activity of the aromatic acids on plants has been studied in detail. In the series of benzoic acid derivatives the greatest activity is shown by compounds containing substituents in the 2,3,6-posi-

Table XXXII. *Repellent properties of diethylamides of some aromatic acids*

Diethylamide of acid	B. P. (°C./mm. Hg)	Repellent activity [a] for mosquitoes on 4-point scale	
		On skin	On clothing
o-Toluic	105/1	4	3
m-Toluic	111/1	4	4
p-Toluic	110/1 [b]	4	4
p-Isopropylbenzoic	106–108/0.1	—	4
o-Methoxybenzoic	100–104/1 [c]	—	3
Veratric	120/0.2	—	2
o-Ethoxybenzoic	115/0.1	4	4
o-Bromobenzoic	138–141/20	3	4
o-Chlorobenzoic	115/0.2	—	4
m-Chlorobenzoic	116/0.2	3	4
2,4-Dichlorobenzoic	125–127/0.4	—	1
3,4-Dichlorobenzoic	192/12	—	1
Cyclohexanecarboxylic	135/22	—	1
Tetrahydrotoluic	79/0.1	—	3

[a] Dash denotes absence of repellent effect.
[b] M. P. 53.3°–55.5° C.
[c] M. P. 36°–38° C.

tions; acids with substitutents in the 2,3,5-positions also are active. Practical use as herbicides has been made of 2,3,6-trichlorobenzoic, 2-methoxy-3,6-dichlorobenzoic, 2,5-dichloro-3-nitrobenzoic, and 2,5-dichloro-3-aminobenzoic acids; 2-bromo-3,5-dichlorobenzoic acid also is highly active. 2,3,5-Triiodobenzoic acid elicits a change in the shape of the leaves of various plants and it also is a defoliant for some crops. Similar properties are shown by some other trihalobenzoic acids containing iodine in the 3-position. Tetrasubstituted benzoic acids also have herbicidal properties, but their action is less selective.

The substituted phenylacetic acids have high physiological activity toward plants. In the series of halogen-substituted phenylacetic acids the following general relationship between activity and structure is observed. Among the monohalogen-substituted phenylacetic acids the greatest activity is shown by the 2-isomers and the least by the 4-isomers; introduction of a second halogen into the molecule of phenylacetic acid does not substantially change the activity; a third halogen atom, depending on the position, either increases the activity of the compound (2,3,6-isomer) or decreases it (2,3,5-isomer). Hydrocarbon radicals lower the activity, while one methoxyl group only slightly affects the activity of the 2,3,6-trisubstituted phenylacetic acid, but the position of the methoxyl group is of considerable importance.

Of the chloro derivatives of phenylacetic acid that have been studied, the greatest herbicidal activity is shown by 2,3,6-trichlorophenylacetic acid, on which the well-known preparation fenac is based.

The naphthoic acids and their halogen derivatives are comparatively weak herbicides, while 1-naphthylacetic acid finds practical use as a plant growth regulator. Replacement of hydrogen in the naphthalene nucleus of 1-naphthylacetic acid by halogens and other radicals in most cases lowers the physiological activity of the compound.

Mention should be made of 2,6-dichlorothiobenzamide and 2,6-dichlorobenzonitrile, which are used as selective herbicides. 3,5-Diiodo-4-hydroxybenzonitrile (ioxynil), its bromine analog, and their esters with various aliphatic acids have also been used as selective herbicides. Some amides of diphenylacetic acid have high herbicidal activity. Practical use has been made of *N,N*-dimethyldiphenylacetamide (diphenamide), the nitrile of which also shows herbicidal properties. The dibasic acid derivatives, *N*-1-naphthylamino acid and dimethyl ester of tetrachloroterephthalic acid, also are used to control weeds.

Derivatives of benzoic acid

As indicated above, the derivatives of benzoic acid have not been used as insecticides, but they show herbicidal and fungicidal properties.

2,6-Dichlorobenzonitrile is a white crystalline substance, m.p. 145° C. Its solubility in water at 25° C. is about 10 mg./l.; it is more soluble in alcohol, acetone, and aromatic hydrocarbons. The LD_{50} for experimental animals lies within the limits 1,000–6,800 mg./kg.

The compound is a promising herbicide; at dosages of 1–2.5 kg./ha. it gives good results as a pre-emergence treatment in the control of many annual weeds, water hyacinths, etc. It is used in the form of a 50% wettable powder or of granules containing 2.5% active ingredient.

The synthesis of 2,6-dichlorobenzonitrile presents certain difficulties. It is made by the joint oxidation of 2,6-dichlorotoluene and ammonia. The reaction takes place at about 360° C., on a vanadium pentoxide catalyst:

$$\text{2,6-Cl}_2\text{C}_6\text{H}_3\text{CH}_3 + \text{NH}_3 + [\text{O}] \xrightarrow{\text{V}_2\text{O}_5} \text{2,6-Cl}_2\text{C}_6\text{H}_3\text{CN} + 3\text{H}_2\text{O}$$

Under optimum conditions the yield of the product is more than 50%. Another method for the synthesis of 2,6-dichlorobenzonitrile is chlorination of 2-chloro-6-nitrobenzonitrile, which is obtained by the action of cuprous cyanide on 2,3-dichloronitrobenzene. The last-named compound is produced in 15–25% yield by direct nitration of o-dichlorobenzene:

[Reaction scheme: o-dichlorobenzene with Cl, Cl, NO₂ → (CuCN) → 2,3-dichloro compound with Cl, CN, NO₂ → (Cl₂) → compound with Cl, CN, Cl]

It also is possible to make 2,6-dichlorobenzonitrile by a Sandmeyer reaction, but this method is only important for laboratory preparation:

[Reaction scheme: O₂N–C₆H₃(NH₂)–NO₂ → (HNO₃, HCl) → O₂N–C₆H₃(N₂⁺Cl⁻)–NO₂ → (CuCN) → O₂N–C₆H₃(CN)–NO₂ → (Cl₂) → Cl–C₆H₃(CN)–Cl]

2,6-Dichlorothiobenzamide is a white crystalline substance, m.p. 151° to 152° C. About 940 mg. of the compound dissolves in a liter of water at 20° C. It is stable to the action of light and to heating at 90° C.; it is rapidly hydrolyzed in an aqueous alkaline solution, but is more resistant in an acid medium. Its LD_{50} is 500–750 mg./kg. It is produced in good yield by the action of hydrogen sulfide on 2,6-dichlorobenzonitrile in the presence of catalytic amounts of pyridine:

[Reaction: Cl–C₆H₃(CN)–Cl + H₂S → Cl–C₆H₃(CSNH₂)–Cl]

2,6-Dichlorothiobenzamide is known as Prefix for the control of weeds in rice plantings and in orchards. At a dosage of 0.75–2 kg./ha. it gives satisfactory results in the eradication of many annual weeds, including miliary plants, Bermuda grass, and cyperus. However, it does not completely destroy quack grass, calistegia, bindweed, silverweed, aquilegia, crowfoot, dock, and some other plants. Prefix is used in the form of a 75% wettable powder and a 7.5% granulated formulation for application to the soil.

2,3,6-Trichlorobenzoic acid is a white crystalline substance, m.p. 126° C., poorly soluble in water and organic solvents. Salts of 2,3,6-trichlorobenzoic acid with the alkali metals, ammonia, and aliphatic amines are highly water-soluble. The LD_{50} for experimental animals is 700–1,500 mg./kg. The compound is used to control various weeds.

The water-soluble salts of pure 2,3,6-trichlorobenzoic acid are used in grain plantings at dosages of 100–200 g./ha. together with MCPA to control weeds that are resistant to the latter. However, in the presence of even small traces of related compounds a negative effect on crop plants is observed. A mixture of isomers can be used to control shoots of weeds of the pink smartweed type. Preparations of this type are known as 2KF and Trisben-200. The mixture that enters into the composition of Trisben-200 melts within the limits 87°–99° C., and consists of 2,3,6-, 2,3,4-, 2,4,5-, 2,4,6-, and 3,4,5-trichlorobenzoic acids. These preparations are dimethylamine salts of the acids mentioned. They are used at dosages of 25–50 kg./ha.

A mixture of isomers of trichlorobenzoic acids is obtained by the chlorination of benzoyl chloride in the presence of a catalyst for substitutive chlorination. Under optimum conditions for the process it is mainly the 2,3,6-isomer that is formed. After the chlorination is ended the isomeric trichlorobenzoyl chlorides are saponified with alkali or organic bases in an aqueous medium:

$$C_6H_5COCl + 3\ Cl_2 \rightarrow C_6H_2Cl_3COCl + 3\ HCl$$
$$C_6H_2Cl_3COCl + 2\ KOH \rightarrow C_6H_2Cl_3COOK + KCl + H_2O$$

A purer product is produced by the following scheme:

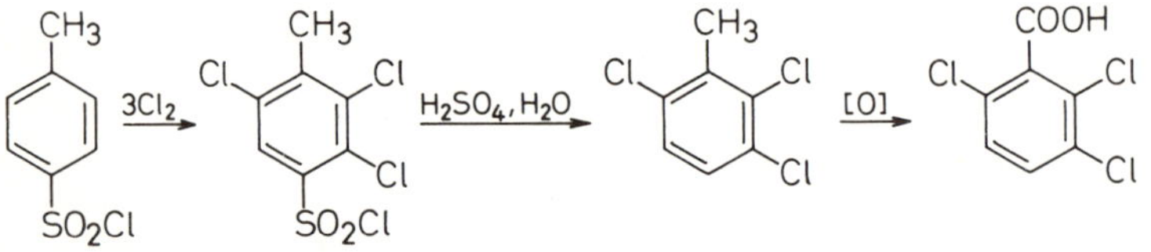

The necessary 2,3,6-trichlorotoluene can be synthesized also by direct chlorination of o-chlorotoluene (a by-product in the production of p-chlorotoluene by direct chlorination of toluene):

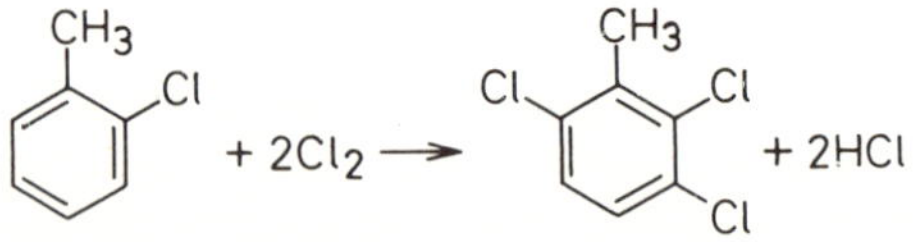

However, this product contains some amount of the other isomers.

Oxidation of trichlorotoluene is best carried out with oxides of nitrogen at a temperature of the order of 180°–190° C., by a continuous process. It is possible to carry out the oxidation also with oxygen in the presence of catalysts. The solution of 2,3,6-trichlorobenzoic acid in trichlorotoluene that is obtained is extracted with alkali. The acid goes over in the aqueous solution to the form of the salt and the trichlorotoluene after appropriate purification is returned to the oxidation process. Since the isomeric acids are precipitated by mineral acids from aqueous alkaline solutions at different pH, the purification of 2,3,6-trichlorobenzoic acid can be accomplished by fractional precipitation.

The esters of 2,3,6-trichlorobenzoic acid have a stronger herbicidal effect, but they have not yet been practically utilized in agriculture, apparently because of their great toxicity to useful plants. The use of 2,3,5,6-tetrachlorobenzoic acid and pentachlorobenzoic acid as herbicides also has been proposed.

2-Methoxy-3,6-dichlorobenzoic acid (dicamba, Banvel-D) is a white crystalline substance, m.p. 114°–116° C., very poorly soluble in water, but highly soluble in most organic solvents. Its salts with the alkali metals and amines are highly water-soluble. Thus, 100 ml. of water at room temperature dissolves 38 g. of the sodium salt and more than 72 g. of the dimethylamine salt of 2-methoxy-3,6-dichlorobenzoic acid. The herbicidal activity and the area of action of 2-methoxy-3,6-dichlorobenzoic acid are similar to those of 2,3,6-trichlorobenzoic acid. 2-Methoxy-3,6-dichlorobenzoic acid is relatively low in toxicity to mammals (LD_{50} 1,000–1,100 mg./kg.). It is used mainly with MCPA against weeds that are resistant to the latter compound, in an amount equal to about 6.6% of the MCPA used. This mixture is effective in the control of fumitory, wild poppy, snakeweed, nipplewort, ragweed, and some other resistant weeds.

2-Methoxy-3,6-dichlorobenzoic acid is produced from the nontoxic isomers of benzene hexachloride by the following process:

It is usually used in the form of the sodium, potassium, or dimethylamine salts, which are highly soluble in water.

An analog of Banvel-D is Banvel-T, 2-methoxy-3,5,6-trichlorobenzoic acid, which is now being investigated as a possible herbicide to control weeds in lawns, in grains, flax, and cruciferous crops.

2,5-Dichloro-3-nitrobenzoic acid (dinoben), m.p. 220°–221° C., is poorly soluble in water and moderately soluble in organic solvents. It is used for the control of weeds in vegetable crops. At a dosage of 5–6 kg./ha. dinoben is toxic to a large number of weeds and relatively safe for carrots, peas, and corn. It is used in the form either of the highly water-soluble amine salts or of granulated formulations. Its toxicity to mammals is low, LD_{50} 3,500 mg./kg.

2,5-Dichloro-3-nitrobenzoic acid is produced either by nitration of 2,5-dichlorobenzoic acid or by oxidation of 2,5-dichloro-3-nitrotoluene:

$$\text{2,5-Cl}_2\text{C}_6\text{H}_3\text{COOH} + \text{HNO}_3 \longrightarrow \text{2,5-Cl}_2\text{-3-NO}_2\text{-C}_6\text{H}_2\text{COOH} + \text{H}_2\text{O}$$

When 2,5-dichloro-3-nitrobenzoic acid is reduced 2,5-dichloro-3-aminobenzoic acid is formed, which is used as a herbicide under the name of amiben, a crystalline substance, m.p. 200°–201° C., LD_{50} for warm-blooded animals 5,500 mg./kg. Amiben is used for pre-emergence control of weeds in plantings of cucurbits, carrots, soy beans, and other crops in the form of amine salts and granulated formulations at dosages of 4–8 kg./ha. The use of dinoben and amiben, however, is still rather limited.

Derivatives of hydroxybenzoic acids

In this class of compounds mention should be made of salicylic acid and its derivatives, which are used as fungicides and disinfectants. Free salicylic acid is used in canning fruits and vegetables. However, salicylanilide has received the widest distribution as a disinfectant for nonmetallic materials and as a fungicide for controlling brown spot (leaf mold) of tomatoes and diseases of some other crops.

Salicylanilide is a white crystalline substance, m.p. 135° C.; about 55 mg. dissolve in a liter of water at 23° C. It is highly soluble in carbon tetrachloride, dichloroethane, and other solvents. Its LD_{50} is more than 5 g./kg. Salicylanilide forms highly water-soluble salts with the alkali metals, ammonia, and amines; its zinc and copper salts are practically insoluble in water and are easily precipitated from aqueous solutions of its alkali salts by reaction with salts of the corresponding metals. Zinc salicylanilide is a more powerful disinfectant and fungicide than the free salicylanilide.

Salicylanilide is produced by the condensation of salicyclic acid with aniline at an elevated temperature, sometimes in the presence of condensing agents:

$$\text{o-HOC}_6\text{H}_4\text{COOH} + \text{C}_6\text{H}_5\text{NH}_2 \longrightarrow \text{o-HOC}_6\text{H}_4\text{CONHC}_6\text{H}_5 + \text{H}_2\text{O}$$

A stronger fungicidal effect is shown by chlorosalicylanilide and especially by dichlorosalicylanilide, which is obtained by direct chlorination of salicylanilide in carbon tetrachloride:

$$\text{o-HOC}_6\text{H}_4\text{CO-NH-C}_6\text{H}_5 + 2\text{Cl}_2 \longrightarrow \text{Cl-C}_6\text{H}_3(\text{OH})\text{CONH-C}_6\text{H}_4\text{-Cl} + 2\text{HCl}$$

Dichlorosalicylanilide is a white crystalline substance, m.p. 232°–234° C. It is practically insoluble in water and poorly soluble in the usual organic solvents. The LD_{50} for experimental animals is close to that of salicylanilide.

High fungicidal and bactericidal activity is characteristic also of the corresponding bromine derivatives, which are used for the treatment of several fungus diseases of the human skin.

The zinc salt of dichlorosalicylanilide also shows a fungicidal effect and is used to protect plants, nonmetallic materials, and textiles. Dichlorosalicylanilide is a good compound for the control of amphibian mollusks that are vectors of fascioliasis. Molluskicidal properties also are possessed by 5,2'-dichloro-4'-nitrosalicylanilide (m.p. 230° C., LD_{50} more than 5,000 mg./kg.), which is produced by the condensation of 2-chloro-4-nitroaniline with 5-chlorosalicylic acid:

$$\text{Cl-C}_6\text{H}_3(\text{OH})\text{COOH} + \text{NH}_2\text{-C}_6\text{H}_3(\text{Cl})\text{-NO}_2 \longrightarrow \text{Cl-C}_6\text{H}_3(\text{OH})\text{CONH-C}_6\text{H}_3(\text{Cl})\text{-NO}_2 + H_2O$$

The copper salt of phenylsalicylic acid [m.p. 148°–150° C. (dec.), LD_{50} 520 mg./kg.] dissolved in organic solvents is used to treat seed boxes and shelves in greenhouses. It is synthesized by the reaction of sodium or potassium phenylsalicylate with cupric sulfate in aqueous solution:

$$2 \, \text{C}_6\text{H}_5\text{-C}_6\text{H}_3(\text{OH})\text{COOK} + CuSO_4 \longrightarrow (\text{C}_6\text{H}_5\text{-C}_6\text{H}_3(\text{OH})\text{COO})_2 Cu + K_2SO_4$$

Some derivatives of hydroxybenzoic acid have a herbicidal effect as well as fungicidal properties. Examples of such compounds are the nitriles of the 3,5-dihalo-4-hydroxybenzoic acids, which are powerful contact herbicides.

3,5-Diiodo-4-hydroxybenzonitrile (ioxynil) is a cream-colored crystalline substance, m.p. 205°–207° C. (dec.), practically insoluble in water, slightly soluble in organic solvents. Its LD_{50} is 120–190 mg./kg. For use in agriculture it is marketed in the form of an aqueous solution of the sodium salt.

The compound is being widely investigated in various European and American countries for the control of weeds in grain fields. Ioxynil eradicates weeds at dosages of 400 g./ha; preparations of it are often used in combination with 2 M-4 CP. In this case the dosage of ioxynil is 150 g./ha. and that of 2 M-4 CP is 375 g./ha.

Ioxynil is produced by direct iodination of 4-hydroxybenzonitrile with iodine:

$$\underset{\underset{CN}{}}{\overset{OH}{\bigcirc}} + 2I_2 \longrightarrow \underset{\underset{CN}{}}{\overset{OH}{I\bigcirc I}} + 2HI$$

When salts of ioxynil react with acid chlorides the esters of the corresponding acids are formed, some of which surpass ioxynil in their herbicidal effect. Weaker herbicidal effect is shown, for example, by the caprylic ester of ioxynil:

$$\underset{\underset{CN}{}}{\overset{ONa}{I\bigcirc I}} + RCOCl \longrightarrow \underset{\underset{CN}{}}{\overset{OCOR}{I\bigcirc I}} + NaCl$$

3,5-Dibromo-4-hydroxybenzonitrile (bromoxynil) is a white crystalline substance, m.p. 194°–195° C. It has a solubility of 130 p.p.m. in water, 7% in ethanol, and 1–2% in xylene at 20°–25° C. Its LD_{50} is 190 mg./kg. Bromoxynil is conveniently formulated as a 25% solution of its potassium salt. It is produced by the action of sodium hypobromite on 4-hydroxybenzonitrile:

$$\underset{\underset{OH}{}}{\overset{CN}{\bigcirc}} + 2NaOH + 2Br_2 \longrightarrow \underset{\underset{OH}{}}{\overset{CN}{Br\bigcirc Br}} + 2NaBr + 2H_2O$$

4-Hydroxybenzonitrile can be prepared by the dehydration of 4-hydroxybenzaldoxime:

$$HO-\bigcirc-CH=NOH \longrightarrow HO-\bigcirc-CN + H_2O$$

3,5-Dibromo-4-octanoyloxybenzonitrile (bromoxynil octanoate) is a cream colored waxy solid, m.p. 45°–46° C. (pure). It is insoluble in water and 70% soluble in xylene. Its LD_{50} is 420 mg./kg. Bromoxynil octanoate is usually formulated as an emulsifiable concentrate. It is produced by the reaction of octanoyl chloride on a salt of bromoxynil:

$$\underset{\underset{CN}{}}{\overset{ONa}{Br\bigcirc Br}} + CH_3(CH_2)_6COCl \longrightarrow \underset{\underset{CN}{}}{\overset{OCO(CH_2)_6CH_3}{Br\bigcirc Br}} + NaCl$$

The amides of benzoic acid and the toluic acids have a rather strong repellent effect on blood-sucking dipterous insects and ixodid ticks. The most interesting of this group is diethyl-m-toluamide.

N,N-Diethyl-m-toluamide (deet, DETA) is a corlorless liquid, b.p. 11° C. at 1 mm. of Hg, d_4^{20} 1.0095, n_D^{20} 1.5206. The compound is insoluble in water, miscible in all proportions with ethyl and isopropyl alcohols, propylene glycol, and cottonseed oil, soluble in halogen derivatives of aliphatic and aromatic hydrocarbons and in aromatic hydrocarbons. The LD_{50} for rats is 2,000 mg./kg. The technical grade compound produced in the United States contains not less than 95% of total isomers of diethyltoluamide, and the content of N,N-diethyl-m-toluamide must be not less than 70%.

Deet, when applied to the skin, repels blood-sucking insects for eight to ten hours, in contrast to dimethyl phthalate which gives absolute protection of the skin from mosquitoes only for one and a half to two hours.

N,N-Diethyl-m-toluamide is produced by two methods:

1. Reaction of m-toluyl chloride with diethylamine in the presence of alkali hydroxides or an excess of amine:

$$m\text{-CH}_3\text{-C}_6\text{H}_4\text{-COCl} + (C_2H_5)_2NH + KOH \rightarrow m\text{-CH}_3\text{-C}_6\text{H}_4\text{-CON}(C_2H_5)_2 + KCl + H_2O$$

2. Reaction of m-toluic acid with diethylamine at an elevated temperature in the presence of dehydration catalysts:

$$m\text{-CH}_3\text{-C}_6\text{H}_4\text{-COOH} + (C_2H_5)_2NH \rightarrow m\text{-CH}_3\text{-C}_6\text{H}_4\text{-CON}(C_2H_5)_2 + H_2O$$

The simplicity of the second method is seductive, but conversion by it does not exceed 70%. In even 10–20 hours of work the catalyst loses activity and the conversion decreases to 30%. As a result, the problem of devising a special process for the regeneration of m-toluic acid, diethylamine, and catalyst arises.

The m-toluic acid necessary for the production of deet is obtained in good yields by the oxidation of m-xylene by the oxygen of the air at an elevated temperature and slight pressure, in the presence of catalysts, usually salts of cobalt with organic acids. In this reaction some isophthalic acid also is obtained. Separation of the mixture is accomplished either by fractional precipitation from aqueous solutions of the alkali salts by acidification with mineral acids, or by extraction of the *m*-toluic acid with organic solvents in which isophthalic acid is slightly soluble.

Since the compound is intended for systematic application to the human skin, it must be quite pure.

High repellent properties also are inherent in hexamethylenebenzamide, which is synthesized by reaction of benzoyl chloride with hexamethylene-amine in the presence of alkalies:

$$C_6H_5COCl + HN\begin{matrix}CH_2CH_2CH_2\\|\\CH_2CH_2CH_2\end{matrix} + KOH \rightarrow C_6H_5CON\begin{matrix}CH_2CH_2CH_2\\|\\CH_2CH_2CH_2\end{matrix} + KCl + H_2O$$

This compound is used in the Soviet Union under the name of benzimine. It enters into the composition of the repellent cream, Taiga. Benzimine does not irritate the skin, is low in toxicity to man, and is effective for a long time.

Dibasic acids and their derivatives

The number of derivatives of dibasic acids used as pesticides is comparatively small. Besides the repellent esters of phthalic acid described, *N*-1-naphthylphthalamic acid and dimethyl tetrachloroterephthalate and its thio-analog have been used to control weeds.

N-1-Naphthylphthalamic acid (Alanap) is a white crystalline substance, m.p. 203° C. (the technical grade compound melts at 175°–180° C.). It is almost insoluble in water, and slightly soluble in most organic solvents; when it is heated it is converted to the slightly active 1-naphthylphthalimide. Its LD_{50} is 8,500 mg./kg. Alanap is produced by the reaction of phthalic anhydride with 1-naphthylamine in organic solvents:

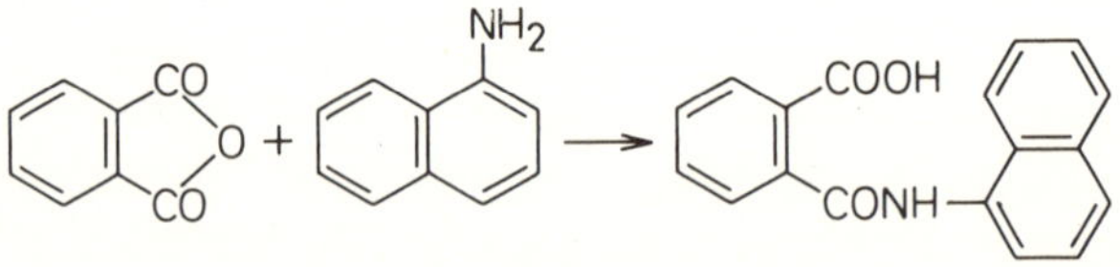

As a pre-emergence herbicide it is used in the form of the sodium salt or a 90% wettable powder at dosages of 2–10 kg./ha.

Dimethyl tetrachloroterephthalate (DAS-893, Dacthal) is a white crystalline substance, m.p. 156° C., LD_{50} more than 3,000 mg./kg. At dosages from 5 to 15 kg./ha., Dacthal gives satisfactory results in the control of creeping weeds in lawns; it also is used to control weeds in plantings of cruciferae, onion, and some other crops. Only the methyl ester has herbicidal properties; the free acid and its salts are inactive. The compound is used in the form of a 50% wettable powder and 2.5 and 5% granules.

Dacthal is produced by esterification of tetrachloroterephthalic acid with methanol or by the action of methanol on the acid dichloride. Tetrachloroterephthalic acid dichloride can be obtained by the following process:

$$\underset{\underset{CCl_3}{|}}{\overset{\overset{CCl_3}{|}}{C_6H_4}} + \underset{\underset{COOH}{|}}{\overset{\overset{COOH}{|}}{C_6H_4}} \longrightarrow 2\,\underset{\underset{COCl}{|}}{\overset{\overset{COCl}{|}}{C_6H_4}} + 2HCl$$

$$\underset{\underset{COCl}{|}}{\overset{\overset{COCl}{|}}{C_6H_4}} + 4Cl_2 \longrightarrow \underset{\underset{COCl}{|}}{\overset{\overset{COCl}{|}}{C_6Cl_4}} + 4HCl$$

This method of preparation is the cheapest and most convenient, since it gives practically no by-products except hydrogen chloride.

O,S-Dimethyl tetrachlorothioterephthalate (Glenbar) is a white crystalline substance, m.p. 161°–162° C. It is practically insoluble in water (0.36 mg./liter) and highly soluble in most organic solvents. The LD_{50} for rats is 3,300 mg./kg.

Glenbar has been proposed as a herbicide to control weeds in rice, legumes, and many other crops. The compound is marketed in the form of emulsive concentrate, wettable powder, and a granulated product containing 12% active ingredient. The dosage is 2–6 kg./ha.

Arylalkylcarboxylic acids and their derivatives

2,3,6-Trichlorophenylacetic acid (fenac) in the pure form melts at 159° to 161° C. In practice, however, a product containing 65–70% 2,3,6-isomer and 30–35% related compounds usually is used. At 30° C., about 200 mg. of the acid dissolves in a liter of water. The LD_{50} is 1,700–1,800 mg./kg.

Fenac is used in the form of the highly water-soluble alkali and amine salts or of emulsions of the esters. It gives good results in the control of field bindweed, quack grass, annual grasses, and many dicotyledonous weeds. At dosages of 15–20 kg./ha., fenac sterilizes the soil for one to two years, thereby controlling vegetation on railways and airfields.

Fenac can be produced by two methods:
1. From 2,3,6-trichlorotoluene by the following process:

$$\text{2,3,6-Cl}_3\text{C}_6\text{H}_2\text{CH}_3 \xrightarrow{Cl_2} \text{2,3,6-Cl}_3\text{C}_6\text{H}_2\text{CH}_2\text{Cl} \xrightarrow{KCN} \text{2,3,6-Cl}_3\text{C}_6\text{H}_2\text{CH}_2\text{CN} \xrightarrow{H_2SO_4} \text{2,3,6-Cl}_3\text{C}_6\text{H}_2\text{CH}_2\text{COOH}$$

2. By the chlorination of phenylacetyl chloride in the presence of catalysts for substitutive chlorination, the best of which is antimonous

chloride. By this method a product is obtained with up to 35% other isomers as impurities.

α-*Cyano-β-2,4-dichlorophenylacrylic acid* (m.p. 197.5°–198.3° C., LD_{50} 50–250 mg./kg.) is used in the form of the diethanolamine salt (m.p. 137.9°–138.9° C.) at 0.1% to retard the growth of plants. It is produced by the condensation of 2,4-dichlorobenzaldehyde with the ethyl ester of cyanoacetic acid with subsequent saponification:

$$\text{Cl-C}_6\text{H}_3(\text{Cl})\text{-CHO} + \text{H}_2\text{C}\genfrac{}{}{0pt}{}{\text{CN}}{\text{COOC}_2\text{H}_5} \longrightarrow \text{Cl-C}_6\text{H}_3(\text{Cl})\text{-CH=C}\genfrac{}{}{0pt}{}{\text{COOC}_2\text{H}_5}{\text{CN}} \longrightarrow$$

$$\text{Cl-C}_6\text{H}_3(\text{Cl})\text{-CH=C}\genfrac{}{}{0pt}{}{\text{COOH}}{\text{CN}}$$

1-Naphthylacetic acid is a white crystalline substance, m.p. 131°–132° C. Its solubility at 25° C. is 41–42 mg. in 100 ml. of water. It is practically nontoxic to mammals. The amide of the acid melts at 183° C.

The three most important methods for production of 1-naphthylacetic acid are the following:

1. Condensation of monochloroacetic acid with an excess of naphthalene at 200°–220° C., in the presence of a ferric bromide catalyst:

$$\text{C}_{10}\text{H}_8 + \text{ClCH}_2\text{COOH} \longrightarrow \text{C}_{10}\text{H}_7\text{-CH}_2\text{COOH} + \text{HCl}$$

This reaction takes 15–20 hours; the yield of 1-naphthylacetic acid is about 70%. Some naphthalenediacetic acid is formed as a by-product, which is easily removed because of its higher solubility in water; it can be separated also by distillation of the methyl esters in vacuum. The condensation of naphthalene with monochloroacetic acid is used for the industrial production of 1-naphthylacetic acid.

2. Hydrolysis of naphthylacetonitrile. The synthesis of this nitrile is accomplished by the reaction of chloromethylated naphthalene with cyanide salts:

$$\text{C}_{10}\text{H}_8 \xrightarrow{\text{CH}_2\text{O, HCl}} \text{C}_{10}\text{H}_7\text{-CH}_2\text{Cl} \xrightarrow{\text{KCN}} \text{C}_{10}\text{H}_7\text{-CH}_2\text{CN} \xrightarrow{\text{Hydrolysis}} \text{C}_{10}\text{H}_7\text{-CH}_2\text{COOH}$$

This method of preparing 1-naphthylacetic acid is more complicated, but it leads to the pure compound in good yields.

3. Rearrangement of 1-naphthyl methyl ketone by the Willgerodt method also yields 1-naphthylacetic acid of high purity. The first intermediate product in this reaction is 1-naphthylacetamide, which can be used without further treatment as a plant growth regulator. The 1-naphthyl methyl ketone necessary for this synthesis is obtained in a 94% yield by a Friedel-Krafts reaction from naphthalene and acetyl chloride in dichloroethane at 34° C. When the optimum conditions are maintained for the reaction, the technical compound contains not more than 3% of the 2-isomer as impurity:

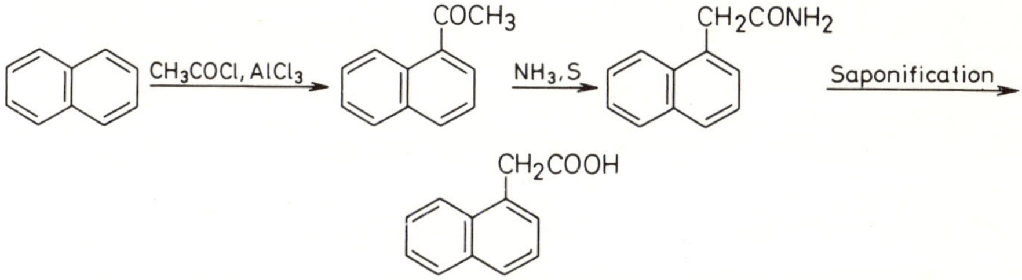

When 1-naphthylacetic acid is esterified with methyl alcohol in the presence of sulfuric acid, the methyl ester of 1-naphthylacetic acid is formed in practically quantitative yield (b.p. 122°–122.5° C. at 1 mm. of Hg). It is interesting to note that a quantitative yield of the ester is obtained only when the reaction is carried out at room temperature; with heating the yield drops sharply.

The methyl ester of 1-naphthylacetic acid is used to retard the sprouting of potatoes in prolonged storage. This compound at a dosage of 100 g. per ton of potatoes (in the form of a 3.5% dust on clay) retards sprouting of the potato tubers for a long time. This compound is known as Compound M-1. 1-Naphthylacetic acid and its amide also are used for thinning the blossoms of apple trees and for retarding the opening of buds in order to protect them from frosts.

2-Chloro-3-(4-chlorophenyl)propionic acid methyl ester (bidisin) has a b.p. 110°–113° C. at 0.1 mm. of Hg. It is soluble in ether, acetone, and aromatic hydrocarbons, but is practically insoluble in water. The acute oral LD_{50} for rats is 1,190 mg./kg. It shows exceptional specificity for the postemergence control of *Avena fatua,* though it has no effect on other weeds. Crops tolerant of bidisin include winter and spring wheat, spring barley, beets, potatoes, peas, and carrots. A dosage of 6–8 l./ha. is used.

N,N-Dimethyldiphenylacetamide (diphenamid) is a white crystalline substance, m.p. 134.5°–135.5° C. Its solubility in water is about 0.024% at 25° C. The LD_{50} is 700 mg./kg. Diphenamid is a herbicide that acts on the roots. It is recommended for the control of annual weeds in tomatoes, strawberries, soy beans, cotton, and some other crops. At dosages up to 4.5 kg./ha. it is safe for the crops listed.

Diphenylacetonitrile (diphenatrile), m.p. 74° C., LD_{50} 3,500 mg./kg., has been proposed for the control of creeping weeds in lawns.

General references

BASKAKOV, YU. A.: Zhur. Vsesoyuz. Khim. Obshchestva im D. I. Mendeleeva 5, 250 (1960).
— Zhur. Vsesoyuz. Obshchestva im. D. I. Mendeleeva 9, 486 (1964).
CRAFTS, A. S.: The chemistry and mode of action of herbicides. New York-London: Interscience (1961).
FISCHER, A.: Herbicide. Limburgerhof: BASF Landw.-Versuchsstation 1966.
HALROYD, J.: Proc. 9th Brit. Weed Control Conf. 1, 74 (1968).
HART, R. D., J. R. BISHOP, and A. R. COOKE: Proc. 7th Brit. Weed Control Conf., p. 3 (1964).
HEYWOOD, B. J., K. CARPENTER, and H. J. COTTRELL: Proc. 7th Brit. Weed Control Conf., p. 10 (1964).
Khimicheskie sredstva zashchity rastenii [Chemical agents for plant protection]. State Scientific and Technical Publishing House for Chemical Literature [USSR] (1961).
MANDEL'BAUM, YA. A., V. M. SAF'YANOVA, and V. I. LOMAKINA: Zhur. Vsesoyuz. Khim. Obshchestva im D. I. Mendeleeva 5, 307 (1960).
—, and KH. E. KHCHEYAN: Khim. Prom. No. 10, 686 (1961).
MEL'NIKOV, N. N., and YU. A. BASKAKOV: Khimiya gerbitsidov i regulyatorov rosta rastenii [Chemistry of herbicides and plant growth regulators]. State Scientific and Technical Publishing House of Chemical Literature [USSR] (1962).
— — Organicheskie insektofungitsidy i gerbitsidy [Organic insectofungicides and herbicides]. State Scientific and Technical Publishing House of Chemical Literature [USSR] (1958).
WAIN, R. L.: Proc. 7th Brit. Weed Control Conf., p. 1 (1964).
ZBIROVSKY, M. J., J. MYSKA, and J. ZEMANEK: Herbicidy. Prague (1960).

XIV. Aryloxalkylcarboxylic acids and their derivatives

General characteristics of pesticidal properties

The insecticidal properties of the aryloxyalkylcarboxylic acids and their derivatives that have been studied are insignificant, but many compounds of this group are herbicides, fungicides, and plant growth regulators. The aryloxyalkylcarboxylic acids have the greatest value as herbicides and plant growth regulators. With respect to their scale of manufacture and use, they occupy first place among compounds of all classes. There is a trend toward further growth in the production of this important group of herbicides.

The physiological activity of a large number of aryloxyalkylcarboxylic acids toward plants has been studied, and general rules for the change of activity with structure have been established, although the mechanism of their action on plants is still not clear.

The physiological activity of phenoxyacetic acid toward plants is increased when halogen is introduced into the molecule. Fluorine and chlorine have the most significant effect; that of bromine and iodine is substantially less. The position of the halogen influences the activity of the compound. The 4-halophenoxyacetic acids have the highest activity. For example, 4-chlorophenoxyacetic acid is almost ten times more active than the 2-isomer. The activity of 3-chlorophenoxyacetic acid also is less than that of the 4-isomer.

The isomeric dichlorophenoxyacetic acids fall in the following order with respect to physiological activity:

$$2,4\text{-} > 2,5\text{-} > 3,4\text{-} > 3,5\text{-} > 2,6\text{-}$$

The herbicidal activity of 2,5-dichloro- and 3,4-dichlorophenoxyacetic acids is great enough so that they, as well as 2,4-dichlorophenoxyacetic acid, have been proposed for use in agriculture. A positive property of 3,4-dichlorophenoxyacetic acid is its high selective activity in comparison with 2,4-D. 3,4-Dichlorophenoxyacetic acid is safer for cotton, alfalfa, potatoes, and sugar beets, but it is very toxic to sunflowers and many dicotyledonous weeds.

Of the isomeric trichlorophenoxyacetic acids highest activity is shown by 2,4,5-trichlorophenoxyacetic acid, which approaches 2,4-D in the strength of its herbicidal action. With respect to activity the isomeric trichlorophenoxyacetic acids fall in the following order:

$$2,4,5- > 2,3,4- > 3,4,5- > 2,3,5- > 2,4,6- > 2,3,6-$$

Among the tetrachlorophenoxyacetic acids only 2,3,4,5-tetrachlorophenoxyacetic acid shows appreciable physiological activity; the 2,3,4,6-tetrachloro- and 2,3,5,6-tetrachloro-isomers are inactive. Also inactive as a plant growth regulator is pentachlorophenoxyacetic acid, which has been proposed, however, for use as a systemic fungicide.

2,4,6-Trichloro-and 2,4,6-tribromophenoxyacetic acids are inactive as herbicides and, according to some data, are even antagonists for 2,4-D. However, 2,4-dichloro-6-fluorophenoxyacetic acid and 2,4-dibromo-6-fluorophenoxyacetic acid approach 2,4-D in activity. Relatively high herbicidal activity is shown by 2,4,6-trifluorophenoxyacetic acid, many other mixed polyhalofluorophenoxyacetic acids, and their derivatives.

Introduction into the molecule of phenoxyacetic acid of an aliphatic hydrocarbon radical in place of one hydrogen atom raises the activity of the compound only slightly. In this connection the activity of 2-alkyl-, 4-alkyl-, 2-halo-4-alkyl-, and 4-halo-2-alkylphenoxyacetic acids decreases with an increase in the size of the alkyl radical. A similar picture is observed also when an aromatic radical is introduced into the phenoxyacetic acid molecule.

Halogenation in the ring of the alkyl- and arylphenoxyacetic acids increases their activity, and the increase is greater in the alkylphenoxyacetic acids with the lower alkyls. 2-Alkyl-4-halophenoxyacetic acids are somewhat more active than 2-halo-4-alkylphenoxyacetic acids. The activity of the halomethylphenoxyacetic acids is especially outstanding. When a second halogen atom is introduced into the molecule of an alkylphenoxyacetic acid, in most cases the activity of the compound falls; the same effect is observed when a third halogen atom is introduced into the molecule. An exception is 2,4-dichloro-5-methylphenoxyacetic acid, which has a rather strong herbicidal effect. This acid is a structural analog of 2,4,5-T.

Introduction into the phenoxyacetic acid molecule of a chloromethyl group preserves the plant-growth stimulating activity, but sharply decreases the herbicidal activity of the compound. Thus, for example, 4-chloro-2-chloromethylphenoxyacetic acid is not of practical interest as a herbicide, while 4-chloro-2-methylphenoxyacetic acid (MCPA) is widely used in agriculture to control dicotyledonous weeds.

The acetylphenoxyacetic acids are inactive; this apparently is associated with the presence of a carbonyl group in their molecules and the possibility in connection with this of forming additional hydrogen bonds. For this reason, probably, the phenylenebis(acetic) acids and the carboxyphenoxyacetic acids also are low in activity.

In the series of aryloxyacetic acids containing an alkoxyl group in the aromatic nucleus the most active compounds are those with the alkoxyl in the 3-position, and the least active are the 2-isomers. An increase in the number of carbon atoms in the alkoxyl radical leads to an increase in activity up to a certain limit, after which it sharply decreases. It is possible that this is associated with a decrease in the rate of diffusion of the compound into the plant cells.

The alkoxyhalophenoxyacetic acids are more active than the acids that do not contain halogen. Thus, the activity of 2-chloro-4-methoxyphenoxyacetic acid with respect to stimulating root formation on bean cuttings approaches that of 3-indolyl-γ-butyric acid.

For weed control a mixture of aryloxyacetic acids has been proposed which is obtained from phenols with b.p. 180°–200° C. that are separated from products of dry distillation of wood. This mixture contains up to 40% chloroguaiacolacetic acids (a mixture of isomers) and about 25% 2,4-D and MCPA.

When hydrogen of phenoxyacetic acid is replaced by nitro, amino, acylamido, alkylamino, and sulfo groups, compounds of low physiological activity are formed. Even 2-nitro-4-chlorophenoxyacetic acid and its analogs have little activity.

Introduction of halogen into the methylene group of 2,4-dichlorophenoxyacetic acid lowers the physiological activity of the compound. Thus, 2,4-dichlorophenoxyfluoroacetic acid is less active than 2,4-D, and the (+)-form of the 2,4-dichlorophenoxyfluoroacetic acid stimulates the growth of cuttings of pea epicotyl more strongly than the (−)-form or the racemic mixture. The (+)-form also is more active as a herbicide. The activity of 2,4-dichlorophenoxydifluoroacetic acid is lower than that of the monofluoro derivatives.

Replacement of hydrogen in the methylene group by a hydrocarbon radical as a rule decreases the activity of the compound, and the larger the number of carbon atoms in the hydrocarbon radical the greater the decrease in activity. The (+)-forms of the stereoisomeric compounds are more active than the racemates; the (−)-forms either are low in activity or are antagonists for plant growth stimulators.

The acid chlorides and anhydrides of the aryloxyacetic acids do not differ in activity from the acids, since on hydrolysis by water they go over to the corresponding acids.

The activity of the amides, anilides, and other similar derivatives in most cases is close to the activity of the corresponding acids, but there have been examples of significant deviations from this behavior.

The aryloxyacetylamino acids have high activity, which depends not only on the structure of the aryloxyacetic acid, but also on the configuration of the starting amino acid. It is possible that this is associated with the ability of various plant enzymes to cleave the aryloxyacetylamino acids to the corresponding aryloxyacetic acids.

The esters of the aryloxyacetic acids more powerful herbicides than the free acids or their salts. The increased activity of the esters of these acids in comparison with the salts and the free acids usually is explained by their higher rates of penetration through the cuticle of the leaves. However, the fact that it requires a considerably smaller dose of the ester than of the salt (almost 2–3 times) to achieve the same effect cannot satisfactorily be explained in this way. The herbicidal activity of the esters of high-molecular alcohols differs from that of the esters of the low-molecular alcohols; this fact in all probability is associated with their lower solubility in lipoids and waxes and their lower rate of diffusion into the plants.

The activity of the salts of the aryloxyacetic acids with various organic bases is 1.2–1.7 times higher than the activity of the salts of these acids with the alkali metals. It is possible that the increase in herbicidal effects of the amine salts results from a change in surface tension of their aqueous solutions. However, this apparently is only one of many causes for their stronger effect. The other factors that determine the heightened activity of the amine salts have not yet been established. Individual cases are known of higher herbicidal activity of oil solutions of the aryloxyacetic acids in comparison with the amine salts and even the esters of these acids.

When hydroxyl oxygen of the aryloxyacetic acids is replaced by sulfur no substantial change occurs in the physiological activity of the compound, while replacement of ether oxygen by sulfur or other bivalent groups lowers the activity.

The activity of the aryloxyacetylhydroxamic acids and other analogous compounds approaches that of the aryloxyacetic acids.

The naphthoxyacetic acids are considerably less active than the phenoxyacetic acids. Of the two isomeric naphthoxyacetic acids the more active is 2-naphthoxyacetic acid. Introduction of halogens and other substituents into the naphthalene nucleus decreases the activity of the compound.

The fluorenoxy-, acenaphthenoxy-, and other polynuclear aryloxyacetic acids are weak herbicides and therefore not of practical interest.

The aryloxyalkylcarboxylic acids containing an uneven number of methylene groups between the ether oxygen and the carbonyl group have herbicidal activity toward many dicotyledonous plants, while the compounds containing an even number of methylene groups between the ether oxygen and the carbonyl do not show a herbicidal effect. This is so because in most plants (except legumes) the aryloxyalkylcarboxylic acids undergo β-oxidation, as a result of which the first group of compounds is converted to the corresponding aryloxyacetic acids, and the second to inactive phenols:

$$ArOCH_2CH_2CH_2COOH + 6[O] \rightarrow ArOCH_2COOH + 2\ CO_2 + 2\ H_2O$$
$$ArOCH_2CH_2COOH + 6[O] \rightarrow ArOCOOH + 2\ CO_2 + 2\ H_2O$$
$$ArOCOOH \rightarrow ArOH + CO_2$$

Since the aryloxy-γ-butyric acids do not undergo β-oxidation to the corresponding aryloxyacetic acids in legume crops (this conversion does occur in other plants) these former acids are used successfully to control dicotyledonous weeds in legume planting. With an increase in the length of the chain between the ether oxygen and the carbonyl with an uneven number of methylene groups the herbicidal activity of the acids is weakened.

In spite of numerous investigations, the mechanism of action of the aryloxyacetic acids and their homologs has not been determined, and there are only isolated hypotheses that lack experimental confirmation.

Aryloxyacetic acids

The aryloxyacetic acids and their various derivatives are used as herbicides to control dicotyledonous weeds in grain fields and bushes and woody growth in meadows, and to eradicate undesirable trees in clearing sites for the construction of hydroelectric power stations, etc.

2,4-Dichlorophenoxyacetic acid (2,4-D) and its derivatives, 4-chloro-2-methylphenoxyacetic acid and its derivatives (MCPA), and 2,4,5-trichlorophenoxyacetic acid and its derivatives (2,4,5-T) are widely used. 3,4-Dichlorophenoxyacetic acid, 2,4-dichloro-5-methylphenoxyacetic acid, 4-chlorophenoxyacetic acid, and others are being used under pilot experimental conditions.

2-Naphthoxyacetic acid is a plant growth regulator. In some countries it is used in small amounts for the production of seedless tomato fruits.

All the aryloxyacetic acids are weak acids. Their dissociation constants range from 5.2×10^{-4} to 25×10^{-4}. The aryloxyacetic acids are completely analogous to other carboxylic acids in chemical properties. Their characteristics are the formation of acid chlorides and anhydrides, amides, esters, and many other derivatives.

When the acid chlorides of the aryloxyacetic acids react with silver phosphate, alkylation of the aromatic nucleus takes place:

$$6\ ArOCH_2COCl + 2\ Ag_3PO_4 \rightarrow 3\ ArOCH_2COOCH_2OAr + 6\ AgCl + P_2O_5 + 3\ CO$$

$$ArOCH_2COOCH_2OAr \rightarrow Ar\!\!<\!\!\begin{array}{l}OH\\CH_2COOCH_2OAr\end{array}$$

The reaction of the silver salts of the aryloxyacetic acids with bromine proceeds by approximately the same route:

$$2\ ArOCH_2COOAg + Br_2 \rightarrow ArOCH_2COOCH_2OAr + 2\ AgBr + CO_2$$

When aluminum chloride acts on the aryloxyacetyl chlorides cyclization occurs and the corresponding derivatives of coumaranone are obtained:

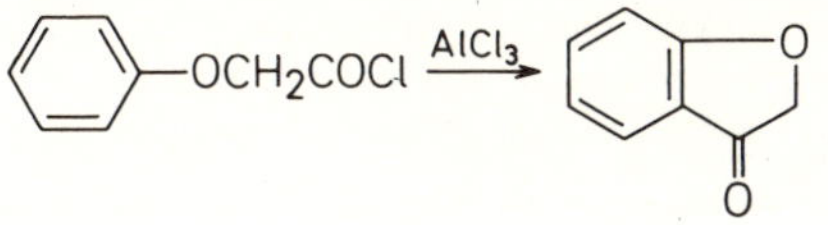

With aldehydes the aryloxyacetic acids enter into a Perkin reaction, forming the corresponding derivatives of cinnamic acid:

$$ArOCH_2COOK + Ar'CHO \rightarrow ArOC(=CHAr')COOK + H_2O$$

When concentrated nitric acid acts on the aryloxyacetic acids, depending on the nitration conditions and the structure of the starting acid, mono- or dinitroaryloxyacetic acids are formed.

Reaction of halogens with the aryloxyacetic acids gives the corresponding halogen derivatives, some of which are obtained quite pure and in good yields. This reaction often is used for the industrial synthesis of 2,4-D, MCPA, and 4-chlorophenoxyacetic acid. Chlorination and bromination proceed most readily; iodination is considerably more difficult. Some compounds are not even iodinated by free iodine.

Hydrolysis of the ether bond takes place comparatively easily when the aryloxyacetic acids are heated with the hydrohalogen acids. The reaction with HI goes rapidly, while that with HCl is considerably slower. The more acid the properties of the phenol obtained on hydrolysis, the more readily this reaction will proceed.

Two groups of general methods for the production of aryloxyacetic acids are known: from compounds containing an ether group, and by reactions in which the formation of an ether bond occurs. The first group of methods for synthesizing aryloxyacetic acids includes the following:

1. Oxidation of aryloxyethanols by various oxidizing agents or by oxygen in the presence of catalysts:

$$ArOCH_2CH_2OH + 2\,O \rightarrow ArOCH_2COOH + H_2O$$

By this reaction only low yields of aryloxyacetic acids are obtained. With catalytic oxidation in the presence of a platinum catalyst the yields are better, but the efficiency of the expensive catalyst is low.

2. Oxidation of unsaturated mixed ethers of phenols, for example aryl allyl ethers, aryloxy-2-chlorobutenes-2, and others:

$$ArOCH_2CH = CH_2 + 5\,O \rightarrow ArOCH_2COOH + CO_2 + H_2O$$

3. Synthesis of aryloxyacetic acids through the aryl chloromethyl ethers, which are obtained by chlorination of the corresponding anisoles in the presence of phosphorus pentachloride:

$$ArOCH_3 + Cl_2 \rightarrow ArOCH_2Cl + HCl$$
$$ArOCH_2Cl + NaCN \rightarrow ArOCH_2CN + NaCl$$
$$ArOCH_2CN + 2\,H_2O \rightarrow ArOCH_2COOH + NH_3$$

By this method 2,4-D is synthesized from anisole in a 60% yield.

The second group of methods for preparing aryloxyacetic acids finds more practical usage; it is employed in all present day processes for the industrial production of these herbicides and plant growth regulators:

1. One of the most widely used methods for the production of aryloxyacetic acids is the reaction of alkali phenolates with salts of monochloroacetic acid:

$$ArONa + ClCH_2COON \rightarrow ArOCH_2COON + NaCl$$
$$ArOCH_2COONa + HCl \rightarrow ArOCH_2COOH + NaCl$$

This reaction can be carried out both in aqueous solution and in organic solvents with the use of an excess of phenolate or in the presence of sodium chloride to decrease the hydrolysis of sodium monochloroacetate, etc. On the basis of a study of the kinetics of the reaction of different phenolates with sodium monochloroacetate it has been established that the optimum temperature for the process when it is carried out without pressure is 105° to 107° C.

2. The aryloxyacetic acids also can be prepared through the esters or amides of the haloacetic acids:

$$\text{ArONa} \xrightarrow[-\text{NaX}]{\text{XCH}_2\text{COOR}} \text{ArOCH}_2\text{COOR} \xrightarrow[-\text{ROH}]{\text{NaOH}}$$
$$\text{ArOCH}_2\text{COONa} \xrightarrow[-\text{NaCl}]{\text{HCl}} \text{ArOCH}_2\text{COOH}$$

$$\text{ArONa} \xrightarrow[-\text{NaCl}]{\text{ClCH}_2\text{CONH}_2} \text{ArOCH}_2\text{CONH}_2 \xrightarrow[-\text{NH}_4\text{HSO}_4]{\text{H}_2\text{O, H}_2\text{SO}_4} \text{ArOCH}_2\text{COOH}$$

This method has preparative importance only for the synthesis of aryloxyacetic acids under laboratory conditions.

3. It is possible to obtain aryloxyacetic acids by the reaction of phenolates with chlorocyanoacetic ester, but this method has only theoretical importance:

$$\text{ArONa} \xrightarrow[-\text{NaCl}]{\text{Cl(CN)CHCOOC}_2\text{H}_5} \text{ArOCH(CN)COOC}_2\text{H}_5 \xrightarrow[-\text{NH}_3,\,\text{C}_2\text{H}_5\text{OH}]{\text{NaOH, H}_2\text{O}}$$
$$\text{ArOCH(COONa)} \xrightarrow[-\text{NaCl, CO}_2]{\text{HCl}} \text{ArOCH}_2\text{COOH}$$

2,4-Dichlorophenoxyacetic acid (2,4-D) is a white crystalline substance, m.p. 141° C., b.p. 160° C. at 0.4 mm. of Hg. Pure 2,4-D has practically no odor; the technical grade compound smells more or less like dichlorophenol. According to international standards the permissible impurity in technical 2,4-D is not more than 0.3% dichlorophenol. At 20° C., 540 mg. of the acid dissolves in 1 liter of water; the solubility at 25° C. in 100 g. of various solvents is ethyl alcohol 130 g., ether 243 g., toluene 0.67 g., and *n*-heptane 0.11 g. 2,4-D is highly soluble also in benzene, carbon tetrachloride, acetone, and tetra- and pentachloroethanes. It is stable in storage both in solutions of different solvents and in the crystalline state. When it is irradiated with ultraviolet light slight decomposition may occur. The dissociation constant of 2,4-D is 23×10^{-4}.

With inorganic and organic bases it forms stable salts. Table XXXIII gives the properties of some salts of 2,4-D.

The salts of the bivalent metals are poorly soluble in water, and consequently when 2,4-D is dissolved in hard water precipitates may separate out. To avoid this, complexons (Trilon B) often are added to technical preparations of 2,4-D.

On prolonged boiling of 2,4-D with HBr or HCl decomposition may occur:

When 2,4-D is chlorinated with gaseous chlorine, 2,4,6-trichlorophenoxyacetic acid is formed, but at 200°–205° C. disintegration products also are found, including 2,4-dichlorophenol, bis(2,4-dichlorophenoxy)methane, and others. When the ethyl ester of 2,4-D is chlorinated at 195°–210° C.,

Table XXXIII. *Properties of salts of 2,4- dichlorophenoxyacetic acid*

Salt	M. P. (°C.)	Solubility in water	
		g./100 g. H_2O	Temp. (°C.)
Allylamine	106–107	1.2	31.5
Ammonium	—	3.5	20
Benzylamine	138–139	1.6	31.5
n-Butylamine	93–94.5	1.8	30.5
Diallylamine	—	710	32
Di-n-butylamine	107–109	1.2	31.5
Dimethylamine	85–87	Highly sol.	20
Diethanolamine	94–94.5	480	30
Diethylamine	129–131	> 50%	20
Potassium	—	7	20
Calcium	—	0.025	20
Magnesium	—	0.17	20
Methylamine	157–159	450	20
Monoethanolamine	145–147	> 50%	20
Morpholine	136–138	220	30
Sodium (monohydrate)	216–218	27.5	0
		33.5	20
		50.6	30
		74.59	45
Piperidine	131–132	230	31
Triethanolamine	—	440	32
Triethylamine	—	340	20

without a catalyst, the esters of 2,4-dichlorophenoxychloro- and dichloroacetic acids and of dichloroacetic acid, and 2,4-dichlorophenyl chloromethyl ether are obtained.

The main product of the nitration of 2,4-D by nitric acid or nitrating mixture is 2,4-dichloro-5-nitrophenoxyacetic acid with a small trace of 2,4-dichloro-6-nitrophenoxyacetic acid:

By the reaction of 2,4-dichlorophenoxyacetyl chloride with amino acids a large number of amides have been obtained that have high herbicidal activity:

$$\underset{\text{Cl}}{\text{Cl}}\text{−C}_6\text{H}_3\text{−OCH}_2\text{COCl} \xrightarrow{\text{H}_2\text{NRCOOH, KOH}}$$

$$\underset{\text{Cl}}{\text{Cl}}\text{−C}_6\text{H}_3\text{−OCH}_2\text{CONHRCOOH} + \text{KCl} + \text{H}_2\text{O}$$

Some species of microorganisms oxidize 2,4-D to 4-chloropyrocatechol (through the stage of formation of a hydroxy derivative and 2,4-dichlorophenol). 2,4-D and especially its esters with phenols act as disinfectants on microorganisms. It is a compound of medium toxicity to mammals. On peroral administration the LD_{50} for experimental animals is within the range 375–1,000 mg./kg. The maximum safe dose for monkeys is 214 mg./kg. It is assumed that for man the lethal dose of 2,4-D is about 15 g. In the United States the standard tolerance for 2,4-D residues in food products has been set at 5 mg./kg.

A large number of methods is known for the preparation of 2,4-D, but only two are of practical importance for industrial production:

1. Condensation of salts of monochloroacetic acid with 2,4-dichlorophenolates of the alkali metals or ammonia in aqueous or anhydrous medium:

$$\underset{\text{Cl}}{\text{Cl}}\text{−C}_6\text{H}_3\text{−ONa} \xrightarrow{\text{ClCH}_2\text{COONa}} \underset{\text{Cl}}{\text{Cl}}\text{−C}_6\text{H}_3\text{−OCH}_2\text{COONa}$$

To decrease hydrolysis of the monochloroacetic acid the reaction may be carried out with an excess of the 2,4-dichlorophenolate and the excess 2,4-dichlorophenol removed by steam distillation after acidification of the reaction medium. In this case the yield of 2,4-D is substantially increased. According to literature data, when 2 moles of sodium 2,4-dichlorophenolate are used per mole of sodium monochloroacetate the yield reaches 94% theoretical, while without the excess phenolate under similar conditions it does not exceed 83%. The 2,4-dichlorophenol distilled off with steam can be returned to the process. To suppress hydrolysis it also has been proposed that sodium chloride be added to the reaction mixture.

It should be noted that for the production of 2,4-D it is advantageous to use the purest possible 2,4-dichlorophenol in order to prevent the consumption of valuable monochloroacetate in the formation of by-products. Furthermore, when pure 2,4-dichlorophenol is used 2,4-D of higher quality is obtained.

Technical grade 2,4-D almost always contains a small amount of 2,4-dichlorophenol and as a result has an unpleasant odor.

2,4-Dichlorophenol is prepared by direct chlorination of phenol with chlorine, with subsequent purification of the chlorination products by fractional distillation. It is most often recommended that the chlorination be carried out at a temperature a little higher than the melting point of phenol. Chlorination at a higher temperature gives a large amount of 2,6-dichlorophenol as a by-product. It is possible to obtain 98% 2,4-dichlorophenol by chlorination of phenol in liquid sulfur dioxide. The chlorination is carried out at a temperature not higher than the boiling point of sulfur dioxide, which hinders the formation of the 2,6-isomer and also practically excludes the possibility of obtaining 2,4,6-trichlorophenol. 2,4-Dichlorophenol can be prepared also from a mixture of isomers of trichlorobenzene (obtained from the nontoxic isomers of benzene hexachloride) by alkaline hydrolysis, but the product formed in this process is a mixture of isomeric dichlorophenols containing not more than 25% of the necessary isomer.

2. Chlorination of phenoxyacetic acid or its esters. Chlorine, sodium hypochlorite, sodium chlorate with HCl, sulfury chloride, and chloramines can be used as chlorinating agents. The chlorination of phenoxyacetic acid with chlorine is carried out either in aqueous medium or in organic solvents or in a fusion:

$$\text{C}_6\text{H}_5\text{OCH}_2\text{COOH} \xrightarrow[-\text{HCl}]{2\text{Cl}_2} \text{2,4-Cl}_2\text{C}_6\text{H}_3\text{OCH}_2\text{COOH}$$

The positive side of this method is that the product obtained by the chlorination of phenoxyacetic acid does not have the characteristic odor peculiar to 2,4-dichlorophenol.

This 2,4-D contains as impurities 4-chlorophenoxyacetic acid, 2-chlorophenoxyacetic acid, 2,4,6-trichlorophenoxyacetic acid, and 2,4,6-trichlorophenol. The yield of 2,4-D calculated on the total chlorophenoxyacetic acids is over 90%, but the product obtained by this method is lower in quality than that prepared from purified 2,4-dichlorophenol.

The phenoxyacetic acid necessary for the production of 2,4-D by the chlorination method is obtained in 80–90% yield by the reaction of sodium phenolate with sodium monochloroacetate at elevated temperature:

$$\text{C}_6\text{H}_5\text{ONa} + \text{ClCH}_2\text{COONa} \rightarrow \text{C}_6\text{H}_5\text{OCH}_2\text{COONa} + \text{NaCl}$$
$$\text{C}_6\text{H}_5\text{OCH}_2\text{COONa} + \text{HCl} \rightarrow \text{C}_6\text{H}_5\text{OCH}_2\text{COOH} + \text{NaCl}$$

Esters of 2,4-dichlorophenoxyacetic acid. The effectiveness of the esters of 2,4-D in controlling various weeds is considerably higher than that of its salts and various other derivatives. The same effect in the eradication of dicotyledonous weeds is obtained when the esters of 2,4-D are used at dosages one-half to one-third (in terms of the acid) the dosage of the sodium salt. In a number of cases the esters are active on weeds against which the salts of 2,4-D have little effect. The esters of 2,4-D are being more and more widely used in agriculture. Their production constitutes more than 50% of the total production of all other derivatives of this acid.

The lower alkyl esters of 2,4-D (ethyl, isopropyl, butyl, etc.) are comparatively volatile and their vapors may damage crops sensitive to 2,4-D that are located next to the treated plots. With an increase in molecular weight the volatility of the esters decreases. Thus, the vapor pressure of the isooctyl ester of 2,4-D is almost 17 times lower than the vapor pressure of the butyl ester. However, the high molecular esters such as the cetyl ester are less effective, apparently because of their poor solubility in the plant juices and slow movement through the plant. Table XXXIV gives the properties of some esters of 2,4-D.

Table XXXIV. *Properties of some esters of 2,4-dichlorophenoxyacetic acid*

Formula of starting alcohol	M. P. (°C.)	B. P. (°C./mm. Hg)
CH_3OH	43	119/1
C_2H_5OH	15.2–15.4	149–150/1
$(CH_3)_2CHOH$	24	183/18
$CH_3(CH_2)_2CH_2OH$	9	146–147/1
$(CH_3)_2CHCH_2OH$	17	133–134/1
$CH_3(CH_2)_3CH_2OH$	15	160/2
$(CH_3)_2CHCH_2CH_2OH$	—	136–138/1
$CH_3(CH_2)_3CH(C_2H_5)CH_2OH$	12	173–174/0.5
$CH_3(CH_2)_6CH_2OH$	—	173–174/1
$CH_3(CH_2)_7CH_2OH$	43	—
$CH_3CCl=CHCH_2OH$	33–34	186–188/1
$2,4\text{-}Cl_2C_6H_3OCH_2CH_2OH$	88	—
⟨O⟩CH_2OH (furfuryl)	—	197–198/2

Of the large number of esters of 2,4-D that have been studied, practical use has been made of the ethyl, isopropyl, butyl, amyl, heptyl, octyl and isooctyl, chlorocrotyl, polypropylene- and polyethyleneglycol esters, and others. These esters of 2,4-D are produced commercially by esterification of the acid with the appropriate alcohols or by chlorination of the esters of phenoxyacetic acid. Esterification usually is carried out in the presence of acid catalysts and the water is distilled off as an azeotropic mixture with an organic solvent.

Chlorination of the esters of 2,4-D in the presence of catalysts for substitutive chlorination proceeds smoothly and gives a pure product. Iron plays a negative role in this reaction.

To prepare an amine salt the ester is treated with the theoretical amount of the appropriate amine in the presence of water:

$$Cl_2C_6H_3OCH_2COOCH_3 + H_2O + N(CH_3)_3 \rightarrow$$
$$Cl_2C_6H_3OCH_2COO^-\overset{+}{N}H(CH_3)_3 + CH_3OH$$

The chlorocrotyl ester of 2,4-D (Crotylin) is obtained by the reaction of dichlorobutene with a salt of 2,4-D:

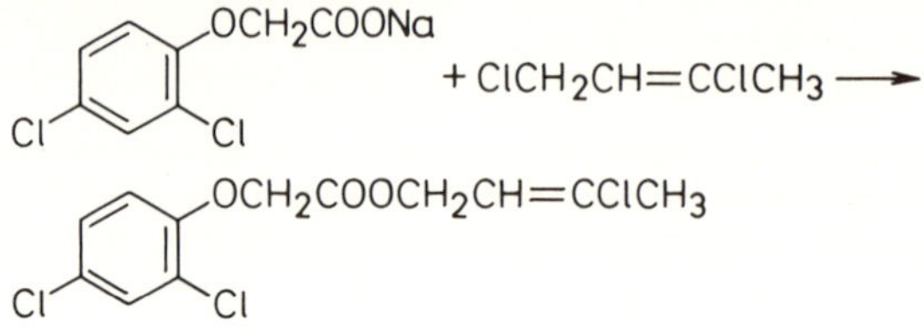

4-Chloro-2-methylphenoxyacetic acid (MCPA, 2M-4C) is a white crystalline substance, m.p. 120°–120.2° C. At 20° C., 0.063 g. of the acid dissolves in 100 ml. of water. It is highly soluble in alcohol, ether, carbon tetrachloride, dichloroethane, benzene, chlorobenzene, and other organic solvents. The dissociation constant, 5.4×10^{-4}, is lower than that of 2,4-D. Pure 4-chloro-2-methylphenoxyacetic acid has practically no odor; the technical grade compound, however, usually has the unpleasant odor of chlorocresol. MCPA is stable in storage both in solutions and in the crystalline state. Its toxicity to mammals, 590–1,000 mg./kg., is close to that of 2,4-D. The technical grade product containing related compounds as impurities is somewhat more toxic than the pure acid. Thus, Decotex has an LD_{50} of about 540 mg./kg. for white mice. The toxicity of MCPA for fish is 200 mg./l.

Table XXXV. *Solubility in water of some salts of 4-chloro-2-methylphenoxyacetic acid (MCPA)*

Salt	Solubility at 20° C. (%)	Salt	Solubility at 20° C. (%)
Ammonium	32	Calcium	0.55
Diethylamine	58	Magnesium	7.6
Diethanolamine	58	Sodium	25
Potassium	48	Triethanolamine	28

4-Chloro-2-methylphenoxyacetic acid is completely analogous to 2,4-D in chemical properties. The solubility of some salts of MCPA is given in Table XXXV.

The production of MCPA under industrial conditions, like that of 2,4-D, is possible by two routes: condensation of 4-chloro-2-methylphenolates with monochloroacetates and chlorination of 2-methylphenoxyacetic acid with various chlorinating agents.

1. The reaction of 2-chloro-4-methylphenol with monochloroacetates is carried out in alkaline medium (pH about 10) with heating of the reaction mixture to 103°–105° C.:

$$\text{Cl-C}_6\text{H}_3(\text{CH}_3)\text{-ONa} + \text{ClCH}_2\text{COONa} \longrightarrow \text{Cl-C}_6\text{H}_3(\text{CH}_3)\text{-OCH}_2\text{COONa} + \text{NaCl}$$

For the reaction it is possible to use either pure 4-chloro-2-methylphenol or the technical grade containing 6-chloro-2-methylphenol and 4,6-dichloro-2-methylphenol as impurities. However, in the latter case a highly contaminated product is obtained that requires special purification.

The 4-chloro-2-methylphenol necessary for the production of MCPA is obtained by chlorination of o-cresol with chlorine, alkali hypochlorites, or sulfuryl chloride. The purest 4-chloro-2-methylphenol is produced when the o-cresol is chlorinated with sulfuryl chloride:

$$\text{C}_6\text{H}_4(\text{CH}_3)\text{-OH} + \text{SO}_2\text{Cl}_2 \longrightarrow \text{Cl-C}_6\text{H}_3(\text{CH}_3)\text{-OH} + \text{SO}_2 + \text{HCl}$$

When o-cresol is chlorinated with chlorine and the other chlorinating agents, a mixture is obtained that contains 65–70% 4-chloro-2-methylphenol with 6-chloro-2-methylphenol and 4,6-dichlorophenol as impurities. To isolate pure 4-chloro-2-methylphenol, the technical product is subjected to fractional distillation.

6-Chloro-2-methylphenol is used for the production of disinfectants for nonmetallic materials, and also for the synthesis of ethers of polyethylene glycol which together with esters of MCPA are active herbicides for the control of weeds in plantings of grains that are resistant to the action of MCPA alone.

2. The second method for the preparation of MCPA, chlorination of 2-methylphenoxyacetic acid with chlorine or a hypochlorite, is of great interest. When 2-methylphenoxyacetic acid is chlorinated, it is predominantly the hydrogen atom in the 4-position of the benzene ring that is replaced. The process is most often carried out in dry organic solvents with or without catalysts. Chlorination in anhydrous solvents is more convenient because corrosion of the apparatus is greatly reduced and the compound obtained is almost free of impurities. Halogen derivatives of the aliphatic hydrocarbons, halogenated ethers, and mixtures of them have been proposed as solvents.

In addition to the principal product and a small amount of 6-chloro-2-methylphenoxyacetic acid, the ester acetal of 4,5,6-trichloro-2-methylcyclohexen-2-one-1 and glycolic acid are formed according to the following scheme:

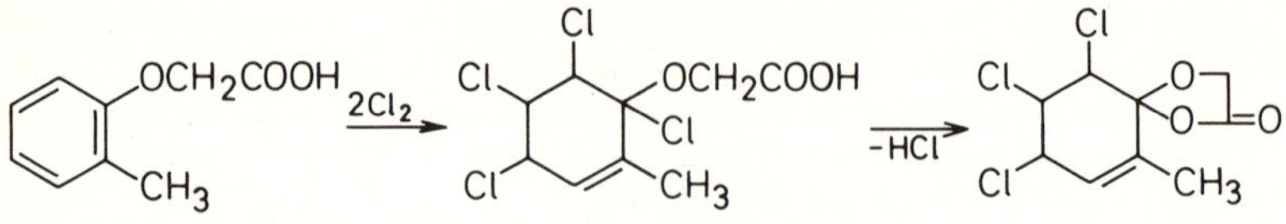

[Reaction scheme: 2-methylphenoxyacetic acid + 2Cl₂ → tetrachloro intermediate with OCH₂COOH and CH₃ groups → (−HCl) → ester acetal with chlorinated cyclohexene ring, CH₃ group, and cyclic O—CH—C(=O)—O acetal]

The ester acetal (m.p. 137° C.) on reaction with caustic alkalies goes over to 4,6-dichloro-2-methylphenol, and on reduction with zinc it goes to 2-methylphenoxyacetic acid. Some conversions of this compound are:

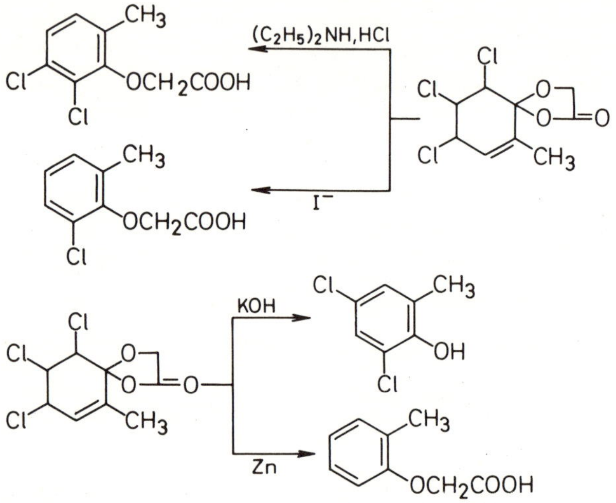

[Reaction scheme showing: 2,3-dichloro-6-methylphenoxyacetic acid ←(C₂H₅)₂NH,HCl— ester acetal intermediate; and 2-chloro-6-methylphenoxyacetic acid ←I⁻— ester acetal intermediate]

[Reaction scheme showing: ester acetal → KOH → 4,6-dichloro-2-methylphenol; and ester acetal → Zn → 2-methylphenoxyacetic acid]

Depending on the solvent used, isolation of the MCPA can be carried out by two methods:

1. MCPA is extracted from hydrophobic solvents by caustic alkalies; after the extraction the solvent is purified and returned to the process, and the aqueous solution of alkali salts of MCPA after appropriate purification is distributed as a commercial product or is evaporated to dryness, or, finally, is acidified with mineral acids to obtain the free acid, which is further reworked into the necessary formulations.

2. When chlorination is carried out in hydrophobic solvents they are distilled off in vacuum, and the acid after suitable purification is used for the preparation of commercial formulations.

Chlorination of 2-methylphenoxyacetic acid is carried out also in aqueous solution or in a fusion. In aqueous solution usually an alkali salt of the 2-methylphenoxyacetic acid is chlorinated:

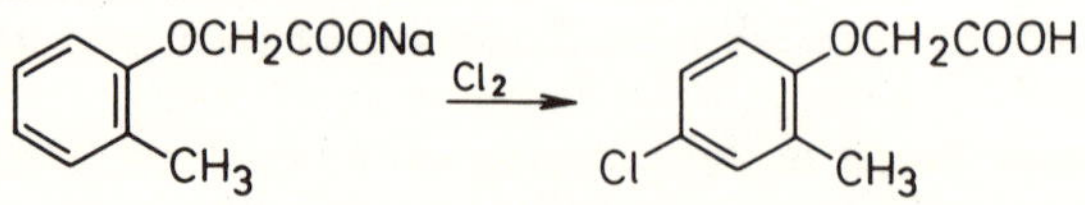

[Reaction: sodium 2-methylphenoxyacetate + Cl₂ → 4-chloro-2-methylphenoxyacetic acid (MCPA)]

The esters of MCPA are more effective herbicides than the salts, and therefore a considerable quantity of this herbicide is put out in the form of esters with various alcohols, the preparation of which is completely analogous to the preparation of the esters of 2,4-D. MCPA is close to 2,4-D in herbicidal activity, but it is safer for a number of crops. Because of this MCPA has been widely used in Europe. This herbicide is especially convenient for the control of weeds in flax and some other crops.

2,4,5-Trichlorophenoxyacetic acid (2,4,5-T) is a white crystalline substance, m.p. 158°–159° C.; at 20° C., 0.189 g. of the acid dissolves in 1 liter of water. It is highly soluble in alcohol, ether, chloroform, and benzene, and moderately soluble in toluene and paraffinic hydrocarbons. The LD_{50} for rats is 500 mg./kg.

The chemical properties of 2,4,5-T recall those of 2,4-D and MCPA. The properties of some salts of 2,4,5-T are given in Table XXXVI.

Table XXXVI. *Properties of some salts of 2,4,5-trichlorophenoxyacetic acid*

Salt	M. P. (°C.)	Sol. in H_2O at 20° C. (g./100 g. H_2O)
Diethanolamine	157–159	> 50
Monoethanolamine	179–181	> 50
Sodium	—	3.5
Trimethylamine	—	330
Triethylamine	104–106	400

The principal method for synthesizing 2,4,5-trichlorophenoxyacetic acid is the reaction of 2,4,5-trichlorophenolates with alkali monochloroacetates in concentrated aqueous solution at pH 10–12 and 103°–107° C.:

$$\text{Cl-C}_6\text{H}_2\text{Cl}_2\text{-ONa} + \text{ClCH}_2\text{COONa} \longrightarrow \text{Cl-C}_6\text{H}_2\text{Cl}_2\text{-OCH}_2\text{COONa} + \text{NaCl}$$

There are reports that it is possible to carry out the process in butyl alcohol with subsequent preparation of the butyl ester, but this method is less convenient because it is more difficult to free the 2,4,5-T of the trichlorophenol that has not reacted with the monochloroacetate.

The esters of 2,4,5-T are produced when the acid is esterified with the appropiate alcohols and the water is distilled off as an azeotropic mixture with organic solvents. In Table XXXVII the properties of some esters of 2,4,5-T are given.

2,4,5-T is used in the form of its esters as a herbicide for the control of undesirable woody growth in clearing pastures and for the destruction of trees in building hydroelectric stations on submerged areas. As a plant growth stimulator (for obtaining parthenocarpic tomato fruits) 2,4,5-T is used in the form of the sodium salt.

In addition to the aryloxyacetic acids described, a large number of other compounds of this class have been synthesized and studied which also are well-known for their practical interest. The properties of some of these compounds are presented in Table XXXVIII.

Table XXXVII. *Properties of some esters of 2,4,5-trichlorophenoxyacetic acid*

Formula of alcohol	M. P. (°C.)	B. P. (°C./mm. Hg)
C_2H_5OH	66–67.5	—
$(CH_3)_2CHOH$	46	—
$CH_3(CH_2)_2CH_2OH$	29	—
$CH_3(CH_2)_3CH_2OH$	15	—
$CH_3(CH_2)_4CH_2OH$	26	—
$CH_3(CH_2)_6CH_2OH$	—	187–189/0.3
$CH_3(CH_2)_7CH_2OH$	43	—
$CH_3OCH_2CH_2OH$	—	145–150/1
$CH_3OCH(CH_3)CH_2OH$	—	148–152/1
C_6H_5OH	116–117	—
C_6Cl_5OH	269–270	—

Table XXXVIII. *Properties of some aryloxyacetic acids*

Acid	M. P. (°C.)	Acid	M. P. (°C.)
4-Chlorophenoxyacetic	159–160	4-Fluorophenoxyacetic	117
3,4-Dichlorophenoxyacetic	156	2,4-Dichloro-5-methyl-	
2,5-Dichlorophenoxyacetic	157	phenoxyacetic	143
2,3,4-Trichlorophenoxy-		2-Naphthoxyacetic	155–156
acetic	141		

Aryloxypropionic acids

As already indicated, the β-aryloxypropionic acids do not have pesticidal properties and, therefore, are not used in agriculture. The aryloxy-α-propionic acids are active herbicides and plant growth regulators.

4-Chloro-2-methylphenoxy-α-propionic acid (mecoprop, MCPP, 2 M-4 CP) is a white crystalline substance, m.p. 94°–95° C. At 20° C., 100 ml. of water dissolves 0.062 g. of the acid, 58 g. of its diethanolamine salt, and at 26° C. 26 g. of the sodium salt, and 4.5 g. of the magnesium salt. The potassium salt of mecoprop is highly soluble and at room temperature gives 32% solutions calculated as the acid. Free mecoprop and its different salts are stable in storage. In other reactions it behaves completely like MCPA. By crystallization of its salts with optically active bases the dextro and levo acids are separated. Only the dextro form, which has the same configuration as *d*-lactic acid, is active. The LD_{50} for mice is 650 mg./kg. For weed control mecoprop is used in the form of its alkali or amine salts (ethanolamine, diethylamine) at dosages of 2–2.5 kg./ha. It is one of the new herbicides used in controlling of bedstraw and chickweed in grain fields.

The synthesis of 4-chloro-2-methylphenoxy-α-propionic acid is accomplished by condensation of salts of α-chloropropionic acid and 4-chloro-2-methylphenol:

$$Cl-\underset{CH_3}{\underset{|}{C_6H_3}}-ONa + \underset{CH_3}{\underset{|}{CHCl}}COONa \longrightarrow NaCl + Cl-\underset{CH_3}{\underset{|}{C_6H_3}}-O\underset{CH_3}{\underset{|}{CH}}COONa$$

2,4,5-Trichlorophenoxy-α-propionic acid (2,4,5-TP, silvex) is a white crystalline substance, m.p. 179°–181° C. At 25° C., 100 ml. of water dissolves 0.014 g. of the acid; its salts with the alkali metals and amines are considerably more soluble. The LD_{50} is 650 mg./kg. Silvex is close to 2,4,5-T in herbicidal activity, but it is safer for cotton, thereby permitting its use for the control of brushwood growth in cotton-growing regions. When it is used to spray pears in the autumn, it gives some increase in yield the following year.

2,4,5-Trichlorophenoxy-α-propionic acid is produced in a manner similar to 2,4,5-T by the reaction of sodium 2,4,5-trichlorophenolate with sodium α-monochloropropionate in aqueous medium. Other aryloxy α-propionic acids also have been synthesized and studied, such as 2,4-dichlorophenoxy-α-propionic (m.p. 116°–118° C.), 3-chloro-2-methylphenoxy-α-propionic (m.p. 131°–132° C.), 4-chloro-2,3-dimethylphenoxy-α-propionic (m.p. 136.5°–137.5° C.), and others. Some of them have proved to be rather active selective herbicides, but they have not yet been practically utilized in agriculture.

Aryloxy-γ-butyric acids

Of the aryloxy-γ-butyric acids only three are used in agriculture for weed control: 4-chloro-2-methylphenoxy-γ-butyric acid, 2,4-dichlorophenoxy-γ-butyric acid, and 2,4,5-trichlorophenoxy-γ-butyric acid.

4-Chloro-2-methylphenoxy-γ-butyric acid (MCPB, 2 M-4 CB) is a white crystalline substance, m.p. 100°–101° C. One liter of water dissolves 44 mg. of the compound; its solubility in ethyl alcohol is 15% and in acetone 20%. Its salts with the alkali metals and aliphatic amines are highly water-soluble. In most plants except legumes it is oxidized to MCPA:

$$Cl-\underset{CH_3}{C_6H_3}-OCH_2CH_2CH_2COOH \xrightarrow{6[O]} Cl-\underset{CH_3}{C_6H_3}-OCH_2COOH + 2H_2O + 2CO_2$$

This permits its use in controlling weeds in legumes and broad-beans at dosages of 2–3 kg./ha. To decrease the amount of the compound used, it is often employed in combination with MCPA.

Under the influence of phosphoric acid or phosphoric anhydride MCPB is cyclized with the formation of 7-chloro-9-methylhomochromanone (m.p. 68° C.):

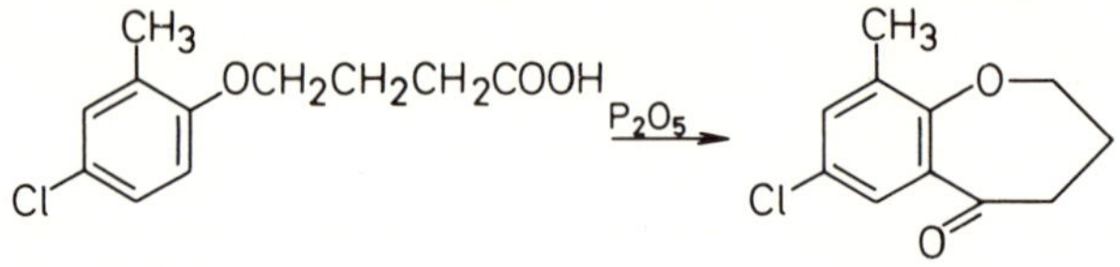

The preparation of MCPB can be carried out by three methods:
1. Through 1,3-chlorobromopropane or 1,3-dibromopropane:

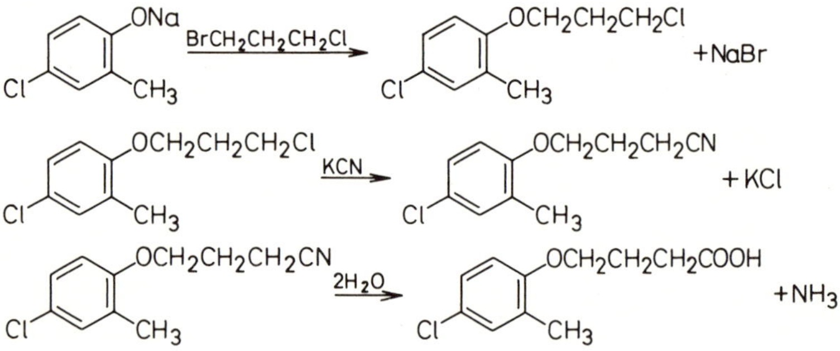

2. By the reaction of sodium 4-chloro-2-methylphenolate with butyrolactone or γ-chlorobutyric acid:

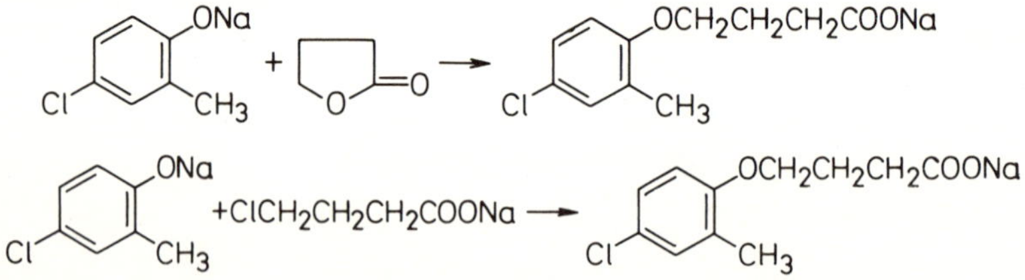

3. By the chlorination of 2-methylphenoxy-γ-butyric acid in aqueous medium or in a hydrophobic organic solvent:

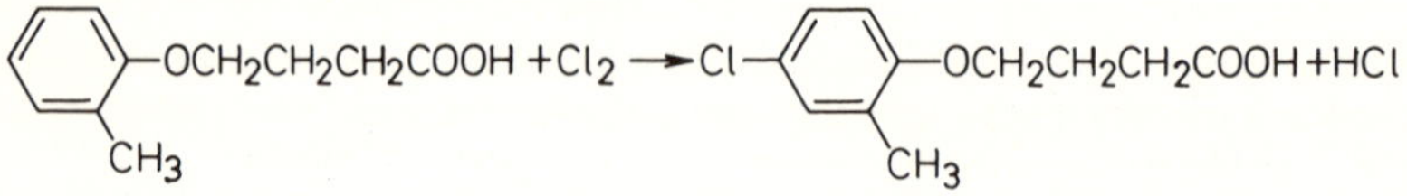

The 2-methylphenoxybutyric acid necessary for this synthesis is easily prepared from a phenolate and butyrolactone. The condensation of the phenolates with butyrolactone usually takes place at an elevated temperature with the use of solid anhydrous phenolate. The solvent may be either an excess of the butyrolactone (b.p. 208° C.) or chlorobenzene (b.p. 131° C.). The yields of the aryloxy-γ-butyric acids in this reaction are good in most cases (80–95%).

2,4-Dichlorophenoxy-γ-butyric acid (2,4-DB) is a white crystalline substance, m.p. 117°–119° C. At room temperature 1 liter of water dissolves 53 mg. of the compound; its solubility in acetone is 10%.

2,4-DB also is capable of undergoing β-oxidation in many plants and forming 2,4-D. However, its use is more limited, mainly to controlling weeds in alfalfa.

2,4,5-Trichlorophenoxy-γ-butyric acid (2,4,5-TB) is a crystalline substance, m.p. 114°–115° C., practically insoluble in water; its sodium salt at 25° C. forms a 20% solution in water.

Unlike the two aryloxy-γ-butyric acids described, 2,4,5-TB undergoes β-oxidation with great difficulty and therefore is not suitable in controlling weeds in legumes, but it has high activity against field bindweed and calystegia. The compound is recommended for the control of these weeds in plantings of black currants and some other crops.

The methods for the synthesis of 2,4-dichlorophenoxy-γ-butyric acid and 2,4,5-trichlorophenoxy-γ-butyric acid are completely analogous to the methods of preparing 4-chloro-2-methylphenoxy-γ-butyric acid.

Formulations of aryloxyalkylcarboxylic acids

The aryloxyalkylcarboxylic acids are used in the form of aqueous solutions of their salts with the alkali metals or aliphatic amines. To increase the effectiveness of the compounds, surface-active agents that improve the wetting of plants by herbicides are added to the aqueous solutions of the salts of the aryloxyalkylcarboxylic acids.

When herbicide solutions are used on crops that are resistant to the compound the addition of wetting agents is advantageous. On the other hand, the addition of wetting agents may cause suppression of the growth of useful crops. Thus, for example, it is not advisable to add detergents when using MCPA to control weeds in flax. However, when MCPA is used to control weeds in wheat the addition of surface-active agents increases the effectiveness of the compound without harming the treated crop.

A wide variety of materials may be used as wetting agents, including ethers of polyethylene glycol, arylalkylsulfonates of the alkali metals and unsubstituted and substituted ammonia, alkylsulfonates, and alkylsulfates, the choice depending on economic considerations and the composition of the preparation being produced. Besides the surface-active agents, usually complexons are added to prevent the precipitation of the calcium, magnesium, and iron salts of the aryloxyalkylcarboxylic acids in hard water. Most often substances of the type of Trilon B (ethylenediaminepolyacetic acids and their salts) are added.

The salts of the aryloxyalkylcarboxylic acids are most often marketed in the solid state or in the form of concentrated aqueous solutions containing 10 to 40% active ingredient calculated as the acid. Their salts with amines are marketed almost exclusively as aqueous solutions containing from 40 to 70% active ingredient.

MCPA and 2,4-D are used on grain crops in combination with 2,3,6-trichlorobenzoic acid or 3,6-dichloro-2-methoxybenzoic acid, which increase the range of action of the preparation. The content of the benzoic acids mentioned usually is from 6 to 10%.

An interesting use of the aryloxyalkylcarboxylic acids is in the form of colloidal solutions, which are fine dispersions of the free acids in water or of aqueous solutions of these acids in hydrophobic organic solvents. According to some data, the effectiveness of such dispersions is even higher than that of solutions of the amine salts.

The esters of the aryloxyacetic acids, as indicated, considerably surpass various salts in effectiveness. They are marketed in the form of emulsive concentrates and solutions in oils for aerial spraying. The concentrates contain, in addition to the active ingredient, a solvent (most often of petroleum or coal-tar origin) and surface-active agents. A mixture of alkylaryl ethers of polyethylene glycol (of the type of OP-7) with a calcium alkylarylsulfonate is used as a detergent. The calcium alkylarylsulfonate is highly soluble in organic solvents, permitting its use in concentrates with a hydrophobic solvent and the active ingredient.

For nonselective eradication of vegetation, preparations are used that contain a combination of several active herbicides from different classes of organic compounds.

The derivatives of the aryloxyacetic acids are used in the form of granulated preparations to control aquatic vegetation and weeds in corn. Preparations for application to the soil contain the amides of aryloxyacetic acids, particularly 2,4-D.

General references

BASKAKOV, YU. A.: Khim. sredstva zashchity rastenii No. 3, 42 (1959).
CRAFTS, A. S.: The chemistry and mode of action of herbicides. New York-London: Interscience 1961.
—, and U. ROBBINS: Khimicheskaya bor'ba s sornyakami [Chemical control of weeds]. Publishing House "Kolos" (1964).
KURT, H.: Chemische Unkrautbekämpfung. Jena (1963).
LUND, H.: Acta Chem. Scand. 12, 793–796 (1958).
MEL'NIKOV, N. N., and YU. A. BASKAKOV: Khimiya gerbitsidov i regulyatorov rosta rastenii [Chemistry of herbicides and plant growth regulators]. State Scientific and Technical Publishing House of Chemical Literature [USSR] (1962).
— — Regulyatory rosta rastenii v sel'skom khozyaistve [Plant growth regulators in agriculture]. Foreign Literature Publishing House [USSR] (1958).
Weed Control Handbook, Vols. I and II. Oxford: Blackwell (1968).
ZBIROVSKY, M., J. MYSKA, and J. ZEMANEK: Herbicidy. Prague (1960).

XV. Derivatives of carbonic acid

General characteristics of pesticidal properties

Esters of carbonic and chlorocarbonic acids as yet have not been practically used in controlling harmful insects, although some of them have high insecticidal or acaricidal activity. In particular, the simplest esters of chlorocarbonic acid such as methyl, ethyl, and isopropyl chlorocarbonates are active fumigants; but they are not used because they are easily hydrolyzed by water, forming HCl. The insecticidal activity of the dialkyl carbonates is slight. The aliphatic-aromatic esters of carbonic acid are more active. Thus, mixed esters of the general formula (I) have powerful acaricidal and

(R = CH_3, CH_3CH_2, C_3H_7, C_4H_9;   R' = *sec*-C_4H_9,   *sec*-C_5H_{11})

fungicidal effects; they suppress the development of powdery mildews. It is possible that the biological activity of this group of compounds should be ascribed to the radicals of dinitro-*sec*-butyl- and dinitro-*sec*-amylphenol, esters which even with other acids show high activity as acaricides and fungicides for powdery mildews.

The acaricidal and fungicidal activity increases going from the mixed esters of carbonic acid to the mixed esters of thio- and dithiocarbonic acids. Compounds of the general formulas (II) and (III) are active fungicides, but compounds of structure (IV) are almost completely inactive.

(R = CH_3, C_2H_5, C_3H_7, C_4H_9;   X = S, O;   Y = Cl, Br, NO_2, CH_3, CH_3O)

Bis(ethylxanthogen) disulfide is an active insecticide for the control of pediculosis and was widely used during World War II.

Phytocidal properties have been discovered in those esters of carbonic acid that on hydrolysis yield herbicidal alcohols or phenols. Thus, for example, herbicidal properties are shown by the carbonates of aryloxyethanols, pentachlorophenol, dinitroalkylphenols, and also by aliphatic-aromatic mixed esters containing physiologically active ester radicals (2-nitro-3,4,6-trichlorophenyl, pentachlorophenyl, etc.). However, a practical use has not yet been found for these compounds.

The alkali metal xanthates and their oxidations products — bis(alkylxanthogen) disulfides and trisulfides — are more active and are used as defoliants, desiccants, and herbicides. A combination of pesticidal properties is shown by the simplest thio derivatives of carbonic acid, carbon sulfoxide and carbon disulfide, which still finds some use as a fumigant.

Esters of carbonic acid

The full esters of carbonic acid are made by the action of phosgene on alcohols or phenols in the presence of inorganic or organic bases. For the synthesis of dialkyl carbonates the alkali metal carbonates can be used as bases, while the aromatic esters of carbonic acid are obtained in better yields from phosgene and the phenolates of the alkali metals:

$$2\ C_2H_5OH + COCl_2 \xrightarrow{CaCO_3} (C_2H_5O)_2CO + 2\ HCl$$
$$2\ C_6H_5OK + COCl_2 \rightarrow (C_6H_5O)_2CO + 2\ KCl$$

In this process the first chlorine of the phosgene reacts the most easily to form esters of chlorocarbonic acid; replacement of the second chlorine takes place under more severe conditions.

It also is possible to prepare esters of carbonic acid from disphogene:

$$CCl_3OCOCl + 4\ C_6H_5OK \rightarrow [CCl_3OCOOC_6H_5 + KCl] \rightarrow$$
$$2\ (C_6H_5O)_2CO + 4\ KCl$$

Mixed esters of carbonic acid are synthesized in two steps through the alkyl or aryl chlorocarbonate:

$$C_2H_5OH + COCl_2 \rightarrow C_2H_5OCOCl + HCl$$
$$C_2H_5OCOCl + C_6H_5OK \rightarrow C_2H_5OCOOC_6H_5 + KCl$$

Derivatives of thio- and dithiocarbonic acids

Derivatives of thio- and dithiocarbonic acids are of practical interest because among them are found effective acaricides, fungicides, defoliants, desiccants, and herbicides.

Three types of mixed esters of thiocarbonic acid are known that have pesticidal properties (V, VI, VII):

$$\underset{ArSCOR}{\overset{O}{\|}} \quad (V) \qquad \underset{ArOCSR}{\overset{O}{\|}} \quad (VI) \qquad \underset{ArOCOR}{\overset{S}{\|}} \quad (VII)$$

Compounds of type (V) can be prepared from the appropriate thiophenol and alkyl chlorocarbonate or from an arylthiol chlorocarbonate and alcohol:

1. $ArSK + ROCOCl \rightarrow ArSCOOR + KCl$
2. $ArSK + COCl_2 \rightarrow ArSCOCl + KCl$
   $ArSCOCl + ROH \rightarrow ArSCOOR + HCl$

The synthesis of compounds of type (VI) starts from either the aryl chlorocarbonate or the aryl thiochlorocarbonate:

$$\left. \begin{array}{l} ArOCOCl + RSK \\ RSCOCl + ArOK \end{array} \right\} \rightarrow ArO\overset{\overset{O}{\|}}{C}SR$$

The second route in this case gives a somewhat higher yield of the product.

Alkylaryl thionocarbonates of type (VII) are prepared from thiophosgene by the following scheme:

$$ROH + CSCl_2 + KOH \rightarrow RO\overset{\overset{S}{\|}}{C}Cl + KCl + H_2O$$

$$RO\overset{\overset{S}{\|}}{C}Cl + ArOK \rightarrow RO\overset{\overset{S}{\|}}{C}OAr + KCl$$

The simplest derivatives of dithiocarbonic acid are the alkali alkylxanthates, of which sodium ethylxanthate is used for the desiccation of cotton and some other crops, and sodium isopropylxanthate is used as a herbicide. The alkylxanthates of the alkali metals are obtained by the reaction of carbon disulfide with alcohols in the presence of caustic alkalies:

$$ROH + CS_2 + NaOH \rightarrow RO\overset{\overset{S}{\|}}{C}SNa + H_2O$$

This reaction proceeds very readily, and the alkylxanthates are formed in quantitative yields.

The synthesis of mixed esters of dithiocarbonic acid is more complicated. Thus, compounds of the general formula (VIII) are

$$ArS\overset{\overset{S}{\|}}{C}OR \qquad \text{(VIII)}$$

obtained by reaction of alkali alkylxanthates with aromatic diazo compounds:

$$ArN_2^+Cl^- + RO\overset{\overset{S}{\|}}{C}SNA \rightarrow ArS\overset{\overset{S}{\|}}{C}OR + N_2 + NaCl$$

To prepare mixed aliphatic esters of dithiocarbonic acid, the alkyl halides are heated with the appropriate alkali alkylxanthates in equimolecular quantities in some organic solvent:

$$ROC(=S)SNa + R'Cl \rightarrow ROC(=S)SR' + NaCl$$

Sodium isopropylxanthate is a light yellow crystalline substance that decomposes on heating below its melting point. At 4° C. the solubility of the compound in water is 30%, at 24° C. 46%, and at 35° C. 54%. Sodium isopropylxanthate is most often used in the form of a 1% aqueous solution as a herbicide to control weeds in cabbage and some other crops. It successfully suppresses such weeds as purslane, goosefoot, bedstraw, and ragwort. It gives good results at dosages of 10–12 kg./ha.

Sodium isopropylxanthate is produced from NaOH, carbon disulfide, and isopropyl alcohol. First, a fine dispersion of the alkali is prepared from a 25% aqueous solution of NaOH and the isopropyl alcohol, and then an equimolecular quantity of carbon disulfide is added:

$$(CH_3)_2CHOH + NaOH + CS_2 \rightarrow (CH_3)_2CHOC(=S)SNa + H_2O$$

The yield of 90% pure sodium isopropylxanthate by this reaction is more than 90%.

The synthesis of sodium ethylxanthate is carried out under similar conditions; this compound can be used at dosages up to 15 kg./ha. as a desiccant for various crops.

Also proposed as herbicides are the xanthates of aryloxyethanols, of which the most active are 4-chloro-2-methylphenoxyethanol and 2,4-dichlorophenoxyethanol. These compounds are prepared by the reaction of equimolecular quantities of the aryloxyethanol, carbon disulfide, and NaOH in some inert solvent, the best being toluene or ligroin.

Bis(methylxanthogen) disulfide (dimethylxanthogen disulfide) is a crystalline substance, m.p. 22.5°–23° C. It is slightly soluble in water but highly soluble in organic solvents. Its LD_{50} is 240 mg./kg. Used as a pre-emergence herbicide and fumigant, at dosages of 8–10 kg./ha. this compound kills many weeds at the time of seed germination. However, it is equally toxic for some crop plants; therefore, precautionary measures should be observed when using it. It usually is employed in the form of emulsive concentrates.

Bis(methylxanthogen) disulfide is produced by the oxidation of sodium methylxanthate with various oxidizing agents in aqueous medium:

$$2\ CH_3OC(=S)SNa + O + 2\ HCl \rightarrow (CH_3OC(=S))_2S_2 + 2\ NaCl + H_2O$$

Sodium nitrite (with acid), hydrogen peroxide, chlorine, hypochlorites, etc., can be used as oxidizing agents.

Bis(ethylxanthogen) disulfide (Herbisan) is a crystalline substance with an unpleasant odor, m.p. 27°–29° C., LD_{50} 600 mg./kg. In addition to its herbicidal effect on weeds, it also is an insecticide and has been used to control pediculosis. On prolonged contact with the skin it may cause strong irritation. In the United States it is used to control weeds in onions. As a herbicide it is employed at dosages of 8–11 kg./ha., in the form of aqueous emulsions containing a solvent and emulsifier.

Bis(ethylxanthogen) disulfide is produced by the oxidation of sodium ethylxanthate with various oxidizing agents, for example sodium nitrite.

Other bis(alkylxanthogen) disulfides also have a herbicidal effect, although a practical use has not yet been found for them.

Bis(ethylxanthogen) trisulfide is a dark-yellow liquid that distills with decomposition even in a high vaccum. Its LD_{50} is about 500 mg./kg. At dosages up to 4–5 kg./ha., it gives satisfactory results in the desiccation of cotton and some other crops.

It is produced by the reaction of dry sodium or potassium xanthate with sulfur dichloride:

$$2\ C_2H_5O\overset{\overset{\displaystyle S}{\|}}{C}SNa + SCl_2 \rightarrow (C_2H_5O\overset{\overset{\displaystyle S}{\|}}{C})_2S_3 + 2\ NaCl$$

or by prolonged heating of bis(ethylxanthogen) disulfide with sulfur:

$$(C_2H_5O\overset{\overset{\displaystyle S}{\|}}{C})_2S_2 + S \rightarrow (C_2H_5O\overset{\overset{\displaystyle S}{\|}}{C})_2S_3$$

Similarly, the reaction of an alkali ethylxanthate with sulfur monochloride produces bis(ethylxanthogen) tetrasulfide, which is close to the trisulfide in physiological activity toward plants:

$$2\ C_2H_5O\overset{\overset{\displaystyle S}{\|}}{C}SNa + S_2Cl_2 \rightarrow (C_2H_5O\overset{\overset{\displaystyle S}{\|}}{C})_2S_4 + 2\ NaCl$$

Carbon disulfide. The simplest derivative of dithiocarbonic acid is its anhydride, carbon disulfide. In the pure form carbon disulfide is a colorless liquid with a weak odor reminiscent of the halogen derivatives of aliphatic hydrocarbons, b.p. 46.2° C., m.p. −108.6° C. At 20° C. 100 g. of water dissolves 0.217 g. of carbon disulfide. It is very explosive when mixed with air. Its spontaneous ignition temperature is 120° C.; the limits of its explosive concentrations when mixed with air are 25 to 1,680 g./m.³, or 0.8 to 53% by volume. The maximum permissible concentration in air is 10 mg. per m.³ Because of its high flammability and explosive hazard, it is used as a fumigant for households and other purposes mainly in a mixture with carbon tetrachloride (3 parts by weight of carbon tetrachloride must be used to 1 part by weight of carbon disulfide). For the treatment of soil to control insect pests or weeds, carbon disulfide is used at dosages from 120 to 500 g./m.² of soil. At present carbon disulfide has only limited use in agriculture.

It is produced either from methane and sulfur or from sulfur and carbon at an elevated temperature:

$$CH_4 + 4 S \rightarrow 2 H_2S + CS_2$$
$$C + 2 S \rightarrow CS_2$$

It is a basic raw material for the production of the many active pesticides that are derivatives of dithiocarbamic acid.

6-Methyl-2,3-quinoxaline dithiocarbonate (Morestan, chinomethionate, quinomethionate) is a yellow crystalline substance, m.p. 172° C. Its LD_{50} is 2,500 mg./kg. It is poorly soluble in water and moderately soluble in most organic solvents. It is used in the form of a wettable powder against the eggs of plant-feeding mites and against true powdery mildew at a concentration of 0.05–0.1% active ingredient. It is synthesized by the reaction of phosgene with 2,3-dimercaptoquinoxaline in the presence of bases to bind the HCl:

Quinoxaline-2,3-trithiocarbonate (Eradex, Bayer 30 686). Active pesticides are found also among the derivatives of trithiocarbonic acid, as exemplified by Eradex. It is a brown substance, m.p. 180° C., vapor pressure at 20° C. 1×10^{-7} mm. of Hg. It is practically insoluble in water as well as most ordinary organic solvents, and poorly soluble in acetone and ethyl alcohol. Its LD_{50} for rats when orally administered is 3,400 mg./kg.; by intraperitoneal administration the LD_{50} is 230 mg./kg.

At room temperature Eradex is resistant to hydrolysis by water, but it is relatively easily oxidized by the oxygen of the air with the formation of S-oxides without loss of biological activity.

Used in the form of a wettable powder, Eradex is an active acaricide for the control of plant-feeding mites in the egg stage. It also holds promise for use in controlling true powdery mildew.

Eradex is produced by the reaction of 2,3-dimercaptoquinoxaline with thiophosgene:

Other cyclic trithiocarbonates also have been described, but they have not yet been used as pesticides.

General references

MEL'NIKOV, N. N.: Khim. v Sel'skom Khozyaistve No. 1, 16 (1966).
PIANKA, M., and C. SMITH: Chem. & Ind. (London), p. 1216 (1965).
Sasse, K.: Höfchen-Briefe 13, 197 (1960).
UNTERSTEINHOEFER, G.: Höfchen-Briefe 13, 207 (1960).

XVI. Derivatives of carbamic acid

General characteristics of pesticidal properties

Derivatives of carbamic acid are widely used as chemical agents for plant protection. The aryl esters of N-methylcarbamic acid are used to control harmful insects. The alkyl esters of N-arylcarbamic acids are powerful herbicides for the control of monocotyledonous weeds. Fungicidal compounds are also found among the derivatives of carbamic acid. The relationship of the biological activity of this group of compounds to their structure has been studied, and for some of the most important compounds their metabolism in living organisms has been investigated.

Only the esters of N-alkylcarbamic acids have insecticidal properties; the insecticidal activity of esters of N-arylcarbamic acids is slight. Replacement of the second hydrogen atom on the nitrogen by an alkyl radical sharply decreases the insecticidal activity. The only exceptions are esters of N,N-dimethylcarbamic acid with complex heterocyclic hydroxy compounds. The maximum insecticidal effect is shown by the aryl esters of N-methylcarbamic acid. Replacement of the methyl group by radicals with a larger number of carbon atoms considerably lowers the activity of the compounds. Replacement of the carbonyl or ester oxygen by sulfur also decreases the insecticidal activity.

All the insecticidal esters of N-alkylcarbamic acids cause inhibition of cholinesterase. The monosubstituted phenyl N-methylcarbamates depress the cholinesterase activity more strongly as the resistance of the compound to hydrolysis increases. *Meta-* and *para-*substituted phenyl esters of N-methylcarbamic acid fall in the following order with respect to their inhibition of fly brain cholinesterase:

$$(CH_3)_3N > \text{tert-}C_4H_9 > (CH_3)_2N > \text{iso-}C_3H_7 > C_2H_5 > CH_3 > Cl > NO_2$$

The order of activity, with the exception of the trimethylammonium group, indicates that the anticholinesterase activity is a function of the ability of the substituent to give up electrons to the N-methylcarbamoyl part of the molecule. When the electron density around the carbonyl is increased, the molecule becomes more resistant to hydrolysis. The carbamates behave as competitive inhibitors of the cholinesterase.

The relationship of the insecticidal and anticholinesterase activity of the carbamates to their structure has a more complex nature than previously indicated, and it still has not been studied sufficiently. There is no doubt that, in addition to the hydrolytic stability of the carbamates, a certain arrangement of the molecule or of its parts with respect to the corresponding active centers of the esterase is necessary for high anticholinesterase activity. The carbamate molecule should fit the esterase molecule like a "key in a lock." Thus, for example, the *l*-isomers of O-*sec*-butylphenyl *N*-methylcarbamate and S-*sec*-butylphenyl *N*-methylcarbamate considerably surpass the corresponding *d*-isomers in insecticidal activity and anticholinesterase effect.

The connection of the insecticidal properties of the phenyl *N*-methylcarbamates containing two or more substituents in the aromatic radical with their structure has a considerably more complex character and the available experimental data still do not permit strict conclusions. In this group of compounds are found a number of very active substances such as 3,4,5-trimethylphenyl *N*-methylcarbamate and its homologs.

In the series of esters of *N*-methylcarbamic acid the greatest activity is shown by 1-naphthyl *N*-methylcarbamate, which is used in agriculture under the names of Sevin, carbaryl, and naphthyl carbamate. The corresponding 2-isomer has little activity. The esters of dihydro- and tetrahydro-1-naphthols show high insecticidal activity.

The keto oximes of esters of *N*-methylcarbamic acid have been proposed as active insecticides. Many esters of methylcarbamic acid are highly toxic to mammals.

All the esters of the arylcarbamic acids are selective herbicides for monocotyledonous plants. Their action on dicotyledonous plants is weaker, and they, therefore, can be used to control monocotyledonous weeds in plantings of such crops as sugar beets, cotton, and carrots. The sensitivity of different weeds to individual compounds varies within wide limits, and it therefore is possible to use them for the control of monocotyledonous weeds in grain fields. For example, 4-chlorobutyn-2-yl *N*-(*m*-chlorophenyl)carbamate is used to control wild oats in wheat. Usually the herbicidal effect of the esters of arylcarbamic acids is evident when they are applied to the soil prior to emergence of sprouts, but some compounds are sufficiently active also for the treatment of growing plants.

The esters of phenylcarbamic acid are active herbicides and, furthermore, different substituents may be present in the aromatic radical. Going from the esters of phenylcarbamic acid to the esters of naphthylcarbamic, diphenylcarbamic, and other polycyclocarbamic acids the herbicidal activity sharply decreases.

Replacing hydrogen of the phenyl radical in isopropyl phenylcarbamate by different functional groups in the *ortho*- and *para*-positions to the amino group sharply lowers the herbicidal activity. All the *para*-derivatives are practically inactive; the only exception is isopropyl *N*-(4-fluorophenyl)-carbamate, which is 20% less effective than isopropyl *N*-phenylcarbamate. The *ortho*-derivatives show weaker activity than isopropyl *N*-phenylcarbamate. It is interesting to note that *N*-(4-chlorophenyl)dimethylurea, which is close in structure to isopropyl *N*-(4-chlorophenyl)carbamate, has high activity and is used as a herbicide. This demonstrates the different mechanism of action on plants of even such structurally similar compounds, although all these substances disrupt the photosynthetic processes and inhibit the Hill reaction. Replacing hydrogen in the phenyl radical by substituents in the *meta*-position to the amino group of the first type increases the activity of the compounds, but substituents of the second type decrease it. Introduction of two substituents into the aromatic radical of isopropyl phenylcarbamate lowers the physiological activity. A similar picture is also observed when three substituents are introduced. A less significant decrease in activity occurs if even one substituent of the first type is in the *meta*-position to the amino group.

When functional groups are introduced into the phenyl radical of isopropyl phenylcarbamate the selective effect on monocotyledons is preserved. The change in herbicidal activity of esters of arylcarbamic acids resulting from the nature and structure of the ester radical has been studied in detail in the examples of the esters of phenyl- and *m*-chlorophenylcarbamic acids. The greatest activity is displayed by the esters of these acids that contain three or four carbon atoms in the ester radical. If the number of carbon atoms in the ester radical is more than five, the activity of the ester decreases.

The arylcarbamic acid esters that are derivatives of unsaturated alcohols as a rule have higher herbicidal activity than the corresponding esters of the saturated alcohols with the same number of carbon atoms. However, in these cases the selective effect of the compounds is somewhat altered. Aromatic esters of phenyl- and *m*-chlorophenylcarbamic acids are only slightly active.

Introduction of halogen into the aliphatic radical affects the activity of the compounds variously. Thus, the 2-chloroethyl ester of phenylcarbamic acid is more active than the ethyl ester of this acid, while the 2-chloroethyl ester of *m*-chlorophenylcarbamic acid is less active than the corresponding ethyl ester. Introduction of a nitro group into the ester radical lowers the herbicidal activity of the compound.

Replacement of hydrogen on the nitrogen of the esters of alkylcarbamic acids by other groups, depending on the nature of the substituent, decreases the activity of the compound to a greater or less degree. The only exception is replacement of hydrogen by hydroxyl, carboxyl, or carboxylic acid groups. The carboxylic compounds after partial hydrolysis are easily decarboxylated and go over to the esters of the arylcarbamic acid.

When oxygen is replaced by sulfur the phytotoxic effect of the compounds decreases, while when the carbonyl oxygen is replaced by an imino group, derivatives of isourea are formed that are active herbicides.

The isopropyl esters of phenyl- and *m*-chlorophenylcarbamic acids (IPC and chloro-IPC), the 4-chlorobutyn-2-yl ester of *m*-chlorophenylcarbamic acid, and some others are used as herbicides.

The herbicidal activity of the carbamates is associated with their ability to form hydrogen bonds with the chlorophyll molecule or with the proteins of enzymes that are implicated in the process of photosynthesis. However, the mechanism of action of the arylcarbamates apparently is not so simple, and the effect of these compounds on the growth of plants is a result of their disruption of several vitally important systems.

In addition to the study of the herbicidal activity of the esters of arylcarbamic acids, their effect in inhibiting sprouting of potatoes in prolonged storage has been investigated. Isopropyl 3-chlorophenylcarbamate has proved to be a good inhibitor of sprouting and is used for this purpose in the German Democratic Republic. Sprouting of potatoes is inhibited for several months when they are dusted with this compound at the rate of 10–30 g./ton.

The esters of arylcarbamic acids with ketoximes of the type ArNHCOON=CRR′ also have a herbicidal effect.

Aryl esters of alkylcarbamic acids

The most important method for producing the aryl esters of alkylcarbamic acids is the reaction of alkyl isocyanates with phenols or aromatic alcohols:

$$ArOH + RNCO \rightarrow ArOCONHR$$

This reaction takes place easily, producing quantitative yields at a comparatively low temperature. Tertiary amines usually are used as catalysts. The greatest difficulties in carrying out this reaction are presented by the preparation of the alkyl isocyanates, which are formed when amines react with phosgene at an elevated temperature:

$$RNH_2 + COCl_2 \rightarrow RNCO + 2\,HCl$$

However, the HCl that is evolved adds upon cooling to the isocyanate, yielding the alkylcarbamoyl chloride:

$$RNCO + HCl \rightleftarrows RNHCOCl$$

This reaction is reversible; depending on the temperature, an equilibrium is established that is characteristic of each alkyl isocyanate. The lower the boiling point of the alkyl isocyanate, the more the equilibrium is shifted to the right. To obtain the alkyl isocyanate it is necessary to remove the HCl from the reaction mixture. This can be accomplished, for example, by boiling a solution of the alkylcarbamoyl chloride in carbon tetrachloride or another suitable solvent in an apparatus of special construction that permits withdrawing the HCl from the sphere of the reaction. By this means even methyl isocyanate is obtained in satisfactory yield.

The alkyl isocyanates are obtained in good yields and a high state of purity by thermal decomposition of diarylalkylureas:

$$(Ar)_2NCONHR \rightarrow (Ar)_2NH + RNCO$$

The diarylalkylureas are prepared by the following reactions:

$$(Ar)_2NH + COCl \rightarrow (Ar)_2NCOCl + HCl$$
$$(Ar)_2NCOCl + 2\,RNH_2 \rightarrow (Ar)_2NCONHR + RNH_3Cl$$

The pure diarylamine produced by the pyrolysis of the diarylalkylurea can be returned to the process.

Another method of synthesizing the aryl esters of alkylcarbamic acids is the reaction of the appropriate phenols with the alkylcarbamoyl chlorides:

$$ArOH + RNHCOCl \rightarrow ArOCONHR + HCl$$

By this process the aryl N-alkylcarbamates are obtained in good yields but with a lower degree of purity than through the alkyl isocyanates.

It is possible to prepare the aryl N-alkylcarbamates also from the aryl carbonates and an amine:

$$ArOCOCl + RNH_2 + NaOH \rightarrow ArOCONHR + NaCl + H_2O$$

This reaction goes quite easily in the presence of HCl acceptors (alkali hydroxides, carbonates, organic amines).

The necessary aryl chlorocarbonates are formed in satisfactory yields by the reaction of phosgene with phenolates or of phosgene with phenol in the presence of tertiary amines:

$$ArOH + (C_2H_5)_3N + COCl_2 \rightarrow ArOCOCl + (C_2H_5)_3NHCl$$

When alkali phenolates are used, the reaction is carried out in aqueous medium, sometimes with the addition of a hydrophobic organic solvent; but when the process is carried out in the presence of tertiary amines it is in an organic solvent. The tertiary amine hydrochloride that is formed is removed by filtration or by washing with water.

The reaction for preparing the aryl chlorocarbonates is carried out at the lowest possible temperature with an excess of phosgene to suppress the formation of the diaryl carbonate, and also to diminish the rate of hydrolysis of the aryl chlorocarbonate if the reaction is carried out in an aqueous medium:

$$ArOCOCl + ArONa \rightarrow (ArO)_2CO + NaCl$$
$$ArOCOCl + H_2O \rightarrow ArOH + CO_2 + HCl$$

The yields of the aryl chlorocarbonates usually do not exceed 75–85% theoretical, calculated on the phenol used, and only in isolated cases do they reach 90% or more.

The aryl esters of N-methylcarbamic acid are also prepared by the action of methylamine on the diaryl carbonates with heating:

$$(ArO)_2CO + CH_3NH_2 \rightarrow ArOCONHCH_3 + ArOH$$

Separation of the carbamate from the phenol that is formed is accomplished by extraction of the phenol with aqueous solutions of alkalies.

In a search for new effective insecticides of this series, a large number of aryl esters of methylcarbamic acid have been synthesized with various phenols as the starting point, including alkyl- and alkenylphenols, haloalkylphenols, dialkyl- and trialkylphenols, alkylthiophenols, alkylthioalkylphenols, dimethylaminophenols, naphthols, dihydro-, tetrahydro-, and hexahydronaphthols, oxindene, hydroxydiphenyls, and many other aromatic hydroxy compounds. However, only a limited number of these are used in agriculture, including such compounds as carbaryl (Sevin), Zectran, Mesurol, and others. Carbaryl has received the widest distribution; the production of carbaryl in the United States is more than 40,000 tons a year.

Heterocyclic esters of dimethylcarbamic acid such as Isolan, Pyrolan, and dimetilan also are used.

The properties of some insecticidal esters of N-methylcarbamic acid are given in Table XXXVIII a.

Naphthyl N-methylcarbamic acid (carbaryl, Sevin, naphthyl carbamate) is a white crystalline substance, m.p. 142° C. At 20° C. less than 0.1% dissolves in water. It is highly soluble in organic solvents. Its vapor pressure at 26° C. is less than 0.005 mm. of Hg. At room temperature it is resistant to the action of water, light, and the oxygen of the air. In alkaline medium it is rapidly hydrolyzed, and therefore it is not compatible with compounds having an alkaline reaction (Bordeaux mixture, calcium and barium polysulfides, etc.):

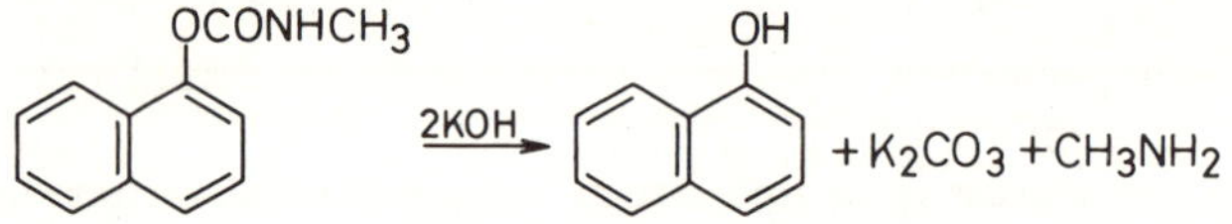

$$\text{Naphthyl-OCONHCH}_3 \xrightarrow{2\text{KOH}} \text{Naphthyl-OH} + \text{K}_2\text{CO}_3 + \text{CH}_3\text{NH}_2$$

The LD_{50} for rats is 500 mg./kg. When rats were fed a diet containing 200 mg. of carbaryl/kg. for two years, no harmful consequences were noted; the LD_{50} for goldfish is 25 mg./kg. The compound does not irritate the human skin. The maximum permissible concentration of carbaryl in air is 1 g./m.^{3}.

For the production of carbaryl 1-naphthol is used that is obtained through tetralin by the following scheme:

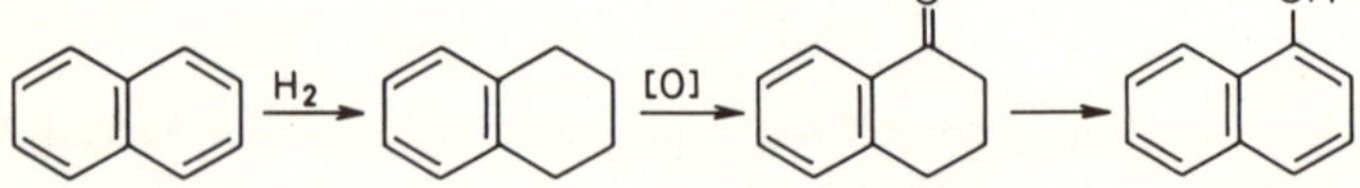

$$\text{naphthalene} \xrightarrow{H_2} \text{tetralin} \xrightarrow{[O]} \text{tetralone} \longrightarrow \text{1-naphthol}$$

The use of 1-naphthol obtained through 1-naphthylamine is not permitted, since the latter contains as a contaminant 2-naphthylamine, which has high cancerogenic activity.

Carbaryl is used to control pests of cotton and fruit crops. Since carbaryl is not toxic to plant-feeding mites, it is advantageous to use it in combination with acaricides. But when apple trees are treated with carbaryl, part of the young fruit falls off. Carbaryl also is used in animal husbandry.

In a study of the decomposition of carbaryl in plants and animals it has been established that the principal metabolites are products of oxidation and hydrolysis. Oxidation is directed primarily at the methyl group attached to the nitrogen atom. Also formed as metabolites are 4-hydroxy- and 5-hydroxynaphthyl N-methylcarbamates, which further break down with rupture of the aromatic ring:

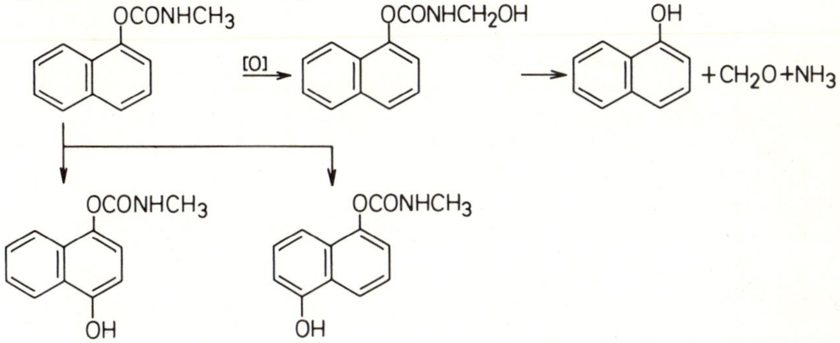

In the United States the maximum permissible amount of carbaryl in food products of seasonal consumption has been set at 10 mg./kg.

Almost all the general methods given for the preparation of carbamates are described in the patent literature; the most important of them apparently is the reaction of 1-naphthol with methylcarbamoyl chloride, which takes place at room temperature over 10–20 hours:

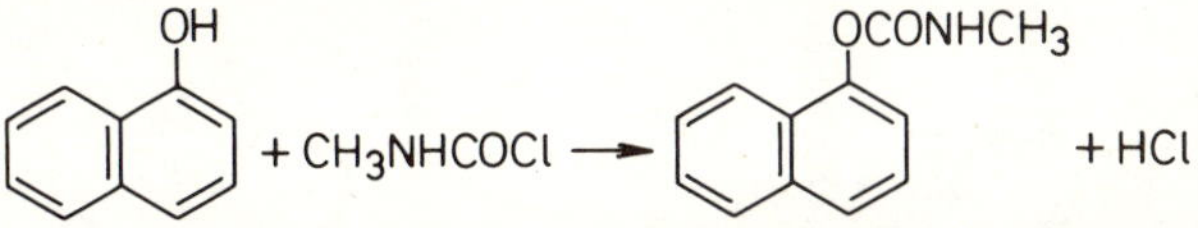

To speed up the reaction it has been proposed that the HCl be blown off with a stream of air or nitrogen. The methylcarbamoyl chloride necessary for the synthesis of carbaryl by this method is obtained by the reaction of phosgene with methylamine at 300°–400° C.; after the reaction mixture has cooled to 50°–60° C., the carbamoyl chloride that has been formed is dissolved in a suitable solvent. The solution obtained is used for the synthesis of carbaryl.

A better method of making carbaryl is the reaction of 1-naphthol with methyl isocyanate. The purest compound is obtained by this method. Methyl isocyanate can be synthesized, in addition to the methods described, by the action of tertiary amines on carbamoyl chloride:

$$CH_3NHCOCl + NR_3 \rightarrow CH_3NCO + R_3N \cdot HCl$$

However, when the tertiary amine is regenerated, contaminated wastewaters are obtained that require special purification. This greatly complicates the production.

The synthesis of carbaryl also is possible through 1-naphthyl chlorocarbonate, which is obtained either by the action of phosgene on sodium

Table XXXVIII a. *Properties of some insecticidal esters of N-methylcarbamic acid*

Trade or common name, chemical name	Structural formula	M.P. (°C.)	LD$_{50}$ (rats) (mg./kg.)	Uses
CPMC 2-chlorophenyl N-methylcarbamate	2-Cl-C$_6$H$_4$-OCONHCH$_3$	90–91	150	Pests of rice
Tumacide (MTMC) 3-methylphenyl N-methylcarbamate	3-CH$_3$-C$_6$H$_4$-OCONHCH$_3$	76–77	268	Pests of rice
MPCM (Meobal) 3,4-dimethylphenyl N-methylcarbamate	3,4-(CH$_3$)$_2$-C$_6$H$_3$-OCONHCH$_3$	79–80	290 [a]	Pests of rice
Banol (carbanolate) 6-chloro-3,4-xylenyl N-methylcarbamate	6-Cl-3,4-(CH$_3$)$_2$-C$_6$H$_2$-OCONHCH$_3$	130–133	293	Agricultural and animal parasites

Derivatives of carbamic acid 191

Maqbarl 3,5-xylenyl N-methylcarbamate	(structure)	99–100.5	245	Pests of rice
H-28 3-sec-butylphenyl N-methylcarbamate	(structure)	32–33.5	340	Pests of rice
H-22 3-tert-butylphenyl N-methylcarbamate	(structure)	143–144.5	470	Pests of rice
Ortho-5353 3-sec-pentylphenyl N-methylcarbamate (mixture of isomers)	(structure)	40–45	87	Agricultural pests
Aminocarb (matacil) 4-dimethylamino-3-tolyl N-methyl- carbamate	(structure)	93–94	50	Insecticide, molluscicide

Table XXXVIII a (continued)

Trade or common name, chemical name	Structural formula	M.P. (°C.)	LD$_{50}$ (rats) (mg./kg.)	Uses
Furodan 2,2-dimethyl-2,3-dihydrobenzofuranyl-7 N-methylcarbamate		150–152	5 [b]	Systemic soil insecticide
Mobam 4-benzothienyl N-methylcarbamate		128	234	Agricultural pests
APC (Hydrol) 4-Diallylamino-3,5-dimethylphenyl N-methylcarbamate		—	48 [c]	Agricultural pests
C 8353 2-(1,3-dioxolanyl-2)-phenyl N-methyl-carbamate		114–115	120	Agricultural pests
Lanoate Methyl thiomethylketoxime N-methyl-carbamate	$CH_3S-C=NOCONHCH_3$ $\quad\quad\ \ |$ $\quad\quad\ \ CH_3$	—	50	Systemic insecticide

[a] Mice, 60.8. [b] Rabbits, 885. [c] Mice.

1-naphtholate in aqueous medium or by the action of a tertiary amine on a mixture of 1-naphthol and phosgene in an organic solvent. The further conversion of the 1-naphthyl chlorocarbonate to carbaryl is accomplished by its reaction with methylamine in the presence of HCl acceptors:

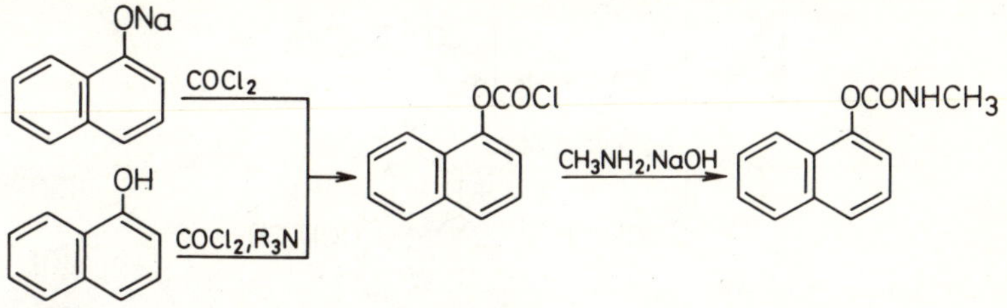

Carbaryl is used in agriculture in the form of wettable powders containing 50 and 85% active ingredient, and also in the form of dusts and granulated preparations at dosages of 0.3 to 3 kg./ha.

3-Isopropylphenyl N-methylcarbamate (UC-10854) forms white crystals, m.p. 53° C. At 30° C., 1 liter of water dissolves 85 mg. of the compound; in xylene it forms a 10% solution, in toluene 20%, in isopropyl alcohol 40%, and in acetone 50%. Its LD_{50} for rats is 41–63 mg./kg. It is similar to carbaryl in chemical properties, and it has almost no acaricidal effect; it is somewhat weaker than carbaryl in insecticidal activity. It has been tested in the form of a 50% wettable powder and a 15% emulsive concentrate.

It is prepared by the reaction of *m*-isopropylphenol with methyl isocyanate in benzene or another suitable solvent in the presence of tertiary amines (triethyl amine):

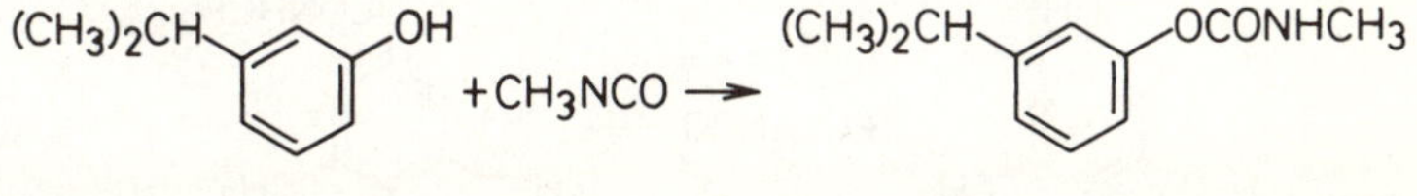

2-Isopropoxyphenyl N-methylcarbamate (Baygon, Bayer-39007) is a white crystalline substance, m.p. 91.5° C.; it is highly soluble in organic solvents and soluble to the extent of about 1% in water; the vapor pressure at 20° C. is 1×10^{-2} mm. of Hg. Its LD_{50} for rats is 100 mg./kg. It is an active insecticide for the control of aphids, Colorado potato beetles, cotton pests, cockroaches, flies, and mosquitoes. It is used in the form of a 50% wettable powder, 20% emulsive concentrate, 5% dust, and 5–10% granulated preparations.

Several methods are known for the preparation of 2-isopropoxyphenyl *N*-methylcarbamate, the most important of which are the reaction of methylcarbamoyl chloride or methyl isocyanate with 2-isopropoxyphenol:

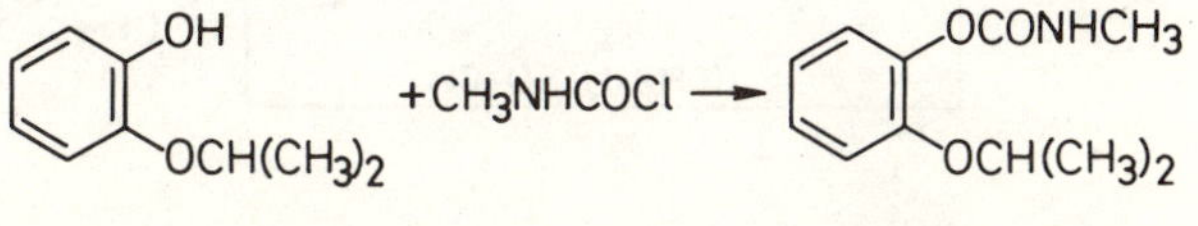

It is possible to obtain this compound also by the reaction of isopropoxyphenyl chlorocarbonate and di(isopropoxyphenyl) carbonate with methylamine:

$$\left[\underset{OCH(CH_3)_2}{\bigcirc}\right]_2 CO + CH_3NH_2 \longrightarrow$$

$$\underset{OCH(CH_3)_2}{\bigcirc}^{OCONHCH_3} + \underset{OCH(CH_3)_2}{\bigcirc}^{OH}$$

4-Methylmercapto-3,5-dimethylphenyl N-methylcarbamate (Mesurol, metmercapturon, Bayer-37344) is a crystalline substance, practically odorless, m.p. 121.5° C. It is highly soluble in most organic solvents, but almost insoluble in water.

The compound is resistant to hydrolysis in neutral and weakly acid media, but is rapidly hydrolyzed by alkalies. It is an active insecticide and acaricide with a wide spectrum of action and relatively long residual activity. It is used at dosages from 0.5 to 2.5 kg./ha. The usual working concentration is 0.05–0.1% active ingredient. Its LD_{50} for rats is 100 mg./kg. It is marketed in the form of a 50% wettable powder and 3 and 5% dusts.

4-Methylmercapto-3,5-dimethylphenyl N-methylcarbamate is synthesized by the reaction of 4-methylmercapto-3,5-dimethylphenol with methyl isocyanate in the presence of triethylamine. The reaction goes in any organic solvent, for example in benzene, with practically quantitative yield:

$$\underset{SCH_3}{\underset{H_3C \quad CH_3}{\bigcirc}^{OH}} + CH_3NCO \longrightarrow \underset{SCH_3}{\underset{H_3C \quad CH_3}{\bigcirc}^{OCONHCH_3}}$$

The necessary 4-methylmercapto-3,5-dimethylphenol is prepared by the action of HCl on a mixture of 3,5-xylenol and dimethyl sulfoxide or by the action of sulfuryl chloride on a mixture of phenol and dimethyl disulfide:

$$\underset{CH_3 \quad CH_3}{\bigcirc}^{OH} \xrightarrow{\substack{(CH_3)_2SO, HCl \\ (CH_3S)_2, SO_2Cl_2}} \underset{(CH_3)_2S^+Cl^-}{\underset{CH_3 \quad CH_3}{\bigcirc}^{OH}} \xrightarrow{H_2O} \underset{SCH_3}{\underset{CH_3 \quad CH_3}{\bigcirc}^{OH}}$$

Metmercapturon is formed also when 4-methylmercapto-3,5-dimethylphenyl chlorocarbonate reacts with methylamine:

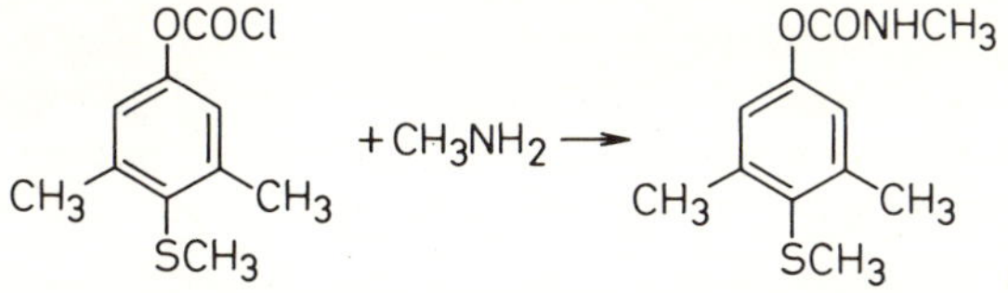

A large number of analogs and homologs of metmercapturon have been obtained in this way. The chlorocarbonates are relatively easily formed by the reaction of the appropriate phenol with phosgene in the presence of dimethylaniline at a temperature below 10° C., in any organic solvent.

4-(N,N-Dimethylamino)-3,5-dimethylphenyl N-methylcarbamate (Zectran) is a white crystalline substance, m.p. 85° C., poorly soluble in water, soluble in most organic solvents. With mineral acids it forms salts that are highly soluble in water. However, in this process it loses its contact insecticidal activity. The compound is resistant to hydrolysis in neutral medium, but it is rapidly hydrolyzed in alkaline medium. The LD_{50} for experimental animals is 16–63 mg./kg.

The metabolism of Zectran in the dog takes the following course: Zectran → 4-dimethylamino-3,5-dimethylphenol → 1,4-dihydroxy-3,5-dimethylphenol → dihydroxybenzene glucuronate. The metabolism of Zectran in broccoli is represented by the following scheme:

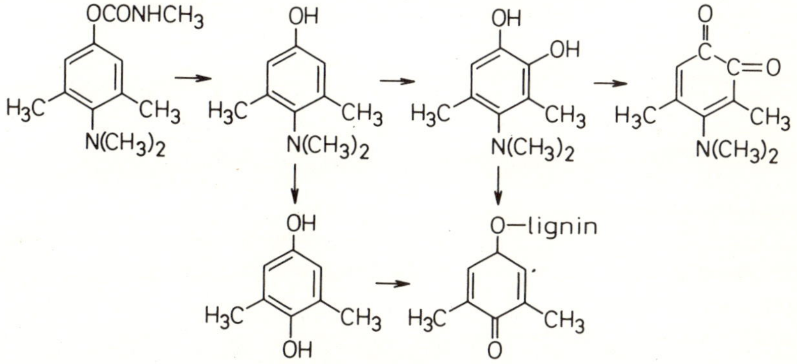

The principal method for preparing Zectran is the reaction of methyl isocyanate with 4-dimethylamino-3,5-dimethylphenol, which takes place in the presence of small amounts of tertiary amines in organic solvents with quantitative yields.

Since Zectran is an insecticide with a broad spectrum of action, a number of its analogs and homologs having insecticidal and acaricidal properties have been synthesized and studied. The synthesis of these compounds has been carried out by all possible methods for the preparation of aryl esters of alkylcarbamic acids, which indicates the possibility of preparing Zectran also by these methods.

5,5-Dimethyldihydroresorcinyl N,N-dimethylcarbamate (dimetan) is a yellow crystalline substance, m.p. 46° C., about 3% soluble in water; its solubility in most organic solvents is high. The LD_{50} for rats is about 150 mg./kg. Dimetan is active against a number of species of aphids in 0.01% concentration; it is moderately toxic to mites and also has a systemic effect.

The compound is easily hydrolyzed by alkali hydroxides with the formation of dimethylamine and this reaction is utilized for its quantitative determination. It is prepared by the reaction of the monosodium salt of dimedon with dimethylcarbamoyl chloride:

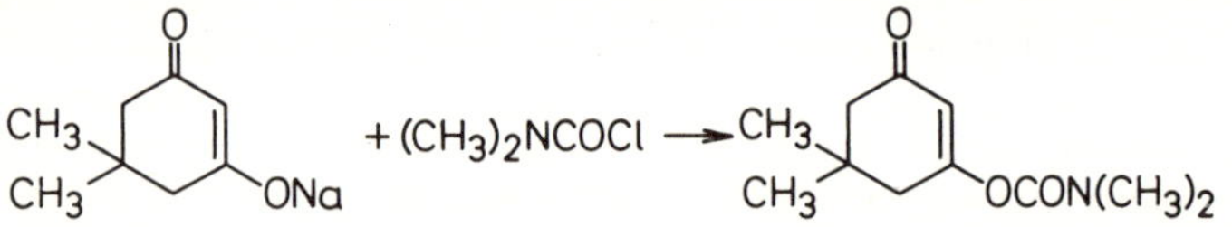

To control cockroaches and other harmful insects under domestic conditions 2-(4',5'-dimethyl-1',3'-dioxolanyl-2')-phenyl *N*-methylcarbamate is used, which also is known as Compound C 10015 (white crystals, m.p. 65°–100° C., LD_{50} for rats 67 mg./kg.):

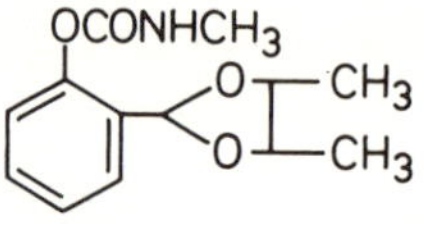

3,4-Dichlorobenzyl N-methylcarbamate (Romate) is a crystalline substance, m.p. 53°–54° C., b.p. 139° C. at 0.5 mm. of Hg. It is practically insoluble in water, but highly soluble in most organic solvents. It is a herbicide for the control of weeds in plantings of cotton, potatoes, tobacco, and some other crops.

Its synthesis is by reaction of 3,4-dichlorobenzyl alcohol with methyl isocyanate in benzene at room temperature in the presence of dibutyltin acetate as a catalyst:

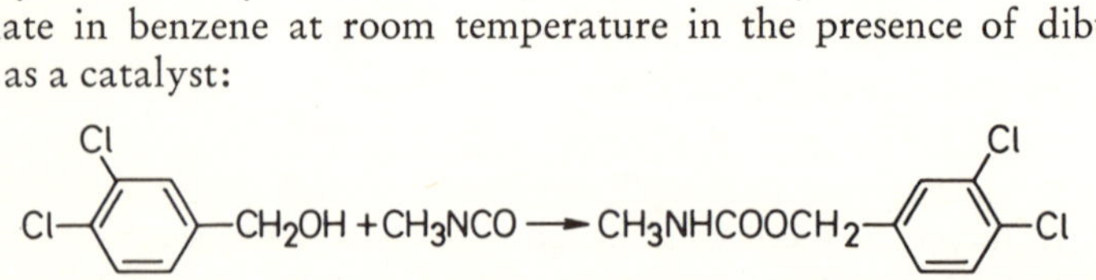

The yield of the carbamate from this reaction is about 83% of theoretical.

The necessary 3,4-dichlorobenzyl alcohol is prepared in 94% yield by the reduction of 3,4-dichlorobenzaldehyde with sodium borohydride at 30° C.

1-Isopropyl-2-methylpyrazolyl-5 dimethylcarbamate (Isolan) is a colorless liquid that steam distills, b.p. 105°–107° C. at 0.3 mm. of Hg. It is miscible with water, alcohol, acetone, and xylene in all proportions, and is up to 3% soluble in paraffinic hydrocarbons.

Isolan is an active aphicide with systemic action. It is employed in the form of 0.005–0.05% aqueous solutions. Its LD_{50} for different experimental animals is 11–23 mg./kg. It is produced by the action of dimethylcarbamoyl chloride on the potassium derivative of the enol form of 1-isopropyl-3-methylpyrazolone:

$$\underset{\underset{C_3H_7-iso}{|}}{\underset{N-N}{CH_3}}\!\!\!\diagdown\!\!OK + (CH_3)_2NCOCl \longrightarrow \underset{\underset{C_3H_7-iso}{|}}{\underset{N-N}{CH_3}}\!\!\!\diagdown\!\!OCON(CH_3)_2$$

Because of its high cost its use is limited.

1-Phenyl-3-methylpyrazolyl-5 dimethylcarbamate (Pyrolan) forms colorless crystals, m.p. 50° C., solubility in water about 0.2%. Although poorly soluble in petroleum oils, it is highly soluble in alcohol, acetone, and aromatic hydrocarbons. Its LD_{50} for rats is 62 mg./kg. Pyrolan has a systemic action against various species of aphids. Its effect on other species of insects is weak.

It is produced by the condensation of the potassium derivative of 1-phenyl-3-methylpyrazolone-5 with dimethylcarbamoyl chloride with heating in any organic solvent:

$$\underset{\underset{C_6H_5}{|}}{\underset{N-N}{CH_3}}\!\!\!\diagdown\!\!OK + (CH_3)_2NCOCl \longrightarrow \underset{\underset{C_6H_5}{|}}{\underset{N-N}{CH_3}}\!\!\!\diagdown\!\!OCON(CH_3)_2$$

2-Dimethylcarbamoyl-3-methylpyrazolyl-5 dimethylcarbamate (dimetilan) is also sometimes used as an agent for the control of flies and parasites of domestic animals:

$$(CH_3)_2NCO\!\!-\!\!\underset{N-N}{\overset{CH_3}{|}}\!\!\!\diagdown\!\!OCON(CH_3)_2$$

Its LD_{50} for rats is 65 mg./kg.

Interesting derivatives of *N*-methylcarbamic acid are the esters of the oximes of aliphatic and cyclic aldehydes and ketones, which have strong insecticidal, acaricidal, and nematicidal effects. At present field scale tests are being carried out on two such compounds:

1. *2-Methyl-2-methylthiopropionaldoxime O-N-methylcarbamate (aldicarb, Temik)* is a white crystalline substance, m.p. 100° C., vapor pressure at 20° C. 0.05 mm. of Hg, solubility in water 0.6%, in acetone 30%, in ethyl alcohol 25%, in chloroform 35%, in toluene 10%, in chlorobenzene 15%. It is insoluble in paraffinic hydrocarbons. Its LD_{50} for rats is 0.93 mg./kg. In spite of its high toxicity, it has been proposed as a systemic

nematicide for introduction into the soil in the form of special granules. It is synthesized by the reaction of the corresponding oxime with methyl isocyanate:

$$CH_3S\underset{CH_3}{\overset{CH_3}{\underset{|}{\overset{|}{C}}}}CH=NOH + CH_3NCO \rightarrow CH_3S\underset{CH_3}{\overset{CH_3}{\underset{|}{\overset{|}{C}}}}CH=NOCONHCH_3$$

2. *3-Chloro-6-cyanonorbornanone-2 oxime O,N-methylcarbamate (Compound UC-20047 A)* is a crystalline substance, m.p. 141°–146° C., solubility in water 0.2%, in acetone 30%, in methyl alcohol 25%, in ethyl alcohol 5%, and in xylene 1%. Its LD_{50} for rats is 26 mg./kg. It is prepared by the reaction of 3-chloro-6-cyanonorbornanone oxime with methyl isocyanate:

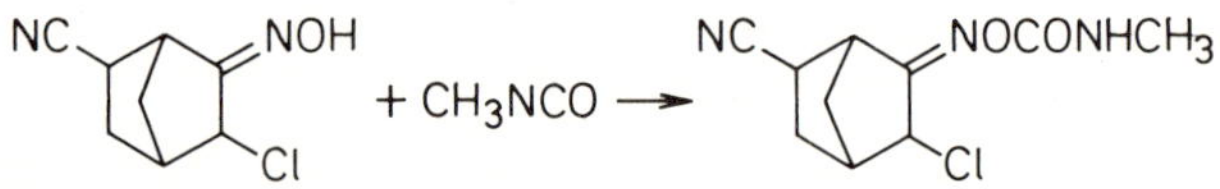

Alkyl esters of arylcarbamic acids

The alkyl esters of arylcarbamic acids have been used exclusively as agents to control weeds, although reports have appeared in the literature that some alkyl and aryl esters of arylcarbamic acids also have a fungicidal effect.

The principal methods of making this group of carbamic acid esters are the following:

1. Reaction of aryl isocyanates with alcohols:

$$ArNCO + ROH \rightarrow ArNHCOOR$$

This reaction proceeds with practically quantitative yields and gives arylcarbamic acid esters of high purity. Depending on the nature of the alcohol, the reaction takes place at room temperature or at an elevated temperature. The aryl isocyanates necessary for this synthesis are obtained in good yields by the action of phosgene on amines or amine hydrochlorides:

$$ArNH_2 + COCl_2 \rightarrow ArNCO + 2\,HCl$$

Aryl isocyanates are most easily prepared from amines that are halogen- and nitro-substituted in the nucleus. This is caused by corresponding isocyanates that react more slowly with the amines, and consequently a smaller amount of diarylurea is formed. For preparation of the arly isocyanates, the reaction of the amines with phosgene is usually carried out at a temperature from 90°–250° C., with chlorobenzene, trichloroethylene, or other halogen derivatives of hydrocarbons as the solvent. It must be pointed out, however, that ferric chloride has a negative effect; even a small trace of it sharply lowers the yield of isocyanate and increases the amount of tar in the reaction mixture.

2. Reaction of alkyl chlorocarbonates with aromatic amines, which takes place in the presence of different organic and inorganic bases or of an excess of the starting amine:

$$ArNH_2 + ClCOOR + NaOH \rightarrow ArNHCOOR + NaCl$$

The yields of arylcarbamic acid esters by this method exceed 90%.

3. Reaction of aryl isothiocyanates with alcohols and subsequent oxidation of the product obtained:

$$ArNCS + ROH \rightarrow ArNHCSOR$$
$$ArNHCSOR + H_2O_2 + NaOH \rightarrow ArNHCOOR + Na_2SO_4 + H_2O$$

The yield of the final alkyl arylcarbamates by this method does not exceed 60–65% calculated on the starting aryl isothiocyanate.

4. Heating a mixture of an amine and alcohol with urea leads to the formation of alkyl arylcarbamates in small yields:

$$ArNH_2 + ROH + CO(NH_2)_2 \rightarrow ArNHCOOR + 2\,NH_3$$

5. Reaction of alcohols with benzoylazides:

$$ArCON_3 + ROH \rightarrow ArNHCOOR + N_2$$

6. By a Hoffman reaction from benzamide:

$$ArCONH_2 + RONa + Br_2 \rightarrow ArNHCOOR + NaBr$$

Preparation of alkyl esters of arylcarbamic acids is possible also by several other methods, but only the first two are of practical importance for industrial production.

The chemical properties of alkyl arylcarbamates are reminiscent of those of the corresponding aryl arylcarbamates. The most important reaction of these compounds that has practical significance is their hydrolysis by alkali hydroxides and acids. The process proceeds with the formation of the free amine or its salt (in the case of acid hydrolysis). Apparently as a result of this reaction, which takes place under the influence of soil microorganisms, a comparatively rapid breakdown of herbicide occurs and the amine that is formed is further oxidized to complete destruction of the molecule.

Isopropyl N-phenylcarbamate (IPC, propham) is a white crystalline substance, m.p. 89°–90° C. Several values are given for the solubility of IPC in water, from 32 to 100 mg./l. IPC is highly soluble in alcohol, acetone, benzene, chlorobenzene, ethyl acetate, and many other organic solvents. Its LD_{50} is 1,000 mg./kg. There are contradictory data in the literature on the chronic toxicity. According to the data of some authors IPC has a blastomogenic effect, while other investigators deny this.

The compound is used as a root herbicide to control monocotyledonous weeds by introducing it into the soil. It is employed at dosages from 6–12 kg./ha. depending on conditions. It also is used to control weeds in sugar beet plantings in combination with endothall (the disodium salt of 3,6-endoxohexahydrophthalic acid). This preparation is known as Endoif and Murbetol. It contains about 16% endothall, 8% IPC, a surface-active agent, a stabilizer, and water. IPC is used most often in the form of aqueous suspensions.

In the soil the compound quickly breaks down and is converted to products of complete destruction. It has an especially high rate of breakdown in plants that are low in sensitivity to the compound. In plants that have high sensitivity to IPC it is broken down slowly.

IPC is produced by the first two methods described. The yield from phenyl isocyanate and isopropyl alcohol is practically quantitative. The phenyl isocyanate is added gradually, with good stirring, to an excess of absolute isopropyl alcohol. Two to 5 moles of the alcohol are used to 1 mole of isocyanate. After the reaction is ended, the mixture is cooled and the IPC that has separated out is filtered off. The pure compound is obtained with melting point not lower than 88°–89° C.; the yield is not less than 96%. The filtrate, which is a saturated solution of IPC in isopropyl alcohol, is diluted with the alcohol and used for the production of subsequent lots of IPC.

Isopropyl N-(3-chlorophenyl)carbamate (chloro-IPC, chloropropham) is a white crystalline substance, m.p. 40°–41° C., b.p. 112°–113° C. at 1 mm. of Hg, d_{20}^{20} 1.1913. The solubility of chloro-IPC in water is 80 mg./l.; it is highly soluble in aromatic hydrocarbons, halogen derivatives of hydrocarbons of the aliphatic and aromatic series, ketones, and esters, and slightly soluble in paraffinic hydrocarbons. The technical grade compound melts at 30°–40° C. and contains small amounts of IPC and diisopropyl carbonate as impurities. Chloro-IPC is more resistant to hydrolysis and oxidation than IPC, and therefore lasts longer in the soil (up to two months). Its LD_{50} for various experimental animals is 3,000–7,500 mg./kg.

Chloro-IPC is widely used for the control of weeds in plantings of onions, carrots, garlic, and other crops. It suppresses the growth and development of weed grasses, and is safe for many dicotyledonous crops. Mixtures of it with various herbicides are especially effective. Thus, good results are obtained in the control of weeds in carrots by the use of Probanil, which contains 85% chloro-IPC and 15% propazine. Good results also have been obtained by the use of chloro-IPC in combination with monuron. Addition of 10–15% of the other compounds makes it possible to decrease the dosage of chloro-IPC by 2–3 times. Chloro-IPC gives good results in the control of starwort, foxtail, and other weeds. It is employed mainly in the form of aqueous dispersions.

For large-scale industrial production of chloro-IPC the first two methods described for the preparation of alkyl arylcarbamates are used. It can be prepared in small quantities by a Curtius reaction from *m*-chlorobenzoyl chloride, sodium azide, and isopropyl alcohol:

$$\text{Cl-C}_6\text{H}_4\text{-COCl} + \text{NaN}_3 + \text{C}_3\text{H}_7\text{OH} \longrightarrow \text{Cl-C}_6\text{H}_4\text{-NHCOOC}_3\text{H}_7 + \text{N}_2 + \text{NaCl}$$

In this case it is not necessary to prepare the *m*-chloroaniline that is required for the synthesis of chloro-IPC by the other methods. The Curtius method presents a danger of explosion and cannot be used in industry.

The *m*-chloroaniline used for the production of chloro-IPC is obtained by the reduction of *m*-chloronitrobenzene. The reduction is best carried out with hydrogen on a chromium-copper catalyst or Raney nickel. Iron filings or alkali sulfides also can be employed as reducing agents. The synthesis of *m*-chloronitrobenzene is accomplished by chlorination of nitrobenzene in the presence of iodine and iron or of antimonous chloride. An important condition for successful chlorination of nitrobenzene is the absence of moisture. When water is present, the chlorination proceeds very slowly. To bind the water that may enter from the air into the reaction mixture chlorosulfonic acid should be added.

By-products that are formed are other isomers of chloronitrobenzene and dichloronitrobenzene, which are converted by further chlorination to pentachloronitrobenzene, used in agriculture as a fungicide.

The *m*-chloronitrobenzene is purified by recrystallization from methanol.

1-Methylpropyn-2-yl N-(m-chlorophenyl)carbamate (Bi-PC) is a white crystalline substance, m.p. 45°–46° C., insoluble in water but highly soluble in organic solvents. The LD_{50} is 250–2,500 mg./kg. Bi-PC is used to control weeds in sugar beet plantings in a mixture with cyclooctyldimethylurea in a ratio of 2:3; this mixture is effective at dosages of 2–10 kg. of the compound/ha.

Bi-PC is prepared by the reaction of *m*-chlorophenyl isocyanate with methylethynylcarbinol or of methylpropynyl chlorocarbonate with *m*-chloroaniline in the presence of bases:

$$HC\equiv C\text{-}CH(CH_3)\text{-}OH + \text{Cl-C}_6\text{H}_4\text{-NCO} \longrightarrow \text{Cl-C}_6\text{H}_4\text{-NHCOOCH}(CH_3)C\equiv CH$$

$$HC\equiv C\text{-}CH(CH_3)\text{-}OCOCl + \text{Cl-C}_6\text{H}_4\text{-NH}_2 \longrightarrow \text{Cl-C}_6\text{H}_4\text{-NHCOOCH}(CH_3)C\equiv CH$$

4-Chlorobutyn-2-yl N-(m-chlorophenyl)carbamate (barban, chlorinat).
This is a white crystalline substance, m.p. 76°–78° C.; its solubility in water is 110 mg./liter. It is slightly soluble in saturated hydrocarbons, highly soluble in aromatic hydrocarbons and halogenated derivatives of hydrocarbons. The LD_{50} for different experimental animals is 240–800 mg./kg. Solutions of the compound in organic solvents and the solid compound irritate the skin. Therefore, gloves should be worn when working with it.

4-Chlorobutyn-2-yl *N*-(*m*-chlorophenyl)carbamate is thermally unstable and when heated it readily breaks down with the evolution of HCl. The decomposition takes place very easily in the presence of ferric chloride; even 0.005% ferric chloride in the compound causes it to break down at or below its melting point. The hydrolytic stability of barban is also lower than of IPC or chloro-IPC. Upon hydrolysis the final products are *m*-chloroaniline and CO_2; this reaction is used for quantitative determination of barban.

Barban is a herbicide of narrowly selective action. It is designated for the control of wild oats in wheat and barley. It also is used to control weeds in vegetables. In contrast to other carbamates, barban acts on green plants and is used to control wild oats at the period of development of the third leaf in dosages from 0.5 to 2 kg./ha., depending on the degree of infestation of the planting. However, higher concentrations of the compound cause damage to useful crops.

Investigation of residues of barban in plants has shown that it completely breaks down in the course of 20–30 days.

The following routes are used for its synthesis:

1. By the reaction of *m*-chlorophenyl isocyanate with butynediol, 4-hydroxybutyn-2-yl *N*-(*m*-chlorophenyl)carbamate is obtained, which is converted to barban by the action of thionyl chloride in the presence of catalytic amounts of pyridine:

$$\text{m-Cl-C}_6\text{H}_4\text{-NCO} \xrightarrow{\text{HOCH}_2\text{C}\equiv\text{CCH}_2\text{OH}} \text{m-Cl-C}_6\text{H}_4\text{-NHCOOCH}_2\text{C}\equiv\text{CCH}_2\text{OH} \xrightarrow{\text{SOCl}_2} \text{m-Cl-C}_6\text{H}_4\text{-NHCOOCH}_2\text{C}\equiv\text{CCH}_2\text{Cl}$$

It should be noted that when *m*-chlorophenyl isocyanate reacts with butynediol, a rather considerable amount of butyn-2-ylidene-1,4 bis[*m*-chlorophenyl)carbamate] is formed:

$$\text{m-Cl-C}_6\text{H}_4\text{-NCO} \xrightarrow{\text{HOCH}_2\text{C}\equiv\text{CCH}_2\text{OH}} \text{m-Cl-C}_6\text{H}_4\text{-NHCOOCH}_2\text{C}\equiv\text{CCH}_2\text{OCONH-C}_6\text{H}_4\text{-Cl-m}$$

Depending on the conditions, the amount of the by-product indicated is from 15–30%. This impurity usually is present in the technical compound as an inert material.

2. By the reaction of 1-chlorobutyn-2-ol-4 with *m*-chlorophenyl isocyanate or by the reaction of 1-chlorobutyn-2-ol-4 with phosgene and subsequent treatment with *m*-chloroaniline:

$$ClCH_2C\equiv CCH_2OH + \underset{Cl}{\underset{|}{C_6H_4}}NCO \longrightarrow$$

$$ClCH_2C\equiv CCH_2OH \xrightarrow[-HCl]{COCl_2} ClCH_2C\equiv CCH_2OCOCl \xrightarrow[-HCl]{ClC_6H_4NH_2}$$

$$\underset{Cl}{\underset{|}{C_6H_4}}NHCOOCH_2C\equiv CCH_2Cl$$

Because of the skin-vesicant action of the intermediate product and the small yields obtained, these methods are less interesting.

Among the analogs of barban that have been studied, there are compounds with herbicidal, fungicidal, and insecticidal properties. The 4-chlorobutynyl esters of alkyl- and dialkylcarbamic acids are interesting.

Isopropyl N-acetoxy-N-phenylcarbamate (acylate). In the Soviet Union a new group of herbicides has been proposed based on *N*-hydroxy-IPC. The simplest derivative of this series is the compound acylate, a white crystalline substance, m.p. 95° C., slightly soluble in water but highly soluble in most organic solvents. Its LD_{50} is 3,000 mg./kg. It is intended for the control of barley-like weeds in plantings of sugar beets or legume crops. At dosages of 6–8 kg./ha. in rather humid weather it gives good results in the control of annual weed grasses and is harmless to legumes and sugar beets.

Acylate is produced by the following method:

$$C_6H_5NHOH + ClCOOC_3H_7 + NaOH \rightarrow C_6H_5\underset{OH}{\underset{|}{N}}COOC_3H_7 + NaCl + H_2O$$

$$C_6H_5\underset{OH}{\underset{|}{N}}COOC_3H_7 + (CH_3CO)_2O \rightarrow C_6H_5\underset{OCOCH_3}{\underset{|}{N}}COOC_3H_7 + CH_3COOH$$

Methyl N-(3,4-dichlorophenyl)carbamate (swep). Swep is being widely investigated as a pre-emergence herbicide for the control of weeds in soy beans and rice. It is a white crystalline substance, m.p. 113°–114° C. The LD_{50} for rats is 550 mg./kg. It is almost insoluble in water but highly soluble in organic solvents. Swep is recommended for use in the form of a wettable powder containing not less than 50% active ingredient. It is prepared by the reaction of 3,4-dichlorophenyl isocyanate with absolute methanol:

$$Cl-C_6H_3(Cl)-NCO + CH_3OH \longrightarrow Cl-C_6H_3(Cl)-NHCOOCH_3$$

The process is best carried out with an excess of methanol, as has been described for the synthesis of IPC. It also is possible to obtain swep by the reaction of methyl chlorocarbonate with 3,4-dichloroaniline in the presence of inorganic or organic bases.

Acetone oxime N-phenylcarbamate (proxypham). In the German Democratic Republic use has been started of the new herbicide proxypham, which is a white crystalline substance, m.p. 109°–109.5° C., vapor pressure at 20° C. 5×10^{-6} mm. of Hg, volatility 0.05 mg./m.^{3}, solubility in water about 500 mg./liter. It is highly soluble in aromatic hydrocarbons and their halogen derivatives. Its LD_{50} for rats is about 3,000 mg./kg.

Upon heating for 1 hour to 130° C., proxypham breaks down. In 30 days about 80% of the compound breaks down in the soil. Hydrolysis by water takes place in the following manner:

$$C_6H_5NHCOON = C(CH_3)_2 + H_2O \rightarrow$$
$$C_6H_5NH_2 + NH_2OH + CO_2 + CH_3COCH_3$$

This reaction can be utilized for quantitative determination of the compound. The main method for making proxypham is the reaction of acetone oxime with phenyl isocyanate:

$$C_6H_5NCO + HON = C(CH_3)_2 \rightarrow C_6H_5NHCOOCN = C(CH_3)_2$$

Proxypham is used to control monocotyledonous weeds in combination with IPC and monuron.

3-Methoxycarbonylaminophenyl N-3'-methylphenylcarbamate (phenmedipham, betanal) is a white crystalline substance, m.p. 143°–144° C., solubility in water 10 p.p.m., acetone 20%, methanol 5%, benzene 0.25%, and hexane 0.05%. Its LD_{50} is 3,000–8,000 mg./kg. The activity of phenmedipham in controlling dicotyledonous weeds after their emergence in green stands of Beta beets has been described. The dosage is 1 kg. of active ingredient per hectare. It is prepared by the reaction of *m*-tolyl isocyanate with 3-hydroxyphenyl O-methylcarbamate:

$$CH_3OCHN(O)-C_6H_4-OH + CH_3-C_6H_4-NCO \rightarrow CH_3O-CHN(O)-C_6H_4-OCONH-C_6H_4-CH_3$$

Methyl N-(4-aminobenzenesulphonyl)carbamate (asulam) is a crystalline substance that decomposes at 143° C.; solubility in water 0.5%, methanol 28%, acetone 34%, hydrocarbons <2% at 20°–25° C. Its LD_{50} is 5,000 mg./kg. (mice).

Methyl N-(4-nitrobenzenesulphonyl)carbamate (MB 8882) is a crystalline substance, m.p. 150°–152° C., solubility in water 0.1%, methanol 4%, acetone 26%, and hydrocarbons <2%. Its LD_{50} is 5,000 mg./kg. (mice). These two recent herbicides have rather different properties from the other carbamate herbicides. They may be taken up by roots or foliage, are translocated within the plant and then interfere with the growing point and cause chlorosis. Possible uses are the control of annual weeds. Dosages are of the order of 2–4 kg./ha.

General references

ALLEN, B. P., and A. K. OSWALD: Proc. 9th Brit. Weed Control Conf. **1**, 471 (1968).
ARNDT, F., and C. KÖTTER: Weed Research **8**, 259 (1968).
BACHMANN, F., and J. B. LEGGE: J. Sci. Agr. Suppl., p. 39 (1968).
BASKAKOV, YU. A.: In book „Novye insektofungitsidy i gerbitsidy" [New insectofungicides and herbicides], p. 185. Foreign Literature Publishing House [USSR] (1960).
BOCHAROVA, L. P.: Khim. v Sel'skom Khozyaistve No. 4, 57 (1964).
COTTRELL, H. J., and B. J. HEYWOOD: Nature **201**, 655 (1965).
DURDEN, J. A., and H. J. WEIDEN: J. Agr. Food Chem. **17**, 94 (1969).
EDWARDS, C. J.: Proc. 9th Brit. Weed Control Conf. **2**, 575 (1968).
FISCHER, A.: Herbicide. Limburgerhof: BASF Land.-Versuchsstation (1966).
GUBLER, K., A. MARGOT, and H. GYSIN: J. Sci. Agr. Suppl., p. 13 (1968).
HACSKAYLO, J.: J. Econ. Entomol. **61**, 1108 (1968).
HOLMES, H. M.: Proc. 9th Brit. Weed Control Conf. **2**, 580 (1968).
KILSHEIMER, J. R., H. A. KAUFMAN, H. M. FOSTER, P. R. DRISCOLL, L. A. GLICK, and R. P. NAPIER: J. Agr. Food Chem. **17**, 91 (1969).
KURT, H.: Chemische Unkrautbekämpfung. Jena: Fischer Verlag (1963).
MEL'NIKOV, N. N.: Khim. v Sel'skom Khozyaistve No. 6, 16 (1964).
— Zhur. Vsesoyusnogo Khim. Obshchestva im. D. I. Mendeleeva **9**, 482 (1964).
—, and YU. A. BASKAKOV: Khimiya gerbitsidov i regulyatorov rosta rastenii [Chemistry of herbicides and plant growth regulators]. State Scientific and Technical Publishing House of Chemical Literature [USSR] (1962).
METCALF, R. L., and T. R. FUKUTO: Farm. Chem. **130** (No. 11), 45 (1967).
—, T. R. FUKUTO, C. COLLINS, R. BORCK, S. ABDEL-AZIZ, R. MUNOZ, and C. C. CASSILL: J. Agr. Food Chem. **16**, 300 (1968).
O'BRIEN, R. D.: Insecticides, action and metabolism. New York: Academic Press (1967).
PEASE, H. L., and J. J. KIRKLAND: J. Agr. Food Chem. **16**, 554 (1968).
SHAPOVALOVA, G. K., Z. I. MAKSIMOVA, and I. A. MEL'NIKOVA: Khim. v Sels'kom Khozyaistve No. 3, 23 (1964).
SOLLKE, K., and K. KOSMANN: Zucker **21**, 183 (1968).
THOMAS, T. M.: Proc. 9th Brit. Weed Control Conf. **2**, 575 (1968).
Weed Control Handbook, Vols. I and II. Oxford: Blackwell (1968).
WEIDEN, M. H., H. H. MOOREFIELD, and L. K. PAYNE: J. Econ. Entomol. **58**, 1954 (1965).
WINNER, C., and W. SCHÄUFELE: Zucker **21**, 139 (1968).
ZBIROVSKY, M., J. MYSKA, and J. ZEMANEK: Herbicidy. Prague (1960).

XVII. Derivatives of thio- and dithiocarbamic acids

General characteristics of pesticidal properties

The physiological activity of derivatives of thio- and dithiocarbamic acids has been the subject of detailed investigation. As a result of systematic study it has been established that most derivatives of the thiolocarbamic acids are active herbicides that easily penetrate into plants and move through the xylem. The most effective herbicides are the S-alkyl N,N-dialkylthiocarbamates, which act selectively on annual grasses and some dicotyledons and can be used successfully in such crops as vegetables, sugar beets, beans, etc. In this series, compounds also have been found that are useful for the control of velvet grass in rice plantings. To control weeds the thiolocarbamates usually are introduced into the soil either before sowing of the seed or prior to emergence of the seedlings. In different countries five or more S-alkyl dialkylthiocarbamates are used in agriculture to some extent.

The insecticidal effect of this group of compounds is slight; the dialkylthiocarbamates show a nematicidal effect, but they have yet to be practically used for this purpose.

Derivatives of dithiocarbamic acid are used as nematicides. The simplest derivative of dithiocarbamic acid with a wide spectrum of action that has been used in agriculture is sodium N-methyldithiocarbamate, known as carbothion or Vapam. It is an active soil sterilant; it destroys not only weed seeds, but nematodes, insects, and plant disease agents.

Among the derivatives of dithiocarbamic acid of more complex structure are found fungicides, nematicides, herbicides, and plant growth regulators; many of them are used in agriculture.

The following general conclusions can be drawn about the relationship between biological activity and structure of derivatives of dithiocarbamic acid.

The nematicidal, fungicidal, and herbicidal activity of the alkali metal salts of alkyldithiocarbamic acids decreases with an increase in the length of the alkyl radical. The maximum activity is shown by the salts of N-methyldithiocarbamic acid. The nature of the cation is not of substantial importance. This situation is true for almost all the water-soluble salts of N-alkyldithiocarbamic acids. Replacement of the second hydrogen atom on the nitrogen by an alkyl or aryl radical also lowers the biocidal activity of the water-soluble salts of dithiocarbamic acid.

The fungicidal activity of the esters of alkyl- and dialkyldithiocarbamic acids is considerably weaker than that of the corresponding water-soluble and water-insoluble salts of these acids.

The nematicidal activity of the esters of alkyl- and dialkylcarbamic acids in a number of cases is higher than the activity of the salts. The most toxic in this connection are the methyl and ethyl esters. With an increase in the size of the ester radical the nematicidal activity falls.

The herbicidal acivity also decreases when the number of carbon atoms in the ester radical is increased to more than five, and when the total number of carbon atoms in the alkyl radicals on the nitrogen is more than six.

The fungitoxicity of the water-insoluble salts of the alkyl- and dialkyldithiocarbamic acids decreases with an increase in the number of carbon atoms in the alkyl radicals. Introduction of an aromatic radical on the nitrogen sometimes increases the fungicidal activity of the compound or its selective action on some species of fungi. In the series of alkylenebis(dithiocarbamates) of the alkali metals or of zinc, the fungicidal activity decreases with an increase in the number of methylene groups between the nitrogen atoms. The maximum activity is shown by salts of ethylenebis(dithiocarbamic) acid and other 1,2-alkylenebis(dithiocarbamic) acids. The disulfides obtained by oxidation of alkyl- and dialkyldithiocarbamic acids are more active than the salts of the starting dithiocarbamic acids, although the difference is not very substantial.

When a carboalkoxyl group is introduced into the ester radical of a dialkyldithiocarbamic acid ester the nematicidal activity falls sharply, but plant growth regulating activity appears.

Compounds of the general formula (I) are the first compounds physiologically active toward plants that do not have a cyclic grouping in their structure. Removal of the sulfur atom from the carbonyl group leads to a loss of the physiological activity.

$$R_2NCSCH_2COOH \quad \text{(with C=S)} \qquad (I)$$

Some substances of this type show systemic fungicidal properties, but these are not strong enough for use under practical conditions.

Esters of thiocarbamic acids

As indicated, the main use of the esters of alkyl- and dialkylthiocarbamic acids is for weed control. Various methods have been developed for the synthesis of this interesting class of compounds. The methods of preparing esters of thiocarbamic acid which are of practical interest for industrial production are given below:

1. Reaction of the mercaptides of alkali metals, ammonia, and amines with carbamoyl chlorides:

$$R_2N\overset{\overset{O}{\|}}{C}Cl + NaSR' \rightarrow R_2N\overset{\overset{O}{\|}}{C}SR' + NaCl$$

This reaction proceeds rapidly in any organic solvent that does not contain active functional groups capable of interacting with the carbamoyl chloride or mercaptide. The best solvents apparently are the aromatic hydrocarbons, although substances like dioxane, formalglycol, etc. also may be used. It is also possible to carry out the reaction in aqueous medium. In this case the process is carried out at the lowest possible temperature to diminish the side reaction of hydrolysis of the carbamoyl chloride. The necessary carbamoyl chloride is synthesized from phosgene and the appropriate amine with the use of special HCl acceptors (tertiary amines) or with an excess of the starting amine:

$$2\,R_2NH + COCl_2 \rightarrow R_2N\overset{\overset{O}{\|}}{C}Cl + R_2NH_2Cl$$

2. Reaction of alkyl thiochlorocarbonates with amines in the presence of acceptors of HCl:

$$R_2NH + R'S\overset{\overset{O}{\|}}{C}Cl \rightarrow R_2N\overset{\overset{O}{\|}}{C}SR' + HCl$$

The thiolochlorocarbonates are prepared in good yields from the corresponding mercaptides and phosgene with the use of organic or inorganic bases as HCl acceptors:

$$RSH + COCl_2 \rightarrow RS\overset{\overset{O}{\|}}{C}Cl + HCl$$

3. Reaction of the dialkylthiocarbamates of alkali metals or ammonia with alkyl halides:

$$R_2N\overset{\overset{O}{\|}}{C}SNa + R'Cl \rightarrow R_2N\overset{\overset{O}{\|}}{C}SR' + NaCl$$

The thiolocarbamates are formed in practically quantitative yields by the reaction of an amine, alkali, and carbonyl sulfide in aqueous solution or in an organic solvent:

$$R_2NH + NaOH + COS \rightarrow R_2N\overset{\overset{O}{\|}}{C}SNa + H_2O$$

Carbonyl sulfide can be obtained from carbon monoxide and sulfur at an elevated temperature:

$$CO + S \rightarrow COS$$

or by decomposition of ammonium thiocyanate with sulfuric acid:

$$NH_4SCN + H_2SO_4 + H_2O \rightarrow (NH_4)_2SO_4 + COS$$

S-Ethyl N,N-di-n-propylthiocarbamate (Eptam, EPTC). The simplest derivative of a dialkylthiocarbamic acid is Eptam, widely used as a pre-emergence herbicide to control weeds in plantings of alfalfa, beans, beets, carrots, cabbage, potatoes, flax, and many other crops. It is a liquid with an unpleasant odor, b.p. 127° C. at 20 mm. of Hg, d_{30}^{30} 0.9543, n_D^{20} 1.4755; its solubility in water at 20° C. is 375 mg./liter; it is highly soluble in most organic solvents, but miscible in all proportions with methanol, isopropyl alcohol, acetone, benzene, toluene, and xylene. The LD_{50} is 1,630 mg./kg.

Eptam is hydrolyzed by the action of alkalies, forming mercaptide and free amine:

$$(C_3H_7)_2NCOSC_2H_5 + 3\,NaOH \rightarrow$$
$$(C_3H_7)_2NH + Na_2CO_3 + C_2H_5SNa + H_2O$$

Upon oxidation it is broken down with the formation of amine and ethanesulfonic acid, which is formed through a series of intermediate compounds. Hydrolysis and oxidation probably are the main processes that lead to a complete breakdown of Eptam in the soil.

It is used in the form of aqueous emulsions and granulated preparations. It is applied to the soil and subsequently covered, the dosage being 5–6 kg./ha.

Eptam is produced by both the first and the second methods for synthesizing esters of dialkylthiocarbamic acids:

$$(C_3H_7)_2NH + C_2H_5SCOCl + NaOH \rightarrow (C_3H_7)_2N\overset{O}{\underset{\|}{C}}SC_2H_5 + NaCl + H_2O$$

$$(C_3H_7)_2NCOCl + C_2H_5SH + NaOH \rightarrow (C_3H_7)_2N\overset{O}{\underset{\|}{C}}SC_2H_5 + NaCl + H_2O$$

The closest homolog of Eptam, *S-n-propyl N,N-di-n-propylthiocarbamate (Vernam, R-1607)* has been marketed as an experimental herbicide for the control of weeds by its introduction into the soil. It is a liquid, b.p. 150° C. at 30 mm. of Hg, d_{20}^{20} 0.954, n_{20}^{20} 1.4736, solubility in water less than 0.01%. It is miscible in all proportions with kerosine, xylene, methyl isobutyl ketone, and some other organic compounds. The LD_{50} for rats is 1,780 mg./kg. It is used at dosages from 2–4 kg./ha. in the form of an emulsion or a granulated preparation and is similar to Eptam in properties and methods of production.

S-Propyl N-ethyl-N-butylthiocarbamate (Tillam) is a clear liquid, b.p. 142.5° C. at 20 mm. of Hg, d_{20}^{20} 0.945. At 21° C. 92 mg. dissolve in 1 liter of water; it is miscible with kerosine, benzene, toluene, xylene, methanol, isopropyl alcohol, and acetone. The LD_{50} for rats is 1,120 mg./kg. This compound has been proposed for the control of weeds in sugar beet plantings by its introduction to the soil with subsequent covering. It is marketed in the form of a 10% granulated preparation and an emulsive concentrate containing more than 60% active ingredient with the dosage being 2–4 kg./ha.

The chemical properties of Tillam are similar to those of Eptam. It is best produced by the first two methods described for the synthesis of esters of dialkylthiocarbamic acid.

S-Ethyl hexahydro-1H-azepine-1-carbothioate (S-ethyl 1-hexamethyleneiminothiocarbamate, molinate, Ordram, Hydram, Yalan, Stauffer R-4572) is a light yellow liquid, b.p. 137° C. at 10 mm. of Hg, solubility in water about 0.1%; miscible in all proportions with kerosine, toluene, xylene, and ketones. The LD_{50} for rats is 720 mg./kg. It is marketed in the form of a 5% granulated preparation and a 60% emulsive concentrate. It has been proposed for use on rice and other crops to control miliary weeds prior to emergence of seedlings; the dosage is 2–6 kg./ha.

The best method for producing this compound is the reaction of ethyl thiochlorocarbonate with hexamethylenimine in the presence of HCl acceptors:

$$C_2H_5SCOCl + HN\hspace{-0.2em}\bigcirc \xrightarrow{NaOH} C_2H_5SCON\hspace{-0.2em}\bigcirc + NaCl$$

It is possible to prepare molinate also by the reaction of sodium hexahydro-1*H*-azepine-1-carbothioate with ethyl chloride or bromide.

S-2,3-Dichloroallyl N,N-diisopropylthiocarbamate (di-allate, Avadex) is a liquid, b.p. 149°–150° C. at 9 mm. of Hg; about 40 mg. dissolve in 1 liter of water at 25° C.; it is miscible in all proportions with most aromatic hydrocarbons, ketones, and halogen derivatives of hydrocarbons. The LD_{50} for rats is 393 mg./kg.

For weed control it is recommended that the compound be covered with soil to a depth of 2.5–5 cm. before sowing of the seed. It acts selectively on monocotyledonous weeds and some dicotyledons, but is comparatively safe for barley and wheat. It is applied under wheat not earlier than 10–15 days before sowing, since damage to the seedlings is possible. It is recommended for the control of wild oats in such crops as flax, barley, corn, peas, lentils, sugar beets, beans, and some others at dosages from 0.6–1.7 kg./ha.

It is stable in storage, but in an alkaline medium it is hydrolyzed with the formation of products of low toxicity to plants:

$$(C_3H_7)_2N\overset{O}{\overset{\|}{C}}SCH_2CCl = CHCl + 3\ NaOH \rightarrow$$
$$(C_3H_7)_2NH + Na_2CO_3 + NaSCH_2CCl = CHCl + H_2O$$

The main method of producing di-allate is the reaction of alkali salts of diisopropylthiocarbamic acid with 1,2,3-trichloropropylene:

$$(C_3H_7)_2NC(O)SNa + ClCH_2CCl=CHCl \rightarrow$$
$$(C_3H_7)_2NCOSCH_2CCl=CHCl + NaCl$$

S-2,3,3-Trichloroallyl N,N-diisopropylthiocarbamate (triallate) is a clear liquid, b.p. 165° C. at 6 mm. of Hg, practically insoluble in water, but miscible with alcohol, benzene, acetone, and halogen derivatives of aliphatic and aromatic hydrocarbons. The LD_{50} for rats is 1,340 mg./kg. Triallate is marketed in the form of an emulsive concentrate containing about 400 g. of active ingredient/liter, and also in the form of a granulated preparation. It is designated for control of wild oats and other weeds in plantings of wheat, barley, flax, beets, and peas at dosages from 1 to 1.5 kg./ha., with covering by soil. It is safer than di-allate for wheat.

Tri-allate is similar to di-allate in chemical properties. It also is produced in a manner similar to di-allate, by the reaction of sodium diisopropylthiocarbamate with 1,2,3,3-tetrachloropropylene-2.

Salts of substituted dithiocarbamic acids

One of the most important groups of derivatives of dithiocarbamic acid is the salts of methyl-, dimethyl-, and ethylenebis(dithiocarbamic) acids. At present they are produced in various countries in tens of thousands of tons and are used to control various plant diseases. Compounds like maneb, zineb, ziram, carbothion, polycarbazine, and tetramethylthiuram disulfide (TMTD) are very widely used, because of their relatively low cost, simplicity of production, and the availability of raw materials.

The alkali salts of the alkyldithiocarbamic acids are produced easily and in practically quantitative yields by the reaction of an amine and alkali hydroxides with carbon disulfide:

$$R_2NH + NaOH + CS_2 \rightarrow R_2NC(S)SNa + H_2O$$

By the reaction of these water-soluble alkali salts with aqueous solutions of salts of zinc, iron, manganese, etc., the corresponding salts of the dithiocarbamic acids are formed, which are practically insoluble in water and therefore precipitate:

$$2 R_2NC(S)SNH_4 + ZnSO_4 \rightarrow (R_2NC(S)S)_2 Zn + (NH_4)_2SO_4$$

To prepare the zinc and other water-soluble salts it is best to use the highly water-soluble ammonium salts of the substituted dithiocarbamic acids. In this case the wastewater contains ammonium salt and after evaporation of the mother liquor commercial fertilizers are obtained. When the sulfates of the appropriate metals are used for the precipitation, rich nitrogenous fertilizers result.

Salts of dithiocarbamic acids are produced also by the reaction of oxides of the corresponding metals with amines and carbon disulfide:

$$2 R_2NH + 2 CS_2 + ZnO \rightarrow \left(R_2N\overset{S}{\overset{\|}{C}}S \right)_2 Zn + H_2O$$

When zineb is prepared by this method the necessity of purifying the wastewater is almost completely avoided, since the mother liquor after filtration of the zinc salt can be used for the subsequent operation of preparing the same salt of dithiocarbamic acid. A deficiency of this method is the fact that the salt obtained in this way contains some zinc oxide as an impurity. If the compound is used as a fungicide to protect plants from diseases, then this impurity is not important, since it does not lower the effectiveness. However, the amount of this impurity should be minimal.

To prepare the disulfides, salts of the dithiocarbamic acid are oxidized with simultaneous acidification by mineral acids:

$$2 R_2N\overset{S}{\overset{\|}{C}}SNa + O + H_2SO_4 \rightarrow \left[R_2N\overset{S}{\overset{\|}{C}}S \right]_2 + H_2O + Na_2SO_4$$

Good results are obtained with the following oxidizing agents: hydrogen peroxide, alkali nitrites, peracids, and the like. Used most often is sodium nitrite with sulfuric acid, or hydrogen peroxide.

Sodium N-methyldithiocarbamate (Vapam, carbothion) in the pure form is a white crystalline substance with a creamy tinge, highly soluble in water (more than 40% at 20° C.), but practically insoluble in hydrocarbons, halogenated hydrocarbons, and other hydrophobic solvents, moderately soluble in methyl and ethyl alcohols. It is unstable in storage and gradually breaks down with the formation of methyl isothiocyanate:

$$CH_3NH\overset{S}{\overset{\|}{C}}SNa \rightarrow CH_3NCS + NaSH$$

It also breaks down in aqueous solution, and in this case the lower the concentration of the solution the faster the breakdown. Thus, for example, at 60° C., in the course of one week in 40% solution 4.2% of the carbothion decomposes, in 20% solution 8.9, in 10% solution 18.3, and in 2.5% solution 68.5. To stabilize solutions of sodium N-methyldithiocarbamate small amounts of the lower tertiary aliphatic amines are added. For agricultural needs carbothion usually is marketed in the form of a 30—40% aqueous solution containing 0.1–1% of a tertiary amine (for example, trimethyl- or triethylamine). In this form the compound is stable and can be stored for many months without decomposition.

Sodium N-methyldithiocarbamate is moderately toxic to animals. The LD_{50} for white mice is 285 mg./kg., and for rats 820 mg./kg. The methyl isothiocyanate formed in its decomposition strongly irritates the mucous membranes of the eyes.

It is obtained in practically quantitative yield by the reaction of equimolecular quantities of carbon disulfide, methylamine, and sodium hydroxide in aqueous solution at 20°–40° C. The aqueous solution of the compound is diluted to an active ingredient content of 30–40%, stabilizers are added, and the mixture is put out as a commercial product.

In the United States and other countries this mixture is widely used as a practical soil sterilant that provides complete destruction of nematodes, disease agents, and weeds in the soil. Usually to obtain a better effect the soil is wet with water before treatment. The dosage of the compound is from 250 to 1,500 kg./ha. Soil is treated with carbothion 1–4 weeks before planting (depending on the nature of the soil, the dosage of the compound, and the crop).

Sodium N-methyldithiocarbamate reacts quantitatively with iodine and this reaction is employed for its analytical determination:

$$2\ CH_3NH\overset{S}{\overset{\|}{C}}SNa + I_2 \rightarrow \left(CH_3NH\overset{S}{\overset{\|}{C}}S\right)_2 + 2\ NaI$$

Oxidation can be carried out also with other oxidizing agents. The dimethylthiuram disulfide obtained, under the name of Tridipam, is being studied as a soil fungicide.

It is assumed that the action of carbothion on different organisms is based on the formation, when it breaks down, of methyl isothiocyanate, which further enters into reaction with the vital systems of plants and animals.

The effect of the salts of the dithiocarbamic acids on different species of microorganisms apparently is based on disruption of the oxidation-reduction processes in their cells. Furthermore, the dithiocarbamates are capable of reacting with the amino groups of proteins to form the corresponding derivatives of thiourea.

Salts of dimethyldithiocarbamic acid. To protect plants from diseases, the zinc (ziram, Zerlate) and iron (ferbam, Fermate) salts of dimethyldithiocarbamic acid, and on a very small scale the manganese (marbam) salt, are used. The highest fungicidal activity is shown by the zinc salt, as can be seen just from a comparison of the LD_{50} for the fungus *Sclerotinia fructicola:* the LD_{50} of ziram is 0.4 mg./liter, of ferbam 1 mg./liter, and of marbam 1 mg./liter.

The zinc salt of dimethyldithiocarbamic acid is a crystalline substance, m.p. 240°–246° C. Its solubility in water at 25° C. is about 0.065 g./liter. It is slightly soluble in most organic solvents, with an LD_{50} of 1,400 mg./kg. Dogs can survive on a diet containing 5 mg./kg. of ziram (calculated on the weight of the dog) for a year without harm. When ziram comes in contact with the mucous membranes it is capable of causing strong irritation, which occurs also upon systematic contact of the compound with sweaty areas of the skin.

Upon prolonged heating to 170°–180° C., violent decomposition of the compound occurs with the formation of a carbonaceous product; this must be considered when drying the compound in industrial dryers, including spray dryers.

Zinc dimethyldithiocarbamate has a chelate structure (II and III), as shown by a study of the absorption spectra in the ultraviolet region. In dilute solution structure (II) predominates:

$$(CH_3)_2NC(=S)(S)Zn^+ \quad (II) \qquad (CH_3)_2NC(=S)(S)Zn(S)(S=)CN(CH_3)_2 \quad (III)$$

Ziram is rather stable in storage. Dilute acids do not decompose it at room temperature, but concentrated sulfuric, phosphoric, and other acids break the compound down completely with the evolution of carbon disulfide and a small amount of hydrogen sulfide. Ziram also decomposes when it is heated with caustic alkalies.

It is used most often in the form of a wettable powder containing from 70–90% active ingredient. This powder is produced by spray drying of the slurry obtained by precipitation of the zinc salt of dimethyldithiocarbamic acid from aqueous solutions of zinc sulfate and ammonium dimethyldithiocarbamate. This method of producing wettable powders of fungicides is often used in industry. By this means a wettable powder of good quality is obtained and the stages of grinding and mixing are eliminated, but the expenditure of energy for the removal of water is increased.

Zinc dimethyldithiocarbamate can be synthesized by two methods:

1. Reaction of ammonium dimethyldithiocarbamate with zinc sulfate or another highly water-soluble salt:

$$2(CH_3)_2N\overset{S}{\overset{\|}{C}}SNH_4 + ZnSO_4 \rightarrow \left[(CH_3)_2N\overset{S}{\overset{\|}{C}}S\right]_2 Zn + (NH_4)_2SO_4$$

Ammonium dimethyldithiocarbamate is obtained by the reaction of equimolecular quantities of carbon disulfide, ammonia, and dimethylamine in aqueous medium:

$$(CH_3)_2NH + NH_3 + CS_2 \rightarrow (CH_3)_2N\overset{S}{\overset{\|}{C}}SNH_4$$

Both the first and the second reactions proceed at room temperature and the target products are obtained in good yields.

2. Reaction of carbon disulfide, zinc oxide, and dimethylamine at 30°–40° C., with a small excess of the dimethylamine:

$$CS_2 + (CH_3)_2NH + ZnO \rightarrow \left[(CH_3)_2N\overset{S}{\overset{\|}{C}}S \right]_2 Zn + H_2O$$

After removal of the ziram from the aqueous mother liquor, the latter is returned to the process for the subsequent operation. This method completely eliminates wastewater.

By both the first and second methods for the synthesis of ziram, small amounts of hydrogen sulfide and zinc dimethylthiocarbamate are obtained because of side reactions. Ziram is used as a protective fungicide for the treatment of growing plants at a concentration of 0.2–0.7% active ingredient.

Ferric dimethyldithiocarbamate (ferbam) is a dark brown crystalline substance that breaks down at 180° C. At 20° C., 1 liter of water dissolves 120 mg. of the compound; it is soluble in chloroform, pyridine, acetonitrile, etc. Like zinc dimethyldithiocarbamate, ferbam has a chelate structure. In the absence of moisture in the cold it is rather stable, but when it is heated in the moist condition it may decompose; it also is decomposed by the action of alkaline agents. The LD_{50} for rats is 4,000 mg./kg. Ferbam is used in the form of a wettable powder containing 60–80% active ingredient to protect plants from diseases.

It is produced by the reaction of sodium dimethyldithiocarbamate with ferric chloride:

$$3(CH_3)_2N\overset{S}{\overset{\|}{C}}SNa + FeCl_3 \rightarrow \left[(CH_3)_2N\overset{S}{\overset{\|}{C}}S \right]_3 Fe + 3\, NaCl$$

A number of mixed preparations containing other substance besides the salts of dimethyldithiocarbamic acid also are used. For example, Vancide-F contains 90% ferbam and 10% mercaptobenzothiazole; Vancide-M contains manganese dimethyldithiocarbamate and mercaptobenzothiazole; and Vancide-Z contains 90% ziram and 10% zinc salt of mercaptobenzothiazole. In France a mixture of ziram with copper oxychloride is used (15% ziram and 35% copper oxychloride calculated as copper).

Tetramethylthiuram disulfide (thiram, Thiuram-D, TMTD) is a white or cream-colored crystalline substance, m.p. 155°–156° C. It is practically insoluble in water and slightly soluble in most organic solvents. TMTD is stable in storage and is not explosive since it is nonvolatile, but in the form of a fine dust it gives explosive mixtures with air. The LD_{50} is 780 mg./kg. Finely powdered TMTD can cause irritation on contact with the skin and mucous membranes. When hens are fed a diet containing 35 mg./kg. of this compound the number of eggs laid is reduced.

Thiram is produced in good yields by the oxidation of the alkali salts of dimethyldithiocarbamic acid with hydrogen peroxide or other oxidizing agents (for example, sodium nitrite).

In agriculture seeds of corn, beans, and many other crops are disinfected with thiram. It also is used to control soil fungi that cause root rot diseases of plants. Thiram is most often used as a seed disinfectant together with insecticides (γ-benzene hexachloride, heptachlor, dieldrin, etc.) for combined protection of seedlings from diseases and soil-inhabiting insect pests. Thiram is sometimes used in mixtures with fungicides and bactericides, since it acts only on some species of fungi and has almost no effect on phytopathogenic bacteria. Examples of this type of preparation are Fenthiuram and Fenthiuram-molybdate. Fenthiuram contains 40% thiram, 10% copper 2,4,5-trichlorophenolate, 20% γ-benzene hexachloride or heptachlor, and a diluent. Fenthiuram-molybdate contains, in addition to the compounds listed, 8% ammonium molybdate as a microfertilizer for legumes.

Table XXXIX. *Fungicidal activity of derivatives of dithiocarbamic acid*

Compound	Min. conc. of compound causing complete inhibition of growth of fungi (mg./l.)			
	Botrytis cinerea	*Penicillium italicum*	*Aspergillus niger*	*Rhizopus nigricans*
$(CH_3)_2NCSSNa$	0.2	0.5	20	2
$(CH_3)_2NCSSSCSN(CH_3)_2$	0.2	0.2	10	2
$(C_3H_7)_2NCSSNa$	200	200	200	1000
$CH_3NHCSSNa$	10	10	50	200
$CH_3NHCSSSCSNHCH_3$	5	5	20	100
$NaSSCNHCH_2CH_2NHCSSNa$	0.1	0.1	0.5	20
$(-SCSNHCH_2CH_2NHCSS-)_n$	0.2	0.2	1	50
$SCNCH_2CH_2NCS$	0.05	0.02	0.05	10
$NaSCSNH(CH_2)_4NHCSSNa$	1	1	5	50
$SCN(CH_2)_4NCS$	0.05	0.05	0.5	50
$NaSCSNH(CH_2)_6NHCSSNa$	2	2	5	100
$SCN(CH_2)_6NCS$	0.05	0.05	1	>5000
$NaSCSNH(CH_2)_8NHCSSNa$	5	2	10	500
$SCN(CH_2)_8NCS$	1	0.5	1	>1000
$NaSCSNH(CH_2)_{10}NHCSSNa$	10	10	100	1000
$SCN(CH_2)_{10}NCS$	>200	>200	>200	>200

Thiram is employed as a wettable powder to protect growing plants from diseases. For example, it gives good results in the control of gray mold of amygdalaceous crops and strawberries, and a number of other diseases. Thiram and ziram are used in the rubber industry as accelerators of the vulcanization of rubber. Tetraethylthiuram disulfide has a considerably weaker fungicidal effect. It is employed in the treatment of chronic alcoholism because in the body it blocks the oxidation of ethyl alcohol at the stage of acetaldehyde.

Tetramethylthiuram monosulfide also shows a fungicidal effect, but a considerably weaker one than that of the disulfide.

The salts of ethylenebis(dithiocarbamic) acid with respect to scale of production and use in agriculture occupy first place among the fungicides employed for the protection of growing plants. Four salts of this acid are marketed: the ammonium (amoben), sodium (nabam), zinc (zineb), and manganese (maneb) salts, and a large number of combinations of these salts with other compounds. The salts of ethylenebis(dithiocarbamic) acid considerably surpass in fungicidal strength the corresponding salts of dimethyldithiocarbamic acid and similar compounds. Table XXXIX gives the minimum concentrations of derivatives of dithiocarbamic acid that suppress the growth of four species of fungi.

The sodium and ammonium salts of ethylenebis(dithiocarbamic) acid are highly phytocidal for green plants and are, therefore, employed only for soil treatment or in a mixture with zinc sulfate. These salts are being displaced more and more by zineb and maneb and their production is gradually being reduced. Diammonium and disodium ethylenebis(dithiocarbamates) have a slight systemic effect.

The diammonium ethylenebis(dithiocarbamate) usually is marketed as an aqueous solution, since it breaks down rapidly in the dry state. Its LD_{50} is about 400 mg./kg. (calculated on the dry basis). Frequently for on-the-spot use it is treated with calcium hypochlorite solution and thiuram disulfide is thus formed which is less phytotoxic for green plants and can be used for spraying potatoes and other crops.

Diammonium ethylenebis(dithiocarbamate) is produced by the reaction of ammonia, carbon disulfide, and ethylenediamine in water at a temperature not higher than 40° C. (preferably not above 25° C.):

$$\begin{array}{c} CH_2NH_2 \\ | \\ CH_2NH_2 \end{array} + 2\,CS_2 + 2\,NH_3 \rightarrow \begin{array}{c} CH_2NHC(S)SNH_4 \\ | \\ CH_2NHC(S)SNH_4 \end{array}$$

This salt also is an intermediate product in the production of zineb.

The solubility of disodium ethylenebis(dithiocarbamate), nabam, in water is about 20%. Practically insoluble in most hydrophobic organic solvents, it crystallizes from water in the form of the hexahydrate; in storage it breaks down easily. The anhydrous salt also is unstable in storage. It is formed by the reaction of NaOH, carbon disulfide, and ethylenediamine in aqueous medium:

$$\begin{array}{c} CH_2NH_2 \\ | \\ CH_2NH_2 \end{array} + 2\,CS_2 + 2\,NaOH \rightarrow \begin{array}{c} CH_2NHC(S)SNa \\ | \\ CH_2NHC(S)SNa \end{array} + 2\,H_2O$$

In the United States nabam is marketed in the form of a 19% aqueous solution (based on the anhydrous salt or 27% on the hexahydrate). For spraying it usually is mixed on-the-spot with zinc sulfate; more rarely it is employed for soil treatment.

The ammonium and sodium salts are reactive and may undergo various conversions. The most important reactions of these compounds are oxidation and reaction with heavy metals with which they form salts that are slightly soluble in water. The principal reactions of the salts of ethylenebis-(dithiocarbamic) acid that are of practical importance are shown in the following diagram:

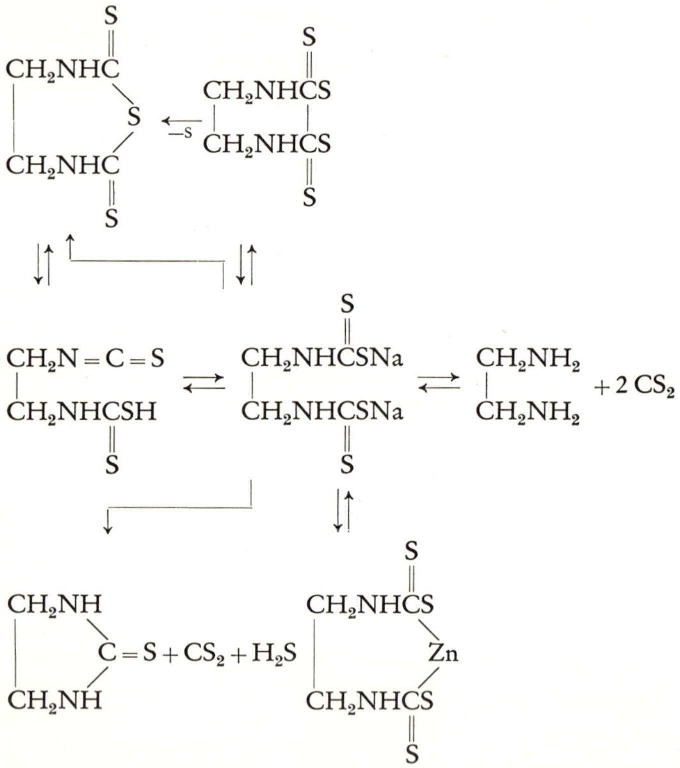

$$
\begin{array}{c}
\text{CH}_2\text{NHC(=S)} \\
| \quad\quad\quad \text{S} \\
\text{CH}_2\text{NHC(=S)}
\end{array}
\xleftarrow{-S}
\begin{array}{c}
\text{CH}_2\text{NHCS} \\
| \\
\text{CH}_2\text{NHCS}
\end{array}
$$

$$
\begin{array}{c}
\text{CH}_2\text{N=C=S} \\
| \\
\text{CH}_2\text{NHCSH} \\
\| \\
\text{S}
\end{array}
\rightleftarrows
\begin{array}{c}
\text{CH}_2\text{NHCSNa} \\
| \\
\text{CH}_2\text{NHCSNa} \\
\| \\
\text{S}
\end{array}
\rightleftarrows
\begin{array}{c}
\text{CH}_2\text{NH}_2 \\
| \\
\text{CH}_2\text{NH}_2
\end{array}
+ 2\,\text{CS}_2
$$

$$
\begin{array}{c}
\text{CH}_2\text{NH} \\
\quad\quad \diagdown \\
\quad\quad\quad \text{C=S} + \text{CS}_2 + \text{H}_2\text{S} \\
\quad\quad \diagup \\
\text{CH}_2\text{NH}
\end{array}
\quad\quad
\begin{array}{c}
\text{CH}_2\text{NHCS} \\
| \quad\quad \diagdown \\
\quad\quad\quad \text{Zn} \\
| \quad\quad \diagup \\
\text{CH}_2\text{NHCS} \\
\|\\
\text{S}
\end{array}
$$

It is the opinion of most investigators that the mechanism of action of the derivatives of ethylenebis(dithiocarbamic) acid differs from that of the derivatives of dimethyldithiocarbamic acid. This difference is associated first with their different structure, since the presence of hydrogen on the nitrogen in the ethylenebis(dithiocarbamates) is strongly reflected in their reactivity, which is considerably higher than that of the dimethyldithiocarbamic acid derivatives. It is assumed that the action of the ethylenebis-(dithiocarbamates), apart from their effect on the oxidation-reduction systems of fungi, is associated with the easy formation of the corresponding isothiocyanates, which can interact with various active groups in protein molecules. This assumption is supported by the higher fungicidal activity of the corresponding alkylenebis(isothiocyanates).

Zineb. One of the most important fungicides used in agriculture is zinc ethylenebis(dithiocarbamate). It is a white crystalline substance that breaks down before melting: decomposition temperature from 140°–160° C. At 20° C., about 0.001 g. dissolves in 1 liter of water; it is almost insoluble in most organic solvents, but moderately soluble in pyridine. The LD_{50} for experimental animals is 2,000—5,000 mg./kg. It is unstable in the presence of moisture and light. The moist compound (containing more than 4% moisture) under unfavorable storage conditions may decompose to the extent of more than 50% in a year. To avoid decomposition with the evolution of explosive carbon disulfide it is recommended that zineb be stored on shelves in well-ventilated places at the lowest possible temperature.

Zineb also breaks down relatively rapidly (10–15 days) on plants with the final formation of easily volatile products. In warmer and clearer weather the breakdown of the compound takes place considerably faster. The final product of this decomposition is zinc sulfite, which may play the role of a zinc micronutrient. In this connection, an increase in yield of a number of crops has been noted when they were treated with zineb.

Methods of determining zineb on plants based on the determination of zinc (e. g., the polarographic method) do not give dependable results, because only the decomposition product, zinc sulfite, is determined.

In agriculture a wettable powder containing 70–90% zineb is most often used. The plants are sprayed with the preparation at 0.2–0.5% active ingredient.

Zineb can be obtained by the following two principal methods:

1. Precipitation of zinc ethylenebis(dithiocarbamate) from aqueous solutions of diammonium or disodium ethylenebis(dithiocarbamate) by zinc sulfate:

$$\begin{array}{c} S \\ \| \\ CH_2NHCSNH_4 \\ | \\ CH_2NHCSNH_4 \\ \| \\ S \end{array} + ZnSO_4 \rightarrow \begin{array}{c} S \\ \| \\ CH_2NHCS \\ | \quad\quad\quad\quad Zn \\ CH_2NHCS \\ \| \\ S \end{array} + (NH_4)_2SO_4$$

This reaction should be carried out in dilute solutions (not higher than 5–7%), because polymeric zineb precipitates from concentrated solutions, and its toxicity for plant disease agents is somewhat lower. The zineb formed is filtered off, washed with water, surface-active agents are added, and the mixture is dried in a spray dryer; after supplementary grinding a very fine powder is obtained that is easily dispersed in water. In the drying it is necessary to remember that prolonged heating above 120° C. may lead to decomposition of the compound with the formation of carbonaceous products.

2. Reaction of zinc oxide, carbon disulfide, and ethylenediamine:

$$\begin{array}{c} CH_2NH_2 \\ | \\ CH_2NH_2 \end{array} + 2\,CS_2 + ZnO \rightarrow \begin{array}{c} CH_2NHCS \\ | \\ CH_2NHCS \end{array}\!\!\!\!\!\!\!\!\overset{S}{\underset{S}{\|}}\!\!\!\!\!\!\!\!Zn + H_2O$$

To decrease side reactions forming zinc 2-aminoethylenedithiocarbamate, it is advisable to add ammonia to the reaction mixture in the amount of 50 mole-% of the ethylenediamine used. In this way zineb is obtained in more than 90% yield and not less than 90% purity. The synthesis is carried out at 20°–30° C., with thorough stirring of the reaction mixture. The reaction takes four to six hours. After removal of the zineb by filtration, the mother liquor is returned to the process for the next operation; then the zineb is washed and dried as described.

Zinc 2-aminoethylenedithiocarbamate also is a good fungicide and even somewhat surpasses zineb in the strength of its action, but it is less stable in storage. It is synthesized by the reaction of ethylenediamine, carbon disulfide, and zinc oxide according to the equation:

$$2\begin{array}{c} CH_2NH_2 \\ | \\ CH_2NH_2 \end{array} + 2\,CS_2 + ZnO \rightarrow \left[NH_2CH_2CH_2NH\overset{S}{\underset{}{\overset{\|}{C}}}S \right]_2 Zn + H_2O$$

In addition to zineb, use has begun in agriculture of one of its compounds with ethylenebis(thiuram polysulfide) that has a stronger fungicidal effect than zineb. This preparation, which has been given the name Polycarbacine or Polyram, is a creamy white substance that decomposes above 120° C. It is insoluble in water and most organic solvents. The LD_{50} is more than 6,400 mg./kg. It is marketed in the form of a wettable powder containing 60–90% active ingredient, which is used as a 0.2–0.5% suspension in water.

Polycarbacine has the following composition:

$$\left[-CH_2NH\overset{S}{\overset{\|}{C}}S\overset{S}{\overset{\|}{C}}NHCH_2- \right]_n \left[-CH_2NH\overset{S}{\overset{\|}{C}}SZnS\overset{S}{\overset{\|}{C}}NHCH_2- \right]_m$$

$$(n:m=1:3)$$

It is obtained by simultaneous oxidation of an aqueous solution of sodium ethylenebis(dithiocarbamate) by hydrogen peroxide (or other suitable oxidizing agent) and precipitation with an aqueous solution of zinc sulfate. It is possible to prepare it from zinc oxide, carbon disulfide, ethylenediamine, and hydrogen peroxide. Less zinc oxide is used for the reaction than for the production of zineb, and the ammonium ethylenebis(dithiocarbamate) remaining in solution is oxidized by hydrogen peroxide with simultaneous acidification of the reaction medium with sulfuric acid.

Combinations of zineb with copper oxychloride and other inorganic and organic compounds of copper also are marketed. For example, preparations are known that contain 15% zineb and 25% copper, 20% zineb and 35% copper, and 40% zineb and 25% copper (in the form of copper oxychloride). A preparation known as Cuprocin also is being investigated that contains 5% copper in the form of copper ethylenebis(dithiocarbamate) obtained from zinc oxide and copper oxide with ethylenediamine and carbon disulfide in the presence of ammonia.

For control of phytophthora of potatoes, rust of grains, and some other plant diseases the manganese salt of ethylenebis(dithiocarbamic) acid, maneb, also is used. Maneb is a yellow crystalline substance, insoluble in water and organic solvents. The pure compound breaks down at about 120° C., and the technical grade product below 100° C. When stored in bulk, maneb is capable of spontaneous decomposition with charring. This property is less marked when the compound is diluted with an inert diluent. However, preparations with a low content of the active ingredient decompose very readily at an elevated temperature in the presence of moisture. According to numerous data, drying of the compound after it has been prepared should be carried out at a temperature not higher than 45° C. The use of urotropin, paraform, and other compounds as stabilizers has been proposed, but such stabilizers do not change the stability very substantially. The LD_{50} for rats is 6,700 mg./kg.

Maneb surpasses zineb in activity against phytophthora in potatoes. It is marketed in the form of wettable powders containing 70–80% active ingredient. It is produced in a manner similar to zineb, by the reaction of manganese sulfate with diammonium or disodium ethylenebis(dithiocarbamate) in aqueous solution:

$$\begin{array}{c} S \\ \| \\ CH_2NHCSNH_4 \\ | \\ CH_2NHCSNH_4 \\ \| \\ S \end{array} + MnSO_4 \rightarrow \begin{array}{c} S \\ \| \\ CH_2NHCS \\ | \\ CH_2NHCS \\ \| \\ S \end{array}\!\!\!Mn + (NH_4)_2SO_4$$

The preparation of maneb by the reaction of manganous oxide with ethylenediamine and carbon disulfide has been patented:

$$\begin{array}{c} CH_2NH_2 \\ | \\ CH_2NH_2 \end{array} + 2\,CS_2 + MnO \rightarrow \begin{array}{c} S \\ \| \\ CH_2NHCS \\ | \\ CH_2NHCS \\ \| \\ S \end{array}\!\!\!Mn + H_2O$$

Interesting field results also are obtained with a preparation containing about 10% zineb and 70% maneb that is known in the United States as Dithane M-45.

To control rust on grains it is recommended that maneb be used with the addition of nickel sulfate (Dithane C-31).

Analogs of zineb and maneb containing various aliphatic and aromatic radicals in place of hydrogen on the nitrogen also have fungicidal activity, but investigation of them still has not gone beyond the framework of laboratory tests.

Practical use has been made in agriculture of zinc 1,2-propylenebis(dithiocarbamate) (propineb), which surpasses zineb in fungicidal effect. It is similar to zineb in properties and methods of preparation.

Of the new compounds based on ethylenebis(dithiocarbamic) acid, mention should be made also of Carbathene, or thioneb, which is a mixture of 80% polyethylenebis(thiuram disulfide) and 20% polyethylenebis(thiuram monosulfide). This preparation is interesting because it leaves no residue on plants. It is low in toxicity to vertebrate animals and is effective in controlling diseases of fruit and berry crops, potatoes, and grapes.

Thioneb can be prepared by oxidation of the alkali metal or ammonium salts of ethylenebis(dithiocarbamic) acid by hydrogen peroxide, chlorine, or other oxidizing agents:

$$\begin{array}{c} S \\ \| \\ CH_2NHCSNH_4 \\ | \\ CH_2NHCSNH_4 \\ \| \\ S \end{array} \xrightarrow{[O]} \left[\begin{array}{c} S \\ \| \\ CH_2NHCS \\ | \\ CH_2NHCS \\ \| \\ S \end{array}\right]_n \longrightarrow \begin{array}{c} S \\ \| \\ CH_2NHC \\ | \\ CH_2NHC \\ \| \\ S \end{array}\!\!\!\!\!\!\!\!\!\!\!\!\diagdown_{\diagup}^{\diagup} S + S$$

The product contains as an impurity a small amount of sulfur, which also has some fungicidal effect.

The compound Lithuram (DPDT), dipyrrolidylthiuram disulfide, which is obtained by the oxidation of sodium or ammonium pyrrolidyldithiocarbamate, also is used as a fungicide:

$$\underset{H}{\underset{|}{\bigcirc_N}} + CS_2 + NaOH \longrightarrow \underset{S=CSNa}{\underset{|}{\bigcirc_N}} \longrightarrow \left[\underset{S=CS-}{\underset{|}{\bigcirc_N}}\right]_2$$

Esters of dithiocarbamic acids

Esters of dithiocarbamic acids substituted on the nitrogen have more limited use, since their pesticidal activity for various species of organisms is lower than that of the salts of these acids.

The methyl (Cystogon) and ethyl (Compound No. 23) esters of dimethyldithiocarbamic acid are used to control nematodes. They are employed in the form of 20 and 10% dusts on kaolin for application to the soil at dosages up to 150 mg./kg. (based on active ingredient). Advantages of these compounds are their low toxicity to mammals and their low phytotoxicity, which permits their use in controlling nematodes during the growing season.

The methyl ester of dimethyldithiocarbamic acid is produced by the reaction of sodium dimethyldithiocarbamate with dimethyl sulfate in aqueous medium in the presence of a small amount of emulsifier at 40° to 50° C.:

$$(CH_3)_2NCSNa + (CH_3)_2SO_4 \rightarrow (CH_3)_2NCSCH_3 + CH_3NaSO_4$$
$$\overset{S}{\underset{\|}{}}\qquad\qquad\qquad\overset{S}{\underset{\|}{}}$$

The ethyl ester of dimethyldithiocarbamic acid is synthesized by alkylation of sodium dimethyldithiocarbamate with ethyl bromide or chloride in an autoclave at about 100° C.:

$$(CH_3)_2NCSNa + C_2H_5Cl \rightarrow (CH_3)_2NCSC_2H_5 + NaCl$$
$$\overset{S}{\underset{\|}{}}\qquad\qquad\qquad\overset{S}{\underset{\|}{}}$$

The persistent and unpleasant odor and relatively high cost of these compounds limit their field of use.

2-Chloroallyl N,N-diethyldithiocarbamate (Vegadex, CDEC) is employed for the control of weeds in vegetable plantings. It is an oily liquid, b.p. 128°–230° C. at 1 mm. of Hg, solubility in water about 0.01%. It is highly soluble in organic solvents and is miscible in all proportions with most aromatic hydrocarbons. Vegadex is moderately toxic to mammals. The LD_{50} for rats is 850 mg./kg.

It is employed as a pre-emergence or preplanting herbicide by application to the soil with subsequent covering to a depth of 2–5 cm. At dosages of 3–9 kg./ha., depending on the nature of the soil, good results have been obtained in the eradication of barnyard millet, annual meadowgrass, henbit, thorn apple, goosefoot, amaranth, starwort, plantain, and wild oat.

Vegadex is stable in storage, but on boiling with caustic alkalies it breaks down with the formation of the diethyldithiocarbamate of the alkali metal. Strong oxidizing agents also decompose the compound to complete breakdown of the molecule.

2-Chloroallyl N,N-diethyldithiocarbamate is produced by the reaction of a 23% aqueous solution of sodium diethyldithiocarbamate with 2,3-dichloropropylene-1 at an elevated temperature and with thorough mixing:

$$(C_2H_5)_2NCSNa + ClCH_2CCl=CH_2 \rightarrow$$
$$\overset{S}{\underset{\|}{}}$$
$$(C_2H_5)_2NCSCH_2CCl=CH_2 + NaCl$$
$$\overset{S}{\underset{\|}{}}$$

The yield by this reaction is about 90% of theoretical.

A large number of other esters of dithiocarbamic acid have been described, but they have not yet been practically utilized in agriculture.

General references

Banki, L., M. Hamran, Gy. Josepovits, and G. Matolcsy: Acta Phytopathol. **1**, 223 (1966).
Baskakov, Yu. A.: In book „Novye insektofungitsidy i gerbitsidy" [New insectofungicides and herbicides] pp. 185–237. Foreign Literature Publishing House [USSR] (1960).
Carter, G. A., J. L. Garraway, D. M. Spencer, and R. L. Wain: Ann. Applied Biol. **51**, 135 (1963).
Golyshin, N. M.: Khim. v Sel'skom Khozyaistve No. 1, 30 (1964).
Heyns, A. J., G. A. Carter, K. Rothwell, and R. L. Wain: Ann. Applied Biol. **57**, 33 (1966).
Khimicheskie sredstva zashchity rastenii [Chemical agents for plant protection] Coll. vol. State Scientific and Technical Publishing House of Chemical Literature [USSR] (1961).
Matolcsy, G., and Gy. Josepovits: Acta Chim. Acad. Hung. **51**, 319 (1967).
Mel'nikov, N. N., and Yu. A. Baskakov: Khimiya gerbitsidov i regulyatorov rosta rastenii [Chemistry of herbicides and plant growth regulators]. State Scientific and Technical Publishing House of Chemical Literature [USSR] (1962).
—, E. M. Sokolova, P. P. Trunov, S. D. Volodkovich, G. M. Dymshakova, A. P. Burdakova, and L. D. Nayanov: Khim. Prom. p. 652 (1967).
Nase, B.: Zeitschr. für. Chem. **8**, 96 (1968).
Pianka, M., J. D. Edwards, and C. B. F. Smith: J. Sci. Food Agr. **17**, 407 (1966).
Polyakov, I. M., M. E. Vladimirskaya, and N. V. Shmettse: Khim. v Sel'skom Khozyaistve No. 12, 27 (1965).
Pronchenko, T. S., E. I. Andreeva, and V. I. Obukhova: Khim. v Sel'skom Khozyaistve No. 7, 30 (1965).
Vasil'ev, I. I., and O. V. Mitrofanova: Khim. v Sel'skom Khozyaistve No. 1, 30 (1964).
Zbarskii, E. A., and E. I. Andreeva: Khim. v Sel'skom Khozyaistve No. 4, 35 (1965).

XVIII. Derivatives of urea and thiourea

General characteristics of pesticidal properties

Urea is one of the products of the life activity of animals. It is used in agriculture as an excellent concentrated fertilizer. Practically all plants tolerate the use of comparatively large dosages of this compound without any harm. But even the simplest derivative of urea, biuret, has appreciable phytocidal effect and at dosages of 40–70 kg./ha. it inhibits the growth of many monocotyledonous and dicotyledonous plants. It is suitable for the control of weeds in crops of cotton, corn, and others. In spite of the large dosages, biuret may have practical importance, since in the soil it undergoes hydrolysis with the formation of ammonia, which is utilized by crop plants as a source of nourishment.

Systematic study of the pesticidal properties of urea derivatives has established that the insecticidal activity of most of the compounds is insignificant; many compounds of this class are active herbicides and a number of them are employed in agricultural practice. The fungicidal properties of the substituted ureas are relatively slight but derivatives of thiourea, and especially of guanidine, have high fungicidal activity and are used in agriculture.

The herbicidal properties of various derivatives of urea have been studied in the greatest detail. In a search for effective herbicides and growth regulators several thousand substituted ureas have been synthesized, including representatives of the aliphatic, alicyclic, aromatic, and heterocyclic series.

Active herbicides have been found among the trialkylureas that contain both the simplest and the more complex hydrocarbon radicals. For example, it has been proposed that for the control of weeds in cotton and corn N-butyl-N',N'-dimethylurea (b.p. 106°–108° C. at 0.7 mm. of Hg) be used; this compound at dosages of 3–4 kg./ha. inhibits the growth of many weeds. According to patent data the same properties are shown also by trialkylureas that contain three different hydrocarbon radicals. Halo- and polyhaloallyl-N',N'-dimethylureas also are active herbicides. All these compounds are obtained in good yields by the reaction of alkyl isocyanates with amines:

$$RNCO + R'_2NH \rightarrow R'_2N\overset{O}{\underset{\|}{C}}NHR$$

The symmetrical tetraalkylureas do not have a herbicidal effect.

Of the aliphatic derivatives of urea, dichloralurea is used as a herbicide for the control of weeds in sugar beets. Active herbicides also are known among the dialkylcycloalkylureas, for example, N-cyclooctyl-N',N'-dimethylurea, N-(chloronorbornyl-2)-N',N'-dimethylurea, N-(tetrahydrodicyclopentadienyl)-N',N'-dimethylurea, and N-(5,6-dihydrodicyclopentadienyl-5)-N',N'-dimethylurea.

Derivatives of urea containing aliphatic and aromatic radicals on the nitrogen have been studied in detail. Among them are found active herbicides that have received wide use in agriculture and industry.

Although there have been no systematic investigations of the relationship of herbicidal effect to structure of the substituted ureas and the available information is concentrated mainly in numerous patents, some generalizations still can be made in this area.

The N-aryl-N',N'-dialkylureas that contain as the aromatic radical a phenyl group in which not more than two hydrogen atoms are replaced by functional groups are active herbicides. In this case at least one of the *ortho*-positions to the amide group should be unsubstituted. The total number of carbon atoms in the two alkyl groups should not be more than five. If the alkyls contain more than five carbon atoms, the activity of the compound decreases. This situation does not extend to the arylcycloalkylureas. An example of such compounds containing more than five carbon atoms in the cycloalkyl radical is N-phenyl-N'-(2-methylcyclohexyl)urea (m.p. 133°–138° C., LD_{50} for rats 5,000 mg./kg.), used as a pre-emergence selective herbicide.

When the number of carbon atoms in the alkyl or cycloalkyl radicals bound to the second nitrogen atom is increased, the strength of the herbicidal effect is decreased, but the selectivity is greater.

Among the haloaryldimethylureas the greatest activity is shown by the chloro derivatives; the bromo derivatives are less active, and the iodo derivatives still less so. The position of the halogen with respect to the amide group also is of considerable importance. The *para*-derivatives are much more active than the *ortho*-derivatives.

Among the dihalo derivatives of aryldimethylureas the maximum activity is shown by the 3,4-dihalo compounds.

When an aryloxy or alkoxy group is introduced into the phenyl radical in the *para*- position to the amide group, active selective herbicides are formed, but they are somewhat less effective than phenyldimethylurea. Thus, for example, N-butoxyphenyl-N',N'-dimethylurea (m.p. 116° to 117.5° C.) is harmless to onions and effectively inhibits the growth of dicotyledonous and monocotyledonous weeds; 4-chlorophenoxyphenyldimethyl urea can be used to control weeds in potatoes, carrots, gladiolas, and other crops at dosages up to 6 kg./ha., while phenyldimethylurea is dangerous for these crops even at a dosage of 3 kg./ha.

Replacement of hydrogen in the phenyl radical by a trifluoromethyl group in the *meta*-position also lowers the activity of the compound with a simultaneous increase in the selectivity of action.

Naphthyldialkylureas have a weaker effect than phenyldialkylureas. Aryltrialkylureas are somewhat less active than aryldialkylureas, and diaryl- and triarylureas show weak herbicidal activity. Introduction of various substituents into an aliphatic radical of aryldialkylureas usually does not substantially decrease the activity of the compounds, but improves their physical properties. Thus, when a carboxyl group is introduced, the solubility in water is increased without lowering the herbicidal activity. Introduction into an aliphatic radical of a nitrile, aldehyde, ether, sulfide, or some other groups also produces no negative effect on the herbicidal activity of the compound. The preparation of such compounds is more complicated, however, and therefore they still are not used in agriculture.

N-Aryl-N'-alkylureas are somewhat less active than their dialkyl analogs, but the N-aryl-N'-alkyl-N'-acylureas have high herbicidal activity. Replacement of one alkyl group on the nitrogen by an alkoxyl does not cause a lowering in activity of the compound but increases its selectivity.

Introduction of a hydroxyl on the nitrogen of the aryl group leads to a small decrease in activity, but strongly increases the selectivity of action of the compound. This last group of compounds was first proposed for use by Soviet investigators. Not only the N-aryl-N-hydroxy-N',N'-dialkylureas, but also the N-aryl-N-hydroxy-N'-alkylureas have high activity.

When the aromatic radical is replaced by a heterocyclic radical, in a number of cases active herbicides are produced. Examples of such compounds are N-methyl-N'-(2-benzothiazolyl)urea (m.p. 287° C., LD_{50} 1,000 mg./kg.) and N-methyl-N'-methyl-N'-(2-benzothiazolyl)urea (m.p. 119° C., LD_{50} 2,500 mg./kg.).

The salts of aryldialkylureas with various herbicidal acids are stronger herbicides than the ureas. The herbicidal activity of the substituted thioureas is lower than that of the corresponding ureas. However, a number of thioureas are highly toxic to rodents and are used to control them. An example of such compounds is 1-naphthylthiourea (krysid) which is widely used to control rats.

Many derivatives of thiourea have high fungicidal and bactericidal activity. In this respect the thiuronium salts containing long aliphatic chains are especially interesting. Some thiuronium salts also show a repellent effect on rodents, and therefore have been proposed for impregnation of containers. A number of thiuronium salts, according to patent data, also have a herbicidal effect. The use of this group of compounds in the national economy still is very limited, but there is no doubt that one can expect to obtain interesting compounds in this class of substances.

The use of aryltetralkylguanidines of general formula (I) as active herbicides has been patented. The activity of the aryldialkylguanidines (II) is lower than that of the corresponding ureas:

$$\text{ArN}=\text{C}\left(\text{N}\begin{smallmatrix}R\\R'\end{smallmatrix}\right)_2 \quad \text{(I)} \qquad \begin{matrix}\text{ArNHC}-\text{N(CH}_3)_2\\\|\\\text{NH}\end{matrix} \quad \text{(II)}$$

Alkylguanidines are eradicatory fungicides, and their fungicidal effect increases with an increase in length of the normal aliphatic radical. The maximum activity is reached at $C_{12}H_{25}$. An increase in length of the hydrocarbon chain beyond 12 carbon atoms lowers the activity of the compound. The most active compounds are the salts of alkylguanidines with acetic acid and other carboxylic acids. Of this group of compounds dodecylguanidine acetate, known as dodine, is used in agriculture.

Substituted ureas of the aliphatic and alicyclic series

Dichloralurea is a white crystalline substance, m.p. 194°–196° C., insoluble in water, somewhat soluble in cyclic ketones. The LD_{50} is 31,600 mg./kg. The compound is resistant to the action of mineral and organic acids, but breaks down rapidly when heated with caustic alkalies:

$$\left(\begin{array}{c} CCl_3CHNH \\ | \\ OH \end{array} \right)_2 CO + 4\ NaOH \rightarrow 2\ CHCl_2COONa + 2\ NaCl + CO(NH_2)_2 + 2\ H_2O$$

A similar process apparently may occur also in the soil under the influence of alkaline agents. In addition to the formation of salts of dichloroacetic acid, trichloroacetic acid also is obtained from dichloralurea on oxidation:

$$\left(\begin{array}{c} CCl_3CHNH \\ | \\ OH \end{array} \right)_2 CO + 2O \rightarrow 2\ CCl_3COOH + CO(NH_2)_2$$

It is possible that these reactions explain the high herbicidal activity of dichloralurea toward monocotyledonous weeds, since it approaches the salts of trichloroacetic acid in the strength of its effect on these plants.

Dichloralurea usually is marketed in the form of a 50% wettable powder. It is used for application to the soil in the form of an aqueous suspension, the dosage being 5–20 kg./ha., and gives satisfactory results in the control of weeds in sugar beets. However, the high rates of application and the presence of an irritating action on the mucous membranes decrease the value of this herbicide.

Dichloralurea is produced in more than 90% yield by the condensation of chloral with urea in the presence of hydrochloric acid with heating:

$$2\ CCl_3CHO + CO(NH_2)_2 \rightarrow \underset{\underset{OH}{|}}{CCl_3CHNH}CONH\underset{\underset{OH}{|}}{CHCCl_3}$$

Two moles of chloral are mixed with a concentrated aqueous solution of 1 mole of urea, a 20% solution of HCl is added, and the mixture is heated with stirring until precipitation ceases. The precipitate is filtered off and dried. This reaction can be carried out also in an organic solvent, increasing the yield of the product from 92–96%.

N-Cyclooctyl-N',N'-dimethylurea (OMU) is a white crystalline substance, m.p. 138° C.; at 20° C., 0.11% dissolves in water, 46.63% in methyl alcohol, 32.28% in ethyl alcohol, 5.53% in benzene, 6.67% in acetone, and 3.76% in ethyl acetate. The LD_{50} is 300 mg./kg. for mice and 1,500 mg./kg. for rats.

The compound is stable at room temperature and when heated to 50° C. in the absence of substances of an alkaline or acid nature. On prolonged heating with aqueous solutions of alkalies or mineral acids it is hydrolyzed with the formation of amides or their salts.

OMU is used to control monocotyledonous weeds at dosages of 0.8–2 kg. per ha., most often in combination with Bi-PC (60% OMU and 40% Bi-PC). Under the name of Alipur this preparation is recommended for the control of weeds in sugar beets.

The following three methods for the synthesis of OMU are described in the literature:

1. Reaction of cyclooctyl isocyanate with dimethylamine:

$$(CH_2)_7CHNH_2HCl + COCl_2 \rightarrow (CH_2)_7CHNCO + 3 HCl$$
$$(CH_2)_7CHNCO + HN(CH_3)_2 \rightarrow (CH_2)_7CHNHCON(CH_3)_2$$

The cyclooctyl isocyanate necessary for this synthesis is obtained in good yields from cyclooctylamine hydrochloride and phosgene in dioxane at 90° C.

2. Reaction of dimethylcarbamoyl chloride with cyclooctylamine in the presence of an HCl acceptor:

$$(CH_3)_2NCOCl + (CH_2)_7CHNH_2 \rightarrow (CH_2)_7CHNHCON(CH_3)_2 + HCl$$

3. Reaction of amines with urea:

$$CO(NH_2)_2 + (CH_2)_7CHNH_2 \rightarrow (CH_2)_7CHNHCONH_2 + NH_3$$
$$(CH_2)_7CHNHCONH_2 + HN(CH_3)_2 \rightarrow (CH_2)_7CHNHCON(CH_3)_2 + NH_3$$

N-(Tetrahydrodicyclopentadienyl)-N',N'-dimethylurea (norea, Herban) is a white crystalline substance, m.p. 168°–169° C. At 20° C., 150 mg. dissolves in a liter of water. It is slightly soluble in benzene, xylene, benzine, and kerosine, but soluble in acetone, cyclohexanone, and ethyl alcohol. The LD_{50} for rats is 2,000 mg./kg.

Norea is marketed in the form of an 80% wettable powder for the control of weeds in cotton, soy beans, peas, and potatoes, the dosage usually being 1.2–4.5 kg./ha. It is similar to other derivatives of urea in chemical properties, and is produced by the reaction of dimethylcarbamoyl chloride with the amine or of tetrahydrodicyclopentadienyl isocyanate with dimethylamine:

A new herbicide, *N*-(5,6-dihydrodicyclopentadienyl-5)-*N'*,*N'*-dimethylurea (m.p. 111°–113° C.), which is similar in properties to norea is obtained by reaction of the corresponding isocyanate with dimethylamine:

$$\text{OCN-[dihydrodicyclopentadienyl]} + (CH_3)_2NH \longrightarrow (CH_3)_2NCONH\text{-[dihydrodicyclopentadienyl]}$$

N-(Chloronorbornyl-2)-*N'*,*N'*-dimethylurea (m.p. 203°–204° C., LD_{50} for rats 1,300 mg./kg.) is being tested under industrial conditions as a soil sterilant:

$$\text{Cl-[norbornyl]-NHCON(CH}_3)_2$$

N-Aryl-*N'*,*N'*-dialkylureas

The large majority of herbicidal derivatives of urea belong to the group of aryldialkylureas. The following three general methods are known for their preparation:

1. Condensation of aryl isocyanates with dialkylamines:

$$ArNCO + R_2NH \rightarrow ArNHCONR_2$$

This method is the most universal and produces various aryldialkylureas with a high degree of purity and in good yields.

2. Reaction of a dialkylcarbamoyl chloride with aniline:

$$R_2NCOCl + ArNH_2 \rightarrow R_2NCONHAr + HCl$$

3. Reaction of a dialkylamine with a diarylurea at an elevated temperature:

$$(ArNH)_2CO + R_2NH \rightarrow R_2NCONHAr + ArNH_2$$

N-Phenyl-N',N'-dimethylurea (fenuron, fenidim) is a white crystalline substance, m.p. 136° C.; at 24° C., 0.29% dissolves in water. It is highly soluble in alcohols, ketones, halogenated hydrocarbons, but barely soluble in saturated hydrocarbons. The LD_{50} for rats is 7,500 mg./kg.

At room temperature fenuron can be stored for an unlimited time without change. Upon boiling with caustic alkalies and mineral acids it breaks down with the formation of amines or their salts:

$$C_6H_5NHCON(CH_3)_2 + 2\ NaOH \rightarrow C_6H_5NH_2 + (CH_3)_2NH + Na_2CO_3$$

A method for the quantitative determination of fenuron in biological materials is based on this reaction.

Fenuron is an active root herbicide that is used to eradicate annual vegetation. The double compounds of phenyldimethylurea with halogen-substituted aliphatic acids are powerful nonselective herbicides. Such a complex with trichloroacetic acid (m.p. 65°–68° C., LD_{50} 4,000 mg. per kg.), known as "Urab", is especially active. It is obtained by melting equimolecular quantities of phenyldimethylurea and trichloroacetic acid.

For the industrial production of fenuron all three of the described methods of synthesis of aryldialkylureas can be used. The diphenylurea necessary for the third method is easily synthesized by heating urea with aniline to 140°–180° C.

N-4-Chlorophenyl-N',N'-dimethylurea (monuron, chlorfenidim) is a white crystalline substance, m.p. 170.5°–171.5° C. The technical grade compound usually melts at 164° C. The solubility of monuron in water at 25° C. is 230 mg./l. It is poorly soluble in kerosine, alcohol, petroleum ether, and benzene, but more soluble in halogenated hydrocarbons and dioxane. The LD_{50} to mammals is 3,500 mg./kg.

Monuron is stable under ordinary conditions of storage and withstands heating to the melting point without decomposition. Upon prolonged boiling with alkalies or mineral acids it breaks down with the formation of dimethylamine, *p*-chloroaniline, or their salts. It is widely used to control weeds in cotton and sugar cane, and in fruit orchards. The compound is employed as a selective herbicide at dosages of 1–3 kg./ha.; at higher dosages it eradicates almost all species of plants. With trichloroacetic acid chlorophenyldimethylurea gives a double compound (m.p. 78°–91° C., LD_{50} for rats 2,300 mg./kg.) that has a rapid herbicidal effect and can be employed to destroy undesirable brushwood.

Monuron is produced by all the methods described. Furthermore, it can be prepared by direct chlorination of phenyldimethylurea:

$$\text{C}_6\text{H}_5\text{-NHCON(CH}_3)_2 + \text{Cl}_2 \rightarrow \text{Cl-C}_6\text{H}_4\text{-NHCON(CH}_3)_2$$

By this reaction the compound is obtained with m.p. 165°–168° C.

N-3,4-Dichlorophenyl-N',N'-dimethylurea (diuron, dichlorfenidim) is a white crystalline substance, m.p. 158°–159° C., vapor pressure at 50° C. 0.31×10^{-5} mm. of Hg. At 25° C., 42 mg. of the compound dissolves in 1 liter of water; it is slightly soluble in most organic solvents. The LD_{50} for rats is 3,400 mg./kg.

On heating to 180°–190° C., the compound breaks down to dimethylamine and 3,4-dichlorophenyl isocyanate. Like fenuron, boiling with solutions of caustic alkalies or mineral acids in water causes diuron to break down with the formation of dimethylamine and 3,4-dichloroaniline or their salts.

In industry diuron is synthesized by the reaction of 3,4-dichlorophenyl isocyanate with dimethylamine. It is used in the form of aqueous suspensions to control weeds in cotton crops. At dosages of 1–2 kg./ha. it gives completely satisfactory results, but at dosages of 1.8–2 kg./ha., damage to cotton is sometimes observed.

N-3,4-Dichlorophenyl-N'-methyl-N'-butylurea (neburon) is a crystalline substance, m.p. 101.5°–103° C. At 24° C., 4.8 mg. dissolves in 1 liter of water; it is slightly soluble in aliphatic hydrocarbons, but more soluble in alcohols, ketones, and halogenated hydrocarbons. The chemical properties of neburon are similar to those of the urea derivatives described. It has low toxicity to mammals, with an LD_{50} of 11,000 mg./kg.

Neburon approaches the compounds described in herbicidal activity, but because of its low solubility in water it is safer for a number of crops. It is recommended for the control of weeds in vegetable crops, peanuts, alfalfa, and perennial grasses, in a dosage of 2–6 kg./ha. It is marketed in the form of a wettable powder.

The main method for the production of neburon is reaction of 3,4-dichlorophenyl isocyanate with methylbutylamine:

$$\text{Cl}_2\text{C}_6\text{H}_3\text{-NCO} + \text{HN(CH}_3\text{)CH}_2\text{CH}_2\text{CH}_2\text{CH}_3 \longrightarrow \text{Cl}_2\text{C}_6\text{H}_3\text{-NHCON(CH}_3\text{)CH}_2\text{CH}_2\text{CH}_2\text{CH}_3$$

N-4-(4'-Chlorophenoxy)phenyl-N',N'-dimethylurea (Tenoran) is a white crystalline substance, m.p. 151°–152° C. Its solubility in water is 37 mg./liter; it is moderately soluble in most organic solvents. The LD_{50} for rats is 1,000 mg./kg. It is marketed in the form of a 50% wettable powder and is recommended for the control of weeds in plantings of carrots, potatoes, asparagus, gladiolas, and other crops at dosages of 6–9 kg./ha.

It is reminiscent of the urea derivatives with respect to its chemical properties. It is produced by the reaction of 4-chlorophenoxyphenyl isocyanate with dimethylamine:

$$\text{Cl-C}_6\text{H}_4\text{-O-C}_6\text{H}_4\text{-NCO} + \text{NH(CH}_3\text{)}_2 \longrightarrow \text{Cl-C}_6\text{H}_4\text{-O-C}_6\text{H}_4\text{-NHCON(CH}_3\text{)}_2$$

or by the reaction of dimethylcarbamoyl chloride with 4-chlorophenoxyaniline. A purer product is obtaind by the first method.

N-4-(4'-Methoxyphenoxy)phenyl-N',N'-dimethylurea (C-3470) is a white crystalline substance, m.p. 136°–138° C. Its solubility in water is 20 ml./liter, in chloroform 1,250 mg./liter; highly soluble in alcohol and acetone. The LD_{50} for rats and mice is >1,000 mg./kg.; the toxicity for fish is 10–20 mg./liter. It is used in the form of a 50% wettable powder at a dosage of 2–3 kg./ha. for control of weeds in onion crops.

C-3470 is synthesized by the reaction of dimethylcarbamoyl chloride with 4-methoxy-4'-aminodiphenyl ether or of the corresponding isocyanate with dimethylamine.

N-(3-Trifluoromethylphenyl)-N',N'-dimethylurea (fluometuron, Cotoran, Herbicide C-2059) is a white crystalline substance, m.p. 163°–164.5° C., solubility in water at 25° C. 90 mg./liter; soluble in ethyl and isopropyl alcohols, acetone, acetonitrile, chloroform, and dimethylformamide. The LD_{50} is 850 mg./kg. for mice, and 6,000 mg./kg. for rats.

Fluometuron is a selective herbicide with prolonged residual action (two to five months) and is used to control weeds in cotton and some other crops at dosages of 1–4 kg./ha. (active ingredient). The compound is marketed in the form of an 80% wettable powder and is used in aqueous suspension applied at 500–1,000 liters/ha. The chemical properties of fluometuron are similar to those of the other derivatives of urea.

Fluometuron can be synthesized by the first two methods that are general for the aryldialkylureas:

$$F_3C\text{-}C_6H_4\text{-}NCO + (CH_3)_2NH \rightarrow F_3C\text{-}C_6H_4\text{-}NHCON(CH_3)_2$$

$$F_3C\text{-}C_6H_4\text{-}NH_2 + (CH_3)_2NCOCl \rightarrow F_3C\text{-}C_6H_4\text{-}NHCON(CH_3)_2$$

The necessary 3-trifluoromethylaniline is obtained by the following route:

$$C_6H_5\text{-}CH_3 \rightarrow C_6H_5\text{-}CCl_3 \rightarrow C_6H_5\text{-}CF_3 \rightarrow 3\text{-}NO_2\text{-}C_6H_4\text{-}CF_3 \rightarrow 3\text{-}NH_2\text{-}C_6H_4\text{-}CF_3$$

In addition to the urea derivatives described, the arylhydantoins (V) have been proposed as herbicides. They are easily produced by the action of water in the presence of HCl on aryl-N'-methyl-N'-cyanomethylureas:

$$\text{ArNHCONCH}_2\text{CN} + H_2O \rightarrow \text{arylhydantoin} + NH_3 \quad (V)$$

with CH₃ on the central N.

In Table XL the properties of some compounds of this type are given. Although the general herbicidal activity of this group of compounds is less than that of the aryldialkylureas, they are more selective, and therefore have a potential use in agriculture.

Of the derivatives of isourea the herbicidal compound trimeturon, N'-4-chlorophenyl-O,N,N-trimethylisourea (m.p. 147°–149° C., solubility in water 0.07%, LD_{50} 1,500 mg./kg.), has been used in agriculture:

$$Cl\text{-}C_6H_4\text{-}N=C(OCH_3)N(CH_3)_2$$

Table XL. *Properties of some arylhydantoin derivatives of the general formula*

R	X	X'	Y	Y'	M. P. (°C.)
CH_3	Cl	H	H	H	108–109
CH_3	H	H	Cl	H	119–120
CH_3	H	Cl	H	H	84–86
CH_3	Cl	H	Cl	Cl	—
CH_3	Cl	H	Cl	H	99–100
CH_3	H	CH_3	H	H	48–52
CH_3	Cl	H	H	Cl	120–121.5
CH_3	H	H	CH_3	H	114–117
CH_3	H	CH_3	CH_3	H	97–99
$(CH_3)_2CH$	H	H	Cl	H	107–109
$(CH_3)_2CH$	H	Cl	Cl	H	77–78
CH_3	H	H	CH_3O	H	117–120
CH_3	CH_3O	H	H	CH_3O	133–136.5
CH_3	H	CF_3	Cl	H	93–100
CH_3	H	H	F	H	135–136
CH_3	H	H	$N(CH_3)_2$	H	154–155

In addition to the aryldialkylureas, the arylalkyloxy- and alkoxyureas containing an alkoxyl or hydroxyl group on one of the nitrogen atoms are being used more frequently. The following herbicides belong to this class of compounds: linuron, monolinuron, Patoran, and other compounds that at the present time are undergoing industrial tests.

N-3,4-Dichlorophenyl-N'-methyl-N'-methoxyurea (linuron, Garnitan) is a white crystalline substance, m.p. 93°–94° C., solubility in water at 25° C. 75 mg./liter. It is soluble in alcohol, acetone, benzene, xylene, and toluene. The LD_{50} for rats is 1,500 mg./kg. It is marketed in the form of a 50% wettable powder and is intended for the control of weeds in corn, cotton, soy beans, and carrots, at a dosage of 1.5–6 kg./ha. On light soils it sometimes damages useful crops; for example cotton foliage is especially sensitive to it. In chemical properties the compound is less stable than its nearest analog, dichlorophenyldimethylurea. Thus, it is hydrolyzed in alkaline, and especially in acid media even at room temperature. Decomposition of linuron in the soil also takes place more rapidly. The most important methods for its synthesis are:

1. Methylation of 3,4-dichlorophenyl-N-hydroxyurea with dimethyl sulfate:

$$\text{Cl}_2\text{C}_6\text{H}_3\text{-NHCONHOH} + 2(CH_3)_2SO_4 + 2NaOH \longrightarrow$$

$$\text{Cl}_2\text{C}_6\text{H}_3\text{-NHCON}(CH_3)(OCH_3) + 2CH_3NaSO_4 + 2H_2O$$

The necessary 3,4-dichlorophenyl-N-hydroxyurea is prepared from 3,4-dichlorophenyl isocyanate and hydroxylamine.

2. Reaction of 3,4-dichlorophenyl isocyanate with O,N-dimethylhydroxylamine:

$$\text{Cl}_2\text{C}_6\text{H}_3\text{-NCO} + HN(CH_3)(OCH_3) \longrightarrow \text{Cl}_2\text{C}_6\text{H}_3\text{-NHCON}(CH_3)(OCH_3)$$

However, the preparation of O,N-hydroxylamine on a large scale presents well-known difficulties.

3. Methylation by dimethyl sulfate of N-3,4-dichlorophenyl-N'-methyl-N'-hydroxyurea, which is obtained from the isocyanate and N-methylhydroxylamine.

N-4-Chlorophenyl-N'-methyl-N'-methoxyurea (monolinuron, Aresin) is a white crystalline substance, m.p. 72° C. The technical grade product usually is a thick oil with crystals. The solubility of monolinuron in water at 20° C. is 580 mg./liter; it is soluble in alcohol, acetone, benzene, and other organic solvents, except saturated hydrocarbons. The LD_{50} for rats is 2,250 mg./kg. The compound is marketed in the form of a 50% wettable powder for the control of weeds in potatoes, and it is similar to linuron in chemical properties and method of preparation.

N-4-Bromophenyl-N'-methyl-N'-methoxyurea (Patoran, C-3126) is a white crystalline substance, m.p. 95.5°–96° C. At 20° C., 330 mg./liter dissolves in water; it is highly soluble in acetone, alcohol, and chloroform. Its LD_{50} for rats and mice is >1,000 mg./kg.; it does not irritate the skin of rabbits.

Patoran is marketed in the form of a 50% wettable powder for the control of weeds in potato fields, the dosage usually being 1.5–2.5 kg./ha. Its properties, like those of N-4-bromo-3-chlorophenyl-N'-methyl-N'-methoxymethylurea (m.p. 94°–96° C.), which has been proposed for the control of weeds in soy beans and carrots, are similar to the properties of linuron.

Meturin. N-Hydroxy(alkoxy)-N-aryl-N',N'-dialkyl- and N-hydroxy-(alkoxy)-N-aryl-N'-alkylureas of the general formula ArN(OH)CONRR'

(VI) have been patented as selective herbicides. The herbicidal properties of this class of compounds were first discovered by Soviet investigators in 1958. An example of this group of herbicides is the compound meturin, *N*-phenyl-*N*-hydroxy-*N'*-methylurea, m.p. 132° C., LD_{50} about 5,000 mg. per kg. It is highly soluble in organic solvents and in aqueous solutions of alkalies. It is produced by the reaction of methyl isocyanate with phenylhydroxylamine:

$$C_6H_5NHOH + CH_3NCO \rightarrow C_6H_5\underset{\underset{OH}{|}}{N}CONHCH_3$$

It is a selective herbicide for the control of weeds in cotton and potatoes.

Investigations on the herbicidal action of urea derivatives have established that aryldimethylureas are easily absorbed by roots and gradually accumulate in the underground part of plants, disrupting the nitrogen metabolism. Urea derivatives are not capable of penetrating into the phloem. When urea derivatives act on the chloroplasts they hinder the Hill reaction. They form complexes with the chloroplasts as a result of the formation of hydrogen bonds and hinder the vital processes of the plants. Evidence of the formation of hydrogen bonds is the fact that the harmful effect of urea derivatives on the chloroplasts can be significantly decreased by treatment with carbohydrates. Furthermore, the inhibition of photosynthetic processes in the chloroplasts by 4-chlorophenyldimethylurea is terminated after it is washed out with water.

However, the entire mechanism of the physiological action of the ureas still is not completely clear, and the solution of many problems is necessary for final elucidation of this process.

Thiourea and its derivatives

Thiourea is a physiologically active compound that disrupts the dormant state of potatoes, and makes possible the planting of potatoes that have just been harvested. It also causes defoliation of cotton, but has not been practically used for this purpose.

The thio analogs of the herbicidal aryldialkylureas are less active as herbicides and more toxic to mammals. Derivatives of isothiourea of the general formula (VII) have been proposed as herbicides:

<pre>
         N   NAr
          \\ ||
   [benzothiazole]—SCN(CH₃)₂   (VII)
</pre>

N,N-Dimethyl-*N'*-3-methylphenylthiourea *(methiuron)* is a white crystalline substance, m.p. 145° C., solubility in water 400 p.p.m., in methanol 7%, and in methylcyclohexanone 10%. Its LD_{50} is 2,200 mg./kg. Methiuron is active both through the foliage and through the soil on germinating annual weeds, and has been used for pre-emergence weed control in sugar beet at doses of 1—3 kg./ha.

It is produced by the reaction of *m*-methylphenyl isothiocyanate with dimethylamine:

[Structure: m-methylphenyl-NCS + (CH₃)₂NH → m-methylphenyl-NH-C(=S)-N(CH₃)₂]

Thiuronium salts of the general formula (VIII)

$$\underset{RNHC=NH_2X}{\overset{SR'}{|}} \quad \text{(VIII)}$$

have powerful fungicidal and bactericidal effects; especially active are compounds containing hydrocarbon radicals with at least 7–10 carbon atoms. These compounds, which also repel rodents, are easily prepared by the reaction of alkyl halides with thiourea or its derivatives:

$$RNHCSNH_2 + R'X \rightarrow \underset{RNHC=NH_2X}{\overset{SR'}{|}}$$

Thiourea derivatives in general are highly toxic to mammals, and are used for rodent control. Such compounds include Promurite, Chloropromurite, and Krysid.

4-Chlorobenzenediazothiourea (Promurite) is a yellow crystalline substance, m.p. 123°–129° C., usually used in the form of the water-soluble sodium salt. The LD_{50} for rats is about 1–1.5 mg./kg.

Promurite is prepared in the following manner: 4-chlorobenzenediazonium chloride is coupled with sodium cyanamide:

[Reaction: Cl-C₆H₄-N₂Cl + NaNHCN → Cl-C₆H₄-N=NNHCN + NaCl]

The 4-chlorobenzenediazocyanamide obtained is converted to the thiourea by the action of sodium hydrosulfide:

[Reaction: Cl-C₆H₄-N=NNHCN + NaSH → Cl-C₆H₄-N=NN(Na)C(=S)NH₂]

This compound is used in the form of food baits.

3,4-Dichlorobenzenediazothiourea (Chloropromurite) is a yellow crystalline substance, m.p. 145° C., used to eradicate rodents. The LD_{50} for rats is 0.5–1 mg./kg. It is prepared in a manner completely analogous to Promurite, except that 3,4-dichloroaniline is used instead of 4-chloroaniline.

1-Naphthylthiourea (antu, Krysid) is a white crystalline substance, m.p. 198° C. The technical product is gray-blue in color with lower melting point. It is slightly soluble in water and most organic solvents. It is used for rat control in the form of food baits containing 0.5–1% active ingredient. Among the disadvantages of this compound is the poor feeding by rodents on the baits when it is used repeatedly. The LD_{50} for gray rats is 6–7 mg.; for other species of rodents this compound is less toxic; the LD_{50} for chicks is 700 mg./kg.

Antu is prepared by prolonged heating of an aqueous solution of equimolecular quantities of 1-naphthylamine hydrochloride and ammonium thiocyanate:

$$\text{naphthyl-NH}_2 \cdot \text{HCl} + \text{NH}_4\text{SCN} \longrightarrow \text{naphthyl-NHCNH}_2\text{(=S)} + \text{NH}_4\text{Cl}$$

To obtain maximum yields (>90%) the duration of heating at 90°–95° C. is 15–20 hours. An important question also is the quality of the raw materials used. The HCl and ammonium thiocyanate should not contain sulfate-ion impurities, since 1-naphthylamine sulfate, which is practically insoluble in water, precipitates and contaminates the 1-naphthylthiourea; the compound contamined with 1-naphthylamine sulfate is poorly fed upon by rodents. After the reaction is ended, the 1-naphthylthiourea is filtered off, washed with water, and dried. The mother liquor is re-used for the next cycle of the process.

For the preparation of baits the finely ground compound with a particle size of less than 0.15 mm. is used. The compound containing not more than 0.2% moisture can be ground to this degree of fineness in a ball mill.

Phenylthiourea is more toxic, but it has a less selective effect.

Substituted arylthioureas not only have a zoocidal effect on rodents, but they also are powerful mothproofing agents. Apparently the substituted thioureas disrupt some enzymatic processes in the moth organism, since they contain potential sulfhydryl groups. However, the simplest thiourea derivatives have not been practically utilized to protect woolen goods from damage by moths.

Fungicidal derivatives of guanidine

Of the substituted guanidines only dodecylguanidine acetate (dodine), which is effective as an eradicative fungicide for the control of apple scab, has been used as yet for protection of plants from diseases.

Dodine is a white crystalline substance, m.p. 136° C., poorly soluble in cold water, more soluble in hot water and alcohol but insoluble in most organic solvents. Because of the presence in its molecule of a radical with a long chain of carbon atoms, dodecylguanidine acetate lowers the surface tension of water. In an alkaline medium, breakdown of the salt occurs and the free base is released; in an acid medium the compound is more stable. Upon boiling with acids or alkalies it decomposes with the formation of ammonia and dodecylamine. This reaction is characteristic of most guanidine derivatives.

The LD_{50} for rats is 1,000–2,000 mg./kg. Dodine strongly irritates the skin and therefore appropriate precautionary measures must be taken in working with it. It is marketed in the form of a 70% wettable powder that is for the treatment of plants at 0.08–0.1%; higher concentrations of the compound are phytotoxic. The compound is capable of penetrating into the leaves of plants and has not only a prophylactic, but also a curative effect.

It is produced by the reaction of cyanamide with dodecylamine:

$$C_{12}H_{25}NH_2 + NH_2CN + CH_3COOH \rightarrow C_{12}H_{25}NHC(=NH)NH_2 \cdot CH_3COOH$$

Other salts of dodecylguanidine have been produced as experimental fungicides, for example the salts with tetrahydrophthalic acid (experimental fungicide 23441) and with sodium acid borate, but these compounds have not yet been practically utilized.

General references

BAGNALL, B. H., and K. U. JUNG: Proc. 9th Brit. Weed Control Conf. 1, 25 (1968).
BASKAKOV, YU. A.: Zhur. Vsesoyuz. Khim. Obshchestva im D. I. Mendeleeva 5, 250 (1960).
BORECKI, Z., E. CZERWINSKA, Z. ECKSTEIN, and R. KOWALIK: Chemiczne srodki grzybobojcze. Warsaw (1965).
CRAFTS, A. S.: The chemistry and mode of action of herbicides. New York: Interscience (1961).
FISCHER, A.: 16th Internat. Symposium, Ghent, p. 719 (1964).
GLENISTER, M., and G. GRIFFITHS: Proc. 9th Brit. Weed Control Conf. 1, 46 (1968).
GREEN, D. H.: Proc. 8th Brit. Weed Control Conf., p. 363 (1966).
HACK, H.: Proc. 9th Brit. Weed Control Conf. 1, 57 (1968).
— Weed Abstr. 17, 870 (1968).
HOGUE, E. G.: Weed Research 16, 185 (1968).
MEL'NIKOV, N. N., and YU. A. BASKAKOV: Khimiya gerbitsidov i regulyatorov rosta rastenii [Chemistry of herbicides and plant growth regulators]. State Scientific and Technical Publishing House of Chemical Literature [USSR] (1962).
OGBORN, J.: Proc. 9th Brit. Weed Control Conf. 2, 731 (1968).
POIGANT, P., D. PILLON, P. CRISINEL, R. CAFFIERO, R. RICHARD, S. GAILARD, and P. DELACE: 2nd Symp. New Herbicides, Paris, p. 12 (1965).
POPA, C., and R. DRIMUS: Chimia Produselor Fitofarmaceutice. Bucharest (1965).
SCHULER, J., and L. EBNER: Proc. 7th Brit. Weed Control Conf. 2, 450 (1964).
SMITH, J. M., and T. G. MARKS: Proc. 9th Brit. Weed Control Conf. 2, 533 (1968).
SOUSADE ALMEIDA, F.: Proc. 9th Brit. Weed Control Conf. 2, 725 (1968).
VOEVODIN, A. V., compiler: Gerbitsidy [Herbicides]. Collection of articles. Publishing House "Kolos" (1964).
WATERSON, H. A.: Proc. 9th Brit. Weed Control Conf. 2, 562 (1968).
ZARN, M. K., T. S. IBRAHIM, and M. S. ELIAN: Proc. 9th Brit. Weed Control Conf. 2, 713 (1968).

XIX. Mercaptans, sulfides, and their derivatives

General characteristics of pesticidal properties

The lower representatives of the aliphatic mercaptan series (up to C_4) are effective fumigants against some species of insects. Beginning with the C_5 alkyl mercaptan the insecticidal properties decrease. The unsaturated mercaptans are somewhat more active than the saturated ones with the same number of carbon atoms.

The thiophenols and aromatic mercaptans have little effectiveness as fumigants because of their slight volatility. They are weak contact insecticides and stronger acaricides for plant-feeding mites. The acaricidal effect of the aromatic mercaptans increases going from the unsubstituted compounds to those that contain halogen or other functional groups in the aromatic nucleus.

The aliphatic disulfides are stronger insecticides than the sulfides.

The aromatic sulfides, both diaryl- and arylbenzyl sulfides, have strong acaricidal properties, but their insecticidal effect is slight. Some mixed sulfides and disulfides of the aromatic series are used in agriculture as specific acaricides.

The chlorine-containing mercaptans of the aliphatic series show a higher insecticidal activity than the unsubstituted analogs; for example, perchloro(methyl mercaptan) is approximately equal to HCN in effectiveness as a fumigant. It has insecticidal, fungicidal, and herbicidal effects. It is a valuable intermediate for the synthesis of important present-day fungicides that contain a trichloromethylmercapto group, such as captan, folpet, and their analogs. Trichloromethyl haloalkyl sulfides obtained by the addition of perchloro(methyl mercaptan) to an unsaturated hydrocarbon and to other compounds containing multiple bonds also have high fungicidal activity. Unfortunately this group of compounds is highly phytocidal.

The aromatic sulfones, especially those containing halogen atoms or other substituents in the aromatic radical, also are active acaricides. Some chlorine-containing aliphatic-aromatic sulfones also show insecticidal activity. An example of such compounds is p-chlorophenyl chloromethyl sulfone which has not withstood competition with DDT because of its higher cost. It is obtained by the reaction of salts of p-chlorobenzenesulfonic acid with dichloroacetic acid:

$$ClC_6H_4SOONa + Cl_2CHCOOH \rightarrow ClC_6H_4SO_2CH_2Cl + NaCl + CO_2$$

The synthesis of the intermediates is very complicated. *p*-Chlorobenzenesulfonates are formed by the reduction of *p*-chlorobenzensulfonyl chloride, and dichloroacetic acid is produced by the dehydrochlorination of chloral hydrate in the presence of sodium cyanide:

$$CCl_3CH(OH)_2 \rightarrow CHCl_2COOH + HCl$$

Sulfides

In the sulfide series use has been made in agriculture of 4-chlorobenzyl 4'-fluorophenyl sulfide and 4-chlorobenzyl 4'-chlorophenyl sulfide. The 4,4'-substituted compounds have the greatest activity; when substituents are introduced into other positions of the benzene ring, less active compounds are produced. The sulfoxides and sulfones have practically identical activity with the starting sulfide. The corresponding phenyl benzyl ethers also are acaricides, with the maximum activity occurring in 4-chlorophenyl 4'-chlorobenzyl ether. 4-Chlorophenyl benzyl ether is more active than 4-chlorobenzyl phenyl ether.

The thioacetals also can be related to a certain extent to the sulfides. Of this class of compounds methylenebis(4-chlorophenyl sulfide) is used in agriculture as a specific acaricide

$$Cl-\langle\rangle-SCH_2S-\langle\rangle-Cl$$

The acaricide Mikazin contains 4,4'-dichlorodiphenyl disulfide (m.p. 69.5°–70.5° C., LD_{50} for mice $>3{,}000$ mg./kg.), 4-chlorophenyl 2,4,5-trichlorophenylazo sulfide, and bis(4-chlorophenoxy)methane.

4-Fluorophenyl 4'-chlorobenzyl sulfide (Fluoroparacide, Fluorosulfacide) is a white crystalline substance, m.p. 36° C., with a characteristic but not strong odor, vapor pressure at 20° C. 8×10^{-5} mm. of Hg. Although insoluble in water, it is highly soluble in organic solvents including petroleum products. The LD_{50} for rats is about 3,000 mg./kg. The compound is used in the form of aerosols obtained by spraying with the aid of Freon, methyl chloride, or special mixtures of Freon and butane for the treatment of greenhouses against plant-feeding mites.

It is chemically stable, and does not break down when heated to 100° C.; it is resistant to hydrolysis, but is relatively rapidly oxidized to form the corresponding sulfoxide and sulfone. These derivatives also have high acaricidal activity.

Fluoroparacide is prepared by the reaction of sodium 4-fluorothiophenolate with 4-chlorobenzyl chloride in alcohol or other suitable organic solvent (e. g., in methyl or isopropyl alcohol):

$$Cl-\langle\rangle-CH_2Cl + NaS-\langle\rangle-F \rightarrow Cl-\langle\rangle-CH_2S-\langle\rangle-F$$

For the synthesis of fluorothiophenol, *p*-fluorobenzenesulfonyl chloride is reduced with iron or zinc. To obtain 4-chlorobenzyl chloride, *p*-chlorotoluene is chlorinated at an elevated temperature and with irradiation by u.v. light.

4-Chlorophenyl 4'-chlorobenzyl sulfide (chlorbenside, Chloroparacide, Chlorosulfacide) is a white crystalline substance, m.p. 72° C., vapor pressure at 20° C. 2.59×10^{-6} mm. of Hg, insoluble in water. The solubility of chlorbenside in organic solvents at 20° C. in g./100 g. of solvent in acetone is 92, benzene 111, toluene 107, xylene 93, methyl alcohol 4, ethyl alcohol 2.8, carbon tetrachloride 49, chloroform 99, acetic acid 63, and kerosine 5–7.5. The LD_{50} for mice is 3,000 mg./kg.; the chronic toxicity is moderate. Feeding rats for three weeks with daily introduction of 250 mg./kg. of the compound with the diet did not affect the growth nor the composition of the animals' blood; only a slight enlargement of the liver was noted, without any pathological changes.

Alkalies and acids weakly affect chlorbenside, but oxidizing agents, including oxygen of the air, rather quickly oxidize it to the sulfoxide (m.p. 124° C.) and sulfone (m.p. 148° C.). However, both the sulfoxide and sulfone have the same acaricidal effect as the sulfide, and therefore upon oxidation no loss in toxicity to mites occurs. It is produced by the condensation of 4-chlorobenzyl chloride with sodium 4-chlorothiophenolate:

$$Cl-\langle\ \rangle-CH_2Cl + NaS-\langle\ \rangle-Cl \rightarrow Cl-\langle\ \rangle-CH_2S-\langle\ \rangle-Cl$$

This reaction proceeds readily in any organic solvent with almost a quantitative yield of the sulfide.

Chlorbenside is marketed in the form of a 20% wettable powder, a 20% emulsive concentrate, and a 10% solution in a readily volatile solvent. It is used as a specific acaricide of prolonged nonsystemic action.

Sulfones

The insecticidal and acaricidal activity of various sulfones has been studied in detail, and active acaricides have been found among them. The insecticidal effect of most of the sulfones, however, is small. The most active are the compounds of the aromatic series, some of which have been used in agriculture.

In Table XLI the acaricidal activity of diaryl sulfones for *Tetranychus urticae* is shown. The most active are the compounds containing chlorine in the 2-, 4-, and 5-positions of one aromatic nucleus and fluorine, chlorine, bromine, or methoxyl in the 4'-position of the second nucleus. A change in position of the substituent both in the first and in the second aromatic nucleus leads to a decrease in acaricidal activity of the compound; the activity drops substantially also when a nitro or amino group is introduced into one of the aromatic radicals.

Table XLI. *Acaricidal activity of some sulfones on Tetranychus urticae*

Compounds	Activity [a] at concentration (mg./l.)		
	1,000	100	10
Diphenyl sulfone	+	±	−
4-Chlorodiphenyl sulfone	+	−	−
4,4′-Dichlorodiphenyl sulfone	±	−	−
4-Chlorophenyl-4′-bromophenyl sulfone	−	−	−
4,4′-Dibromodiphenyl sulfone	−	−	−
2,5-Dichlorodiphenyl sulfone	−	−	−
2,5,2′,5′-Tetrachlorodiphenyl sulfone	−	−	−
2,5,4′-Trichlorodiphenyl sulfone	±	−	−
2,5-Dichloro-4′-methyldiphenyl sulfone	−	−	−
3,4-Dichlorodiphenyl sulfone	−	−	−
3,4,4′-Trichlorodiphenyl sulfone	+	−	−
3,4,2′-Trichlorodiphenyl sulfone	−	−	−
3,4,3′,4′-Tetrachlorodiphenyl sulfone	−	−	−
3,4-Dichloro-4′-methyldiphenyl sulfone	−	−	−
2,3,4,5,4′-Pentachlorodiphenyl sulfone	−	−	−
4,4′-Dichloro-3,3′-dinitrodiphenyl sulfone	±	−	−
4,4′-Dichloro-3,3′-diaminodiphenyl sulfone	±	−	−
2,4,5-Trichlorodiphenyl sulfone	+	±	−
2,4,5,4′-Tetrachlorodiphenyl sulfone	+	+	+
2,4,5-Trichloro-4′-fluorodiphenyl sulfone	+	+	+
2,4,5-Trichloro-4′-bromodiphenyl sulfone	+	+	+
2,4,5-Trichloro-4′-methyldiphenyl sulfone	+	±	−
2,4,5-Trichloro-4′-methoxydiphenyl sulfone	±	−	−
2,4,5-Trichloro-4′-nitrodiphenyl sulfone	+	±	−
2,4,5,2′-Tetrachlorodiphenyl sulfone	−	−	−
2,4,5,3′-Tetrachlorodiphenyl sulfone	−	−	−
2,4,5-Trichloro-4′-aminodiphenyl sulfone	−	−	−
2,4,5-Trichloro-4′-hydroxydiphenyl sulfone	−	−	−

[a] + sign indicates 100% mortality of larvae and death of eggs of the mites, ± sign indicates 50–100%, and − sign indicates less than 50%.

Upon cyanethylation of substituted phenyl benzyl sulfones the insecticidal activity of the compound is increased:

$$ArSO_2CH_2Ar' + 2\ CH_2 = CHCN \rightarrow ArSO_2C(CH_2CH_2CN)_2Ar'$$

The acaricidal activity of the substituted aromatic sulfones is somewhat higher than that of the corresponding sulfoxides and sulfides, but the difference is not very substantial. The most widely used has been 2,4,5,4′-tetrachlorodiphenyl sulfone, which is employed as an active selective acaricide to protect plants; diphenyl sulfone and 4-chlorodiphenyl sulfone are used on a more limited scale.

2,4,5,4′-*Tetrachlorodiphenyl sulfone (tetradifon, Tedion, tetradichlone)* is a white crystalline substance, m.p. 146.5°–147.5° C. At room temperature it is insoluble in water, methyl alcohol, and petroleum ether. The solubility of tetradifon in other solvents in g./100 g. of solvent is: carbon tetrachloride 1.6, acetic acid 1.6, methyl acetate 6.5, ethyl acetate 7.1, acetone 8.2, methyl ethyl ketone 10.5, benzene 14.8, toluene 13.5, xylene 11.5, dioxane 22.3, chloroform 25.5. The LD_{50} for rats is 5,000 mg./kg. When rats were fed a diet containing 0.05% of the compound for two months, no harmful effect on the animals was observed.

Tetradifon is resistant to the action of mineral acids and alkalies even upon prolonged heating. It is produced by a Friedel-Crafts reaction of 2,4,5-trichlorobenzenesulfonyl chloride with chlorobenzene:

$$\text{Cl}_3\text{C}_6\text{H}_2\text{-SO}_2\text{Cl} + \text{C}_6\text{H}_4\text{Cl} \xrightarrow{AlCl_3} \text{Cl}_3\text{C}_6\text{H}_2\text{-SO}_2\text{-C}_6\text{H}_4\text{-Cl} + \text{HCl}$$

This reaction proceeds readily in the presence of anhydrous aluminum chloride or ferric chloride. The 2,4,5-trichlorobenzenesulfonyl chloride is synthesized by sulfochlorination of trichlorobenzene with chlorosulfonic acid:

$$\text{C}_6\text{H}_3\text{Cl}_3 + 2 HOSO_2Cl \rightarrow \text{C}_6\text{H}_2\text{Cl}_3(\text{SO}_2\text{Cl}) + H_2SO_4 + HCl$$

When trichlorobenzene reacts with *p*-chlorobenzenesulfonyl chloride, not the sulfone but tetrachlorobenzene and *p*-chlorobenzenesulfonic acid are produced:

$$\text{C}_6\text{H}_3\text{Cl}_3 + \text{C}_6\text{H}_4(\text{SO}_2\text{Cl})\text{Cl} \rightarrow \text{C}_6\text{H}_2\text{Cl}_4 + \text{C}_6\text{H}_4(\text{SO}_2\text{H})\text{Cl}$$

Tetradifon is marketed in the form of a wettable powder and an emulsive concentrate containing 20–30% active ingredient. It is employed as an acaricide of prolonged action to control the larval stages of various plant-feeding mites, although it is not very active against adult mites.

Diphenyl sulfone (phenyl sulfone, DPS) is a white crystalline substance, m.p. 123°–124° C., slightly soluble in water, but highly soluble in aromatic hydrocarbons and their halogenated derivatives. The LD_{50} for rats is more than 2,000 mg./kg. The compound is used in the form of a suspension with summer oils as an ovicide against the eggs and larval stage of mites; in the form of a wettable powder its effectiveness is low.

It is produced by sulfonation of benzene with sulfuric acid, and also is formed as a by-product in the production of benzenesulfonyl chloride and benzenesulfonic acid:

$$2\ C_6H_6 + H_2SO_4 \rightarrow (C_6H_5)_2SO_2 + 2\ H_2O$$

4-Chlorodiphenyl sulfone (Sulphenone) is a white crystalline substance, m.p. 98° C., insoluble in water. Its solubility in organic solvents in g./100 g. of solvent is: acetone 74.4, dioxane 65.6, benzene 44.4, toluene 29.4, xylene 18.2, carbon tetrachloride 4.9, isopropyl alcohol 2.1, and hexane 0.4. The LD_{50} for rats is 1,400–3,500 mg./kg. Sulphenone is employed mainly in the form of 2.25–0.5% aqueous suspensions or emulsions to control the two-spotted mite. It is gradually being displaced by more effective compounds.

Sulphenone is synthesized industrially by the reaction of benzenesulfonic acid with chlorobenzene or by the reaction of *p*-chlorobenzenesulfonic acid with benzene at a temperature of 200°–250° C.:

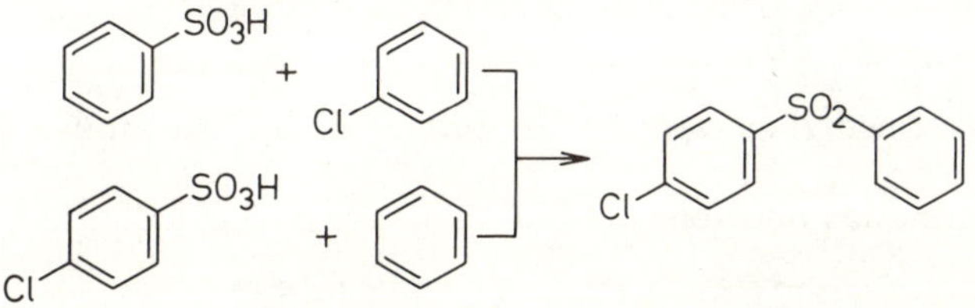

The second method has a better raw-materials basis, since *p*-chlorobenzenesulfonic acid is a by-product in the manufacture of DDT.

Perchloro(methyl mercaptan) and its derivatives

Perchloro(methyl mercaptan), because of its high toxicity for man and animals and its strong corrosive effect, is not used in agriculture. It is the basic raw material for the synthesis of trichloromethylthioimides and trichloromethylthioimides of various acids that have fungicidal properties.

Perchloro(methyl mercaptan) is a red oily liquid, b.p. 148° C. at 760 mm. of Hg (dec.) and 73° C. at 50 mm. of Hg, *d* 1.722. It is so toxic to mammals that in the World War I it was proposed as a chemical warfare agent, but it was hardly ever used for that purpose.

Metals and other reducing agents convert perchloro(methyl mercaptan) to thiophosgene:

$$CCl_3SCl \rightarrow CSCl_2$$

Oxidizing agents oxidize it to trichloromethanesulfonyl chloride:

$$CCl_3SCl + 2\ O \rightarrow CCl_3SO_2Cl$$

Alkalies hydrolyze perchloro(methyl mercaptan) slowly in the cold and rapidly when heated, with the formation of a mixture of salts.

Perchloro(methyl mercaptan) reacts relatively easily with unsaturated compounds to give trichloromethyl chloroalkyl sulfides:

$$RCH=CH_2 + CCl_3SCl \rightarrow RCHClCH_2SCCl_3 \rightarrow RCH=CHSCCl_3 + HCl$$

These sulfides as a rule are strong fungicides and kill many species of phytopathogenic microorganisms, but most of the compounds are also dangerous to crop plants because they cause severe burning of the leaves.

The chlorine atoms in perchloro(methyl mercaptan) have different reactivity in reactions of nucleophilic substitution, thereby enabling its utilization for the introduction of a trichloromethyl group into the molecules of organic compounds. As an example of such reactions, one may cite the interaction of perchloro(methyl mercaptan) with the metal derivatives of amides and imides of carboxylic and other acids:

$$R\underset{CO}{\overset{CO}{\diagup}}NK + ClSCCl_3 \rightarrow R\underset{CO}{\overset{CO}{\diagup}}NSCCl_3 + KCl$$

Perchloro(methyl mercaptan) is produced by the following basic methods:

1. Chlorination of carbon disulfide in the presence of iodine as a catalyst:

$$2\ CS_2 + 5\ Cl_2 \rightarrow 2\ CCl_3SCl + S_2Cl_2$$

Sulfur monochloride is formed as a by-product in this reaction and its removal presents certain difficulties. The simplest laboratory method for the purification of perchloro(methyl mercaptan) is steam distillation. In this process the sulfur monochloride is completely decomposed by the water and the mercaptan distills off. The yield of the mercaptan is decreased by 10–15% as a result of hydrolysis by the water. However, this method of purification is difficult to carry out under industrial conditions because of the strongly corrosive medium. A second method for freeing perchloro(methyl mercaptan) of sulfur monochloride is fractional distillation in vacuum, but in this case the purified product contains up to 3% sulfur monochloride and a second fractionation results in substantial losses of the product.

It has been proposed that perchloro(methyl mercaptan) be purified by washing out the sulfur monochloride with sulfurous acid at a temperature not higher than 10° C. The sulfur monochloride is converted to water-soluble tetrathionic acid:

$$S_2Cl_2 + 2\ H_2SO_3 \rightarrow H_2S_4O_6 + 2\ HCl$$

the calcium salt of which is used to control true powdery mildews. A deficiency of this purification method is the occurrence of side reactions that take place when both the temperature and the time of contact between the mercaptan and the aqueous solution of sulfurous acid is increased. At temperatures above 35° C., the perchloro(methyl mercaptan) is reduced by the sulfurous acid to thiophosgene (b.p. 78° C.), which is highly volatile and more toxic than the mercaptan:

$$CCl_3SCl + H_2SO_3 + H_2O \rightarrow 2\ HCl + H_2SO_4 + CSCl_2$$

2. Chlorination of carbon disulfide in the presence of a small amount of water at a low temperature:

$$CS_2 + 5\ Cl_2 + 4\ H_2O \rightarrow CCl_3SCl + H_2SO_4 + 6\ HCl$$

The yield of pure perchloro(methyl mercaptan) by both the first and the second method is 65–70%. However, the expenditure of chlorine by this second method is almost twice as great.

3. Chlorination of methyl mercaptan and dimethyl disulfide:

$$(CH_3S)_2 + 7\ Cl_2 \rightarrow 2\ CCl_3SCl + 6\ HCl$$

According to patent data perchloro(methyl mercaptan) is obtained in good yield and rather high quality by this method.

The technical perchloro(methyl mercaptan) contains, in addition to sulfur monochloride, a small trace of carbon tetrachloride that is formed by further chlorination of the mercaptan by chlorine.

Of the large number of trichloromethylthioamides and trichloromethylthioimides that have been studied, only three compounds are used in agriculture, the other compounds having proven to be either too unstable in storage or phytocidal in fungicidal concentrations.

N-Trichloromethylthiotetrahydrophthalimide (captan) in the pure state is a white crystalline substance, almost odorless, m.p. 172° C. The technical grade product is yellow or gray, with the characteristic odor of perchloro(methyl mercaptan) and thiophosgene, m.p. 164° C. It is practically insoluble in water and slightly soluble in most organic solvents. In the moist condition captan is hydrolyzed:

$$\text{[structure]}\ N-SCCl_3 + 2H_2O \rightarrow \text{[structure]}\ NH + CO_2 + 3HCl + S$$

This reaction takes place very rapidly in the presence of alkalies at an elevated temperature; therefore, captan is incompatible with all compounds of an alkaline nature.

The LD_{50} for rats is 9,000 mg./kg. When the compound was added to the food of animals in the course of more than a year no deviations from normal behavior were observed. Systematic contact with the skin may cause slight irritation. Captan is employed as a broad-spectrum protective fungicide for the control of various diseases of agricultural crops, including seed disinfection. When grapes are treated with captan, there is a negative effect on alcoholic fermentation during the production of wine, since captan inhibits the development of yeast fungi. It is marketed in the form of a 50% wettable powder and a 75% preparation for seed disinfection; for plant protection it is used in 0.3–0.5% concentration.

Captan is produced by the reaction of tetrahydrophthalimide with perchloro(methyl mercaptan) in aqueous-alkaline medium with thorough stirring at the lowest temperature possible to avoid hydrolysis both of the captan and of the perchloro(methyl mercaptan). The reaction takes place in the presence of excess NaOH:

[structure] NH + NaOH + CCl_3SCl → [structure] N—$SCCl_3$ + NaCl + H_2O

An excess of NaOH is necessary because a part of the perchloro(methyl mercaptan) is unavoidably hydrolyzed and ties up alkali. When the conditions for the process are strictly observed, the yield of captan reaches 90%.

The tetrahydrophthalamide necessary for the synthesis of captan is produced from ammonia and tetrahydrophthalic anhydride at a temperature above 200° C., and under conditions similar to those for the preparation of phthalimide:

[structure] O + NH_3 → [structure] NH + H_2O

Tetrahydrophthalic anhydride is formed in practically quantitative yield by the condensation of divinyl with maleic anhydride at 100°–160° C. This process can be carried out industrially very easily by a continuous operation:

CHCO
‖ >O + CH_2=CH—CH=CH_2 → [structure]
CHCO

N-Trichloromethylthiophthalimide (folpet, Phaltan, phthalan) is a white crystalline substance, m.p. 177° C., insoluble in water, and poorly soluble in the usual organic solvents; it is slowly hydrolyzed by water, especially in an alkaline medium. The rate of hydrolysis by water is higher than that of captan. In producing folpet, therefore, it is necessary to dry the compound carefully and also to use dry diluents. The presence of 1% water in a preparation can hydrolyze about 10% of the folpet:

[structure] $NSCCl_3$ + 2H_2O → [structure] NH + 3HCl + S + CO_2

Because a large amount of HCl is evolved on hydrolysis, folpet like captan cannot be packed in paper containers, since they are very rapidly destroyed.

The LD_{50} for rats is more than 10,000 mg./kg. It is marketed in the form of a 50% wettable powder and is used in 0.2–0.3% concentration of active ingredient.

Folpet is produced similarly to captan by the reaction of perchloro(methyl mercaptan) with phthalimide in the presence of sodium hydroxide in aqueous medium. When sufficiently pure mercaptan is used, the yield of folpet is about 90%. The main contaminant in the technical grade product is the starting phthalimide that has not reacted and that has been formed as a result of hydrolysis. The product also contains a small amount of sulfur.

N-1,1,2,2-Tetrachloroethylthiotetrahydrophthalimide (Difolatan) is a white crystalline substance, m.p. 160°–161° C., practically insoluble in water, poorly soluble in most of the usual organic solvents, but somewhat more soluble than captan. The LD_{50} for rats is more than 6,000 mg./kg. The compound is slowly hydrolyzed by water, more rapidly by caustic alkalies:

$$\text{[cyclohexene-CO-N(SCCl}_2\text{CHCl}_2\text{)-CO]} + 2H_2O \longrightarrow \text{[cyclohexene-CO-NH-CO]} + S + CHCl_2COOH + HCl$$

It is more resistant to hydrolysis than captan and folpet.

Difolatan is marketed in the form of a 50% wettable powder and is used in a manner similar to captan and folpet. The principal crops for which it is recommended are potatoes and grapes.

Difolatan is produced by the reaction of tetrahydrophthalimide with pentachloroethyl mercaptan in the presence of NaOH:

$$\text{[cyclohexene-CO-NH-CO]} + CHCl_2CCl_2SCl + NaOH \longrightarrow \text{[cyclohexene-CO-N(SCCl}_2\text{CHCl}_2\text{)-CO]} + NaCl$$

Pentachloroethyl mercaptan in turn is synthesized from sulfur monochloride and trichloroethylene with subsequent chlorination of the bis(tetrachloroethyl) disulfide to pentachloroethyl mercaptan:

$$CHCl = CCl_2 + S_2Cl \rightarrow (CHCl_2CCl_2)_2S_2$$
$$(CHCl_2CCl_2)_2S_2 + Cl_2 \rightarrow 2\ CHCl_2CCl_2SCl$$

The preparation of this compound is less complicated than that of perchloro(methyl mercaptan), and there is a basis for assuming that *N-1,1,2,2-tetrachloroethylthiotetrahydrophthalimide* will be a serious competitor to captan. A large number of other analogous compounds also have been synthesized, but they have not yet been used in agriculture.

Also close to captan in fungicidal activity are the trichloromethylamides of the aliphatic sulfo acids, while the trichloromethylamides of the aromatic sulfo acids are considerably less active. Introduction of the trichloromethyl group does not always increase the activity of the amides of the sulfo acids. Thus, when an amide hydrogen is replaced by a trichloromethyl group in the 4-chloroanilide of methanesulfonic acid a powerful fungicide is obtained, while replacement of hydrogen on the nitrogen by this group in the 4-thiocyanoanilide of methanesulfonic acid gives a compound of low activity.

N-Trichloromethylthio-4-chloroanilide of methanesulfonic acid (Mesulfan, captan analog No. 6) is a white crystalline substance, m.p. 113° to

114° C., insoluble in water, but moderately soluble in organic solvents. Its chemical properties recall those of captan. Thus, Mesulfan is unstable in alkaline medium and is hydrolyzed by water:

$$CH_3SO_2NC_6H_4Cl + 2\ H_2O \rightarrow CH_3SO_2NHC_6H_4Cl + S + CO_2 + 3\ HCl$$
$$|$$
$$SCCl_3$$

A trace of ferric chloride has a negative effect on storage of the compound. The LD_{50} for rats and mice is $>5,000$ mg./kg.

Mesulfan is produced in good yield by the reaction of perchloro(methyl mercaptan) with 4-chlorophenylmethanesulfonamide in the presence of NaOH:

Cl–C₆H₄–NHSO₂CH₃ + CCl₃SCl + NaOH → Cl–C₆H₄–N(SCCl₃)SO₂CH₃

The sulfamide is synthesized by the following scheme:

$$Na_2S_2O_3 + (CH_3)_2SO_4 \rightarrow CH_3NaS_2O_3 + CH_3NaSO_4$$
$$CH_3NaS_2O_3 + H_2O + Cl_2 \rightarrow CH_3SO_2Cl + H_2SO_4 + NaCl + HCl$$

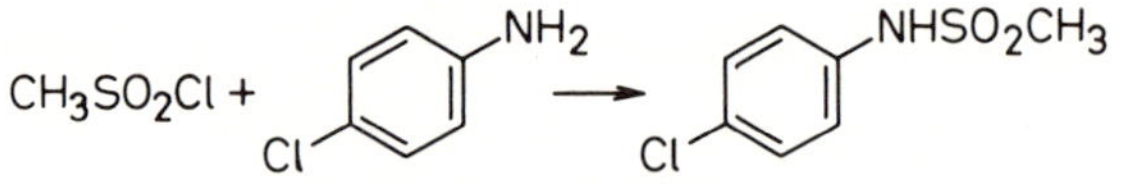

CH₃SO₂Cl + Cl–C₆H₄–NH₂ → Cl–C₆H₄–NHSO₂CH₃

The preparation of methanesulfonyl chloride is carried out at the lowest temperature possible to avoid hydrolysis, since this reaction is carried out in aqueous medium. The synthesis of methanesulfonyl chloride and its homologs also is accomplished by oxidative chlorination of mercaptans, disulfides, thiocyanates, and S-alkylthioureas:

$$(CH_3S)_2 + 4\ H_2O + 5\ Cl_2 \rightarrow 2\ CH_3SO_2Cl + 8\ HCl$$

The yield of methanesulfonyl chloride and its homologs by the method given is 70–80%.

Mesulfan is marketed in the form of a 50% wettable powder, but it is less active than captan.

The N-trichloromethylthioamides of the homologs of methanesulfonic acid are somewhat less active; the only exceptions are the derivatives of ethanesulfonic acid, which are equal to the corresponding derivatives of methanesulfonic acid. Introduction of chlorine into the methyl radical of methanesulfonic acid does not increase the activity of the compound.

In addition to the trichlorothioamides of various acids, a large number of trichloromethylthiosulfonates of the aliphatic and aromatic series have been studied; many of them have a powerful fungicidal effect. However, all compounds of this type are phytotoxic and therefore not used in agriculture.

Trichloromethyl thioethers are obtained in good yields by the reaction of the sodium salts of sulfinic acids with perchloro(methyl mercaptan):

$$RSO_2Na + CCl_3SCl \rightarrow RSO_2SCCl_3 + NaCl$$

The fungicidal activity of compounds containing a trichloromethylthio group, in the opinion of many investigators, is associated with the high reactivity of this group, which by interacting with various sulfhydryl groups in the fungus cell disrupts vital biochemical processes.

General references

BROWN, A.: Insect control by chemicals. New York-London: Wiley (1951).
HUISMAN, H. A., J. H. UHLENBROCK, and J. MELTZER: Rec. trav. chim. 77, 103–122 (1958).
MÄSSING, W.: Mitt. Biol. Bundesanst. Land- u. Forstwirtschaft 83, 59–68 (1955).
MEL'NIKOV, N. N., E. M. SOKOLOVA, P. P. TRUNOV, and G. I. BRUSENINA: Zhur. Priklad. Khim. 34, 2586 (1961).
—, V. I. ZETKIN, B. YA. LIBMAN, E. M. SOKOLOVA, E. V. ZOKHAROV, A. I. PARFENOV, P. P. TRUNOV, and N. M. GOYSHIN: Khim. Prom. No. 10, 692 (1961).
METCALF, R. L.: Organic insecticides. New York: Interscience (1955).
MISTRA, G. S., and R. S. ASTHANA: J. prakt. Chem. 3 (4), 4–12 (1956).
UHLENBROCK, J. H., M. J. KOOPMANS, and H. O. HUISMAN: Rec. trav. chim. 76, 129 (1957).
WAEFFLER, R., R. GASSER, A. MARGOT, and H. GYSIN: Experientia 11, 265 (1955).
ZBIROVSKY, M., and J. MYSKA: Insekticidy, fungicidy, rodenticidy. Prague: Nakl. Ceskosl. Akad. Ved. (1957).

XX. Thiocyanates and isothiocyanates

General characteristics of pesticidal properties

The high physiological activity of various derivatives of thiocyanic acid has long been known. The salts of thiocyanic acid, which are obtained in considerable amounts in purifying coke oven gas of HCN, were some of the first used practically in agriculture.

Ammonium thiocyanate was previously used widely as a desiccant and nonselective contact herbicide. Ammonium thiocyanate and other salts of thiocyanic acid have not yet completely lost their importance. However, it is being displaced by other more effective compounds, because its herbicidal action is comparatively short-lived, and in many cases after the use of ammonium thiocyanate a vigorous secondary growth of weeds has been observed. Ammonium thiocyanate also is employed to disrupt the dormant state of potatoes.

Ammonium thiocyanate is a white hygroscopic substance, m.p. 149° C., very soluble in water and alcohols; at 20° C. its solubility in water is 62%.

In the coal-tar chemical industry ammonium (or sodium) thiocyanate is put out in the form of aqueous solutions containing salts of sulfuric and thiosulfuric acids as impurities. It is used at dosages up to 300 kg./ha. for the control of weeds; at higher dosages it is capable of sterilizing the soil of all species of plants. The defoliation and desiccation of various crops is carried out with salts of thiocyanic acid at dosages of 10–40 kg./ha.

Of considerably greater interest as pesticides are the organic derivatives of thiocyanic acid, which show good insecticidal and fungicidal properties. More than a thousand different aliphatic and aromatic thiocyanates have been synthesized. However, only a few systematic studies on the relationship of the insecticidal activity to the structure of the organic thiocyanates have been published.

In the homologous series of p-alkyl thiocyanates the insecticidal activity toward aphids and the acaricidal activity toward the spider mite rise with an increase in the length of the hydrocarbon chain, reach a maximum at C_{10}—C_{12}, and then fall. In Table XLII data are presented on the insecticidal activity of some alkyl thiocyanates toward the green chrysanthemum aphid.

The thiocyanates with branched chains of carbon atoms have less insecticidal activity than compounds with a normal structure of the carbon chain having the same molecular weight. Introduction into the thiocyanate molecule of various functional groups and halogen atoms somewhat strengthens the action of the compounds. Simultaneously with the increase in toxicity to insects the phytotoxicity also increases. When various functional groups are introduced into the molecule of an aliphatic thiocyanate, often not only the insecticidal activity, but also the fungicidal activity of the compound is increased. The most active atoms and groups are the halogens, a second thiocyano group, sulfide, and ester groupings. Thus, for example, trichloromethyl thiocyanate (b.p 41°–45° C. at 11 mm. of Hg, m.p. 2.5° C.), obtained by the reaction of perchloro(methyl mercaptan) with cyanide salts at a temperature from $-10°$ to $-20°$ C., is so highly phytotoxic that it has been proposed for use as a herbicide. This same compound also has considerable fungicidal effect.

Table XLII. *Insecticidal activity of alkyl thiocyanates toward the green chrysanthemum aphid*

Compound	LC_{50} (%)	Compound	LC_{50} (%)
n-Hexyl thiocyanate	0.08	n-Dodecyl thiocyanate	0.025
n-Octyl thiocyanate	0.032	n-Tetradecyl thiocyanate	0.030
n-Decyl thiocyanate	0.028	n-Hexadecyl thiocyanate	0.050

Simultaneous fungicidal and insecticidal effects are shown also by such compounds as dithiocyanomethane, 1,2-dithiocyanoethane, and others. However, these compounds are toxic to mammals and irritate the skin and mucous membranes.

The esters of thiocyanoalkylcarboxylic acids such as thiocyanoacetic, thiocyanobutyric, and thiocyanopropionic acids have high insecticidal activity. Their insecticidal effect also increases as the carbon chain of the ester radical becomes larger; the maximum activity occurs in compounds containing 6 carbon atoms. Of the acids mentioned the esters of thiocyanoacetic acid have the greatest activity. Many compounds of this series are phytotoxic.

The insecticidal activity of the thiocyanates increases when a hydroxyl group is introduced on the first carbon atom in the molecule. When such thiocyanoalcohols are esterified with various carboxylic acids the activity is multiplied several times.

The thiocyanates containing one or even two ether oxygens in the molecule considerably surpass the alkyl thiocyanates in insecticidal effectiveness. Examples of such compounds are β-butoxy-β'-thiocyanodiethyl ether, which is known as Lethane 384, and β-chloro-β'-thiocyanodiethyl ether (Compound 47). Also effective as insecticides are the ω-thiocyano-ω-aryldialkyl ethers (more than a hundred of which are described in the patent literature), but they all burn plants.

A study of the insecticidal properties of a number of β-thiocyanoethyl- and γ-thiocyanopropyl aryl ethers toward the house fly has shown that substituents in the aromatic radical affect the insecticidal activity. The most active are the compounds in which one hydrogen is replaced by a methyl, allyl, or methoxyl group. Replacement of hydrogen by halogen usually lowers the activity of the compound and the position of the substituent is not of substantial significance. An increase in the number of groups on the phenyl radical considerably increases the activity of the ether. Of the ethers studied, the optimum with respect to the total properties proved to be γ-thiocyanopropyl-2,4,6-trimethylphenyl ether, which does not have an unpleasant odor and is highly active as an insecticide.

Introduction of a thiocyano group into the molecule of ketones lowers the contact insecticidal effect, although there are reports that thiocyanoacetone is more toxic to flies than methyl thiocyanoacetate.

The amides of thiocyanoacetic acid have not only an insecticidal, but also a herbicidal effect. However, they irritate the skin and therefore are not very suitable for practical use in agriculture.

Going from compounds of the aliphatic series to the alicyclic ones the insecticidal activity decreases. For the cycloalkyl esters of thiocyanoacetic acid a decrease in irritating properties and in unpleasant odor has been observed. Investigation of a large number of cycloalkyl thiocyanoacetates has shown that the most active compound of this series is bornyl thiocyanoacetate, which is used to control parasites of domestic animals. The corresponding fenchyl thiocyanoacetate is less insecticidal.

The aromatic thiocyanates containing a thiocyano group in the benzene nucleus show insecticidal properties, but weaker ones than those of compounds of the aliphatic series with the same number of carbon atoms. The insecticidal activity of a compound depends greatly on the substituents in the aromatic radical. When a nitro group or halogen atoms are introduced into the aromatic radical, the activity of the compounds with respect to plant cells is increased. Thus, 2,4-dinitrothiocyanobenzene is an active fungicide that is used in agriculture to protect plants from diseases, and thiocyanohalobenzoic acids have a herbicidal effect.

Introduction of a thiocyano group into the aniline molecule sharply increases the insecticidal and fungicidal properties. Over a number of years 4-thiocyanoaniline has been tested with positive results as a seed disinfectant to control grain smut. In the opinion of I. M. Polyakov the effect of thiocyanoaniline is associated with immunization of plants by means of this compound. Fungicidal activity has been noted for some sulfamides of thiocyanoaniline, which furthermore are devoid of the irritating properties characteristic of thiocyanoaniline. The homologs of thiocyanoaniline, and also the alkylthiocyanoanilines containing a hydrocarbon radical on the nitrogen are less active both as insecticides and as fungicides. Dimethylthiocyanoaniline has been proposed as an active defoliant for cotton; at dosages of 1.5–3 kg./ha. it gives satisfactory leaf removal.

The isothiocyanates also are active pesticides with a broad spectrum of action. Probably the isothiocyanates are the basic active agents of a number of the derivatives of dithiocarbamic acid that are used in agriculture, such as the salts of ethylenebis(dithiocarbamic) acid, the compound carbothion, and others.

The pesticidal activity of a large number of aliphatic and aromatic isothiocyanates has been studied and many of them have proved to be active insecticides, fungicides, and soil sterilants. As yet only methyl isothiocyanate has been used in agriculture as a soil fumigant to control nematodes and phytopathogenic fungi. This compound also acts on some weed seeds.

One of the important reactions of the thiocyanates with respect to their mechanism of action on living organisms is their reduction, in which HCN is formed:

$$RSCN + 2H \rightarrow RSH + HCN$$

HCN is also obtained from the reaction of thiocyanates with various sulfhydryl compounds:

$$RSCN + R'SH \rightarrow RSSR' + HCN$$

There also takes place in the animal organism the production of HCN from alkyl thiocyanates, forming with the blood a characteristically colored cyanohemoglobin; the animals die with characteristic symptoms of HCN poisoning. The tissues of the liver produce a significant amount of HCN from methyl and ethyl thiocyanates, while the higher molecular weight compounds are considerably more stable. The insecticidal effect of the thiocyanates is also based on the formation of HCN in insects.

The high insecticidal activity of thiocyanoacetic acid esters of the type of bornyl thiocyanoacetate can be explained by the following reactions:

$$NCSCH_2COOR + H_2O \rightarrow ROH + NCSCH_2COOH$$
$$NCSCH_2COOH \rightarrow CO_2 + CH_3SCN$$
$$CH_3SCN \xrightarrow{[H]} HCN + CH_3SH$$

The mechanism of action of the thiocyanates on phytopathogenic fungi apparently has a different character, although HCN is produced also in this case.

The pesticidal activity of the isothiocyanates is explained by the high reactivity of the isothiocyano group which can enter into reactions with hydroxyl, thiol, amino, amido, and other similar groups forming corresponding thio- and dithiocarbamates and substituted thioureas:

$$RNCS + R'SH \rightarrow RNHCSSR'$$
$$RNCS + R'OH \rightarrow RNHCSOR'$$
$$RNCS + R'NH_2 \rightarrow RNHCSNHR'$$

Thiocyanates of the aromatic series

At present four thiocyanates of the aliphatic series are being used in agriculture: Lethane 60, Lethane 384, Thanite, and Compound 47. The first three of these compounds because of their high phytotoxicity are employed exclusively in animal husbandry and in households for the control of flies and other insect pests. Compound 47 has been proposed for the treatment of plants in the leafless state and for eradicative fungicidal spraying of the soil, although it is no less phytotoxic.

A general method for the production of the aliphatic thiocyanates is the reaction of alkyl halides with salts of thiocyanic acid, which takes place rather easily when equimolecular quantities of the compounds are boiled in methyl, ethyl, or isopropyl alcohol:

$$RCl + KSCN \rightarrow RSCN + KCl$$

The reaction proceeds most easily with alkyl iodides, less so with alkyl bromides, and still more difficultly with alkyl chlorides. Exceptions are compounds with a labile halogen atom, as in the esters of monochloroacetic acid, haloketones, etc. However, for the production of the alkyl thiocyanates the cheaper chlorine derivatives are used.

Another method of producing aliphatic and aromatic thiocyanates is the reaction of the cyanogen halides (usually cyanogen chloride) with mercaptans or thiophenols, which occurs rapidly in the presence of HCl acceptors:

$$RSH + ClCN \rightarrow RSCN + HCl$$

This reaction is carried out in any inert solvent at the lowest possible temperature depending on the properties of the starting mercaptan. Tertiary amines or inorganic compounds may be used as HCl acceptors.

β-Butoxy-β′-thiocyanodiethyl ether (Lethane 384) is a colorless liquid, b.p. 124° C. at 0.25 mm. of Hg, almost insoluble in water, highly soluble in most organic solvents including kerosine, with which Lethane 384 is miscible in all proportions, d^{25} 0.915. The LD_{50} for experimental animals is about 550 mg./kg. Lethane 384 is a component of some aerosol formulations for the control of flies. Preparations for the treatment of animals contain pyrethrins, Lethane 384, DDT, and other insecticides. In aerosol bombs, for example, a mixture of the following composition is used: Lethane 384 4%, DDT 3%, pyrethrins 0.2%, Freon 92.8%. Lethane 384 is similar to the pyrethrins in rapidity of action on flies. It is not suitable for the treatment of green plants, since it causes severe burning, but it can be used for spraying plants in the dormant state. The usual market form of the compound is a 50% solution in purified kerosine.

Lethane 384 is produced by the following scheme:

$$C_4H_9OH + 2\ H_2C\!\!-\!\!CH_2 \rightarrow C_4H_9OCH_2CH_2OCH_2CH_2OH$$
$$\diagdown\!\!\diagup$$
$$O$$

$$C_4H_9OCH_2CH_2OCH_2CH_2OH + HCl \rightarrow$$
$$C_4H_9OCH_2CH_2OCH_2CH_2Cl + H_2O$$

$$C_4H_9OCH_2CH_2OCH_2CH_2Cl + NaSCN \rightarrow$$
$$C_4H_9OCH_2CH_2OCH_2CH_2SCN + NaCl$$

The first reaction takes place in the presence of catalysts of acid or basic nature at an elevated temperature. Butyl Cellosolve and higher molecular products of hydroxyethylation are formed as by-products. All these compounds are easily separated by fractional distillation in vacuum. The second reaction goes at an elevated temperature, and it is best to use thionyl chloride as the chlorinating agent. The third reaction, replacement of chlorine by a thiocyano group, requires prolonged heating in isopropyl alcohol. To accelerate the process it is carried out at a high temperature under pressure. Note that the thiocyanates upon prolonged heating at a temperature of 150° C. are isomerized to isothiocyanates. This may cause a decrease in yield.

β-Thiocyanoethyl laurate (Lethane 60) is a brown liquid that does not distill without decomposition, is practically insoluble in water, and highly soluble in organic solvents. The LD_{50} for rats is about 300 mg./kg. The compound is used in animal husbandry to treat dairy cattle. The technical grade product is a mixture of β-thiocyanoethyl esters of $C_{10} - C_{18}$ saturated fatty acids in which the β-thiocyanoethyl ester of lauric acid predominates. It is marketed in the form of a 50% solution in kerosine. Under the name of Lethane 384 Special a solution of 37.5% Lethane 60, 12.5% Lethane 384, and 50% kerosine (by volume) is sold.

Lethane 60 is synthesized by the reaction of a mixture of 2-chloroethyl esters of $C_{10} - C_{18}$ fatty acids with salts of thiocyanic acid in an organic solvent at an elevated temperature:

$$C_{11}H_{23}COOCH_2CH_2Cl + NH_4SCN \rightarrow$$
$$C_{11}H_{23}COOCH_2CH_2SCN + NH_4Cl$$

It is best to use alcohols as solvent.

Bornyl thiocyanoacetate (Thanite) is a yellow oily liquid with a specific terpene odor, practically insoluble in water, highly soluble in organic solvents, not distillable without decomposition, d_4^{25} 1.1465, n_D^{25} 1.512. Thanite is used in aerosol compositions to control various species of flies and also to treat dairy cattle. The LD_{50} for rats is >1,000 mg./kg. On prolonged contact it may cause slight irritation of the skin, especially the mucous membranes. The technical grade product is a mixture of 82% isobornyl thiocyanoacetate and not more than 18% related compounds.

Bornyl thiocyanoacetate is produced by the following scheme:

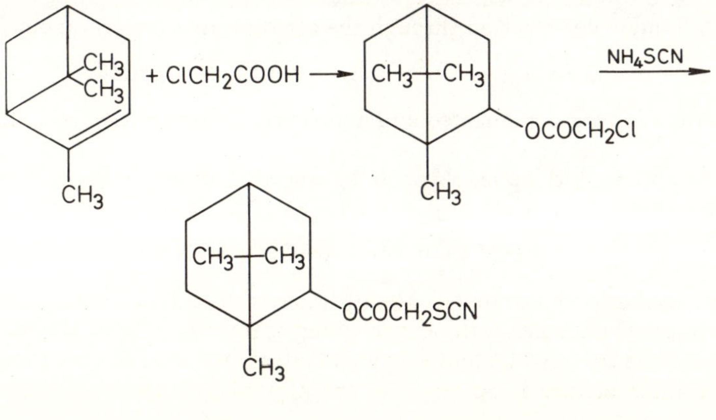

First the monochloroacetic acid and the pinene form bornyl monochloroacetate, which is converted by the action of the salts of thiocyanic acid in alcohol to bornyl thiocyanoacetate in very good yields. After removal of the salts and alcohol, preparations are made up without further purification for use as insecticides. The technical product is corrosive to metals, especially in the presence of moisture.

β-Chloro-β'-thiocyanodiethyl ether (Compound 47), b.p. 100° C. at 1 mm. of Hg, is slightly soluble in water but highly soluble in organic solvents. It is used in the form of a homogenized emulsion for eradicative spraying and treatment of plants in the leafless state. The LD_{50} is about 400 mg./kg.; it strongly irritates the skin.

Compound 47 is produced by the reaction of 2,2'-dichlorodiethyl ether with ammonium thiocyanate in ethyl or isopropyl alcohol:

$$O(CH_2CH_2Cl)_2 + NH_4SCN \rightarrow NCSCH_2CH_2OCH_2CH_2Cl + NH_4Cl$$

Thiocyanates of the aromatic series

The thiocyanates of the aromatic series that contain a thiocyano group in the aromatic nucleus can be prepared by the following methods:

1. Reaction of halogen derivatives with salts of thiocyanic acid, in a manner analogous to the compounds of the aliphatic series. However, this reaction is suitable only for those halogen derivatives of the aromatic series that have a rather labile halogen, such as that in 2,4-dinitrochlorobenzene and similar compounds.

2. Direct thiocyanization:

$$ArNH_2 + (SCN)_2 \rightarrow NCSArNH_2 + HSCN$$

This method is usable for the preparation of thiocyanates of phenols and anilines. The reaction is carried out with free thiocyanogen or with salts of thiocyanic acid that give off thiocyanogen when acted on by chloroamines, chlorine, bromine, sulfuryl chloride, etc. It is possible to carry out the reaction also by electrochemical substitution at the anode.

3. A Sandmeyer reaction through the appropriate diazo compounds:

$$2\ ArN_2Cl + Cu(SCN)_2 \rightarrow 2\ ArSCN + CuCl_2$$

The method is very complicated and expensive to carry out on an industrial scale.

4. Reaction of thiophenols with cyanogen chloride in the presence of HCl acceptors:

$$ArSH + ClCN \rightarrow ArSCN + HCl$$

The aromatic thiocyanates are of interest mainly as fungicides and bactericides. Thus, when a thiocyano group is introduced into the molecule of a phenol its bactericidal and fungicidal effects are considerably increased, and a similar picture is observed in the alkylphenol and naphthol series.

In agriculture so far only 2,4-dinitrothiocyanobenzene is used; 4-thiocyanoaniline (a seed disinfectant against smut) is undergoing tests under industrial conditions, and N,N-dimethyl-4-thiocyanoaniline is being studied as a defoliant.

2,4-Dinitrothiocyanobenzene (DNTB, Nirite) is a yellow crystalline substance, m.p. 138° C., practically insoluble in water, highly soluble in alcohol and aromatic hydrocarbons, slightly soluble in petroleum products that are rich in paraffinic hydrocarbons. It is used in the form of a wettable powder on colloidal sulfur with a 25% content of 2,4-dinitrothiocyanobenzene. Under unfavorable weather conditions it may cause burning of plants. The LD_{50} for experimental animals is about 500 mg./kg.; the maximum permissible concentration in air is 2 mg./m³. The use of 2,4-dinitrothiocyanobenzene stimulates the growth and accelerates the ripening of grapes and increases their yield.

DNTB is produced by the reaction of ammonium thiocyanate with 2,4-dinitrochlorobenzene in water in the presence of dispersing agents at an elevated temperature:

$$O_2N-C_6H_3(NO_2)-Cl + NH_4SCN \longrightarrow O_2N-C_6H_3(NO_2)-SCN + NH_4Cl$$

The product should be carefully washed free of salts of thiocyanic acid, since they increase the phytocidal effect of the compound.

The presence in this molecule of two nitro groups renders it explosive if it is not diluted with an inert diluent.

4-Dimethylaminothiocyanobenzene (Defoliant 2929 RP) is a white crystalline substance, m.p. 73° C., almost insoluble in water, soluble in methyl and ethyl alcohols, acetone, ether, benzene, toluene, and xylene, and also in halogen derivatives of aliphatic and aromatic hydrocarbons. The LD_{50} for rats is 150–200 mg./kg. It is used as a solution containing 150 g. of the active ingredient in 1 liter. For defoliation of cotton it is recommended that 1.5–2 kg./ha. of the active ingredient be used; at these dosages the compound approaches butiphos in effectiveness.

4-Dimethylaminothiocyanobenzene can be synthesized in good yields by direct thiocyanation of dimethylaniline. The most convenient method of preparing it under industrial conditions is the reaction of salts of thiocyanic acid with chlorine and dimethylaniline:

$$C_6H_5-N(CH_3)_2 + NaSCN + Cl_2 \longrightarrow$$
$$NCS-C_6H_4-N(CH_3)_2 + NaCl + HCl$$

This reaction is carried out in methyl alcohol. To avoid the formation of explosive nitrogen trichloride, not ammonium thiocyanate but sodium thiocyanate should be used for the synthesis. It is possible to prepare the compound by an electrochemical method, and also by the action of sulfuryl chloride on a mixture of sodium thiocyanate and dimethylaniline in a suitable solvent.

4-Thiocyanoaniline (Rhodan) is a cream-colored crystalline substance, m.p. 142° C.; its solubility in water is 0.2 g./liter; it is more soluble in organic solvents. The LD_{50} for rats is 130 mg./kg.; Rhodan irritates the skin and sometimes causes the appearance of small sores. As a seed disinfectant against smut of grains it is used as a concentrate with OP-7 containing about 25% active ingredient. Seed disinfection with an aqueous suspension of the compound usually is carried out at an elevated temperature to combine the thermal and chemical methods of destroying the smut. 4-Thiocyanoaniline is produced similarly to 4-dimethylaminothiocyanobenzene.

Isothiocyanates

Two methods are used for the synthesis of isothiocyanates:

1. Isomerization of thiocyanates at a temperature from 100°–250° C., depending on the structure of the starting thiocyanate:

$$RSCN \rightarrow RNCS$$

Thus, for example, allyl thiocyanate at room temperature is slowly isomerized to the isothiocyanate, while methyl thiocyanate is stable even at 120° C.

2. Reaction of primary amines with carbon disulfide, which proceeds through the stage of formation of the dithiocarbamic acid or thiourea:

$$RNH_2 + CS_2 \rightarrow \left[RNH\overset{S}{\overset{\|}{C}}SH \right] \rightarrow RNCS + H_2S$$

By this method it is possible to prepare isothiocyanates of any structure.

Methyl isothiocyanate, used for the control of agricultural pests, is a colorless liquid with an irritating odor and a lachrymatory effect, b.p. 143° C. at 760 mm. of Hg. Pure methyl isothiocyanate is poorly soluble in water, highly soluble in most organic solvents; with many of them it is miscible in all proportions. The LD_{50} for warm-blooded animals is 50 to 80 mg./kg. It is used as a soil fumigant in the form of a 20% solution in a mixture of dichloropropylene and dichlorobutylene (d 1.15, LD_{50} for rats 300 mg./kg., calculated on the formulation).

Methyl isothiocyanate is very reactive and enters into reactions with alcohols, amines, water, and other similar compounds, forming derivatives of thiocarbamic acid or thiourea.

General references

GUSTAFSON, C., T. LIES, and T. WAGNER-JAUREGG: J. Econ. Entomol. 46, 620 (1953).
MEL'NIKOV, N. N., and M. S. ROKITSKAYA: Zhur. Priklad. Khim. 23, 1115 (1950).
—, S. I. SKLYARENKO, and E. M. CHERKASOVA: Zhur. Obshchei Khim. 9, 1819 (1939); 10, 1373 (1940); 14, 113 (1944); 16, 1025 (1946).
—, and N. D. SUKHAREVA: In book "Reaktsii i metody issledovaniya organicheskih soedinenii" [Reactions and methods of investigation of organic compounds], vol. 8. State Scientific and Technical Publishing House of Chemical Literature [USSR] (1959).
— —, and M. L. FEDDER: Doklady Akad. Nauk SSSR 31, 612 (1941).
— — — Zhur. Priklad. Khim. 16, 568 (1943).
METCALF, R. L.: Organic insecticides. New York-London: Interscience (1955).
POPOV, P. V., and M. A. IL'INSKAYA: Trudy Nauch. Inst. po Udobreniyam i Insektofungitsidam im. Ya. V. Samoilova No. 135, 156 (1939).

XXI. Derivatives of sulfuric and sulfurous acids

Derivatives of sulfuric acid

Sulfuric acid was used previously as a herbicide; at present it is used for the desiccation of potato plant tops for the purpose of mechanizing the harvesting. However, because treatment with sulfuric acid greatly increases the acidity of the soil and requires subsequent liming, its scale of use is rapidly declining.

Ammonium sulfamate (AMS, Ammate) is a white crystalline, hygroscopic substance, m.p. 125° C. It is slightly soluble in hydrophobic organic solvents. At 30° C., 232 g. of ammonium sulfamate dissolve in 100 g. of water. It corrodes metals, especially in the presence of water and oxygen of the air.

The technical grade product contains 70–90% of the principal compound with 10–30% impurities (water, ammonium salt of iminodisulfonic acid, and ammonium sulfate). The LD_{50} for rats is about 3,900 mg./kg.

Ammonium sulfamate surpasses all the other salts of this acid in the strength of its herbicidal effect. It is used to eradicate weeds at dosages of 100–200 kg./ha., and for temporary sterilization of the soil at dosages of 400–600 kg./ha. An advantage of ammonium sulfamate is its ability to hydrolyze in the soil with the formation of ammonium sulfate, a good ammoniacal fertilizer:

$$NH_4SO_3NH_2 + H_2O \rightarrow (NH_4)_2SO_4$$

Thus, there is a double utilization of the compound, first as a herbicide and then as a fertilizer.

Ammonium sulfamate is synthesized by neutralization with ammonia of sulfamic acid that is obtained by careful heating of urea with oleum:

$$(NH_2)_2CO + SO_3 + H_2SO_4 \rightarrow 2\ HSO_3NH_2 + CO_2$$
$$HSO_3NH_2 + NH_3 \rightarrow NH_4SO_3NH_2$$

The yield of ammonium sulfamate by this reaction is not less than 90% with a content of not less than 90% of the principal product. Ammonium sulfamate is obtained more cheaply by direct synthesis from ammonia and sulfuric anhydride at an elevated temperature:

$$SO_3 + NH_3 \rightarrow HSO_3NH_2$$
$$HSO_3NH_2 + NH_3 \rightarrow NH_4SO_3NH_2$$

However, purification of the compound obtained in this way is more complicated.

Calcium sulfamate, which is formed by neutralization of sulfamic acid with milk of lime, has been proposed for use to control rust on grains:

$$2\,HSO_3NH_2 + Ca(OH)_2 \rightarrow Ca(SO_3NH_2)_2 + 2\,H_2O$$

Sulfuryl fluoride at room temperature is a colorless and odorless gas, b.p. $-55.2°$ C. at 760 mm. of Hg, vapor pressure 18.3 kg./cm.^{2} at 25° C., critical temperature 96° C. At 25° C., 0.75 g./liter of sulfuryl fluoride dissolves in water. It is chemically stable and does not cause corrosion of metals or destruction of various materials. With caustic alkalies it forms a mixture of fluorides and sulfates.

Daily exposure of animals to an atmosphere containing 100 mg./l. of sulfuryl fluoride (seven hours a day for five days a week) did not cause death of the experimental animals.

Sulfuryl fluoride is produced by thermal decomposition of barium fluorosulfonate:

$$Ba(SO_3F)_2 \rightarrow BaSO_4 + SO_2F_2$$

The compound is used as a fumigant and somewhat exceeds methyl bromide in toxicity to various insect pests.

Tetramine (tetramethylenedisulfotetramine). This zoocide is prepared by condensation of formalin with sulfamide in acid medium:

$$2(NH_2)_2SO_2 + 4\,CH_2O \rightarrow \text{[tetramethylenedisulfotetramine]} + 4\,H_2O$$

It is a white crystalline substance that decomposes at 255°–260° C., is slightly soluble in water, more soluble in acetone and acetic acid. It has been proposed for the treatment of seeds and seedlings of coniferous trees to decrease damage by rodents when forests are planted by direct seeding. It is very toxic to most animals (LD_{50} 0.1–0.3 mg./kg.), hindering its wide use as a zoocide.

A number of mixed amides of sulfuric acid have been studied as fungicides; the most active are the amides containing a trichloromethylthio group on one of the nitrogens. According to patent data, N,N-dimethyl-N'-4-chlorophenyl-N'-trichloromethylthiosulfamide shows fungicidal activity close to that of captan. It still has not been practically used in agriculture. For the synthesis of this compound the diethylamide of chlorosulfonic acid is used, which is formed in good yield when dimethylchloramine reacts with sulfurous anhydride:

$$(CH_3)_2NCl + SO_2 \rightarrow (CH_3)_2NSO_2Cl$$

By the action of *p*-chloroaniline on the dimethylamide of chlorosulfonic acid in the presence of bases N,N-dimethyl-N'-4-chlorophenylsulfamide is obtained:

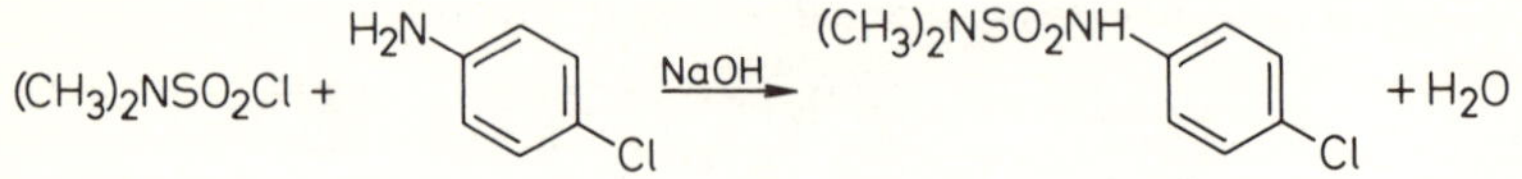

$$(CH_3)_2NSO_2Cl + H_2N\text{-}C_6H_4\text{-}Cl \xrightarrow{NaOH} (CH_3)_2NSO_2NH\text{-}C_6H_4\text{-}Cl + H_2O$$

and from the latter and perchloro(methyl mercaptan) N,N-dimethyl-N'-4-chlorophenyl-N'-trichloromethylthiosulfamide is formed:

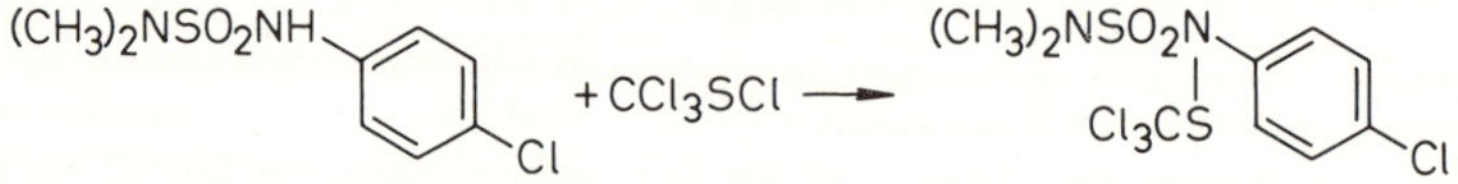

$$(CH_3)_2NSO_2NH\text{-}C_6H_4\text{-}Cl + CCl_3SCl \longrightarrow (CH_3)_2NSO_2N(SCCl_3)\text{-}C_6H_4\text{-}Cl$$

N,N-Dimethyl-N'-phenyl-N'-fluorodichloromethylthiosulfamide (*Eparen*) is a crystalline substance, m.p. 105.5°–105.6° C.; its vapor pressure at 20° C. is 1×10^{-6} mm. of Hg and at 45° C. 4×10^{-5} mm. of Hg. It is practically insoluble in water; 1.5 g. of the compound dissolves in 100 ml. of methyl alcohol, and 7.0 g. dissolves in 100 ml. of xylene. The LD_{50} for rats by different methods of administration is 500–1,000 mg./kg.

Its chemical properties recall those of the corresponding derivatives of perchloro(methyl mercaptan), but it is somewhat more stable.

Eparen can be produced by the following scheme:

$$(CH_3)_2NSO_2Cl + C_6H_5NH_2 \rightarrow (CH_3)_2NSO_2NHC_6H_5 + HCl$$
$$(CH_3)_2NSO_2NHC_6H_5 + CFCl_2SCl \rightarrow (CH_3)_2NSO_2NC_6H_5 + HCl$$
$$\qquad\qquad\qquad\qquad\qquad\qquad\qquad\quad |$$
$$\qquad\qquad\qquad\qquad\qquad\qquad\quad SCFCl_2$$

It is marketed in the form of a 50% wettable powder for use as a 0.075–0.1% aqueous suspension and a 7.5% powder for dusting. Eparen gives a good effect when it is used to control apple scab, and gray rot on strawberries and stone fruits.

The esters of sulfuric acid do not have insecticidal or fungicidal properties, although they are very toxic to vertebrates.

Low fungicidal and acaricidal activity are shown by the salts of thiosulfuric acid, especially in acid medium; this fact is associated with the production of sulfur, which is fungicidal and acaricidal.

Calcium tetrathionate is a white crystalline substance that easily breaks down in alkaline medium but is more stable in acid medium. It is produced by precipitation from a solution of tetrathionic acid with calcium chloride in HCl medium:

$$CaCl_2 + H_2S_4O_6 \rightarrow CaS_4O_6 + 2 HCl$$

A 0.5–1% suspension of calcium tetrathionate gives satisfactory results in the control of powdery mildew of roses and cucumbers. Calcium tetrathionate somewhat exceeds colloidal sulfur in the strength of its fungicidal action. Sodium tetrathionate has little activity.

Derivatives of sulfurous acid

Free sulfurous anhydride, obtainable by burning sulfur on the spot where it was to be used, was previously employed for rodent control and as a fumigant in enclosed premises. When it is used as a rodenticide, sulfur dioxide is produced in special cartridges with a combustible mixture. Recently, however, sulfurous anhydride has had little use. Sodium bisulfite, which is obtained by the action of sulfurous anhydride on sodium hydroxide or sodium carbonate, is used as an agent for the preservation of fodder in animal husbandry.

Table XLIII. *Acaricidal activity of mixed dialkyl sulfites against the spider mite (Tetranychus bimaculatus)*

ROSOR' $\parallel$ O		LC_{95} (mg./l.)
R	R'	
C_2H_5	$C_{10}H_{21}$	2500
C_4H_9	$C_{10}H_{21}$	5000
$ClCH_2CH_2$	$C_{10}H_{21}$	170
$BrCH_2CH_2$	$C_{10}H_{21}$	125
CCl_3CH_2	$C_{10}H_{21}$	1000
$ClCH_2CH(CH_3)$	$C_{10}H_{21}$	500
$ClCH_2CH_2CH_2$	$C_{10}H_{21}$	600
$ClCH_2CH_2CH_2CH_2$	$C_{10}H_{21}$	400
$Cl_2C_3H_5$	$C_{10}H_{21}$	50
$ClCH_2CH_2$	$C_{12}H_{25}$	125
$Cl_2CHCH_2CH_2$	$C_{12}H_{25}$	75
$ClCH_2CH_2CH_2$	$C_{12}H_{25}$	125
$ClCH_2CH_2$	C_4H_9	10000
$ClCH_2CH_2$	$n\text{-}C_7H_{15}$	1000
$ClCH_2CH_2$	$n\text{-}C_8H_{17}$	500
$ClCH_2CH_2$	$n\text{-}C_{11}H_{23}$	150
$ClCH_2CH_2$	$n\text{-}C_{14}H_{29}$	400
$ClCH_2CH_2$	$n\text{-}C_{16}H_{33}$	600
$ClCH_2CH_2$	$C_6H_5CH_2CH_2$	1250
$ClCH_2CH_2$	$4\text{-}ClC_6H_4OCH_2CH_2$	250
$ClCH_2CH_2$	$4\text{-}ClC_6H_4OCH(CH_3)CH_2$	60
$ClCH_2CH_2$	$4\text{-}(CH_3)_3CC_6H_4OCH_2CH_2$	50
$ClCH_2CH_2$	$4\text{-}(CH_3)_3CC_6H_4OCH_2CH_2CH_2$	20
$ClCH_2CH_2$	$4\text{-}(CH_3)_3CC_6H_4OCH_2CH(CH_3)$	10
$ClCH_2CH_2$	$4\text{-}CH_3C_6H_4OCH_2CH(CH_3)$	100
$ClCH_2CH_2$	$4\text{-}(CH_3)_2CHC_6H_4OCH_2CH(CH_3)$	20
$ClCH_2CH_2$	$2\text{-}(CH_3)_2CHC_6H_4OCH_2CH(CH_3)$	20
$ClCH_2CH_2$	$4\text{-}C_2H_5CHC_6H_4OCH_2CH(CH_3)$ $\quad\quad\quad\;\, |$ $\quad\quad\quad CH_3$	10
$ClCH_2CH_2$	$4\text{-}(CH_3)_2CC_6H_4OCH_2CH(CH_3)$ $\quad\quad\quad\;\, |$ $\quad\quad\quad C_2H_5$	16
$ClCH_2CH_2$	$4\text{-}n\text{-}C_5H_{11}C_6H_4OCH_2CH(CH_3)$	25
$ClCH_2CH_2$	$2,4\text{-}Cl_2C_6H_3OCH_2CH(CH_3)$	20
$ClCH_2CH_2$	$2,4,5\text{-}Cl_3C_6H_2OCH_2CH(CH_3)$	50

The insecticidal, acaricidal, fungicidal, and herbicidal properties of a number of esters of sulfurous acid have been studied. The most effective compounds have been found among the mixed aliphatic esters of sulfurous acid. In Table XLIII the toxicities of compounds of this type for the spider mite are shown to illustrate their acaricidal activity.

The acaricidal activity of the mixed aliphatic esters of sulfurous acid increases with an increase in the length of the carbon chain of one of the alkyl groups; the maximum is reached at the compound with a C_{12} radical and thereafter the acaricidal activity decreases. A change in the acaricidal properties also takes place when the second ester radical is modified. The most active compounds are those containing C_{10} or C_{12} radicals and dichloroethyl. The accumulation of a large number of chlorine atoms on the ethyl radical lowers the activity of the compound. Compounds with γ-chloropropyl or dichloropropyl radicals also have good acaricidal activity. Especially active are the compounds having 2-chloroethyl and aryloxypropyl radicals. As aryl radicals the 4-*tert*-butylphenyl radical and its close homologs give the best results. Of this group of compounds, Aramite is used in agriculture.

2-(4-tert-Butylphenoxy)propyl-2-chloroethyl sulfite (Aramite) is a pale yellow liquid, b.p. 175° C. at 0.1 mm. of Hg, practically insoluble in water, miscible with most organic solvents, but slightly soluble in petroleum products that are rich in paraffinic hydrocarbons. It is easily hydrolyzed by caustic alkalies, an aqueous solution of ammonia, and aqueous solutions of mineral acids. In sunlight it breaks down relatively rapidly with the evolution of sulfurous anhydride. To stabilize it, polypropylene glycol is added.

The technical grade product is dark amber in color and contains 90% of the principal compound. It is marketed in the form of a 20–40% emulsifiable concentrate, a 15% wettable powder, and a preparation for the protection of greenhouse crops from mites, employed by covering the heating pipes in the greenhouses with it. The LD_{50} is 2,000 mg./kg. for mice, 3,900 mg./kg. for rats. Aramite has a carcinogenic effect, and therefore its use in agriculture has been discontinued. It is produced by the following scheme:

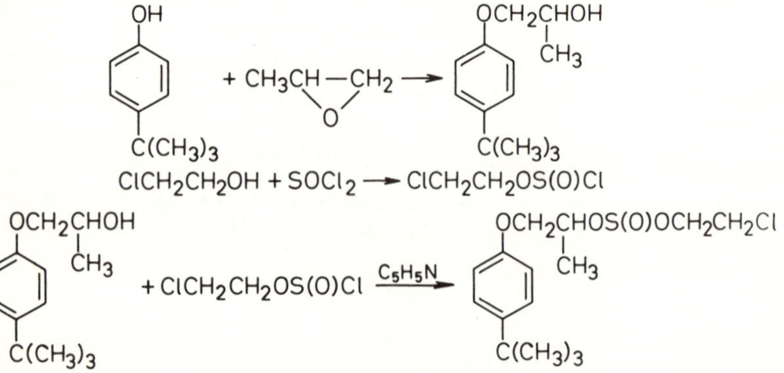

Some cyclic esters of sulfurous acid are of great interest.

1,2,3,4,7,7-Hexachlorobicyclo[2.2.1]heptene-2,5,6-bis(methylene) sulfite (endosulfan, Thiodan) is a white crystalline substance that exists in two forms, m.p. 108°–109° C. and m.p. 296°–298° C. The ratio of these forms in the technical grade product (m.p. 70°–100° C.) is 4:1. It is practically insoluble in water, but highly soluble in organic solvents. The LD_{50} for rats is from 40 to 110 mg./kg.

Endosulfan is marketed in the form of dusts, wettable powders, and emulsive concentrates. It gives good results in the control of various beetles, caterpillars, and aphids on fruit and berry crops at a concentration on the order of 0.2%. Endosulfan is particularly active against the Colorado potato beetle, woolly apple aphid, and a number of other pests. In spite of its relatively high acute toxicity, endosulfan has less chronic toxicity for mammals than other compounds that are products of the diene synthesis.

It is unstable to the action of alkalies, but more stable in sunlight than Aramite.

Endosulfan is produced from hexachlorocyclopentadiene, butenediol, and thionyl chloride. The process consists of two stages: synthesis of endosulfan-alcohol and its reaction with thionyl chloride. Endosulfan-alcohol is synthesized by two routes: reaction of butene-2-diol-1,4 diacetate with hexachlorocyclopentadiene in xylene solution and subsequent hydrolysis in acid medium:

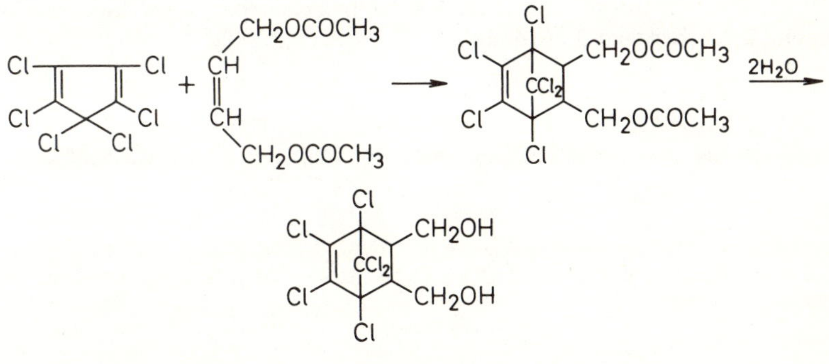

or direct condensation of butene-2-diol-1,4 with hexachlorocyclopentadiene in such solvents as dioxane, tetrahydrofuran, etc.:

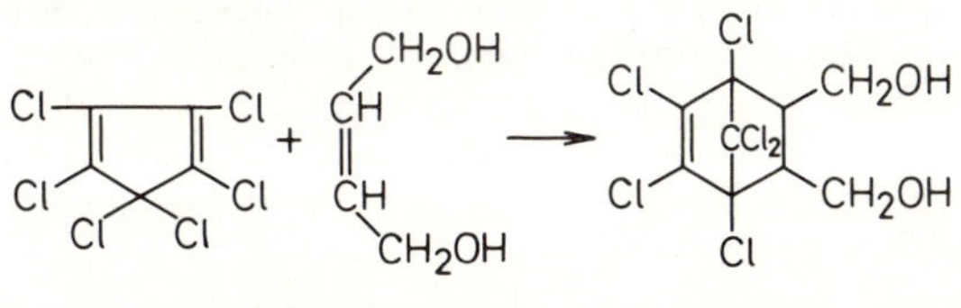

If the condensation is carried out without a solvent, the main product of the reaction is not endosulfan-alcohol, but 4,5,6,7,10,10-hexachloro-4,7-endomethylene-4,7,8,9-tetrahydrophthalan:

The butene-2-diol-1,4 necessary for the synthesis of the endosulfan-alcohol is prepared by selective hydrogenation of butynediol:

$$HOCH_2C \equiv CCH_2OH + H_2 \rightarrow HOCH_2CH = CHCH_2OH$$

Of the two isomeric butynediols only the *cis*-isomer is useful. Because of this the hydrogenation of butynediol is carried out under conditions that guarantee a maximum content of *cis*-butene-2-diol-1,4 in the product.

The final stage of the process of preparing endosulfan consists in the treatment of endosulfan-alcohol with an excess of thionyl chloride:

A by-product in this process is 1,2,3,4,7,7-hexachloro-5,6-bis(chloromethyl)-bicyclo[2.2.1]heptene-2 (Alodan):

Other derivatives of sulfurous acid have not yet been used in agriculture.

General references

Bayer-Compendium Pflanzenschutzmittel, Leverkusen (1964).
METCALF, R. L.: Organic insecticides. New York-London: Interscience (1955).
ZBIROVSKY, M., and J. MYSKA: Insekticidy, fungicidy, rodenticidy. Prague: Nakl. Ceskosl. Akad. Věd. (1957).

XXII. Sulfonic acids and their derivatives

General characteristics of pesticidal properties

The pesticidal properties of a large number of different sulfonic acids, their salts, esters, amides, and other derivatives have been studied, and among them have been found compounds with high physiological activity. However, only a few compounds of this class are used in agriculture, industry, and public health. The free sulfonic acids and their salts with metals and nitrogenous bases do not show contact insecticidal and acaricidal properties, but some sulfonic acids of the triphenylmethane series and substituted arylureas are active mothproofing agents. They are used to protect wool from moths by treating it at the time of dyeing or by other means; the presence of a sulfo group in the molecule facilitates fixation of these compounds on wool. The salts of sulfanilic acid and related compounds have a fungicidal effect against rust of grains, but the mechanism of this action is still not entirely clear. The alkyl- and arylsulfonic acids have herbicidal properties, but they are not strong enough to be of practical importance.

The derivatives of the sulfonic acids are in all respects more active; their properties have been studied in detail and a number of them are used in agriculture. Of these derivatives one should note primarily the esters with aliphatic alcohols and especially with phenols. The esters of aliphatic and aromatic sulfonic acids have weak insecticidal and stronger acaricidal activity. The strongest acaricides are the esters of the sulfonic acids with halogen-substituted phenols. The phenyl and chlorophenyl esters of benzene- and *p*-chlorobenzenesulfonic acids are used in agriculture. The esters of the aliphatic sulfonic acids, although they also show acaricidal properties, are expensive, preparation of these compounds being more complex.

The esters of the aromatic sulfonic acids also have a herbicidal effect. For example, 4-chlorophenyl 4'-chlorobenzenesulfonate when introduced into the soil acts selectively on creeping weeds, wild oats, and other weed plants.

High fungicidal and bactericidal activity is shown by esters of the thiosulfonic acids of both the aliphatic and the aromatic series. These compounds, furthermore, stimulate the growth of a number of plants when used for preplanting treatment of seeds.

The acaricidal, insecticidal, and fungicidal properties of the amides of the sulfonic acids also have been studied in detail, but as yet only a limited number of compounds of this class is in use.

Sulfonic acids

The free sulfonic acids and their salts are used as mothproofing agents for prophylactic treatment of wool. In the textile industry even before World War II the Eulans of different brands were used. The simplest representative of this group of compounds is Eulan-N (I), which is obtained by the condensation of benzaldehyde-2-sulfonic acid with 2,4-dichlorophenol:

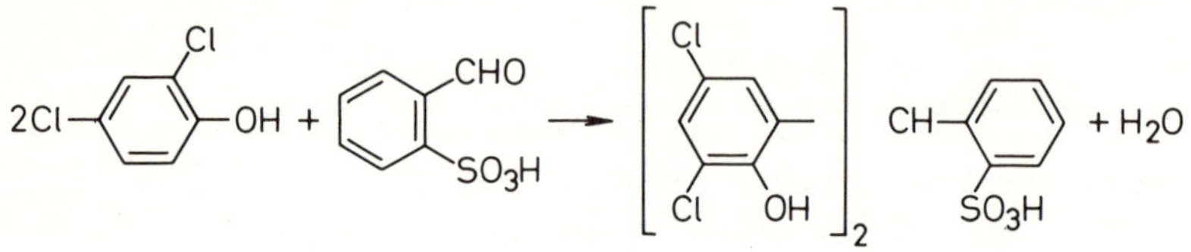

$$2Cl-\underset{Cl}{\underset{|}{C_6H_3}}-OH + \underset{SO_3H}{\underset{|}{C_6H_4}}-CHO \longrightarrow \left[\underset{Cl\quad OH}{\underset{|\quad\;|}{C_6H_2Cl}}\right]_2 CH-\underset{SO_3H}{\underset{|}{C_6H_4}} + H_2O \quad (I)$$

Eulan-CN is 2′,2″-dihydroxy-3′,3″,5′,5″,4-pentachlorotriphenylmethanesulfonic acid-2 (II); it is obtained by a completely analogous method to Eulan-N, with 4-chlorobenzaldehydesulfonic acid-2 instead of benzaldehydesulfonic acid-2. A mothproofing effect is shown by the esters and ethers of the triphenylmethane series, Eulan-SN (III) and the butyl ether of Eulan-CN (IV):

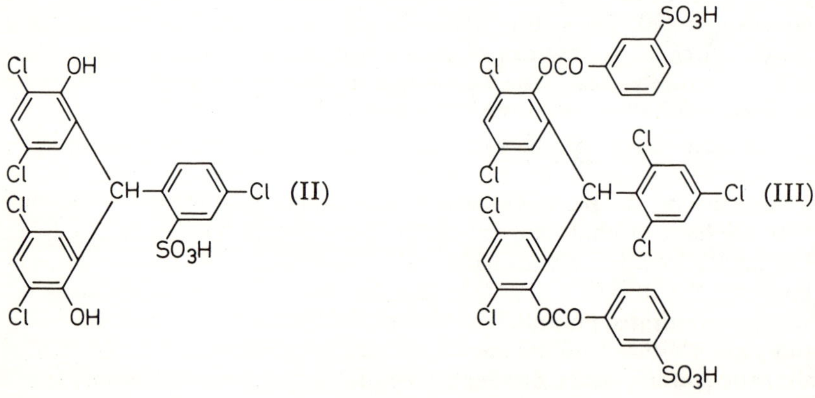

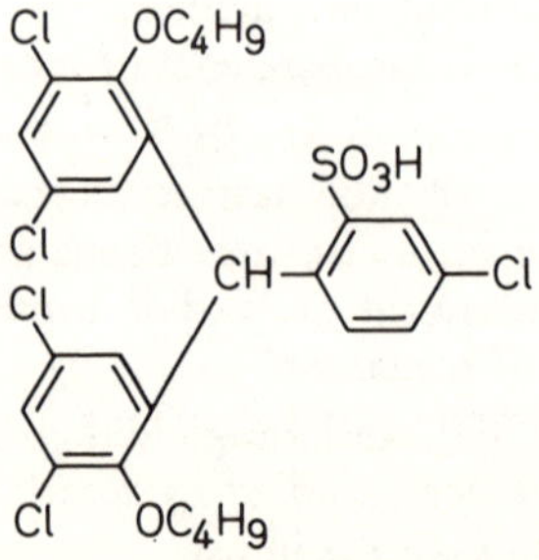

(II) 2′,2″-dihydroxy-3′,3″,5′,5″,4-pentachlorotriphenylmethanesulfonic acid-2

(III) ester derivative

(IV) butyl ether of Eulan-CN

Derivatives of urea also are used as mothproofing agents. The synthesis of one of them, known as Mitin-FF (V), can be carried out by the following scheme:

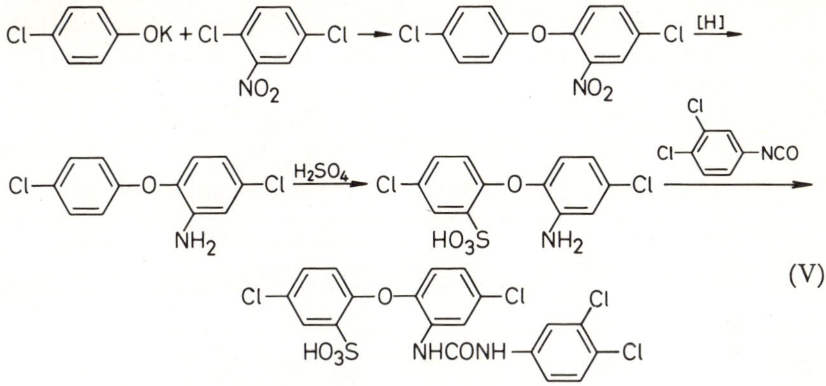

[Reaction scheme showing synthesis of Mitin-FF (V):

Cl—C₆H₄—OK + Cl—C₆H₃(NO₂)—Cl → Cl—C₆H₄—O—C₆H₃(NO₂)—Cl —[H]→

Cl—C₆H₄—O—C₆H₃(NH₂)—Cl —H₂SO₄→ Cl—C₆H₂(HO₃S)—O—C₆H₃(NH₂)—Cl —(Cl—C₆H₃(Cl)—NCO)→

Cl—C₆H₂(HO₃S)—O—C₆H₃(NHCONH—C₆H₃(Cl)—Cl)—Cl]   (V)

In the literature other sulfonic acids are described that have mothproofing properties, but they have not yet been used in industry. The sulfonic acids discussed are used to treat wool in the form of the sodium salts in an amount up to 1% of the weight of the wool.

Esters of sulfonic acids

In agriculture use is made of aromatic esters of the simplest arylsulfonic acids to protect plants from mites. These esters are active not only against the larval stages, but also against the mite eggs. In Table XLIV data are given on the ovicidal activity of aromatic esters of arylsulfonic acids. 4-Chlorophenyl 4'-chlorobenzenesulfonate and the corresponding derivative of 4-bromophenol are the most active; all the others are considerably less active.

The sulfonic acid esters are produced in good yields by the reaction of the acid chlorides of the appropriate sulfonic acids with alkali metal phenolates:

$$ArSO_2Cl + Ar'ONa \rightarrow ArSO_2OAr' + NaCl$$

This reaction is carried out at a low temperature in aqueous medium with thorough stirring of the reaction mixture. Separation of the reaction products does not present difficulties because the esters of the sulfonic acids are insoluble in water. 4-Chlorophenyl 4'-chlorobenzenesulfonate, 2,4-dichlorophenyl benzenesulfonate, and 4-chlorophenyl benzenesulfonate are used in agriculture.

4-Chlorophenyl 4'-chlorobenzenesulfonate (ovex, ester sulfonate, Ovotran, chlorfenson, PCPCBS) is a white crystalline substance, m.p. 86.5° C., insoluble in water. Its solubility in organic solvents in g./100 g. of solvent is: acetone 130, dichloroethane 110, kerosine 2, xylene 78, cyclohexanone 110, carbon tetrachloride 41, ethyl alcohol 1 (25° C.).

The compound as marketed in various countries contains 80–90% of the required isomer and melts at 80° C. In the Soviet Union a product is manufactured that has a lower melting point because of a high content (in excess of 40%) of 2-chlorophenyl 4′-chlorobenzenesulfonate. The LD_{50} for experimental animals is 2,000–2,500 mg./kg.; the maximum permissible concentration in air is 2 mg./m³.

Table XLIV. *Ovicidal effect of aryl esters of substituted benzenesulfonic acids*

		Toxicity for	
R	R′	Tetranychus telarius (LD_{50}, %)	Tetranychus bimaculatus (L_{100}, %)
H	H	0.051	> 0.36
H	4-Cl	0.014	0.36
4-Cl	4-Cl	0.033	0.007
4-Cl	H	0.67	0.36
H	2,4-Cl$_2$	0.053	—
4-Cl	2,4-Cl$_2$	0.32	> 0.36
2,4-Cl$_2$	4-Cl	> 0.5	—
2,4-Cl$_2$	H	0.3	—
4-Cl	2,4,5-Cl$_3$	—	> 0.36
2,4-Cl$_2$	2,4-Cl$_2$	> 0.5	> 0.36
4-Cl	2,3,4,6-Cl$_4$	—	> 0.36
4-Cl	Cl$_5$	—	0.36
4-Cl	4-Br	—	0.03
4-Br	4-Cl	—	0.06
4-Br	4-Br	—	> 0.06
4-Cl	4-CH$_3$	—	> 0.36
4-Cl	4-CH$_3$O	—	> 0.36
4-Cl	4-NO$_2$	—	> 0.36
3-NO$_2$	4-Cl	—	> 0.36
4-Cl	4-(CH$_3$)$_3$C	—	> 0.36
4-Cl	2-C$_6$H$_5$	—	> 0.36

Ovex is chemically stable; it is hydrolyzed by caustic alkalies on heating, forming 4-chlorophenol and 4-chlorobenzenesulfonic acid.

The compound is used to control plant-feeding mites on fruit and other crops; it has a long residual effect. Sometimes, because of the fact that upon hydrolysis it splits off chlorophenol, it imparts an unpleasant taste to fruits. Treatment of plants with ovex, therefore, should be carried out as early as possible, and in any case not later than 30–35 days before the crop is harvested. It is marketed in the form of a 50–80% wettable powder, a 20% emulsive concentrate, and dusts (rarely used).

Ovex is produced by the reaction of 4-chlorobenzenesulfonyl chloride with 4-chlorophenol in the presence of NaOH in aqueous medium:

$$Cl-\underset{}{\underset{}{\bigcirc}}-ONa + Cl-\underset{}{\underset{}{\bigcirc}}-SO_2Cl \longrightarrow$$

$$Cl-\underset{}{\underset{}{\bigcirc}}-OSO_2-\underset{}{\underset{}{\bigcirc}}-Cl + NaCl$$

The 4-chlorophenol necessary for this synthesis is produced in good yield by the chlorination of phenol with sulfuryl chloride:

$$\underset{}{\underset{}{\bigcirc}}\text{-OH} + SO_2Cl_2 \longrightarrow \underset{Cl}{\underset{}{\bigcirc}}\text{-OH} + SO_2 + HCl$$

Direct chlorination of phenol with chlorine leads to a mixture of chlorophenols in which the content of the 4-isomer does not exceed 50%. The process is carried out at as low a temperature as possible. Raising the temperature of chlorination promotes formation of the 2-isomer. Separation of the isomer is accomplished by fractional distillation on a column of not fewer than 30 theoretical plates.

4-Chlorobenzenesulfonyl chloride is synthesized by the action of chlorosulfonic acid on 4-chlorobenzenesulfonic acid, a by-product of DDT manufacture:

$$\underset{SO_3H}{\underset{}{\bigcirc}}\text{-Cl} + ClSO_3H \longrightarrow \underset{SO_2Cl}{\underset{}{\bigcirc}}\text{-Cl} + H_2SO_4$$

Another method of preparing ovex is the chlorination of phenyl 4-chlorobenzenesulfonate with chlorine in the presence of catalysts for substitutive chlorination. In this case the chlorine is directed predominantly into the 4-position:

$$\underset{}{\underset{}{\bigcirc}}-OSO_2-\underset{}{\underset{}{\bigcirc}}-Cl + Cl_2 \longrightarrow Cl-\underset{}{\underset{}{\bigcirc}}-OSO_2-\underset{}{\underset{}{\bigcirc}}-Cl + HCl$$

The content of the necessary isomer in the products is not less than 70–75%. Manufacture of a product with a large amount of the impurity 2-chlorophenyl 4'-chlorobenzenesulfonate not only leads to a loss of raw material, but also causes additional contamination of the treated fruits.

2,4-Dichlorophenyl benzenesulfonate (Genite) melts at 54°–55° C.; the technical grade product of 97% purity solidifies at 42° C. It is insoluble in water, but highly soluble in most organic solvents. The vapor pressure at 30° C. is 2.7×10^{-4} mm. of Hg.

Genite is stable when heated and in prolonged storage. It is hydrolyzed by caustic alkalies, forming 2,4-dichlorophenol and benzenesulfonic acid:

$$\underset{\text{Cl}}{\text{Cl}}\text{C}_6\text{H}_3(\text{OSO}_2\text{C}_6\text{H}_5) \xrightarrow{\text{NaOH}} \underset{\text{Cl}}{\text{Cl}}\text{C}_6\text{H}_3(\text{OH}) + \text{C}_6\text{H}_5\text{SO}_3\text{Na}$$

The LD$_{50}$ for rats is 1,400–2,000 mg./kg. Genite is marketed in the form of a 50% emulsive concentrate; for spraying plants a 0.1% emulsion (active ingredient) is used.

The compound is produced by the reaction of benzenesulfonyl chloride with 2,4-dichlorophenol in the presence of NaOH in aqueous medium at 10°–30° C.:

$$\text{Cl}-\text{C}_6\text{H}_3(\text{Cl})-\text{ONa} + \text{C}_6\text{H}_5-\text{SO}_2\text{Cl} \longrightarrow$$

$$\text{Cl}-\text{C}_6\text{H}_3(\text{Cl})-\text{OSO}_2-\text{C}_6\text{H}_5 + \text{NaCl}$$

It also can be synthesized by chlorination of phenyl benzenesulfonate at an elevated temperature.

4-Chlorophenyl benzenesulfonate (fenson, CPB, PCPB, Murvesco) is a white crystalline substance, m.p. 61°–62° C., insoluble in water, but highly soluble in most organic solvents, especially ketones, aromatic hydrocarbons, and their halogen derivatives. The compound is thermally stable, but is hydrolyzed by alkalies. The LD$_{50}$ for different experimental animals is 1,300–1,800 mg./kg.

The technical grade product melts at 50°–59° C. It is used to control mites on apple and pear trees and also in greenhouses (aerosols). For spraying in orchards, fenson is used at a concentration of 0.25–0.05% active ingredient. It is phytotoxic to cucurbit crops in the early stages of development. It is marketed in the form of a 20–50% wettable powder and in aerosols.

Fenson is prepared in a manner similar to ovex either by the chlorination of phenyl benzenesulfonate with chlorine in the presence of a catalyst for substitutive chlorination at a temperature of about 100° C., or by the condensation of 4-chlorophenol with benzenesulfonyl chloride in alkaline medium. The quality of the 4-chlorophenol used for the production of fenson is very important.

Amides of sulfonic acids

Among the amides of sulfonic acids compounds with varied pesticidal effects have been found. Thus, the 3,4-dichloroanilide of methanesulfonic acid and the 2,4,5-trichloroanilide of chloromethanesulfonic acid under the names of Eulan-BL and Eulan-AVA have been proposed as mothproofing agents for wool by impregnation with 0.5–2% preparations.

The N,N-dimethylamide of p-chlorobenzenesulfonic acid has a strong acaricidal effect and is not inferior in this respect to ovex; 4-thiocyanophenyl methanesulfonamide and its homologs show powerful fungicidal properties and can be used to protect plants from diseases.

The dialkylamides of p-chlorobenzenesulfonic acid strengthen the effect of DDT and its analogs on flies that are resistant to these insecticides. This occurs apparently as a result of inhibition of DDT-dehydrochlorinase by the sulfonamides. Addition to DDT of N,N-dibutyl-4-chlorobenzenesulfonamide permits use of this insecticide even against resistant insects. N,N-Dibutyl-4-chlorobenzenesulfonamide is produced under the name of "antiresistant" [WARF antiresistant]. It is a white substance, almost odorless, m.p. 37°–37.5° C., highly soluble in most of the solvents in which DDT dissolves. The LD_{50} for experimental animals is more than 1,000 mg./kg. It is used together with DDT (and its analogs) in ratios of 0.025–10:1 depending on the object of treatment. The compound is marketed in the form of a technical product, m.p. 33° C.; it is mixed on the site of use with DDT preparations, and also is added to aerosol formulations for household use. In the Soviet Union this compound is not yet in use. It is produced from 4-chlorobenzenesulfonyl chloride and dibutylamine. Some derivatives of the amides of sulfonic acids are used for disinfection, for example, chloramines T and B and the dichloramide of methanesulfonic acid, which contains about 80% active chlorine and surpasses the chloroamides of the aromatic sulfonic acids in strength of action on many microorganisms.

The dichloroamide of methanesulfonic acid is produced by chlorination of the amide of methanesulfonic acid in the presence of weak bases, for example magnesium oxide:

$$CH_3SO_2NH_2 + Cl_2 \rightarrow CH_3SO_2NCl_2 + 2\,HCl$$

A deficiency of this compound is its poor solubility in water, causing it to be used either as solutions in organic solvents or as fine suspensions.

Herbicidal properties have been discovered in the sulfonamide derivatives methyl N-(4-nitrobenzenesulfonyl)carbamate (compound MB-8882), methyl N-(4-aminobenzenesulfonyl)carbamate (compound MB-9057), and methyl N-(4-methylcarbamoylbenzenesulfonyl)carbamate (compound MB-9555) (VI):

$$R-\!\!\left\langle\!\!\bigcirc\!\!\right\rangle\!\!-SO_2NHCOOCH_3 \qquad (VI)$$

$(R = NO_2, NH_2, CH_3OCONH)$

These compounds are crystalline substances with prolonged effects lasting from five to six weeks. They are used at dosages on the order of 2–3 kg./ha. At a concentration 10^{-3} molar they inhibit the Hill reaction and their action on plants causes chlorosis. Their LD_{50}'s vary from 1,000 to

5,000 mg./kg. These compounds have been investigated as growth inhibitors for monocotyledonous weed plants in alfalfa, potatoes, and sugar cane.

Other derivatives of sulfonic acids

In this group of compounds the acid fluorides and esters of thiosulfonic acids should be noted. The former have a strong insecticidal effect and the latter are fungicidal and bactericidal.

When the chlorine in the acid chlorides of the aliphatic sulfonic acids is replaced by fluorine, the toxicity for insects sharply increases, but at the same time the toxicity for mammals also increases almost proportionately. The acid fluorides of the aromatic sulfonic acids are less toxic to mammals, but they also are less toxic to insects.

In Table XLV data are presented on the toxicity of sulfonyl fluorides as fumigants against adult grain weevils. The highest activity against various species of insects is shown by methanesulfonyl fluoride.

Table XLV. *Toxicity of some fumigants for adult grain weevils*

Fumigant	Conc. of compound causing 98–99% mortality (mg./l.)
Methanesulfonyl fluoride	0.003–0.005
Ethanesulfonyl fluoride	0.3–0.7
Benzenesulfonyl fluoride	2.0
Dichloroethane	45–55
Chloropicrin	0.8

Methanesulfonyl fluoride is a liquid, b.p. 124.2° C. at 754 mm. of Hg, d_4^{20} 1.347, solubility in water about 5%. It is highly soluble in most organic solvents. The LD_{50} for rats, dogs, and rabbits is 3.5 mg./kg. It is less toxic by inhalation; thus, a concentration of 0.023 mg./m.^{3} does not cause a harmful effect in experimental animals on prolonged exposure. It has been proposed as a fumigant to control pests of stored products and ectoparasites of animals. However, the possibility of employing methanesulfonyl fluoride to control ectoparasites of domestic animals seems doubtful.

Methanesulfonyl fluoride is produced, like most sulfonyl fluorides of other acids, by the reaction of ammonium or potassium fluoride with methanesulfonyl chloride:

$$CH_3SO_2Cl + NH_4F \rightarrow CH_3SO_2F + NH_4Cl$$

The yield of methanesulfonyl fluoride is 75–80%. Its chemical properties are close to those of the acid chlorides of the sulfonic acids, but it is more resistant to hydrolysis by water than the acid chlorides. In all other reactions it is fully analogous to the acid chlorides.

A second group of sulfonic acid derivatives, the esters of thiosulfonic acids, are not yet used in agriculture but they deserve investigation. These esters are synthesized by the reaction of sulfonic acid chlorides with mercaptans in the presence of bases:

$$RSO_2Cl + NaSR' \rightarrow RSO_2SR' + NaCl$$

Many esters of thiosulfonic acids have high fungicidal and bactericidal activity with relatively low toxicity to mammals. Essential shortcomings of these compounds, however, are their low stability and a strong unpleasant odor, hindering their practical use.

General references

METCALF, R. L.: Organic insecticides. New York-London: Interscience (1955).
Organicheskie insektofungitsidy i gerbitsidy [Organic insectofungicides and herbicides]. State Scientific and Technical Publishing House of Chemical Literature [USSR] (1958).
SUTER, C. M.: The organic chemistry of sulfur: Tetracovalent sulfur compounds. New York (1948).

XXIII. Derivatives of hydrazine and azo compounds

General characteristics of pesticidal properties

Free hydrazine and its salts with various organic and inorganic acids have slight pesticidal activity and are not used in agriculture. Only the double salt of copper sulfate with dihydrazine sulfate, which has been proposed for control of powdery mildew and black spot of roses, has some practical importance. The action of this compound apparently is based on reduction of copper sulfate by hydrazine to cuprous oxide and the free metal. The compound is a light blue powder, slightly soluble in water. The LD_{50} for rats is 590 mg./kg. It is marketed in the form of a 50% wettable powder that is used at a concentration of 0.3–0.4%.

Table XLVI. *Toxicity of nitrogen compounds to the mite Metatetranychus ulmi*

R	R′	Mortality from 0.1% conc. of compound (%)	
		Summer eggs of mites	Adult mites
H	H	98.6	75.7
4-Cl	H	98.6	90.6
4-Cl	4-Cl	8.0	35.7
4-CH$_3$	H	77.3	59.3
4-CH$_3$	4-CH$_3$	14.3	0.0
4-OH	H	0	0
4-H$_2$N	H	68.8	9.5
3-CH$_3$	3-CH$_3$	7.5	27.3
3-OH-4-CH$_3$	H	0	6
Azoxybenzene		90.5	100
Hydrazobenzene		96.3	67.6

The substituted hydrazines show stronger pesticidal activity; the compounds of the aromatic series are the more active.

Table XLVI shows the acaricidal activity of substituted azo-, azoxy-, and hydrazobenzenes toward mites and their summer eggs. Azobenzene, which is used to control mites in greenhouses, has the highest acaricidal activity. Introduction of hydroxy, amine, and methyl groups into the azobenzene molecule sharply decreases the activity of the compound. The substituted hydrazones of aldehydes, ketones, and quinones have high fungicidal activity. Thus, for example, the phenylhydrazone of acrolein and analogous compounds have been proposed for the protection of plants from rust, but because of their high cost they have not been used in agriculture. More promising are quinone derivatives, one of which, the benzoylhydrazone of quinone oxime, is used as a seed disinfectant for sugar beets and a number of other crops. The aromatic azosulfides also effectively eradicate mites.

Investigation of the fungicidal properties of a large number of quinone and quinoxime hydrazones has made it possible to demonstrate general rules for the change in activity of this group of compounds with changes in their structure.

The unsubstituted hydrazones of oximes (I) show comparatively weak fungicidal properties; replacement of hydrogen in the amino group by a carboxylic acid group considerably strengthens their fungicidal effect. The sulfonate (II) also has a fungicidal effect:

$$NH_2N=\!\!\!\bigcirc\!\!\!=O \quad (I) \qquad NaSO_3NHN=\!\!\!\bigcirc\!\!\!=O \quad (II)$$

Fungicidal properties are exhibited by almost all the compounds that are capable of tautomeric conversions according to the scheme:

$$RCONHN=\!\!\!\bigcirc\!\!\!=O \rightleftharpoons RCON=N-\!\!\!\bigcirc\!\!\!-OH \quad (III)$$

Hydrogenation of these compounds to the corresponding 4-hydroxyarylhydrazines (IV) leads to complete loss of fungicidal activity:

$$NH_2NH-\!\!\!\bigcirc\!\!\!-OH \quad (IV)$$

The acyl derivatives of quinone hydrazones are easily prepared from the quinones and hydrazides of carboxylic acids:

$$O=\!\!\!\bigcirc\!\!\!=O + RCONHNH_2 \rightarrow RCONHN=\!\!\!\bigcirc\!\!\!=O + H_2O$$

The fungicidal activity of the benzoylhydrazones of quinones is strengthened by the introduction of halogen atoms into the benzoyl group and into the quinone nucleus. Thus, the 4-chlorobenzoylhydrazone of benzoquinone is more active than the benzoylhydrazone of the quinone and less active than the benzoylhydrazone of chlorobenzoquinone. The same effect is demonstrated when a methyl group is introduced into the quinone nucleus. The most active compound of this group is the benzoylhydrazone of 2-chloro-6-methylquinone.

Replacement of a second hydrogen in a quinone by a hydroxylamine group changes the spectrum of action of the compound. Hydrazonoquinone oximes are more effective than hydrazones of quinones in inhibiting the growth of the phytopathogenic fungi *Phoma, Pythium, Rhizoctonia*, etc., which infect seedlings of sugar beet, cotton, and corn.

Replacement of the benzoyl group in the hydrazones of quinone oximes slightly changes the fungicidal activity of the compounds, while replacement of hydrogen in the quinone nucleus by halogens and other groups sharply decreases the activity of the compound; a similar phenomenon is observed when the oxime hydrogen is replaced by any acyl or alkyl radicals.

Reduction of benzoylhydrazonoquinone oxime with ammonium sulfite to benzoylhydrazinoaniline also yields an active compound:

$$C_6H_5CONHN=\!\!\!\left\langle\!\!\!\!\!\!\right\rangle\!\!\!=NOH \longrightarrow C_6H_5CONHNH-\!\!\!\left\langle\!\!\!\!\!\!\right\rangle\!\!\!-NH_2$$

Discovery of an antitubercular effect in a number of thiosemicarbazones of different aldehydes led to an intensive study of the bactericidal, fungicidal, and bacteriostatic properties of this group of compounds. This group has not yet been practically used for plant protection, although such compounds as the derivatives of the thiosemicarbazones of benzoquinone monoguanidinehydrazone (V) (m.p. 192°–194° C.) inhibit the growth of some microorganisms at a dilution of 1:1,000,000:

$$\underset{NH}{NH_2\overset{\|}{C}NHN}=\!\!\!\left\langle\!\!\!\!\!\!\right\rangle\!\!\!=\underset{S}{NNH\overset{\|}{C}NH_2} \qquad\qquad (V)$$

The preparation of this compound is carried out by reaction of thiosemicarbazide with benzoquinone monoguanidinehydrazone:

$$\underset{NH}{NH_2\overset{\|}{C}NHN}=\!\!\!\left\langle\!\!\!\!\!\!\right\rangle\!\!\!=O +\underset{S}{NH_2NH\overset{\|}{C}NH_2} \longrightarrow \underset{NH}{NH_2\overset{\|}{C}NHN}=\!\!\!\left\langle\!\!\!\!\!\!\right\rangle\!\!\!=\underset{S}{NNH\overset{\|}{C}NH_2}$$

The latter compound is obtained in good yield from quinone and aminoguanidine:

$$\underset{NH}{NH_2\overset{\|}{C}NHNH_2} + O=\!\!\!\left\langle\!\!\!\!\!\!\right\rangle\!\!\!=O \longrightarrow \underset{NH}{NH_2\overset{\|}{C}NHN}=\!\!\!\left\langle\!\!\!\!\!\!\right\rangle\!\!\!=O$$

Many derivatives of semicarbazide have considerable phytotoxicity.

The azo compounds and hydrazines have not yet been used as herbicides and plant growth regulators.

Azobenzene (azobenzide, diphenyldiimide) is an orange-red crystalline substance, m.p. 68° C., b.p. 297.4° C. at 760 mm. of Hg, d 1.203. It is practically insoluble in water, but highly soluble in most organic solvents, and highly volatile at room temperature.

Azobenzene is one of the oldest acaricides of selective action. It was proposed for control of mites in 1945 and is used for treatment of greenhouse plants. For this purpose the steam-heating system of greenhouses is covered with a special paste containing azobenzene. It also can be used in the form of special smoke grenades. The LD_{50} is more than 1,000 mg./kg., but after nine weeks dogs fed with a diet containing 600 mg./kg. of azobenzene displayed disruptions in their liver functions and high mortality was recorded.

Azobenzene is produced by reduction of nitrobenzene:

$$2\ C_6H_5NO_2 + 4\ H_2 \rightarrow C_6H_5N = NC_6H_5 + 4\ H_2O$$

4-Chlorophenyl 2,4,5-trichlorophenylazosulfide is a white crystalline substance, m.p. 123.5°–124° C. (dec.), insoluble in water, barely soluble in acetone and more soluble in alcohol, benzene, and some petroleum products. It is resistant to the action of dilute acids and alkalies at room temperature. The LD_{50} for mice is 3,000 mg./kg.

In Japan use has been begun of mixtures of 4-chlorophenyl 2,4,5-trichlorophenylazosulfide with 4,4'-dichlorodiphenylmethylcarbinol (Milbex) and with 4,4'-dichlorodiphenyl disulfide and bis(4-chlorophenoxy)methane (Mikazine) as selective acaricides. Milbex and Mikazine are marketed in the form of 50% wettable powders of a mixture of the indicated components; they are used at a concentration of 0.05% active ingredient (calculated on total acaricides).

4-Chlorophenyl 2,4,5-trichlorophenylazosulfide is produced by the reaction of a 2,4,5-trichlorophenyldiazonium salt with 4-chlorothiophenol:

Sodium p-dimethylaminobenzenediazosulfonate (Dexon, Bayer 22,555) is being tested as a seed disinfectant and soil fungicide to control microorganisms causing root rots of various crops. It is a yellow-brown powder that decomposes upon heating to 200° C., with a solubility in water of about 3%; it is slightly soluble in hydrophilic solvents and insoluble in hydrophobic ones. Dexon is used in the form of a 70% wettable powder and a 5% granulated formulation. The LD_{50} for experimental animals is 60–150 mg./kg. It is produced by the reaction of 4-dimethylaminophenyldiazonium salts with sodium sulfite:

Benzoylhydrazone of quinone oxime (Cerenox, Seredon) is a yellow-brown substance that decomposes at about 195° C. Its solubility in water is about 5 mg./l. It dissolves readily in dilute NaOH solution and is highly soluble in formamide and some organic solvents. The LD_{50} for rats is 100 mg./kg. It is used as a seed disinfectant for sugar beets and turnips.

It is marketed in the form of a 20% wettable powder for wet treatment of seeds and as a dry disinfectant containing 10% active ingredient.

The seed disinfectant Seredon Special contains 10% benzoylhydrazone of quinone oxime and 5% phenylmercuric acetate. This disinfectant has been recommended for the treatment of seeds of sugar beet, cotton, flax, and corn at a calculated dosage of 1–3 kg. of the compound/ton of seed.

The benzoylhydrazone of quinone oxime is produced in practically quantitative yield when freshly prepared nitrosophenol reacts with benzoylhydrazine in acid medium:

$$NO-C_6H_4-OH + NH_2NHCOC_6H_5 \longrightarrow C_6H_5CONHN=C_6H_4=NOH + H_2O$$

The compound obtained in this way, without further purification and after drying, is used for the preparation of seed disinfectant.

Dimethylalkylhydrazonium salts have been proposed as plant growth regulators:

$$\begin{bmatrix} CH_3 \\ | \\ HN_2NR \\ | \\ CH_3 \end{bmatrix}^+ Cl^-$$

They promote thickening of the stems of some crops and can be used to control lodging of plants.

General references

BROWN, A.: Insect control by chemicals. New York-London: Wiley (1951).
JUNG, J.: Z. Acker- u. Pflanzenbau **125**, 124 (1967).
— Landwirtsch. Forschung **20**, 221 (1967).
PETERSEN, S., W. GAUSS, and E. URBSCHAT: Angew. Chem. **67**, 217 (1955).

XXIV. Organic mercury compounds

General characteristics of pesticidal properties

Mercuric chloride and other inorganic compounds of mercury, because of their high microbiological activity, have been used for many decades in medicine. The fungicidal properties of mercuric chloride with respect to phytopathological fungi were discovered the end of the nineteenth century, and it has since been used to protect plants from diseases. However, its high toxicity to man, animals, and plants made it necessary to seek more active and safer compounds for use in agriculture. Investigation of various mercury derivatives of the aliphatic, alicyclic, aromatic, and heterocyclic series has shown that organomercury compounds considerably surpass the inorganic compounds of this metal in bactericidal and fungicidal effect. Also, organomercury compounds have a favorable chemotherapeutic index, *i.e.*, the concentration that kills plant disease organisms is many times less than the concentration dangerous to the plants. Moreover, many organomercury compounds at the concentrations used not only are harmless to plants, but even stimulate their growth. Sometimes the stimulation is so great that it leads to a substantial increase in yield.

The main use of organomercury compounds is for seed disinfection of various crops. From a fractional gram to 2–3 g. of Hg is used to treat the amount of seed used per hectare; this is quite economical in spite of the high cost of mercury. Recently in a number of countries use of organomercury compounds for plant protection (mainly for apple trees) has been started, since they have not only prophylactic but also curative effects; they also have been used to control some weeds. The insecticidal effect of organomercury compounds is negligible.

Study has shown that strong microbiological activity occurs only in mixed organomercury compounds of the general formula (I), where R is a hydrocarbon radical either containing or not containing functional groups, and X is an organic or inorganic acid group:

$$RHgX \quad (I) \qquad R_2Hg \quad (II)$$

The symmetrical compounds of mercury (II) are practically inactive as fungicides and bactericides.

In the aliphatic series the fungicidal activity of a compound decreases with an increase in size of the hydrocarbon radical bound to the mercury. Maximum activity is shown by ethylmercury derivatives, but the difference is not very substantial. A change in structure of the acid group also affects the toxicity of the compound, since its physical properties (solubility in water and organic solvents) are changed. Compounds of the aliphatic series are more powerful fungicides, and those of the aromatic series are stronger bactericides.

Introduction into an aliphatic radical of functional groups, depending on their nature, changes the fungicidal activity of the compound. Thus, for example, a methoxyl group in the 2-position slightly decreases the fungicidal effect of the compound; when a carbonyl group is introduced into the same position the decrease in fungicidal activity is more significant.

Introduction of functional groups into an aromatic radical bound with mercury also affects the toxicity of the compound both to microorganisms and to animals. The phenyl and tolyl derivatives have the greatest activity. An increase in size of the hydrocarbon radical also lowers the activity of the compound. Polycyclic derivatives of mercury are less active.

The simplest heterocyclic mercury compounds have high bactericidal activity and are less active as fungicides. Examples of such compounds are pyridylmercuric acetate and pyridylmercuric chloride.

Addition of various acid groups to the mercury atom has a substantial effect on the toxicity of the compounds to mammals. The chemical properties of mixed organomercury compounds also depend greatly on the structure of the acid radical.

The water-soluble mixed organomercury compounds give salts with halide ions, thiocyanates, and HCN that are barely soluble in water. This applies to compounds of the aliphatic and aromatic series that do not contain substituents in the hydrocarbon radical. Such compounds can be quantitatively titrated in aqueous solution by the Volhard method:

$$RHgX + KSCN \rightarrow RHgSCN + KX$$

Mercury salts that are slightly soluble in water are precipitated by hydrogen sulfide:

$$2\,RHgX + H_2S \xrightarrow[-2\,HX]{} (RHg)_2S \rightarrow R_2Hg + HgS$$

The action of reducing agents produces symmetrical compounds and metallic mercury separates:

$$2\,RHgX + 2\,H \rightarrow R_2Hg + Hg + 2\,HX$$

This reaction occurs in the soil under the influence of microorganisms, and as a result the duration of the fungicidal effect of organomercury compounds is not very great. Depending on the nature of the soil and the concentration of the organomercury compounds, their decomposition takes place over the course of one to four months.

Oxidizing agents also are capable of breaking down organomercury compounds with the splitting off of the hydrocarbon radical and formation of inorganic mercury compounds:

$$RHgX + O + H_2O \rightarrow ROH + HgXOH$$

In the soil the final product of this reaction is mercuric oxide, which is further converted to mercury sulfide. Mercury sulfide is one of the least toxic compounds for both animals and microorganisms.

Under certain conditions organomercury compounds react with free halogens to give practically quantitative yields of the corresponding halogen derivatives:

$$RHgX + X_2 \to RX + HgX_2$$

This reaction is even used sometimes for the synthesis of halogen derivatives not readily available.

When mixed organomercury compounds are heated to 150°–250° C., they break down with the formation of metallic mercury and other products, whose nature depends on the structure of the hydrocarbon radical on the mercury. For compounds of the aliphatic series this decomposition can be represented by the following general scheme:

$$RCH_2CH_2HgX \to RCH=CH_2 + Hg + HX$$
$$2\,RHgX \to R-R + Hg + HgX_2$$

At room temperature the mixed organomercury compounds are stable and can be kept without decomposition for an unlimited time. Only when there is intense illumination may disproportionation and decomposition similar to thermal breakdown occur to a small degree.

On prolonged heating with concentrated mineral acids the organomercury compounds break down with the formation of the corresponding hydrocarbon:

$$RHgX + HAc \to RH + HgXAc$$

However, this reaction takes place slowly.

The alkylmercury salts with salts of some amines yield double salts that are highly soluble in water. An example of these compounds is the double salt of ethylmercuric chloride with ethylenediamine hydrochloride.

Mercury compounds of the aliphatic series

With respect both to the number of compounds used and to the scale of their production and use in agriculture, the organomercury compounds of the aliphatic series occupy first place. They are used as seed disinfectants for a large number of crops, as disinfectants for nonmetallic materials, and also to combat slime formation in the paper industry. For the last purpose they are gradually being displaced by organotin compounds. Various methyl- and ethylmercury derivatives that are used in industry and agriculture have the highest fungicidal activity. Methoxyethylmercury derivatives, used as seed disinfectants, are close to them in activity.

Aliphatic mercury compounds of the general formula (I) are produced by the reaction of a dialkylmercury with appropriate mercury salts:

$$R_2Hg + Hg(Ac)_2 \to 2\,RHgAc \qquad (I)$$

The dialkylmercury necessary for this synthesis is synthesized industrially by two methods:

1. Reaction of alkyl halides with amalgams of the alkali metals (most often sodium amalgam):

$$2\,RBr + Hg + 2\,Na \rightarrow R_2Hg + 2\,NaBr$$

This reaction goes better with alkyl bromides than with alkyl chlorides. In the latter case a considerable part of the alkyl halide is used up in forming hydrocarbons by a Wurtz reaction. Important conditions for a successful process are the absence of moisture and thorough stirring. To obtain good yields of the dialkylmercury the sodium content of the amalgam should not exceed 0.6%, since more concentrated amalgams have a high melting point and mixing such an amalgam with the alkyl halide is difficult. The process is carried out in the presence of catalysts and at a low temperature. Raising the temperature increases the rate of side reactions. The yield of dialkylmercury by this method is 65–90%, calculated on the alkyl halide; the amalgam is used in large excess.

2. Reaction of the appropriate organomagnesium compounds with mercuric chloride:

$$RX + Mg \rightarrow RMgX$$
$$2\,RMgX + HgCl_2 \rightarrow R_2Hg + MgX_2 + MgCl_2$$

Both the first and the second methods of preparing dialkylmercuries are rather complicated.

G. A. Razuvaev has suggested a new method for preparing alkylmercury salts by initiated breakdown of the mercury salts of carboxylic acids:

$$Hg(OCOCH_3)_2 \rightarrow CH_3HgOCOCH_3 + CO_2$$

Hydrogen peroxide or organic peroxides are used as the initiator. This method is promising for the industrial production of alkylmercury salts.

It also has been proposed that ethylmercury salts be prepared from tetraethyllead and mercuric chloride:

$$(C_2H_5)_4Pb + 2\,HgCl_2 \rightarrow (C_2H_5)_2PbCl_2 + 2\,C_2H_5HgCl$$

The ethylmercury salts formed in this process are contaminated with organolead compounds, separation from which presents great difficulties. Pure ethylmercuric sulfate can be obtained from tetraethyllead by the following reaction:

$$(C_2H_5)_4Pb + 2\,H_2SO_4 + 2\,HgO \rightarrow (C_2H_5Hg)_2SO_4 + (C_2H_5)_2PbSO_4 + 2\,H_2O$$

Methoxyethylmercury derivatives are obtained by the reaction of the appropriate water-soluble salts with methoxyethylmercuric acetate, which is produced in good yield from mercuric acetate and ethylene in methyl alcohol:

$$CH_3OH + CH_2=CH_2 + Hg(OCOCH_3)_2 \rightarrow CH_3OCH_2CH_2HgOCOCH_3 + CH_3COOH$$

$$CH_3OCH_2CH_2HgOCOCH_3 + NaAc \rightarrow CH_3OCH_2CH_2HgAc + NaOCOCH_3$$

Ethylmercuric chloride is a white crystalline substance, m.p. 192° C., solubility in water about 1.5 mg./liter. It is moderately soluble in most organic solvents; its vapor pressure at 20° C. is 8.4×10^{-4} mm. of Hg, volatility at 20° C. 12 mg./m.^{3} Ethylmercuric chloride is highly poisonous. The LD_{50} is 18–30 mg./kg. On prolonged contact with the skin it causes the appearance of blisters that resemble a thermal burn; however, the blisters quickly disappear. The maximum permissible concentration in air is 0.005 mg./m.^{3}

It enters into the composition of a large number of seed disinfectants. In the Soviet Union three preparations based on ethylmercuric chloride are used: Granosan (2% ethylmercuric chloride, 1% dye, 1% mineral oil, and diluent); Mercuran (2% ethylmercuric chloride, 12–14% γ-benzenehexachloride, and diluent); and Mercurhexan (1% ethylmercuric chloride, 20% hexachlorobenzene, 20% γ-benzenehexachloride, and diluent). Ethylmercuric chloride is used as a component of seed disinfectants in various countries.

Ethylmercuric chloride with caustic alkalies in alcoholic solution yields ethylmercuric hydroxide, which is used for the production of various ethylmercury salts. It is resistant to the action of acids but reacts readily with reducing agents, forming diethylmercury and metallic mercury. It also is used as a disinfectant to combat slime formation in the paper industry. In this case the highly water-soluble double salt of ethylmercuric chloride with ethylenediamine·HCl is used, which is formed by simple mixing of the components in equimolecular quantities.

Ethylmercuric chloride is prepared by one of the previously described methods, but it is best prepared from mercuric chloride and diethylmercury:

$$(C_2H_5)_2Hg + HgCl_2 \rightarrow 2\ C_2H_5HgCl$$

This reaction takes place both in organic solvents, for example ethyl or isopropyl alcohol, and in aqueous medium, as well as in the solid state when talc, mercuric chloride, and diethylmercury are ground together. Diethylmercury is synthesized in up to 90% yield from sodium amalgam and ethyl bromide in the presence of ethyl acetate or other compounds as a catalyst. Preparation of diethylmercury from organomagnesium compounds is less convenient, because the reaction of ethylmagnesium bromide with mercuric chloride proceeds slowly and is not completed in less than 30–40 hours if it is carried out in diethyl ether. Use of higher boiling solvents shortens the time of the reaction, but not significantly.

Ethylmercuric phosphate is a white crystalline substance, m.p. 178° C., highly soluble in water and most hydrophilic organic solvents, but barely soluble in hydrocarbons and their halogen derivatives. With water ethylmercuric phosphate gives crystal hydrates with one and two molecules of water, which upon heating are easily lost. However, the anhydrous compound, after standing in moist air, takes on moisture to form the crystal hydrate with one molecule of water (m.p. about 110° C.; the compound melts over a wide interval because of the rapid change in composition due to drying).

Ethylmercuric phosphate surpasses ethylmercuric chloride in fungicidal activity and is widely used as a disinfectant for wood and for casein and albumin glues, to combat slime formation in the paper industry, and to disinfect seeds of various crops. It is most effective for the wet disinfection of seeds; in this process the dosage of the compound does not exceed 0.3 g./ha. Usually the disinfectant contains water-soluble salts of phosphoric acid as well as ethylmercuric phosphate. The total mercury content in such a preparation does not exceed 10%.

By the action of phosphoric acid on ethylmercuric phosphate it is possible to obtain acid monobasic (m.p. 115° C.) and dibasic (m.p. 174° C.) ethylmercuric phosphates:

$$(C_2H_5Hg)_3PO_4 + 2\ H_3PO_4 \rightarrow 3(C_2H_5Hg)H_2PO_4$$
$$2(C_2H_5Hg)_3PO_4 + H_3PO_4 \rightarrow 3(C_2H_5Hg)_2HPO_4$$

When ethylmercuric phosphate reacts with solutions of chlorides, bromides, iodides, thiocyanates, and other similar anions, a precipitate of the corresponding slightly water-soluble ethylmercury salt separates out. The reaction with salts of thiocyanic acid is employed for quantitative determination of ethylmercuric phosphate by Volhard titration.

The compound is produced by heating to 100°–115° C. a mixture of equimolecular quantities of diethylmercury and phosphoric acid:

$$3(C_2H_5)_2Hg + Hg_3(PO_4)_2 \rightarrow 2(C_2H_5Hg)_3PO_4$$

This reaction proceeds practically quantitatively only when a small amount of water is present (about 10% by weight of the reaction mixture). The reaction does not take place in the absence of water or in organic solvents.

Ethylmercuric phosphate has an LD_{50} for rats of 30 mg./kg. When it comes in contact with the skin it causes painful burns that appear many hours after contact. In working with the compound it is necessary to wash the hands as often as possible, and if contact with the skin should occur it should be washed off quickly with a large amount of warm water without soap. With soap the compound forms ethylmercury salts of the fatty acids, which are insoluble in water and difficult to wash off. The maximum permissible concentration in air is 0.005 mg./m.^{3} Ethylmercuric phosphate is of medical interest because its bactericidal activity is ten times that of mercuric chloride.

By a similar method it is possible to synthesize ethylmercuric sulfate, which is an intermediate for the production of the ethylmercuric hydroxide used for the synthesis of a great variety of ethylmercury salts.

N-Ethylmercuri-3,4,5,6,7,7-hexachloro-3,6-endomethylene-1,2,3,6-tetrahydrophthalimide (EMMI) is a white crystalline substance, m.p. 140° to

141° C. It is soluble in water, slightly soluble in most organic solvents, especially paraffinic hydrocarbons, and soluble in acetone. Its chemical properties resemble those of ethylmercuric chloride, but in contrast to the latter it is practically nonvolatile. The LD_{50} for rats is 148 mg./kg.

EMMI has been proposed as an agent for the control of true powdery mildew and apple scab by spraying during the growing season. It is marketed in the form of a 10% wettable powder and an emulsive concentrate. It is safe for apples and pears and some other crops, but it burns the leaves of roses.

It is synthesized by condensation of the sodium derivative of the imide of chlorendic acid (3,4,5,6,7,7-hexachloro-3,6-endomethylene-1,2,3,6-tetrahydrophthalic acid) with ethylmercuric chloride in acetone or other polar solvent:

$$C_2H_5HgCl + \underset{\text{Cl}}{\text{chlorendic imide Na}} \longrightarrow \underset{\text{Cl}}{\text{chlorendic imide-NHgC}_2H_5} + NaCl$$

This reaction takes place during more or less prolonged heating of a solution of equimolecular quantities of the starting compounds in an organic solvent. The sodium chloride formed is removed either by filtration or by washing with water after distillation of the solvent.

The imide of chlorendic acid is obtained by the action of ammonia on the anhydride of chlorendic acid at an elevated temperature, and the latter compound is produced in good yield when hexachlorocyclopentadiene is condensed with maleic anhydride:

$$\text{hexachlorocyclopentadiene} + \text{maleic anhydride} \longrightarrow \text{chlorendic anhydride} \xrightarrow{NH_3} \text{chlorendic imide}$$

N-(Ethylmercuri)-p-toluenesulfonanilide (Ceresan-M). In the United States Ceresan-M is widely used a grain seed disinfectant. Its active ingredient is N-(ethylmercuri)-p-toluenesulfonanilide, a white crystalline substance, m.p. 157° C. It is insoluble in water, slightly soluble in alcohol and benzene, but highly soluble in chloroform and acetone. The LD_{50} for rats is about 50–70 mg./kg. It burns the skin less than ethylmercuric phosphate. It is marketed in the form of preparations for dry and wet disinfection of seeds, which contain 7.7% or 3.2% active ingredient, calculated as mercury. It is produced by the reaction of ethylmercury salts with p-toluenesulfonanilide in the presence of NaOH:

$$CH_3-\langle\ \rangle-SO_2NH-\langle\ \rangle + C_2H_5HgOCOCH_3 \xrightarrow{NaOH}$$

$$CH_3-\langle\ \rangle-\underset{HgC_2H_5}{SO_2N}-\langle\ \rangle + H_2O + CH_3COONa$$

When ethylmercuric chloride or ethylmercuric acetate is used for this reaction, the yield is about 85%. The process can be carried out in aqueous medium.

The preparation of this compound also has been patented in the following way. By the action of acetic acid on tetraethyllead and mercuric oxide an aqueous solution of ethylmercuric acetate is obtained which, after removal of the ethyllead acetate by filtration, is reacted with p-toluenesulfonanilide. The yield of ethylmercuri-p-toluenesulfonanilide by this method is 80%, calculated as mercury:

$$(C_2H_5)_4Pb + HgO + 2\ CH_3COOH \rightarrow C_2H_5HgOCOCH_3 + (C_2H_5)_3PbOCOCH_3 + H_2O$$

Besides the ethylmercury derivatives described, the sodium salt of ethylmercurithiosalicylic acid (Merthiolate, LD_{50} for rats 40 mg./kg.), obtained by the reaction of sodium thiosalicylate with ethylmercuric chloride, is also used in agriculture:

$$\underset{SNa}{\overset{COONa}{\langle\ \rangle}} + C_2H_5HgCl \rightarrow \underset{SHgC_2H_5}{\overset{COONa}{\langle\ \rangle}} + NaCl$$

It is also obtained by mixture of ethylmercuri-2,3-dihydroxypropyl mercaptide with ethylmercuric acetate (LD_{50} 50–60 mg./kg.) employed for wet disinfection of seeds of grains, rice, and cotton.

Methylmercuridicyanodiamide (Panogen) is a white crystalline substance, m.p. 156° C., vapor pressure at 35° C. 6.5×10^{-5} mm. of Hg, solubility in water 21.7 g./liter at 20° C. It is highly soluble in alcohol, acetone, and ethylene glycol, but insoluble in benzene and other hydrocarbons. Panogen is one of the compounds widely used in Europe and America for disinfecting seeds of various crops. It is employed in the form of solutions at the rate of several liters of liquid per ton of seeds to be disinfected, which does not require supplementary drying of the seeds after disinfection. This method permits mechanization of the process of seed disinfection and almost completely eliminates hand labor in this operation.

In the United States methylmercuridicyanodiamide is marketed in two forms: Panogen-15 (2.2% methylmercuridicyanodiamide) and Panogen-42 (6.3% methylmercuridicyanodiamide). The latter preparation is diluted before use with water to the required concentration. The LD_{50} for rats and mice is 45 mg./kg.; like ethylmercuric chloride, when methylmercuridicyanodiamide comes in contact with the skin it causes painful inflammation and formation of blisters.

Its chemical properties are similar to those of the ethylmercury salts, but like most methylmercury salts, it is more soluble in water than the corresponding ethylmercury salt. It is produced by the action of methylmercuric hydroxide on dicyanodiamide:

$$CH_3HgOH + NH_2-\underset{\underset{NH}{\|}}{C}-NHCN \rightarrow CH_3HgNH-\underset{\underset{NH}{\|}}{C}-NHCN + H_2O$$

Methylmercuric hydroxide is prepared in satisfactory yields by the action of KOH on methylmercuric chloride in alcohol:

$$CH_3HgCl + KOH \rightarrow CH_3HgOH + KCl$$

or by the reaction of methylmercuric sulfate and barium or calcium hydroxide in aqueous medium:

$$(CH_3Hg)_2SO_4 + Ba(OH)_2 \rightarrow 2\ CH_3HgOH + BaSO_4$$

Methylmercuric sulfate also has independent importance as a disinfectant for seeds and for nonmetallic materials. First investigations for use as a fungicide were started by N. N. MEL'NIKOV, M. S. ROKITSKA, and Z. É. BEKKER in 1937; then in Western Europe its use began for the disinfection of grain seeds. It is a white crystalline substance that decomposes when it is heated to 260° C. It is highly soluble in water but insoluble in most organic solvents. Its LD_{50} for rats is about 50 mg./kg. Methylmercuric sulfate is marketed in two formulations containing 0.8 and 1.2% Hg. It is obtained in good yield by the reaction of dimethylmercury and mercuric sulfate:

$$(CH_3)_2Hg + HgSO_4 \rightarrow (CH_3Hg)_2SO_4$$

This reaction takes place under conditions similar to those for the synthesis of ethylmercuric phosphate.

It also is possible to produce methylmercury salts by the Razuvaev reaction from mercuric acetate:

$$(CH_3COO)_2Hg \rightarrow CH_3HgOCOCH_3 + CO_2$$

This method is promising for the production of methylmercury salts.

Methylmercuric cyanide (Chipcote) is a white crystalline substance, m.p. 95° C., highly soluble in water and alcohol, but slightly soluble in hydrocarbons. The LD_{50} is about 15 mg./kg. It is considerably more dangerous to handle than Panogen. It is marketed in aqueous solutions with ethylene glycol as an additive and containing 1.53 and 4.5% Hg.

It can be prepared by the reaction of methylmercuric hydroxide with HCN or of dimethylmercury with mercuric cyanide:

$$(CH_3)_2Hg + Hg(CN)_2 \rightarrow 2\ CH_3HgCN$$

Methylmercuric cyanide, manufactured in the United States in 1957, is similar to Panogen in fungicidal effect.

Methylmercury β-hydroxyquinolate (Metazol) is a golden-yellow crystalline substance, m.p. 133°–137° C., slightly soluble in water, but highly soluble in alcohol. The LD_{50} for mice is 72 mg./kg. The compound is marketed in the form of a water-alcohol solution containing 1.3% Hg with a red dye added. An effective seed disinfectant, it is synthesized by the usual method for this type of salt from sodium 8-hydroxyquinolate and methylmercuric chloride.

N-(Methylmercuri)-3,4,5,6,7,7-hexachloro-3,6-endomethylene-1,2,3,6-tetrahydrophthalimide (MEMMI) is a white crystalline substance, m.p. 225°–226° C., poorly soluble in organic solvents and insoluble in water. The LD_{50} for rats is 160 mg./kg. MEMMI is marketed in the form of a wettable powder and is used mainly to control diseases of gladioli and some other crops.

It is produced in a manner entirely analogous to the corresponding ethyl derivative by the action of methylmercuric chloride on the sodium derivative of the imide of chlorendic acid:

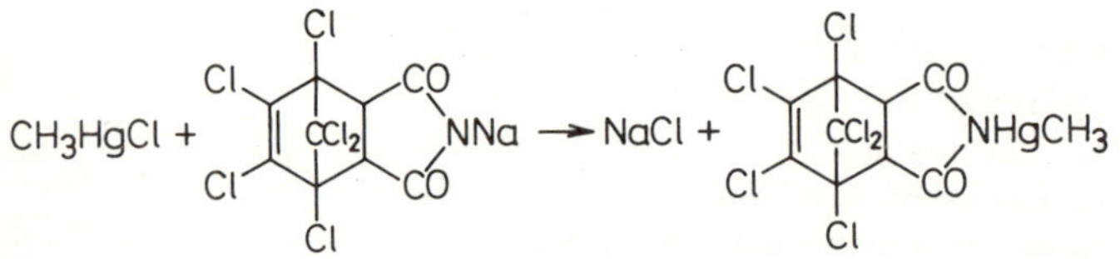

Other fungicides based on methylmercury salts also are known, but their use is somewhat more limited than that of the compounds described.

Methoxyethylmercuric chloride (Ceresan-Universal Nassbeize, Aretan, Agalol) is a white crystalline substance, m.p. 65° C., vapor pressure at 35° C. 1×10^{-3} mm. of Hg. Its solubility in water is about 5%; it is highly soluble in alcohol and acetone. The LD_{50} for rats is about 50 mg./kg.; like ethylmercuric chloride, when it comes in contact with the skin it causes painful burns. The compound is relatively stable in neutral and weakly alkaline media. It is decomposed by acids with the evolution of ethylene:

$$CH_3OCH_2CH_2HgCl + HCl \rightarrow HgCl_2 + CH_3OH + CH_2 = CH_2$$

It is an effective seed disinfectant for grains, beets, and other crops, and is marketed in preparations containing 2.5% and 3.5% Hg. For disinfection it is employed at 0.5–5% concentration (of the formulation) applied at the rate of 1–4 liters/100 kg. of seeds, depending on the crop.

It is produced by precipitation with chloride salts from solutions of methoxyethylmercuric acetate:

$$CH_3OCH_2CH_2HgOCOCH_3 + NaCl \rightarrow$$
$$CH_3OCH_2CH_2HgCl + CH_3COONa$$

Methoxyethylmercuric acetate is formed in practically quantitative yield from ethylene oxide and mercuric acetate in methyl alcohol.

Methoxyethylmercuric silicate (Ceresan-Universal Trockenbeize) is a white crystalline substance, almost insoluble in water, vapor pressure at 35° C. 3.3×10^{-3} mm. of Hg. The LD_{50} for rats is 50 mg./kg. It is marketed as a component of a preparation containing 1.75% mercury for dry disinfection of seeds.

Its chemical properties are similar to those of methoxyethylmercuric chloride. Methoxyethylmercuric silicate is obtained in quantitative yield by precipitation from a solution of methoxyethylmercuric acetate with sodium silicate:

$$3\ CH_3OCH_2CH_2HgOCOCH_3 + Na_3HSiO_4 \rightarrow$$
$$(CH_3OCH_2CH_2Hg)_3HSiO_4 + 3\ CH_3COONa$$

After drying it is used in the preparation of the seed disinfectant.

In the manufacture of organomercury compounds, apparatus or parts constructed of nonferrous metals should not be used, because they form reactive amalgams with mercury. For example, aluminum amalgam is so active that it frees hydrogen from water.

Mercury compounds of the aromatic series

In addition to the compounds previously described, aromatic mercury compounds are also used in agriculture and industry which are not only fungicides and disinfectants, but also herbicides.

The first mercury compounds of the aromatic series used in agriculture for seed disinfection were products of the mercuration of phenol, cresols, chlorophenol, and other similar compounds. Such compounds include Uspulun (III), Germisan (IV), Semesan (V), Semesan-Special (VI), and Semesan-Bel (VII):

Some of them find limited use even now.

As data have accumulated on the fungicidal and bactericidal activity of the different organomercury compounds, the mercuriphenols have been replaced by other arylmercury derivatives. The various phenylmercury derivatives have the greatest effectiveness of all the aromatic mercury compounds that have been studied, and they are widely used in agriculture. The preparation of compounds of this type is relatively simple in comparison with other organomercury compounds and they are completely practicable to manufacture. A number of methods are known for the synthesis of mercury compounds of the aromatic series that are suitable for industrial production.

A principal method of producing phenylmercury derivatives is the direct mercuration of benzene which takes place when mercuric acetate or other mercury salts of carboxylic acids are heated with benzene, with a yield of more than 90%:

$$C_6H_6 + (CH_3COO)_2Hg \rightarrow C_6H_5HgOCOCH_3 + CH_3COOH$$

Small amounts of polymercurated benzene derivatives are formed as by-products, which are easily removed from phenylmercuric acetate because of their slight solubility in hot water and hydrocarbons. However, by carrying out the reaction under mild conditions with a large excess of benzene, these side processes can be reduced to a minimum. Usually the mercuration of benzene is accomplished at a temperature not higher than 130° C. (preferably not above 100° C.); in the presence of some catalysts, for example boron trifluoride, the temperature of the reaction may be lowered to 20°–40° C.

A second method of synthesizing phenylmercury derivatives (especially phenylmercuric halides) is the Nesmeanov reaction, in which the yield of phenylmercuric halides amounts to 70–85%:

$$C_6H_5N_2X + HgX_2 \rightarrow C_6H_5N_2^+X^- \cdot HgX_2$$
$$C_6H_5N_2^+X^- \cdot HgX_2 + Cu_2X_2 \rightarrow C_6H_5HgX + N_2 + 2\,CuX_2$$

The optimum conditions for the process are the presence of cuprous chloride as a reducing agent and carrying out the reaction in aqueous medium at a temperature of about 0° C.

The synthesis of phenylmercury derivatives by other production methods on a mass scale is not of substantial importance. Of these methods the only mention should be made of the reaction of mercuric oxide with phenylhydrazine, and of halides of mercury with organometallic compounds.

Phenylmercuric acetate (Ceresan, PMAC, FMA, Tag, Kwiksan) is a white crystalline substance, m.p. 148°–153° C., vapor pressure at 35° C. 9×10^{-6} mm. of Hg. At 20° C., its solubility in water is 24.7 g./liter; it is more soluble in alcohol, acetone, benzene, and ethyl Carbitol. The compound is resistant to the action of mineral acids and alkalies at room temperature. When heated to 200°–250° C. it breaks down to yield metallic mercury, biphenyl, and phenyl acetate. It is marketed in various formulations both for the control of creeping weeds in lawns (10% wettable powder) and for the disinfection of seeds of grains, cotton, and other crops (1.5% mercury).

Mixed seed disinfectants containing 1% ethylmercuric chloride and 1% of phenylmercuric acetate also are used. The LD_{50} for experimental animals is 17–20 mg./kg.; like most mixed organomercury compounds, when phenylmercuric acetate comes in contact with the skin it causes painful, blistering burns. Phenylmercuric acetate is produced almost exclusively by the first of the methods described.

Table XLVII. *Properties of some phenylmercury derivatives used in agriculture*

Chemical name and formula	M.P. (°C.)	LD_{50} for rats (mg./kg.)	Trade name of preparation	Mercury content (%)
Phenylmercuric chloride C_6H_5HgCl	271	50	Ceredon-Special	3
Phenylmercuriurea $C_6H_5HgHNCONH_2$	—	50	Argox, Leutosan, Mergamma [a]	6.8 7.2 3.3
Phenylmercuric bromide C_6H_5HgBr	275	55	Agronal	1.6–1.8
Phenylmercuripyrocatechin	161	50–100	Germisan	2.5–3
N-(Phenylmercuri)-1,4,5,6,7,7-hexachlorobicyclo[2.2.1]-heptene-5-dicarboximide	228–230	122	PIMM	8

[a] Preparation also contains 40 percent γ-BHC.

Phenylmercuritriethanolammonium lactate (Puratized). The phenylmercury salts of various organic acids react with amines to form complexes in which the phenylmercury group enters into the makeup of the cation. Such compounds are easily obtained both with the simplest amines and with more complex ones. The compounds of phenylmercuric lactate with mono-, di-, and triethanolamines have been practically used as curative fungicides and as preparations for the protection of textiles. The salt with triethanolamine is a white crystalline substance, m.p. 126° C. It is highly soluble in water and hydrophilic organic solvents. Its LD_{50} for rats is 30 mg./kg.

This compound reacts comparatively easily with reducing agents to form diphenylmercury and metallic mercury, and it also breaks down quickly in sunlight. It is obtained by the reaction of phenylmercuric lactate with triethanolamine:

$$C_6H_5HgOCOCHCH_3 + N(CH_2CH_2OH)_3 \rightarrow$$
$$|$$
$$OH$$

$$[C_6H_5HgN(CH_2CH_2OH)_3]^+CH_3CHCOO^-$$
$$|$$
$$OH$$

Some other phenylmercury derivatives also have been used in agriculture and industry as antiseptics and seed disinfectants. The properties of such compounds are given in Table XLVII.

General references

ANDREEVA, E. I.: Khim. v Sel'skom Khozyaistve No. 7, 26 (1964).
—, N. N. MEL'NIKOV, A. V. SKALOZUBOVA, A. V. OBOLENSKOVA, and R. A. PETRENKO: Khim. v Sel'skom Khozyaistve No. 3, 27 (1965).
MAKAROVA, L. G., and A. N. NESMEANOV: Metody élementoorganicheskoi khimii, Rtut' [Methods of heteroorganic chemistry, Mercury]. Publishing House "Nauka" (1965).
MEL'NIKOV, N. N.: Zh. Obshchei Khim. 16, 2065 (1964).
—, and M. S. ROKITSKAYA: Prom. Org. Khim. 7, 387 (1940); Zh. Priklad. Khim. 14, 446 (1941); Zh. Obshchei Khim. 7, 2383, 2518 (1937); 11, 592 (1941); Zh. Priklad. Khim. 12, 1802 (1939).
Novye pestitsidy [New pesticides]. Publishing House "Mir" (1964).
OL'DEKOP, YU. A., and N. A. MAIER: Zh. Obshchei Khim. 30, 612, 3017, 3472, 299 (1960); Vestnik Akad. Nauk S.S.S.R., ser. fiz. tekh. nauk No. 2, 37 (1960); Zh. Obshchei Khim. 30, 275 (1960); Zh. Obshchei Khim. 28, 3008 (1958); Doklay Akad. Nauk S.S.S.R. 6, 503 (1962); Zh. Obshchei Khim. 34, 317 (1964).
Organicheskie insektofungitsidy [Organic insectofungicides]. State Scientific and Technical Publishing House of Chemical Literature [USSR] (1955).
RAZUVAEV, G. D., YU. A. OL'DEKOP, and N. A. MAIER: Doklady Akad. Nauk S.S.S.R. 98, 613 (1954); Zh. Obshchei Khim. 25, 697 (1955).

XXV. Organotin compounds

General characteristics of pesticidal properties

Organotin compounds have been used in various fields of technology. For example, they are used as stabilizers for polyvinyl chloride, antioxidants for rubber, and catalysts for the polymerization of olefins by the Ziegler method. They are employed in agriculture and industry as biologically active agents for controlling plant diseases, for combatting slime formation in the paper industry, as disinfectants for nonmetallic materials, for combating marine growths on boats, and for controlling some species of intestinal worms in poultry and agricultural animals. The organotin compounds are so important in industry and agriculture that in spite of the relatively high cost of tin, thousands of tons yearly of its various derivatives are manufactured industrially.

The first tests of the biocidal activity of organotin compounds were carried out toward the end of the last century (1886), but their insecticidal properties were only discovered in 1929; systematic investigation of their fungicidal properties was begun more than 10 years ago. On the basis of the available data on the fungicidal activity of this class of compounds, the following general conclusions can be drawn about the relationship of their fungitoxicity to structure.

Stannous and stannic chlorides do not have fungicidal properties. When the chlorine atoms in stannic chloride are replaced by alkyl or aryl radicals, the fungicidal effect gradually increases, with the trialkyl- and triaryltin acyls having the maximum activity. Tetraalkyl- and tetraaryltins are not toxic to fungi. This is well illustrated by the data of Table XLVIII for the ethyl derivatives.

Increasing the length of the hydrocarbon chain of the radicals attached to the tin in the trialkyltin acyls leads to an increase in fungicidal activity of the compound up to a certain limit, after which the activity decreases. The compounds containing a total of 12 carbon atoms have the maximum activity. The most active compounds are tributyltin chloride and fluoride. The activity of the cyclic compounds is somewhat higher than that of the aliphatic compounds with the same number of carbon atoms in the molecule. However, the difference is not very great. Triphenyltin acetate is somewhat less active than tricyclohexyltin acetate.

Table XLIX presents the activity of some trialkyltin acetates against four species of microorganisms. As a comparison, the activity of some organomercury compounds is given. The trialkylstannanes containing dif-

Table XLVIII. *Fungicidal activity of tin compounds*

Compound	Min. conc. entirely inhibiting growth (mg./l.)			
	Botrytis allii	Penicillium italicum	Aspergillus niger	Rhizopus nigricans
Stannous chloride ($SnCl_2 \cdot H_2O$)	> 1000	> 1000	> 1000	> 1000
Stannic chloride	> 1000	> 1000	> 1000	> 1000
Ethyltin trichloride	> 1000	> 1000	> 1000	> 1000
Diethyltin dichloride	100	100	500	200
Triethyltin chloride	0.5	2	5	2
Tetraethyltin	50	> 1000	100	100

Table XLIX. *Fungicidal activity of trialkyltin acetates*

Compound	Total no. C atoms in alkyl radicals	Min. conc. entirely inhibiting growth of fungi (mg./l.)			
		Botrytis allii	Penicillium italicum	Aspergillus niger	Rhizopus nigricans
$(CH_3)_3SnOCOCH_3$	3	20	20	200	200
$(C_2H_5)_3SnOCOCH_3$	6	1	2	2	2
$(CH_3)_2C_4H_9SnOCOCH_3$	6	1	2	5	5
$(C_2H_5)_2C_4H_9SnOCOCH_3$	8	0.1	1	0.5	0.5
$(C_3H_7)_3SnOCOCH_3$	9	0.1	0.1	1	1
$(iso-C_3H_7)_3SnOCOCH_3$	9	0.1	0.1	1	0.5
$(CH_3)_2C_8H_{17}SnOCOCH_3$	10	0.5	0.5	0.2	2
$(C_2H_5)_2C_6H_{13}SnOCOCH_3$	10	0.5	0.5	0.1	0.1
$(C_4H_9)_3SnOCOCH_3$	12	0.1	0.1	0.5	0.5
$(C_2H_5)_2C_8H_{17}SnOCOCH_3$	12	0.02	0.5	0.1	5
$(C_5H_{11})_2C_2H_5SnOCOCH_3$	12	0.2	1	1	0.5
$(CH_3)_2C_{12}H_{25}SnOCOCH_3$	14	0.2	1	0.2	10
$(C_5H_{11})_3SnOCOCH_3$	15	0.2	5	5	10
$(C_2H_5)_2C_{12}H_{25}SnOCOCH_3$	16	0.2	5	0.5	50
$(C_6H_{13})_3SnOCOCH_3$	18	1	10	20	100
$(C_8H_{17})_3SnOCOCH_3$	24	100	100	100	100
$(cyclo-C_6H_{11})_3SnOCOCH_3$	18	0.5	0.5	5	20
$(C_6H_5)_3SnOCOCH_3$	18	2	1	0.5	10
C_6H_5HgBr		0.1	0.1	0.1	2
$C_6H_5HgOCOCH_3$		0.5	0.5	0.5	5

ferent hydrocarbon radicals have a greater specificity of action on individual species of fungi, while compounds with identical radicals inhibit the growth of all fungi at similar concentrations. Introduction into an aromatic radical of triphenyltin acetate of any substituents, including halogens, lowers the fungicidal activity of the compound. For example, tri(*p*-chlorophenyl)tin acetate is 10–100 times less active than triphenyltin acetate.

The effect of the anion in the trialkyltin acyls on the fungicidal activity of the compound is less appreciable. Table L gives the activity of some

Table L. *Fungicidal activity of some triethyltin salts*

Compound	Min. conc. entirely inhibiting growth of fungi (mg./l.)			
	Botrytis allii	Penicillium italicum	Aspergillus niger	Rhizopus nigricans
$(C_2H_5)_3SnOH$	0.2	5	0.5	0.5
$(C_2H_5)_3SnCl$	0.5	2	5	2
$(C_2H_5)_3SnBr$	0.5	2	1	1
$(C_2H_5)_3SnI$	0.5	1	5	2
$[(C_2H_5)_3Sn]_2S$	0.2	1	1	1
$[(C_2H_5)_3Sn]_2SO_4$	0.2	0.2	5	5
$(C_2H_5)_3SnOCOCH_3$	1	2	2	2
$(C_2H_5)_3SnOCOC_5H_{11}$	1	5	2	2
$(C_2H_5)_3SnOCOC_{11}H_{23}$	0.2	0.2	5	5
$(C_2H_5)_3SnOCOC_6H_5$	2	10	5	5
$(C_2H_5)_3SnOCOCH=CHCOOSn(C_2H_5)_3$	0.5	5	0.5	0.5
$[(C_2H_5)_3SnOCO]_2CH_2$	0.5	5	2	1
$(C_2H_5)_3SnOC_6H_5$	0.5	1	2	1
$(C_2H_5)_3SnOC_6H_4NO_2$	0.5	2	2	2
$(C_2H_5)_3SnNHSO_2C_6H_4CH_3(4)$	1	5	2	5
$(C_2H_5)_3SnNHSO_2CH_3$	0.5	5	2	2
$(C_2H_5)_3SnOSO_2C_6H_4CH_3(4)$	1	5	2	2
$(C_2H_5)_3SnN(CO)_2C_6H_4$	0.2	5	1	2

triethyltin salts. The toxicity of the trialkyltin salts decreases with an increase in the number of carbon atoms in the molecule. While the LD_{50} of triethyltin chloride is less than 10 mg./kg., that of tributyltin acetate is about 500 mg./kg. The toxicity of the aromatic compounds of tin is greater for mammals, and the LD_{50} of triphenyltin acetate is 125 mg./kg. The anion has a great influence on the toxicity of the trialkyltin salts. Thus, the toxicity of triphenyltin hydroxide is almost four times less than that of the acetate.

Because of the relatively high toxicity of the organotin compounds for animals, only the tributyltin and triphenyltin salts are used as fungicides in agriculture and industry. Dibutyltin dilaurate and dibutyltin maleate are used as anthelminthics in veterinary practice; they enter into the composition of preparations put out by various companies. Triphenyltin oxide and acetate are being tested as agents to arrest feeding by some chewing insects (antifeeding agents). The herbicidal effect of a number of organotin compounds on various species of plants also has been studied. Several compounds of this type have been proposed for weed control. Apparently the unsaturated compounds are especially active as herbicides.

General methods of making fungicidal tin compounds

The main method of preparing trialkyl- and triaryltin salts is the reaction of a tetraalkyltin [or tetraaryltin] with the halides or other salts of tetravalent tin. This reaction proceeds most easily when the tetraalkyl- or tetraaryltin is heated with stannic chloride.

$$3 R_4Sn + SnCl_4 \rightarrow 4 R_3SnCl$$

From the trialkyl- and triaryltin halides good yields of the corresponding hydroxides are obtained, which can be converted by the action of free organic and inorganic acids to the corresponding trialkyl- or triaryltin salts:

$$R_3SnCl + KOH \rightarrow R_3SnOH + KCl$$
$$R_3SnOH + HAc \rightarrow R_3SnAc + H_2O$$

Trialkyltin derivatives are also synthesized by the action of acids or free halogens on a tetraalkyltin, but these methods are uneconomical because of the loss of one radical as hydrocarbon or alkyl halide:

$$R_4Sn + HX \rightarrow RH + R_3SnX$$
$$R_4Sn + X_2 \rightarrow RX + R_3SnX$$

The tetraalkyltins are obtained in satisfactory yields by a Grignard reaction through the organomagnesium compounds:

$$4 RMgX + SnX_4 \rightarrow R_4Sn + 4 MgX_2$$

However, this method has a number of disadvantages, the most important its long reaction time, the necessity of working with inflammable ethers, and the large number of technical steps: preparation of the organomagnesium compound, its reaction with stannic halide, decomposition of the complex that is obtained, separation from water, drying, distillation and regeneration of the solvent, and reworking of the magnesium chloride. A simpler process is carried out in one step in a hydrocarbon solvent with small amounts of anisole or other high-boiling ether as a catalyst:

$$4 RX + SnX_4 + 4 Mg \rightarrow R_4Sn + 4 MgX_2$$

In this way the reaction can be carried out at a higher temperature and easily controlled by addition to the reactor of a mixture of the stannic chloride and alkyl halide. Tetraalkyltins can also be synthesized by starting with an alloy of tin with metallic sodium or magnesium. In this case a large excess of the alloy must be used and tin must be recovered from the residues:

$$2 SnNa + 2 RX \rightarrow 2 NaX + Sn + R_2Sn$$
$$2 R_2Sn \rightarrow R_4Sn + Sn$$

The tetraethyltins are produced in yields up to 85% from alkyl halide, stannic halide, and metallic sodium:

$$4 RX + SnX_4 + 8 Na \rightarrow 8 NaX + R_4Sn$$

For this reaction both alkyl chlorides and bromides can be used. Tetraalkyltins are obtained in good yields by the reaction of a dialkyltin dihalide with alkyl halide and sodium:

$$R_2SnX_2 + 2 RX + 4 Na \rightarrow R_4Sn + 4 NaX$$

The dialkyltin dihalides are synthesized from alkyl halides and metallic tin by prolonged heating at a temperature of 200°–250° C., under pressure:

$$Sn + 2\,RX \rightarrow R_2SnX_2$$

All the methods listed for the production of tetraalkyltins yield products with trialkyltin halide as an impurity, which varies in amount depending on the ratio of the reagents used. All the methods except the last can be used for the synthesis of tetraaryltins.

A general method for synthesizing tetraalkyltins is the reaction of organoaluminum and organolithium compounds with a stannic halide. This reaction gives good yields:

$$4\,RLi + SnX_4 \rightarrow R_4Sn + 4\,LiX$$

The organolithium derivatives are used also for the synthesis of aromatic tin derivatives.

Triphenyltin acetate is used in agriculture to control diseases of sugar beets, potatoes, and a number of other crops. All the rest of the tin compounds are phytotoxic and are used only for seed treatment and as disinfectants. In particular, tributyltin hydroxide and acetate are widely employed as disinfectants.

Tributyltin hydroxide is a liquid, b.p. 186°–190° C. at 5 mm. of Hg. It is moderately soluble in water, and highly soluble in organic solvents. The LD_{50} is about 500 mg./kg. The compound is used to combat slime formation in the paper industry at a dosage from 3 to 5 g./m.³ of water, to treat wood for preservation from blue-stain, yellowing, and other fungal diseases, as an additive to emulsion-type paints to protect them from mold, and to disinfect textile goods.

In aqueous solution tributyltin hydroxide quickly reacts with the appropriate acids to form salts; with halogens it is capable of splitting off all its hydrocarbon radicals successively in the form of butyl halide. It breaks down relatively rapidly in the soil, and also under the influence of sunlight in the presence of oxygen.

Tributyltin hydroxide is produced by the action of caustic alkalies on tributyltin chloride. The reaction is carried out in aqueous solution; however, it is better if the tributyltin chloride is in some organic solvent that is immiscible with water, for a purer compound is obtained.

Tributyltin acetate is a white crystalline substance, m.p. 84.5°–85° C., slightly soluble in water, more soluble in organic solvents. The LD_{50} is 500 mg./kg. When there is long contact or inhalation of the compound, headaches and gastritis are observed as with other organotin compounds. When action of the compound is prolonged, liver complications are possible. It is used for the same purposes as the hydroxide. It also is produced similarly by the reaction of alkali acetates with tributyltin chloride.

Triphenyltin acetate (Brestan) is a white crystalline substance, m.p. 124°–125° C., solubility in water about 20 mg./l., slightly soluble in most organic solvents. The LD_{50} for rats is 125 mg./kg.; it causes irritation of the mucous membranes. The compound is marketed in the form of 20% and 60% wettable powders. It is strongly phytotoxic, but in fungicidal concentrations it is safe for sugar beets, potatoes, rice, and legume crops. It is similar in fungicidal activity to the copper compounds, but it is used at dosages ten times smaller. Recently it has been used successfully in combination with maneb.

In the dry form triphenyltin acetate is stable and can be stored for an unlimited time. In the presence of moisture it is quickly hydrolyzed and converted to triphenyltin hydroxide:

$$(C_6H_5)_3SnOCOCH_3 + H_2O \rightarrow (C_6H_5)_3SnOH + CH_3COOH$$

Triphenyltin acetate is produced by one of the previously described methods.

Triphenyltin hydroxide is a white crystalline substance, m.p. 120° C., practically insoluble in water and most organic solvents. The LD_{50} is about 500 mg./kg. It is a promising fungicide for controlling diseases of the same crops as triphenyltin acetate.

General references

ASCHER, K. R. S., and S. NISSIN: World Rev. of Pest Control 3, 188 (1964).
CZERWINSKA, E., Z. ECKSTEIN, Z. EJMOCKI, and R. KOWALIK: Bull. l'Acad. Polon. S. Chim. 15, 335 (1967).
HÄRTEL, K.: Angew. Chem. 76, 304 (1964).
IVANOVA, S. N.: In book "Novye insektofungitsidy i gerbitsidy" [New insectofungicides and herbicides], pp. 170–184. Foreign Literature Publishing House [USSR] (1960).
JUGHAM, R. K., S. D. ROSENBERG, and H. GILMAN: Chem. Rev. 60, 459 (1960).
KOCHESHKOV, K. A.: Sinteticheskie metody v oblasti metallorganicheskikh soedinenii elementov IV gruppy [Synthetic methods in the field of organometallic compounds of elements of group IV]. Publishing House of the Academy of Sciences of USSR (1947).
VAN DER KERK, G. J. M., and J. G. A. LUIJTEN: J. Applied Chem. 4, 314 (1954).

XXVI. Organophosphorus compounds

General characteristics of pesticidal properties

One of the most important classes of present day pesticides is the organophosphorus compounds, of which more than 80 are used in agriculture. Substances with a great variety of pesticidal properties are found among the organophosphorus compounds, including insecticides, acaricides, nematicides, herbicides, defoliants, and fungicides.

Organophosphorus compounds are widely used as agents to combat plant pests and ectoparasites and, in part, endoparasites of domestic animals.

The scale of use of the organophosphorus compounds as pesticides approaches that of the organochlorine compounds in agriculture. In the United States the production of such compounds as metaphos [methyl parathion], thiophos [parathion], carbophos [malathion], and certain others alone exceeds 40,000 tons a year.

The most important advantages of the organophosphorus compounds as pesticides are presented below:

1. High insecticidal and acaricidal activity.
2. Wide spectrum of action on plant pests.
3. Low persistence and breakdown to form products nontoxic to man and animals.
4. Systemic action of a number of the compounds.
5. Low dosage of compound per unit of area treated.
6. Relatively rapid metabolism in vertebrate organisms and absence of accumulation in their bodies, and also comparatively low chronic toxicity.
7. Rapidity of action on plant pests.

A disadvantage of many organophosphorus compounds is their relatively high toxicity to vertebrates, requiring suitable protective measures when using them. However, a large number of organophosphorus compounds that have moderate or low toxicity for mammals have been synthesized; the use of such compounds in agriculture is entirely safe.

Investigation of the mechanism of action of organophosphorus compounds on mammals and insects has shown that in animals they phosphorylate vitally important esterases, inhibiting their normal functions. The action of these compounds supposedly is directed mainly toward inhibition of cholinesterase, the physiological functions of which are very important in the animal organisms. Cholinesterase hydrolyzes acetylcholine:

$$(CH_3)_3\overset{+}{N}CH_2CH_2OCOCH_3 \xrightarrow{ChE} (CH_3)_3\overset{+}{N}CH_2CH_2OH + CH_3COOH$$

The mechanism of inhibition of cholinesterase by organophosphorus compounds may represented by the following general scheme:

$$\text{ChEH} + \text{ROP(O)(OR')}_2 \rightarrow \begin{bmatrix} & \text{O}^- & \\ & | & \\ \text{RO} - & \text{P(OR')}_2 & \\ & | & \\ & \text{H} - \text{ChE}^+ & \end{bmatrix} \rightarrow$$

$$\begin{bmatrix} \text{O}^- \\ | \\ \text{P(OR')}_2 \\ \| \\ \text{ChE}^+ \end{bmatrix} + \text{ROH} \rightarrow \text{ChEP(O)(OR')}_2$$

The esterase apparently first forms with the organophosphorus compound a complex that further breaks down to give a phosphorylation product of the esterase and the corresponding hydroxy compound. The phosphorylated esterase may be gradually hydrolyzed by water, and by this process the esterase activity is restored. For some compounds the rate of dephosphorylation is so slow that it is not significant. For example, the dephosphorylation of cholinesterase that has been inhibited by tetraethyl pyrophosphate proceeds very slowly in water; in 28 days only 50% of the initial activity is restored:

$$\text{ChEP(O)(OR')}_2 + \text{H}_2\text{O} \rightarrow \text{ChEH} + \text{HOP(O)(OR')}_2$$

The activity of organophosphorus compounds depends greatly on the structure of the ester groups that enter into the makeup of the molecule of the phosphoric acid ester; the effect of the structure of each of the radicals cannot be ignored in this connection. Thus, for example, O,O-diisopropyl O-4-nitrophenyl thiophosphate is only about 1/100 as active as O,O-diethyl O-4-nitrophenyl thiophosphate with respect to the cholinesterase of bees.

To obtain the maximum effect the molecule of the organophosphorus compound should fit the active centers of the esterase like a "key in a lock".

It is fully evident that the activity of a compound will differ with respect to the cholinesterase of different insects and animals. The differential toxicity of the organophosphorus compounds with respect to insects and mammals is explained in a number of cases by their metabolic routes in insects and in warm-blooded animals. For example, malathion in the house fly undergoes the usual type of conversions, while in rats it gives malathionic acid, which has little toxicity:

$$(\text{CH}_3\text{O})_2\text{P} \underset{\text{SCHCOOH}}{\overset{\text{S}}{\diagup\!\!\!\diagdown}} \atop \phantom{(\text{CH}_3\text{O})_2\text{P}}\;\;\;|\atop \phantom{(\text{CH}_3\text{O})_2\text{P}}\;\;\;\text{CH}_2\text{COOH}$$

but the octamethyltetraamide of pyrophosphoric acid in animals goes over to a more toxic product than the starting compound.

For a correct understanding of the reasons for the toxicity of any organophosphorus compound to individual species of animals, it is necassary to know not only the general mechanism of action, but also the mechanism in the given species. Very often the activity with respect to cholinesterase is connected with the rate of hydrolysis of the organophosphorus compound, but this is not always so, and the activity also depends on the steric characteristics of the molecular structure.

Various phosphorus compounds are used as pesticides, including derivatives of phosphorous, thiophosphorous, phosphoric, thiophosphoric, dithiophosphoric, phosphonous, and thiophosphonous acids.

Unlike most other classes of compounds, the nomenclature of the organophosphorus compounds still is not well developed, and different countries use their own naming systems. In most cases the organophosphorus compounds are considered as derivatives of the corresponding acids or hydrogen phosphide (phosphine).

Derivatives of phosphorous acid have the ending "ite" and those of phosphoric acid the ending "ate". For example, the diethyl ester of phosphorous acid (I) carries the name diethyl phosphite, and the diethyl ester of phosphoric acid (II) is called diethyl phosphate:

$(C_2H_5O)_2P\diagup^O_H$ (I) $(C_2H_5O)_2P\diagup^O_{OH}$ (II)

When there is a sulfur atom in the molecule, the names of the compounds will be derived in the following way: the diethyl ester of thiophosphorous acid, depending on the position of the sulfur, is called O,O-diethyl thiophosphite (III) or O,S-diethyl thiophosphite (IV):

$(C_2H_5O)_2P\diagup^S_H$ (III) $\begin{matrix}C_2H_5O\\C_2H_5S\end{matrix}\!\!P\diagup^O_H$ (IV)

The names of the derivatives of thiophosphoric acids are derived in a similar manner, as can be seen from the following examples:

$(C_2H_5O)_2P\diagup^S_{OH}$ O,O-diethyl thiophosphate

$\begin{matrix}C_2H_5O\\C_2H_5S\end{matrix}\!\!P\diagup^O_{OH}$ O,S-diethyl thiophosphate

$\begin{matrix}C_2H_5O\\C_2H_5S\end{matrix}\!\!P\diagup^S_{OH}$ O,S-diethyl dithiophosphate

$(C_2H_5S)_2P\diagup^O_{OH}$ S,S-diethyl dithiophosphate

The amides of the phosphorus acids are named by the same system. Thus, compound (V) will be called O,O-diethyl N-methylamidophosphate and compound (VI) O-ethyl N-methyl N'-n-propyl-diamidothiophosphate:

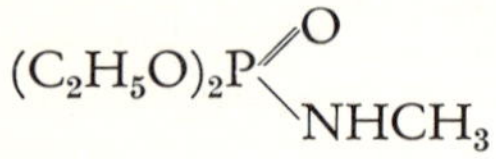

$(C_2H_5O)_2P\begin{smallmatrix}\nearrow O \\ \searrow NHCH_3\end{smallmatrix}$ (V) 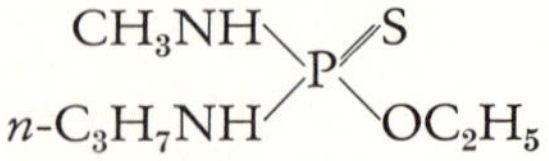 $\begin{smallmatrix}CH_3NH\searrow \\ n\text{-}C_3H_7NH\nearrow\end{smallmatrix}P\begin{smallmatrix}\nearrow S \\ \searrow OC_2H_5\end{smallmatrix}$ (VI)

The names of the full amides of phosphorous acid sometimes are derived also from phosphine. In this case compound (VII) can be called hexamethyltriaminophosphine.

$$[(CH_3)_2N]_3P \qquad (VII)$$

The names of the phosphorus acids having a C–P bond are derived from the following acids:

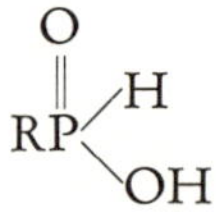

$RP\begin{smallmatrix}\nearrow O \\ | \\ \searrow OH\end{smallmatrix}\quad \text{(with H)}$ alkylphosphonous

$R_2P\begin{smallmatrix}\nearrow O \\ \searrow H\end{smallmatrix}$ dialkylphosphinous

$R\overset{O}{\underset{\|}{P}}(OH)_2$ alkylphosphonic

$R_2P\begin{smallmatrix}\nearrow O \\ \searrow OH\end{smallmatrix}$ dialkylphosphinic

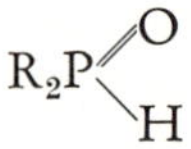

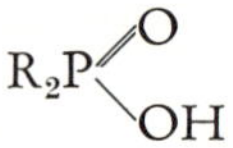

The nomenclature of the derivatives of the acids listed is apparent from the following examples:

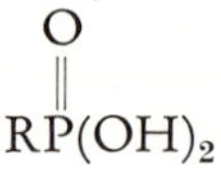

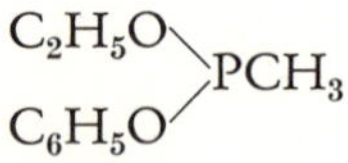

$\begin{smallmatrix}C_2H_5O\searrow \\ C_6H_5O\nearrow\end{smallmatrix}PCH_3$ O-ethyl O-phenyl methylphosphinite

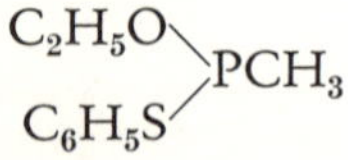

$\begin{smallmatrix}C_2H_5O\searrow \\ C_6H_5S\nearrow\end{smallmatrix}PCH_3$ O-ethyl S-phenyl methylthiophosphinite

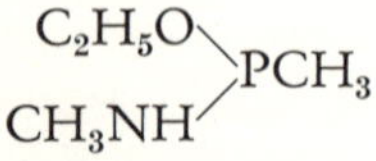

$\begin{smallmatrix}C_2H_5O\searrow \\ CH_3NH\nearrow\end{smallmatrix}PCH_3$ O-ethyl N-methylamidomethylphosphinite

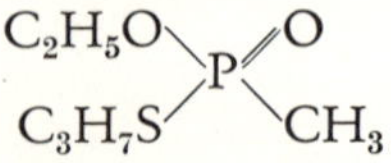

$\begin{smallmatrix}C_2H_5O\searrow \\ C_3H_7S\nearrow\end{smallmatrix}P\begin{smallmatrix}\nearrow O \\ \searrow CH_3\end{smallmatrix}$ O-ethyl S-propyl methylthiophosphonate

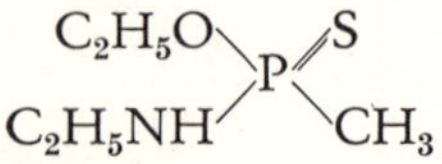

$\begin{smallmatrix}C_2H_5O\searrow \\ C_2H_5NH\nearrow\end{smallmatrix}P\begin{smallmatrix}\nearrow S \\ \searrow CH_3\end{smallmatrix}$ O-ethyl N-ethylamidomethylthiophosphonate

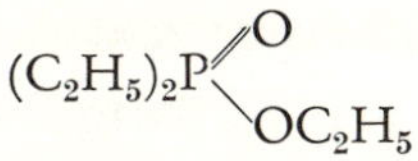

$(C_2H_5)_2P\begin{smallmatrix}\diagup O\\ \diagdown OC_2H_5\end{smallmatrix}$ O-ethyl diethylphosphinate

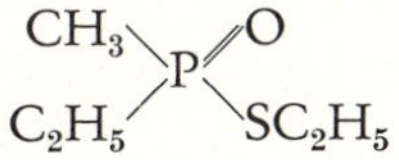

$\begin{smallmatrix}CH_3\diagdown\\ C_2H_5\diagup\end{smallmatrix}P\begin{smallmatrix}\diagup O\\ \diagdown SC_2H_5\end{smallmatrix}$ S-ethyl ethylmethylthiophosphinate

The names of the acid chlorides of the alkyl- and arylphosphinous acids may be derived from phosphine; for example, CH_3PCl_2 (VIII) is called methyldichlorophosphine.

Compounds of the general formula (IX) are named as phosphines, and compounds of formula (X) are named as phosphine oxides:

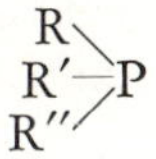

$\begin{matrix}R\diagdown\\ R'-P\\ R''\diagup\end{matrix}$ (IX) $\begin{matrix}R\diagdown\\ R'-P=O\\ R''\diagup\end{matrix}$ (X)

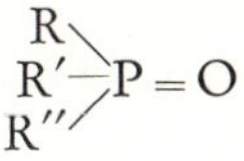

Tetrasubstituted phosphonium salts (XI) are named similarly to the corresponding ammonium derivatives:

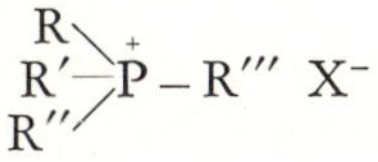

$\begin{matrix}R\diagdown\\ R'-\overset{+}{P}-R''' \quad X^-\\ R''\diagup\end{matrix}$ (XI)

Derivatives of phosphorous acid

Investigation of the pesticidal properties of a large number of derivatives of phosphorous acid has established that many compounds of this series have weak insecticidal and acaricidal activity, but some derivatives of phosphorous and thiophosphorous acids have high herbicidal activity. The herbicidal activity of the esters of phosphorous acid increases with an increase in the number of carbon atoms in the aliphatic ester radical. The greatest activity is shown by the full esters of phosphorous acid with haloaryloxyethanols; of this class the compound Falone is used in agriculture.

Tris(2,4-dichlorophenoxyethyl) phosphite (Falone) is a thick oily liquid with a weak odor that does not distill without decomposition in an ordinary vacuum and is practically insoluble in water, but highly soluble in organic solvents. The LD_{50} for rats is 850 mg./kg.

Upon oxidation the compound forms the corresponding phosphate:

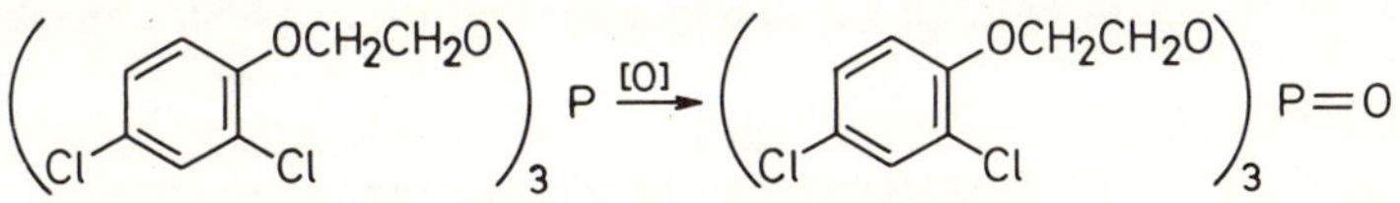

$\left(\underset{Cl}{\underset{|}{\bigcirc}}\text{—}\underset{Cl}{\underset{|}{\bigcirc}}\text{—OCH}_2\text{CH}_2\text{O}\right)_3 P \xrightarrow{[O]} \left(\underset{Cl}{\underset{|}{\bigcirc}}\text{—}\underset{Cl}{\underset{|}{\bigcirc}}\text{—OCH}_2\text{CH}_2\text{O}\right)_3 P=O$

When Falone reacts with water it first is converted to the corresponding dialkyl phosphite, which further is hydrolyzed and oxidized to yield 2,4-chlorophenoxyethanol and phosphoric acid:

$$\left(\underset{Cl}{\underset{|}{\bigcirc}}\text{—OCH}_2\text{CH}_2\text{O}\right)_3 P \xrightarrow{H_2O} \left(\underset{Cl}{\underset{|}{\bigcirc}}\text{—OCH}_2\text{CH}_2\text{O}\right)_2 POH \xrightarrow{H_2O + [O]}$$

$$\underset{Cl}{\underset{|}{\bigcirc}}\text{—OCH}_2\text{CH}_2\text{OH} + H_3PO_4$$

The action of Falone in the soil apparently is based on its conversion to 2,4-D. Falone is recommended for the control of weeds in corn, potatoes, strawberries, and some other crops at dosages from 4–8 kg./ha. For treatment of the soil it usually is used in the form of aqueous emulsions.

Falone is produced by the reaction of phosphorus trichloride with 2,4-dichlorophenoxyethanol in the presence of pyridine, dimethylaniline, or other tertiary amines:

$$PCl_3 + 3 \underset{Cl}{\underset{|}{\bigcirc}}\text{—OCH}_2\text{CH}_2\text{OH} \xrightarrow{3C_5H_5N} \left(\underset{Cl}{\underset{|}{\bigcirc}}\text{—OCH}_2\text{CH}_2\text{O}\right)_3 P$$

The reaction product after removal of the pyridine hydrochloride is used without further purification for the preparation of emulsive concentrates.

Bis(2,4-dichlorophenoxyethyl) phosphite also has herbicidal properties but it is inferior to Falone in activity.

Tributyl trithiophosphite (Merphos, Folex) is a light, oily liquid, b.p. 150°–152° C. at 2 mm. of Hg, almost insoluble in water, but highly soluble in organic solvents. The LD_{50} for rats is 350 mg./kg. The compound is used in the form of aqueous emulsions at dosages of 1–2 kg./ha. for the defoliation of cotton. With respect to its rapidity of action and the effect obtained, Merphos is one of the best defoliants.

Merphos is slowly oxidized by the oxygen of the air to tributyl trithiophosphate; this reaction is accelerated by heating and it can serve as a preparatory method for tributyl trithiophosphate. Water slowly hydrolyzes tributyl trithiophosphite with the formation of butyl mercaptan. In the presence of caustic alkalies the rate of hydrolysis is increased.

Tributyl trithiophosphite is obtained in good yields by the reaction of butyl mercaptan with phosphorus trichloride at an elevated temperature:

$$PCl_3 + 3\ C_4H_9SH \rightarrow (C_4H_9S)_3P + 3\ HCl$$

In the patent literature there are also reports of the high defoliant activity of some homologs of tributyl trithiophosphite, but they have not been practically used.

Derivatives of phosphoric acid

Going from the phosphites to the phosphates the insecticidal and acaricidal activity of the compounds increases. Especially active are the mixed esters of phosphoric acid in which one of the ester radicals has an acid nature. The higher the dissociation constant of such an alcohol or phenol (or acid), the more toxic the compound is to insects and animals. For example, in the series O,O-diethyl O-4-chlorophenyl phosphate, O,O-diethyl O-2,4-dichlorophenyl phosphate, O,O-diethyl O-2,4,5-trichlorophenyl phosphate the greatest insecticidal activity is shown by the last-named ester and the dissociation constants of 4-chlorophenol, 2,4-dichlorophenol, and 2,4,5-trichlorophenol have the following values, respectively: 4.1×10^{-10}, 3.1×10^{-8}, 4.26×10^{-8}. The dialkyl fluorophosphates and also the amides

Table LI. *Biological activity of O,O-diethyl O-aryl phosphates*

X_n	I_{50} [a] (moles) Cholinesterase of fly brain	LD_{50} (%) House flies	Thrips	Citrus red mite
2,4-$(NO_2)_2$	$3.0 \cdot 10^{-9}$	15.5	0.01	0.01
4-NO_2	$2.6 \cdot 10^{-8}$	0.05	0.0001	0.001
2-NO_2	$5.0 \cdot 10^{-8}$	0.70	0.001	0.003
3-NO_2	$5.0 \cdot 10^{-8}$	0.98	0.005	0.03
2,4,5-Cl_3	$6.0 \cdot 10^{-9}$	0.80	0.03	0.0003
2,4,6-Cl_3	$3.3 \cdot 10^{-6}$	17.5	0.03	0.03
2,4-Cl_2	$5.0 \cdot 10^{-7}$	1.5	0.003	0.1
2-Cl	$2.0 \cdot 10^{-5}$	25.0	0.01	0.1
4-Cl	$3.0 \cdot 10^{-5}$	15.0	0.01	0.1
4-CH_3SO_2	$2.5 \cdot 10^{-7}$	0.25	0.0001	0.001
4-CH_3SO	$3.1 \cdot 10^{-6}$	0.15	0.0001	0.0008
4-CH_3S	$3.3 \cdot 10^{-5}$	0.20	0.0001	0.0002
3-$N(CH_3)_2$	$4.0 \cdot 10^{-7}$	2.5	0.1	0.1
4-CN	$1.3 \cdot 10^{-7}$	0.35	0.00002	0.0002
3-CH_3O	$1.3 \cdot 10^{-4}$	> 50	> 0.02	0.3
4-CH_3O	$1.0 \cdot 10^{-3}$	> 50	1.0	1.0
4-CH_3	$1.0 \cdot 10^{-3}$	> 50	> 0.1	> 0.1
H	$1.0 \cdot 10^{-3}$	> 50	> 0.1	> 0.1

[a] I_{50} is the concentration of the compound causing 50% inhibition of the enzyme activity.

of fluorophosphoric acid are highly toxic. However, when the length of the alkyl radicals in the esters and amides of phosphoric acid is increased, the toxicity of the compounds to animals decreases. The maximum toxicity for many mixed esters of phosphoric acid occurs at the diethyl derivatives, but also its position have great influence on the toxicity of a mixed ester of considerably less toxic. This apparently is associated with their high alkylating capability with respect to various nitrogen and sulfur compounds that are present in biological substrates and also with their high rate of hydrolysis.

The biological activity of O,O-diethyl O-aryl phosphates is shown in Table LI. Not only the nature of the substituent in the aromatic radical, but also its position have great influence on the toxicity of a mixed ester of phosphoric acid for insects and mites. The insecticidal properties are increased most when nitro and methylmercapto groups are introduced into the aromatic radical. In general the activity of the aryl phosphates substituted in position 4 is higher than in positions 2 and 3. However, in spite of their high insecticidal activity, the mixed alkyl aryl esters of phosphoric acid have not been practically utilized in agriculture because of their toxicity to animals and man. Certain aliphatic halogen-containing esters of phosphoric acid and enol phosphates that are used both in agriculture and public health are somewhat less toxic.

O,O-Dimethyl O-2,2-dichlorovinyl phosphate (dichlorvos, DDVP, Nuvan) is the first representative of the enolphosphate series. It is a colorless liquid, b.p. 74° C. at 1 mm. of Hg, vapor pressure at 20° C. 1.2×10^{-2} mm. of Hg, volatility at this temperature 145 mg./m.^{3}, d_4^{20} 1.420, n_D^{20} 1.4541. Dichlorvos is highly soluble in most organic solvents; the solubility in water is about 1%. The LD_{50} is 80 mg./kg.

In pure water 50% dichlorvos is hydrolyzed at 20° C. in 61.5 days, and at 70° C. in 25 minutes. The hydrolysis proceeds by the following scheme:

$$(CH_3O)_2\overset{O}{\underset{\|}{P}}OCH = CCl_2 + H_2O \rightarrow (CH_3O)_2\overset{O}{\underset{\|}{P}}OH + CHCl_2CHO$$

Dimethylphosphoric acid is practically harmless and dichloroacetaldehyde rapidly decomposes and evaporates. Therefore, when dichlorvos is applied to plants and other materials no toxic residues remain.

Dichlorvos adds chlorine or bromine at the double bond of the dichlorovinyl group:

$$(CH_3O)_2\overset{O}{\underset{\|}{P}}OCH = CCl_2 + Br_2 \rightarrow (CH_3O)_2\overset{O}{\underset{\|}{P}}OCHBrCCl_2Br$$

and in this process compounds are formed that have a powerful insecticidal effect (see "naled").

Two methods are used for the production of dichlorvos:

1. Splitting out HCl from trichlorfon (chlorophos) by the action of caustic alkalies in aqueous solution at 40°–50° C.:

$$(CH_3O)_2\overset{O}{\underset{\|}{P}}CH(OH)CCl_3 + KOH \rightarrow (CH_3O)_2\overset{O}{\underset{\|}{P}}OCH = CCl_2 + KCl + H_2O$$

The yield of dichlorvos in this process does not exceed 60%.

2. Reaction of chloral with trimethyl phosphite:

$$(CH_3O)_3P + CCl_3CHO \rightarrow CH_3Cl + (CH_3O)_2\overset{O}{\underset{\|}{P}}OCH = CCl_2$$

This process is carried out either in a solvent (e. g., in benzene) or without one (in the latter case cooling is necessary). By the first and second methods dichlorvos of 92–93% purity can be produced by either a batch or a continuous process.

In the presence of traces of moisture dichlorvos on standing breaks down with the formation of acidic products that catalyze further decomposition of the compound. To stabilize the technical grade product 2–4% epichlorohydrin is added, which ties up the acidic substances and improves the conditions for storage of the compound. Dichlorvos keeps well in a glass container.

The homologs of dichlorvos are considerably less toxic to insects. Thus, O,O-dipropyl O-2,2-dichlorovinyl phosphate is two and a half times less toxic to flies than dichlorvos, and O,O-dibutyl O-2,2-dichlorovinyl phosphate is five times less toxic.

O,O-Dimethyl O-2,2-dichloro-1,2-dibromoethyl phosphate (naled, Dibrom) is a crystalline substance, m.p. 25° C., b.p. 110° C. at 0.5 mm. of Hg. It is insoluble in water, slightly soluble in paraffinic hydrocarbons, but highly soluble in most other organic solvents. The LD_{50} for rats is 430 mg./kg. It is an active insecticide for the control of both sucking and chewing insects. Naled also has some fungicidal effect.

With respect to hydrolysis, naled is more stable than dichlorvos. When naled hydrolyzes, dimethylphosphoric acid, HBr, and dichlorobromoacetic acid are formed, and the last of these goes to oxalic acid.

O,O-Diethyl O-(2,2-dichloro-1-β-chloroethoxyvinyl) phosphate (Phosthenon, Phosphinon) is the closest analog of dichlorvos. It is a mobile liquid, b.p. 124° C. at 4 mm. of Hg. However, in spite of its good insecticidal properties, Phosthenon is not used on an appreciable scale because of its high toxicity to animals; the LD_{50} for rats is 6.8–9.7 mg./kg.

Phosthenon is produced by the reaction of triethyl phosphite with 2-chloroethyl trichloroacetate:

$$(C_2H_5O)_3P + CCl_3COOCH_2CH_2Cl \rightarrow (C_2H_5O)_2\overset{\overset{O}{\|}}{P}OC=CCl_2$$
$$\underset{OCH_2CH_2Cl}{|}$$

O,O-Dimethyl O-(1-methyl-2-carbomethoxyvinyl) phosphate (mevinphos, Phosdrin) is a colorless liquid, highly soluble in water, b.p. 76° C. at 0.2 mm. of Hg, vapor pressure at 20° C. 2.2×10^{-3} mm. of Hg, volatility at 20° C. 27 mg./m.³. It is slightly soluble in petroleum ether, but highly soluble in acetone and benzene, d_4^{20} 1.25, n_D^{20} 1.4494.

In neutral medium mevinphos is resistant to hydrolysis (50% hydrolyzed in 30–35 days), but in alkaline medium hydrolysis is appreciably accelerated (at pH 11 50% hydrolyzed in 1.4 hours). Mevinphos exists in the form of *cis-* and *trans-*isomers, of which the *trans-*isomer is the more stable. The *cis-*isomer has insecticidal properties and exceeds the *trans-*isomer in insecticidal activity by approximately 100 times. The technical grade product usually contains 60% *cis-* and 40% *trans-*isomers.

Mevinphos is toxic to warm-blooded animals (LD_{50} for rats 3.7–6.1 mg./kg.). It is a good insecticide of short-term action for sucking and chewing insects. It is marketed in the form of a 50% emulsive concentrate which is used at 0.1% concentration.

The hydrolysis of mevinphos by water proceeds according to the following scheme:

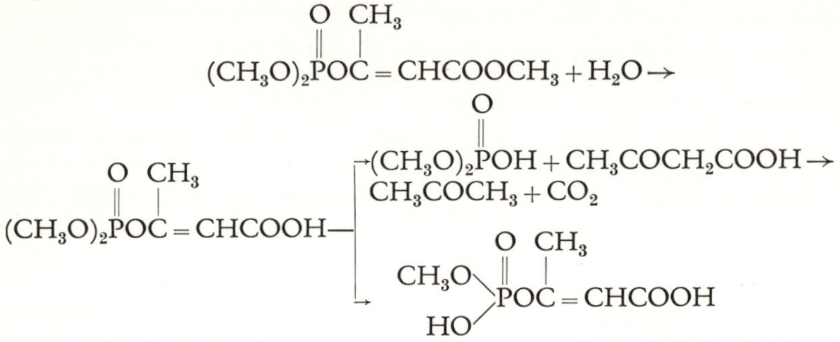

$$(CH_3O)_2\overset{O}{\overset{\|}{P}}O\overset{CH_3}{\overset{|}{C}}=CHCOOCH_3 + H_2O \rightarrow$$

$$(CH_3O)_2\overset{O}{\overset{\|}{P}}O\overset{CH_3}{\overset{|}{C}}=CHCOOH \begin{bmatrix} \rightarrow (CH_3O)_2\overset{O}{\overset{\|}{P}}OH + CH_3COCH_2COOH \rightarrow \\ CH_3COCH_3 + CO_2 \\ \rightarrow \overset{CH_3O}{\underset{HO}{\diagup}}\overset{}{\diagdown}\overset{O}{\overset{\|}{P}}O\overset{CH_3}{\overset{|}{C}}=CHCOOH \end{bmatrix}$$

phosdrinic acid *desmethylphosdrinic acid*

Analogous products are formed when mevinphos breaks down in plant tissues. It has been established that in plants 90% breaks down in 0.8–4.2 days.

It can be produced by two methods:

1. By reaction of the sodium enolate of methyl acetoacetate with dimethylchlorophosphate:

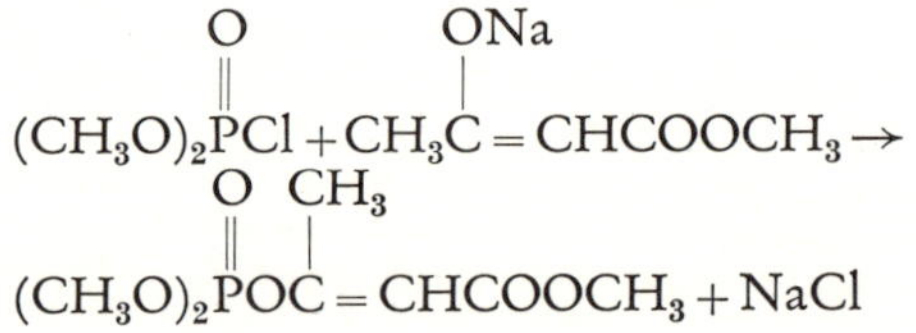

$$(CH_3O)_2\overset{O}{\overset{\|}{P}}Cl + CH_3\overset{ONa}{\overset{|}{C}}=CHCOOCH_3 \rightarrow$$

$$(CH_3O)_2\overset{O}{\overset{\|}{P}}O\overset{CH_3}{\overset{|}{C}}=CHCOOCH_3 + NaCl$$

2. By a Perkov reaction from trimethyl phosphite and chloroacetoacetic ester:

$$(CH_3O)_3P + CH_3COCHClCOOCH_3 \rightarrow$$

$$(CH_3O)_2\overset{O}{\overset{\|}{P}}O\overset{CH_3}{\overset{|}{C}}=CHCOOCH_3 + CH_3Cl$$

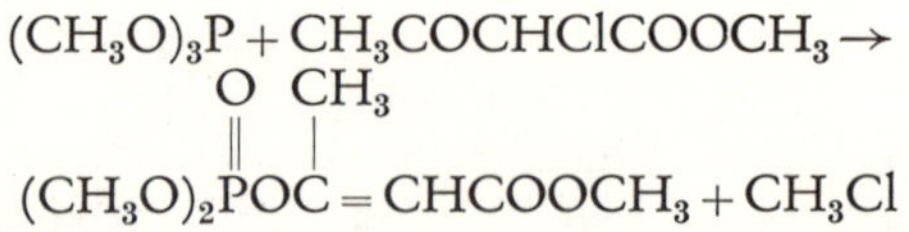

The second method gives better yields of the product, but chloroacetoacetic ester is a requisite.

The corresponding thio-mevinphos is several times less toxic for vertebrates than mevinphos.

Dimethylphosphate of alpha-methylbenzyl-3-hydroxy-cis-crotonate (Ciodrin, crotoxyphos) is a colorless liquid, b.p. 135° C. at 0.03 mm. of Hg; vapor pressure 1.4×10^{-5} mm. of Hg at 20° C., 3.9×10^{-5} at 30° C., 9.8×10^{-5} at 40° C.; half-life in water solution at 38° C. and pH 9, 35 hours and pH 1, 87 hours; solubility in water about 0.1%, miscible with xylene.

The LD$_{50}$ is 125 mg./kg. Ciodrin is used for the control of external parasites on livestock.

Two methods can be used for its production:
1. Reaction of chloroacetoacetic acid ester with trimethyl phosphite:

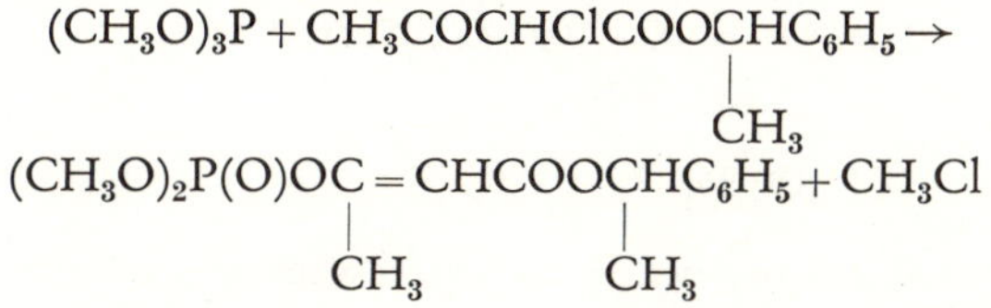

$$(CH_3O)_3P + CH_3COCHClCOOCHC_6H_5 \rightarrow$$
$$\qquad\qquad\qquad\qquad\qquad\qquad\quad |$$
$$\qquad\qquad\qquad\qquad\qquad\qquad\quad CH_3$$
$$(CH_3O)_2P(O)OC=CHCOOCHC_6H_5 + CH_3Cl$$
$$\qquad\qquad\quad |\qquad\qquad\quad\;\; |$$
$$\qquad\qquad\; CH_3\qquad\qquad CH_3$$

2. Reaction of the sodium enolate of phenylethyl acetoacetate with dimethylchlorophosphate:

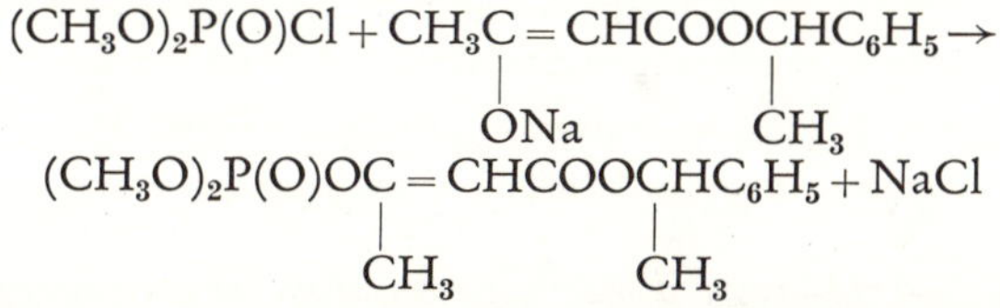

$$(CH_3O)_2P(O)Cl + CH_3C=CHCOOCHC_6H_5 \rightarrow$$
$$\qquad\qquad\qquad\qquad\quad |\qquad\qquad |$$
$$\qquad\qquad\qquad\qquad ONa\qquad\; CH_3$$
$$(CH_3O)_2P(O)OC=CHCOOCHC_6H_5 + NaCl$$
$$\qquad\qquad\quad |\qquad\qquad\quad\;\; |$$
$$\qquad\qquad\; CH_3\qquad\qquad CH_3$$

O,O-Dimethyl O-(2-chloro-2-N,N-diethylcarbamoyl-1-methylvinyl, phosphate (phosphamidon) is a colorless liquid, highly soluble in water, alcohol, and acetone, and slightly soluble in saturated hydrocarbons, b.p. 70° C. at 0.01 mm. of Hg, vapor pressure at 20° C. 2.5×10^{-5} mm. of Hg, volatility 0.41 mg./m.³ The LD$_{50}$ for rats by oral administration is 7.5–15 mg./kg., and by dermal application 125 mg./kg.

Phosphamidon is stable in neutral and weakly acid aqueous solutions; in alkaline medium it is rapidly hydrolyzed. In plants phosphamidon breaks down according to the following scheme:

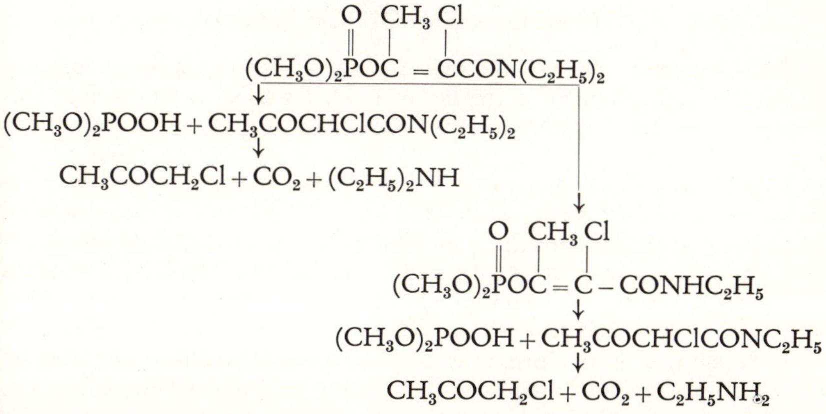

$$\qquad\qquad\qquad O\;\; CH_3\;\, Cl$$
$$\qquad\qquad\qquad \|\quad |\quad\;\, |$$
$$\qquad (CH_3O)_2POC = CCON(C_2H_5)_2$$
$$\qquad\qquad\quad \downarrow$$
$$(CH_3O)_2POOH + CH_3COCHClCON(C_2H_5)_2$$
$$\qquad\qquad\quad \downarrow$$
$$CH_3COCH_2Cl + CO_2 + (C_2H_5)_2NH$$
$$\qquad\qquad\qquad\qquad\qquad\qquad\quad \downarrow$$
$$\qquad\qquad\qquad\qquad O\;\; CH_3\;\, Cl$$
$$\qquad\qquad\qquad\qquad \|\quad |\quad\;\, |$$
$$\qquad\qquad (CH_3O)_2POC = C - CONHC_2H_5$$
$$\qquad\qquad\qquad\qquad\qquad \downarrow$$
$$\qquad (CH_3O)_2POOH + CH_3COCHClCONC_2H_5$$
$$\qquad\qquad\qquad\qquad\qquad \downarrow$$
$$\qquad\qquad CH_3COCH_2Cl + CO_2 + C_2H_5NH_2$$

In spite of its relatively high toxicity for animals, it is used as a systemic insecticide for the control of sucking pests on cotton and other crops. It is effective also in controlling some chewing pests (e. g., the Colorado potato beetle).

The best method of producing phosphamidon is the reaction of trimethyl phosphite with the diethylamide of dichloroacetoacetic acid:

$$(CH_3O)_3P + CH_3COCCl_2CON(C_2H_5)_2 \rightarrow$$
$$(CH_3O)_2\overset{O}{\underset{\|}{P}}OC(\overset{CH_3}{\underset{|}{}}) = CClCON(C_2H_5)_2 + CH_3Cl$$

The reaction goes readily when an equimolecular quantity of trimethyl phosphite is added to a boiling solution of the diethylamide of dichloroacetoacetic acid in chlorobenzene. The yield of phosphamidon by this method is 83%. The diethylamide of dichloroacetoacetic acid that is necessary for the synthesis of phosphamidon is prepared in more than 90% yield by the chlorination of diethylacetoacetamide with sulfuryl chloride:

$$CH_3COCH_2CON(C_2H_5)_2 + 2\ SO_2Cl_2 \rightarrow$$
$$CH_3COCCl_2CON(C_2H_5)_2 + 2\ HCl + 2\ SO_2$$

O,O-Dimethyl O-(N,N-dimethylcarbamoyl-1-methylvinyl) phosphate (Bidrin) is an active systemic insecticide. It is a liquid, miscible with water in all proportions, d_{60}^{60} 1.22. The LD_{50} for rats is 22 mg./kg. At pH 9 and 37° C., 50% is hydrolyzed in 50 days, and at pH 1 in 100 days.

Bidrin is recommended for the control of various plant pests at dosages from 0.25 to 1 kg./ha., and also for the control of some chewing plant pests.

It is synthesized by a Perkov reaction from the dimethylamide of monochloroacetoacetic acid and trimethyl phosphite:

$$(CH_3O)_3P + CH_3COCHClCON(CH_3)_2 \rightarrow$$
$$(CH_3O)_2\overset{O}{\underset{\|}{P}}OC(\overset{CH_3}{\underset{|}{}}) = CHCON(CH_3)_2 + CH_3Cl$$

O,O-Dimethyl O-(2-N-methylcarbamoyl-1-methylvinyl) phosphate (SD91-29) is prepared in a similar manner. Its LD_{50} is 21 mg./kg. This compound gives satisfactory results in the control of some soil-inhabiting pests.

O,O-Diethyl O-[2-chloro-1-(2′,4′-dichlorophenyl)vinyl]phosphate (Birlane) is an oily liquid with a weak odor. At 20° C. 145 mg. of the compound dissolves in 1 liter of water; it is miscible with aromatic hydrocarbons, d^{20} 1.36, vapor pressure at 25° C. 1.7×10^{-7} mm. of Hg. The LD_{50} for rats is 10–39 mg./kg. In aqueous solution 50% is hydrolyzed in 400 hours at pH 9, and in 700 hours at pH 1.1.

Birlane is used in the form of emulsions, wettable powders, and granules applied to the soil to control cabbage, carrot, and onion flies. When it is used on these crops at planting time, the residues of the compound in the harvested vegetables do not exceed 0.05 mg./kg. It is synthesized by the condensation of triethyl phosphite with 2,4,1′,1′-tetrachloroacetophenone:

$$(C_2H_5O)_3P + Cl-\underset{Cl}{\underset{|}{C_6H_3}}-COCHCl_2 \longrightarrow (C_2H_5O)_2\overset{O}{\overset{\|}{P}}OC=CHCl + C_2H_5Cl$$

with the aryl group being 2,4-dichlorophenyl on the phosphate product.

The technical grade product obtained by this method is a mixture of *cis*- and *trans*-isomers.

2-Chloro-1-(2,4,5-trichlorophenyl)vinyl dimethyl phosphate (Gardona) is a white crystalline substance, m.p. 97°–98° C., vapor pressure 4.2×10^{-8} mm. of Hg at 20° C., solubility in water 11 p.p.m. at 20° C. The LD_{50} is 4,000 mg./kg. In aqueous solution 50% is hydrolyzed in 1,300 hours at pH 3, 1,060 hours at pH 7, and 80 hours at pH 10.5 at 50° C. The initial half-life of Gardona is about 1 day on leaves and 4–5 days in medium loam soil. On foliage and in soils breakdown occurs by hydrolysis of the P – O – C bond to give initially 2,4,5-trichlorophenacyl chloride and desmethyl Gardona followed by reaction of the former involving reduction of the ketone group and reductive dechlorination to form several eight-carbon compounds. On foliage and on apple fruit the major breakdown products are conjugates of 1-(2′,4′,5′-trichlorophenyl)ethanol-1 with sugars other than β-D-glucose.

Gardona is synthesized by the condensation of trimethyl phosphite with 2,4,5,1′,1′-pentachloroacetophenone:

$$(CH_3O)_3P + Cl-Ar-COCHCl_2 \longrightarrow CH_3Cl + (CH_3O)_2\overset{O}{\overset{\|}{P}}-OC=CHCl$$

(where Ar = 2,4,5-trichlorophenyl)

It is used in the form of emulsion, wettable powders, and granules to control pests on cabbage, apple, rice, and other crops.

Of possible interest are the analogs of Gardona that are synthesized by the condensation of O,O-dialkyl chlorothiophosphates with chloroacetophenones in the presence of NaOH:

$$(C_2H_5O)_2\overset{S}{\overset{\|}{P}}Cl + Cl-Ar-COCH_2Cl \xrightarrow{NaOH} (C_2H_5O)_2\overset{S}{\overset{\|}{P}}OC=CHCl$$

(where Ar = 2,4,5-trichlorophenyl)

A large number of other similar compounds also have been studied, but they are not yet used in agriculture.

In addition to the use of esters of enolphosphoric acid in agriculture as systemic insecticides, the amides of fluorophosphoric acid are used. Dimefox and mipafox belong to this group.

Tetramethylphosphorodiamidic fluoride [bis(dimethylamido)fluorophosphate, dimefox, Pestox, Hanane, Terra-Sytam] is a colorless, mobile liquid with a weak odor, highly soluble in water and in many polar organic solvents, b.p. 80° C. at 10 mm. of Hg, d_4^{20} 1.1151, n_D^{20} 1.4267. It is stable in aqueous alkaline solutions and is rapidly hydrolyzed in acid medium. It is highly toxic to mammals, for the LD_{50} for rats is 1–1.5 mg./kg. Dimefox is a powerful systemic insecticide and is used to control sucking pests of hops and vectors of a virus disease of cacao trees. For greater safety the compound often is placed in the soil in gelatine ampoules.

Dimefox can be produced by three methods:

1. Action of salts of HF on bis(dimethylamido) chlorophosphate:

$$[(CH_3)_2N]_2\overset{\overset{O}{\|}}{P}Cl + NaF \rightarrow [(CH_3)_2N]_2\overset{\overset{O}{\|}}{P}F + NaCl$$

2. Reaction of dimethylamine with dimethylamido difluorophosphate:

$$(CH_3)_2\overset{\overset{O}{\|}}{N}PF_2 + 2(CH_3)_2NH \rightarrow [(CH_3)_2N]_2\overset{\overset{O}{\|}}{P}F + (CH_3)_2NH \cdot HF$$

3. Reaction of phosphorus fluorooxychloride with dimethylamine:

$$\overset{\overset{O}{\|}}{F}PCl_2 + 4(CH_3)_2NH \rightarrow [(CH_3)_2N]_2\overset{\overset{O}{\|}}{P}F + 2(CH_3)_2NH \cdot HCl$$

N,N'-Diisopropylphosphorodiamidic fluoride [bis(isopropylamido) fluorophosphate, mipafox, Pestox-15, Isopestox] is a white crystalline substance, m.p. 61°–62° C., b.p. 125° C. at 2 mm. of Hg, vapor pressure at 25° C. 0.0025 mm. of Hg. It is highly soluble in most organic solvents except paraffinic hydrocarbons; its solubility in water is about 8%. The LD_{50} for rats is 50 mg./kg. Mipafox is stable in aqueous neutral or weakly acid solutions; 50% is hydrolyzed in 200 days at pH 6 and in six days at pH 8. It shows a strong systemic and contact action against aphids and other plant pests. However, when vapors of mipafox are inhaled by animals or man, it causes irreversible paralysis; the compound, therefore, is not used at the present time.

Mipafox is produced by the following route:

$$POCl_3 + 4\ iso\text{-}C_3H_7NH_2 \rightarrow (iso\text{-}C_3H_7NH)_2\overset{\overset{O}{\|}}{P}Cl + 2\ C_3H_7NH \cdot HCl$$

$$(iso\text{-}C_3H_7NH)_2\overset{\overset{O}{\|}}{P}Cl + KF \rightarrow (iso\text{-}C_3H_7NH)_2\overset{\overset{O}{\|}}{P}F + KCl$$

O,O-Dimethyl N-(isopropoxycarbamoyl) phosphate (dimethyl ester of isopropylurethanephosphoric acid, avenin, K-69-79) is a light oily liquid, not distillable without decomposition, slightly soluble in water, soluble in most organic solvents. When it is heated it breaks down with the formation of the methyl ester of isocyanatophosphoric acid. The LD_{50} for experimental animals is more than 5,000 mg./kg. It has a systemic selective action on the common sugar-beet weevil. Treatment of sugar-beet seeds before planting protects the seedlings from the sugar-beet weevil for 10–12 days. Avenin is an interesting insecticide of low toxicity with narrowly selective action.

It can be synthesized by the action of isopropyl alcohol on the methyl ester of isocyanatophosphoric acid:

$$OCNP(OCH_3)_2 + (CH_3)_2CHOH \rightarrow C_3H_7OCNHP(OCH_3)_2$$

O-Methyl O-2-chloro-4-tert-butylphenyl N-methylamidophosphate (Ruelene) is a white crystalline substance, m.p. 57.4°–59.8° C., b.p. 117° to 118° C. at 0.01 mm. of Hg. Its solubility in water is about 0.5%, in hexane 3.3%; it is highly soluble in most polar organic solvents. The LD_{50} for rats is 770–950 mg./kg. Ruelene is used to control intestinal worms in domestic animals and warble flies by its introduction into the feed. The dosages for different animals range within the limits 37–150 mg. of compound per kg. body weight of the animal. Good results in the control of warble flies are also obtained by spraying the animals with 0.25–0.37% solutions. The compound does not accumulate in animal organs, and it breaks down rapidly.

It can be produced by the reaction of 2-chloro-4-*tert*-butylphenol with O-methyl N-methylamidochlorophosphate in the presence of bases or by the reaction of 2-chloro-4-*tert*-butylphenyl dichlorophosphate with methanol and methylamine:

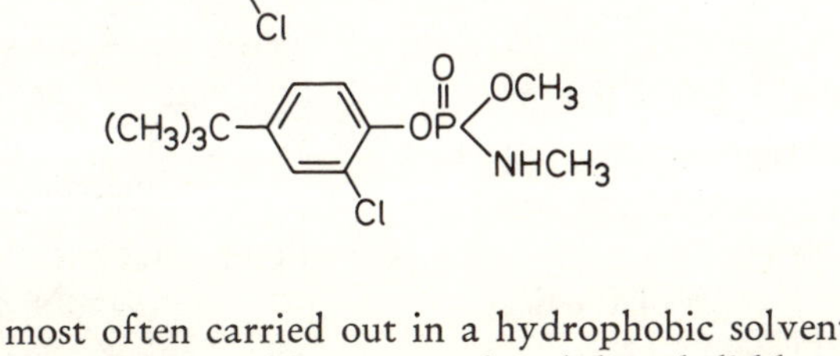

This reaction is most often carried out in a hydrophobic solvent with anhydrous methylamine. The 2-chloro-4-*tert*-butylphenyl-dichlorophosphate necessary for the reaction is obtained in yields up to 80% by the reaction of 2-chloro-4-*tert*-butylphenol with an excess of phosphorus oxychloride in the presence of potassium chloride or anhydrous magnesium chloride:

$$(CH_3)_3C-\underset{Cl}{\underset{|}{C_6H_3}}-OH + POCl_3 \longrightarrow (CH_3)_3C-\underset{Cl}{\underset{|}{C_6H_3}}-O\overset{O}{\underset{\|}{P}}Cl_2 + HCl$$

This reaction takes place on prolonged boiling. 2-Chloro-4-*tert*-butylphenol can be obtained in satisfactory yield by the chlorination of *tert*-butylphenol or by the condensation of isobutylene with o-chlorophenol in the presence of various catalysts. The compound usually sold is 92% pure.

Diethyl N-1,3-dithiolanyl-2-imino phosphate (cyolane, dithiolane, Iminophosphate) is a colorless-to-yellow liquid (m.p. 37°–45° C.) b.p. 115°–118° C. at 0.001 mm. of Hg; soluble in acetone, benzene, toluene, ethanol, water, cyclohexanone; very slightly soluble in hexane. The LD_{50} is 8.9 mg./kg., dermal toxicity on rabbits (LD_{50}) 23.0 mg./kg. Cyolane is a contact and stomach insecticide with systemic activity in plants through root and foliar absorption. It is used in the form of emulsions (25% E.C.) and granules to control pests on cotton and other crops. It is synthesized by the condensation of 2-iminodithiolane with diethyl chlorophosphate:

$$(C_2H_5O)_2\overset{O}{\underset{\|}{P}}-Cl + HN=C\underset{S-CH_2}{\overset{S-CH_2}{\diagup\diagdown}} \longrightarrow (C_2H_5O)_2\overset{O}{\underset{\|}{P}}-N=C\underset{S-CH_2}{\overset{S-CH_2}{\diagup\diagdown}} + HCl$$

In addition to the use of derivatives of phosphoric acid as insecticides and acaricides, certain amides of phosphoric acid have been proposed for the sexual sterilization of insects. Those studied in most detail are the derivatives of ethylenimine, such as tris(1-aziridinyl) phosphine oxide [tris-(ethylenimido)phosphate, tepa, aphoxide], its methyl homolog (metepa, metaphoxide), tris(1-aziridinyl)phosphine sulfide [tris(ethylenimido) thiophosphate, thiotepa], its methyl homolog (methiotepa), hexakis(1-aziridinyl)phosphonitrile [hexa(ethylenimido)cyclotriphosphazene, apholate], and hexamethyl triamidophosphate (hempa, HMPA):

$$\left(\underset{H_2C}{\overset{H_2C}{\diagdown\diagup}}N\right)_3 P=O \qquad \left(\underset{H_2C}{\overset{CH_3-HC}{\diagdown\diagup}}N\right)_3 P=O$$
$$\text{tepa} \qquad\qquad\qquad \text{metepa}$$

$$\left(\underset{H_2C}{\overset{H_2C}{\diagdown\diagup}}N\right)_3 P=S \qquad \left(\underset{H_2C}{\overset{CH_3-HC}{\diagdown\diagup}}N\right)_3 P=S$$
$$\text{thiotepa} \qquad\qquad\qquad \text{methiotepa}$$

$$\left(\begin{array}{c}H_2C\text{---}CH_2\\ \diagdown N\diagup\end{array}\right)_2$$

$$\left(\begin{array}{c}H_2C\\ |\\ H_2C\end{array}\hspace{-4pt}\diagdown N\right)_2 \hspace{8pt} \begin{array}{c}P\\ N\hspace{8pt}N\\ |\hspace{8pt}\|\\ P\hspace{8pt}P\\ \diagdown N\diagup\end{array} \hspace{8pt}\left(N\hspace{-4pt}\diagdown\hspace{-4pt}\begin{array}{c}CH_2\\ |\\ CH_2\end{array}\right)_2 \hspace{20pt} [(CH_3)_2N]_3P=O$$

apholate   hempa

When insects are fed these compounds they do not produce normal progeny. These compounds should be used to control only those species of insects that have many generations in one season. The compounds mentioned are toxic and dangerous for vertebrate animals and man; their use, therefore, is very limited and in fact has not gone beyond the stage of field experiments.

Derivatives of thiophosphoric acid

Replacement of one of the oxygen atoms by sulfur in derivatives of phosphoric acid leads to a considerable decrease in toxicity of the compound for mammals without substantial change in the insecticidal and acaricidal activity, although there are exceptions to the general rule. Consequently, the derivatives of thiophosphoric acid are widely used in agriculture as agents to control plant pests.

It is well known that the derivatives of thiophosphoric acid may have a thiono (I) or thiolo (II) structure:

$$(RO)_2P\diagup^{\displaystyle S}_{\displaystyle OR'} \hspace{20pt}(I) \hspace{60pt} (RO)_2P\diagup^{\displaystyle O}_{\displaystyle SR'} \hspace{20pt}(II)$$

The thiolo derivatives of thiophosphoric acid are the more toxic to mammals.

The thiono derivatives when heated or when acted on by certain reagents rearrange to the thiolo isomers. This reaction is known as a Pishchemuka rearrangement. Depending on the conditions and the reagents, the simplest derivatives (e. g., salts) of thiophosphoric acid may form either thiolo or thiono isomers; this fact is associated with the existence of a dual reactive capacity in the thiophosphoric acid derivatives as a result of tautomeric conversions:

$$\begin{array}{c}RO\\ R'O\end{array}\hspace{-4pt}\diagdown P\diagup^{\displaystyle S}_{\displaystyle OH} \rightleftarrows \begin{array}{c}RO\\ R'O\end{array}\hspace{-4pt}\diagdown P\diagup^{\displaystyle O}_{\displaystyle SH}$$

The compounds used as pesticides are mainly the mixed esters of thiophosphoric acid of the general formulas (III), (IV), (V), (VI), and (VII), where R and R' are a lower aliphatic radical, Ar is an aromatic or heterocyclic radical containing various substituents in the aromatic or heterocyclic nucleus, and R'' is an aliphatic, aromatic, or heterocyclic radical:

$$\begin{array}{c}RO\\R'O\end{array}\!\!\!>\!\!P\!\!<\!\!\begin{array}{c}S\\OAr\end{array} \quad (III) \qquad \begin{array}{c}RO\\R'O\end{array}\!\!\!>\!\!P\!\!<\!\!\begin{array}{c}O\\SR''\end{array} \quad (IV)$$

$$\begin{array}{c}RO\\R'O\end{array}\!\!\!>\!\!P\!\!<\!\!\begin{array}{c}S\\OCH_2CH_2SR''\end{array} \quad (V) \qquad \begin{array}{c}RO\\R'O\end{array}\!\!\!>\!\!P\!\!<\!\!\begin{array}{c}O\\SCH_2CH_2SR''\end{array} \quad (VI)$$

$$\begin{array}{c}RO\\R'O\end{array}\!\!\!>\!\!P\!\!<\!\!\begin{array}{c}O\\SCH_2COOR''\end{array} \quad (VII) \qquad \begin{array}{c}RO\\R'HN\end{array}\!\!\!>\!\!P\!\!<\!\!\begin{array}{c}S\\OAr\end{array} \quad (VIII)$$

Mixed esteramides of phosphoric acid of the general formula (VIII) are used as pesticides, but this group of compounds is considerably less numerous.

For the series of mixed esters of thiophosphoric acid of the general formula (III) the following conclusions can be drawn about the relationship of insecticidal activity to structure:

1. The most powerful insecticidal and acaricidal action is shown by the mixed esters of thiophosphoric acid in which R and R' are lower aliphatic radicals with the total number of carbon atoms not exceeding four. The maximum activity is reached in compounds containing two ethyl radicals or one methyl and the other ethyl. The minimum toxicity for mammals is shown by O,O-dimethyl O-aryl esters of thiophosphoric acid.

2. The mixed aliphatic-aromatic esters of thiophosphoric acid that do not contain functional groups in the aromatic radical show low insecticidal activity.

3. Among the mixed O,O-dialkyl O-aryl esters of thiophosphoric acid, the maximum insecticidal activity is shown by the esters containing a nitro group in the 4-position. The esters containing a cyano, sulfide, sulfoxide, or sulfonyl group in the 4-position are rather highly active, but most of the insecticidal compounds of this series are also relatively highly toxic for mammals. Compounds with other substituents are less effective. The corresponding 2- and 3-substituted derivatives show a weaker insecticidal effect.

4. Introduction of a second substituent in the aromatic radical somewhat decreases the toxicity of the compound for mammals without substantially lowering the insecticidal activity. In this case the position of the substituents has a great influence. Thus, the introduction of an alkyl or halogen in position 3 of a starting 4-nitro- or 4-methylphenol lowers the toxicity for vertebrates and does not decrease the insecticidal activity of the compound, while introduction of these substituents in position 2 lowers the biological activity of the compounds in all respects.

Satisfactory insecticidal activity is shown also by compounds containing substituents in the 2,4,5-positions, for example, O,O-dimethyl O-2,4,5-trichlorophenyl thiophosphate and O,O-dimethyl O-2,5-dichloro-4-bromophenyl thiophosphate, while the toxicity of these compounds for mammals is low.

5. The presence in the aromatic radical of more than three substituents lowers the insecticidal activity of the compound; sometimes the nature of the action of the compound also is altered. Thus, O,O-dialkyl O-2,3,4,5,6-pentachlorophenyl thiophosphate shows fungicidal activity, although it is insufficient for practical use.

6. Replacement of the aromatic radical by a heterocyclic one in most cases also leads to active insecticides; examples of this are diazinon, Dursban, Zinophos, Asunthol, and others.

7. Introduction of an amido group in place of one aliphatic ester radical usually gives biologically active compounds that have insecticidal (Dowco-109) or herbicidal (Zytron) effects.

8. O,O-Dialkyl S-aryl thiophosphates in a number of cases show not only insecticidal but also fungicidal properties.

9. A fungicidal effect is observed in the mixed O-alkyl diamidothiophosphates.

In the series of compounds with the general formulas (V) and (VI) the following general relationships of activity to structure are observed:

1. Increasing the total number of carbon atoms in R and R' to more than four decreases the compound's insecticidal activity.

2. Increasing the number of methylene groups between the phosphorus and sulfur atoms sharply lowers the activity of the compound. The maximum activity corresponds to the number of 1- or 2-methylene groups.

3. Oxidation of the sulfur to sulfoxyl or the formation of a sulfonium compound increases the toxicity of the compound, especially in the latter case.

4. Introduction into R'' of various substituents sometimes gives active compounds.

5. Practically all compounds of structures (V) and (VI) have a systemic effect; the contact effect of the thiono isomers is somewhat weaker than that of the thiolo isomers.

6. Replacement of sulfur by nitrogen leads to compounds that are toxic both for insects and mites as well as for mammals.

7. Increasing the number of carbon atoms in R'' to more than three lowers the insecticidal activity of the compound.

8. Compounds containing aromatic radicals as the R'' show a lower insecticidal activity than compounds of the aliphatic series.

The activity of compounds of the general formula (VIII) changes almost analogously to compounds of structure (III). The highest activity is shown by compounds with $R = R' = R'' = C_2H_5$.

The principal methods of producing thiophosphates of the general formulas (III) and (V) are the following reactions:

$$\underset{R'O}{\overset{RO}{\diagdown}}\overset{S}{\underset{\|}{P}}Cl + HOAr \rightarrow \underset{R'O}{\overset{RO}{\diagdown}}\overset{S}{\underset{\|}{P}}OAr$$

$$PSCl_3 + ArOH \xrightarrow{KCl} Ar\overset{\overset{S}{\|}}{O}PCl_2 + HCl$$

$$Ar\overset{\overset{S}{\|}}{O}PCl_2 + 2 ROH \rightarrow (RO)_2\overset{\overset{S}{\|}}{P}OAr$$

The first of these reactions goes easily when organic or inorganic bases are used as HCl acceptors. Of the organic bases, it is best to use tertiary amines, e. g., triethylamine or pyridine. The reaction occurs with high yields, also, in the presence of caustic alkalies or carbonates of the alkali metals. The best results are obtained when potassium carbonate is used, especially when the process is carried out in solution in aliphatic ketones (acetone, methyl ethyl ketone). The dialkyl chlorothiophosphates necessary for these syntheses are obtained in high yields by the following route:

$$PSCl_3 + ROH \rightarrow R\overset{\overset{S}{\|}}{O}PCl_2 + HCl$$

$$R\overset{\overset{S}{\|}}{O}PCl_2 + R'OH \rightarrow \overset{RO}{\underset{R'O}{}}{\Large>}\overset{\overset{S}{\|}}{P}Cl$$

The reaction of phosphorus thiotrichloride with alcohols proceeds readily on slight warming and the yield of alkyl dichlorothiophosphates is not <90%. The reaction of alkyl dichlorothiophosphates with alcohols is carried out from −5° to 0° C., with vigorous stirring. Both tertiary amines and alkali hydroxides are employed to bind the HCl and in preparing dimethyl chlorothiophosphate it is possible to use a 40% aqueous solution of NaOH. The mixed dialkyl chlorothiophosphates, one of whose ester radicals is methyl, are produced in a yield close to 90%.

The dialkyl chlorothiophosphates are obtained in high yields also by the action on alkyl dichlorothiophosphates of sodium or magnesium alcoholates:

$$R\overset{\overset{S}{\|}}{O}PCl_2 + R'ONa \rightarrow \overset{RO}{\underset{R'O}{}}{\Large>}\overset{\overset{S}{\|}}{P}Cl + NaCl$$

$$2 R\overset{\overset{S}{\|}}{O}PCl_2 + (R'O)_2Mg \rightarrow 2 \overset{RO}{\underset{R'O}{}}{\Large>}\overset{\overset{S}{\|}}{P}Cl + MgCl_2$$

For the reaction with a sodium alcoholate the optimum temperature is from −10° to −15° C. When the temperature is raised the yield decreases and a considerable amount of trialkyl thiophosphates is formed. The synthesis of dialkyl chlorothiophosphates from magnesium alcoholates is comparatively simple and proceeds at a higher temperature. A third method of synthesis of dialkyl chlorothiophosphates employs chlorination of dialkyl dithiophosphoric acids or of dialkyl thiophosphoro disulfides:

$$2\,(AlkO)_2P(S)SH + 2\,Cl_2 = 2\,(AlkO)_2PSCl + 2\,HCl + 2\,S$$
$$(AlkO)_2P(S)SS(S)P(OAlk)_2 + 2\,Cl_2 = 2\,(AlkO)_2PSCl + S_2Cl_2$$

The methods listed can be used for the industrial production of the dialkyl chlorothiophosphates.

The aryl dichlorothiophosphates are formed on prolonged (not <10 hours) heating to 100°–115° C. of the appropriate phenol with an excess of phosphorus thiotrichloride in the presence of small amounts of potassium chloride or anhydrous magnesium chloride. When the reaction is over, the excess phosphorus thiotrichloride is distilled off in vacuum on a water bath and the aryl dichlorothiophosphate is used for the subsequent process.

To produce the dialkyl aryl thiophosphates the aryl dichlorothiophosphate is introduced into reaction with the alcoholates of the appropriate alcohols; it also is possible to use an alcohol with HCl acceptors, but the yields of dialkyl chlorothiophosphates are lower in the latter case.

The O-aryl O-alkyl N-alkylamidothiophosphates are synthesized by the reaction of amines with aryl dichlorothiophosphates in the appropriate alcohol:

$$ArOP(S)Cl_2 + ROH + 3\,R'NH_2 \rightarrow \begin{array}{c} ArO \\ RO \end{array}\!\!\!P\!\!\!\begin{array}{c} S \\ NHR' \end{array} + 2\,R'NH_2 \cdot HCl$$

For the production of thiolophosphates of formulas (IV), (VI), and (VII) use is most often made of the reaction of salts of the dialkyl thiophosphoric acids with halogen derivatives:

$$\begin{array}{c} RO \\ R'O \end{array}\!\!\!P\!\!\!\begin{array}{c} O \\ SNa \end{array} + R''Cl \rightarrow \begin{array}{c} RO \\ R'O \end{array}\!\!\!P\!\!\!\begin{array}{c} O \\ SR'' \end{array} + NaCl$$

This reaction occurs readily with most dialkyl thiophosphates. An exception is dimethyl thiophosphate. In reactions with it a methylation process occurs, and a considerable amount of trimethyl thiolophosphate is formed as a by-product.

Throughout the world more than 30 different pesticides are used that are derivatives of thiophosphoric acid, the production of which exceeds 50,000 tons a year.

O,O-Diethyl O-4-nitrophenyl thiophosphate (parathion, thiophos, E-605) is a clear oily liquid, m.p. 6.1° C., b.p. 113° C. at 0.05 mm. of Hg, vapor pressure at 20° C. 0.57×10^{-5} mm. of Hg, volatility 0.09 mg./m.³, d_4^{25} 1.265. Its solubility in water is 24 mg./l.; it is highly soluble in most organic solvents, with the exception of the paraffinic hydrocarbons. The LD_{50} is 6–12 mg./kg.

When heated to 100° C., parathion is gradually converted to the thiolo isomer:

$$(C_2H_5O)_2\overset{S}{\underset{\|}{P}}-O-\underset{}{\bigcirc}-NO_2 \xrightarrow{100°C} \begin{matrix} C_2H_5O \\ C_2H_5S \end{matrix}\!\!>\!\!\overset{O}{\underset{\|}{P}}-S-\underset{}{\bigcirc}-NO_2$$

At a higher temperature the reaction may take place very violently with an explosion. Parathion is slowly hydrolyzed by water with the formation of *p*-nitrophenol and diethylthiophosphoric acid:

$$(C_2H_5O)_2\overset{S}{\underset{\|}{P}}-O-\underset{}{\bigcirc}-NO_2 \xrightarrow{H_2O} (C_2H_5O)_2P\!\!\underset{OH}{\overset{S}{\diagup}} + HO-\underset{}{\bigcirc}-NO_2$$

At pH 1–5 about 50% is hydrolyzed in 690 days at 20° C., and in 17 to 20 hours at 70° C. Hydrolysis is more rapid in alkaline medium, thereby enabling the quantitative analysis of parathion by colorimetric determination of the *p*-nitrophenol that is split off.

Oxidizing agents convert parathion to O,O-diethyl O-4-nitrophenyl phosphate (paraoxon), which is more toxic for vertebrates than the starting parathion:

$$(C_2H_5O)_2\overset{S}{\underset{\|}{P}}-O-\underset{}{\bigcirc}-NO_2 \xrightarrow{2O_2, H_2O} (C_2H_5O)_2\overset{O}{\underset{\|}{P}}-O-\underset{}{\bigcirc}-NO_2 + H_2SO_4$$

Reducing agents (for example, metals in acid medium) convert it to the corresponding amino compound (aminoparathion):

$$(C_2H_5O)_2\overset{S}{\underset{\|}{P}}-O-\underset{}{\bigcirc}-NO_2 \xrightarrow{6[H]} (C_2H_5O)_2\overset{S}{\underset{\|}{P}}-O-\underset{}{\bigcirc}-NH_2 + 2H_2O$$

O,O-Diethyl O-4-aminophenyl thiophosphate is nontoxic to animals and does not have an insecticidal effect.

Study of the metabolism of parathion in insects and man has shown that in different species of living organisms the metabolism may proceed in several directions. The principal metabolic product of parathion in man is *p*-nitrophenol and other hydrolysis products, while in the bovine organism parathion is reduced to aminophenol, which is derivatized in the form of the aminophenyl glucuronide. The general scheme of metabolism of parathion can be represented by the cycle of reactions:

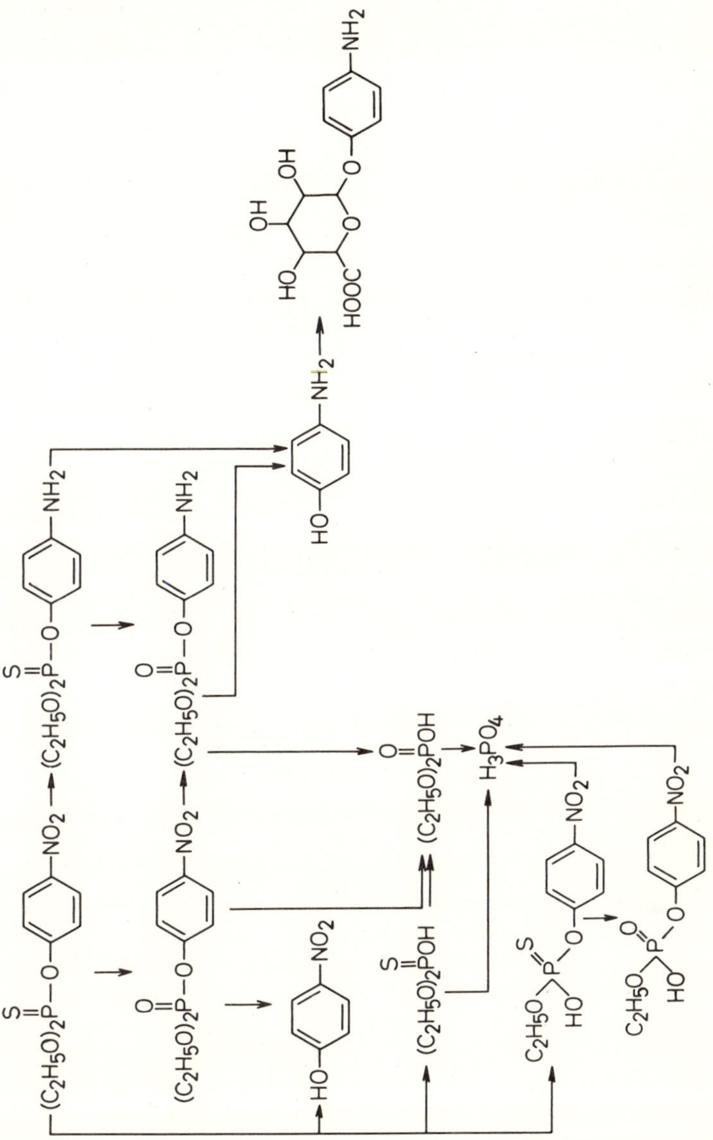

In connection with the toxicity of parathion the tolerance for residues of it in food products in the Soviet Union is 1 mg./kg., including its hydrolysis products. The waiting time [minimum interval after treatment] is determined by the crop that is treated, since the rate of metabolism of parathion depends on the species of the plant. The maximum permissible concentration in air is 0.05 mg./m.³ Parathion is widely used to control various plant pests in the form of emulsions, wettable powders, dusts, and formulations in oil, at dosages ranging from 0.2–1 kg./ha.

Any of the reactions previously described can be used for making parathion, but for industrial synthesis the reaction most often used is diethyl chlorothiophosphate with sodium *p*-nitrophenolate:

$$(C_2H_5O)_2\overset{\overset{S}{\|}}{P}Cl + NaO-\!\!\left\langle\!\!\bigcirc\!\!\right\rangle\!\!-NO_2 \longrightarrow (C_2H_5O)_2\overset{\overset{S}{\|}}{P}-O-\!\!\left\langle\!\!\bigcirc\!\!\right\rangle\!\!-NO_2 + NaCl$$

This process is carried out in chlorobenzene or in aqueous medium in the presence of emulsifiers. It is obtained in good yields also when diethyl chlorothiophosphate reacts with *p*-nitrophenol in the presence of potassium carbonate (HCl acceptor) in acetone or in other solvents (the acceptor is a tertiary amine):

$$(C_2H_5O)_2\overset{\overset{S}{\|}}{P}Cl + HO-\!\!\left\langle\!\!\bigcirc\!\!\right\rangle\!\!-NO_2 + K_2CO_3 \longrightarrow$$

$$(C_2H_5O)_2\overset{\overset{S}{\|}}{P}-O-\!\!\left\langle\!\!\bigcirc\!\!\right\rangle\!\!-NO_2 + KHCO_3 + KCl$$

The diethyl chlorothiophosphate necessary for the synthesis of parathion is prepared either by the general methods described or from phosphorus pentasulfide:

$$4 C_2H_5OH + P_2S_5 \longrightarrow 2(C_2H_5O)_2\overset{\overset{S}{\|}}{P}SH + H_2S$$

$$2(C_2H_5O)_2\overset{\overset{S}{\|}}{P}SH + 3Cl_2 \longrightarrow 2(C_2H_5O)_2\overset{\overset{S}{\|}}{P}Cl + 2HCl + S_2Cl_2$$

The yield of diethyl chlorothiophosphate from diethyldithiophosphoric acid is >80%, but this method apparently is less economical than one using sodium or magnesium ethylate and ethyl dichlorothiophosphate, or alcohol, ethyl dichlorothiophosphate, and NaOH, although in the latter case the yield does not exceed 60% calculated on the ethyl dichlorothiophosphate.

Usually by the aqueous method technical grade parathion is obtained that contains 85–92% O,O-diethyl O-4-nitrophenyl thiophosphate; by the acetone method it is possible to obtain the compound in 98% purity. The main impurities in the technical grade product are O-ethyl O,O-bis(4-nitrophenyl)thiophosphate, nitrophenol, and a small amount of triethyl thiophosphate.

O,O-Dimethyl O-4-nitrophenyl thiophosphate (methyl parathion, metaphos, metacide, Folidol-80, Wofatox) is a white crystalline substance, m.p.

$35°-36°$ C., b.p. $109°$ C. at 0.05 mm. of Hg, vapor pressure at $20°$ C. 0.97×10^{-5} mm. of Hg, volatility 0.14 mg./m.³, d_4^{20} 1.358, n_D^{35} 1.5515. Its solubility in water at $25°$ C. is about 55 mg./liter; it is slightly soluble in paraffinic hydrocarbons but highly soluble in aromatic hydrocarbons and most organic solvents.

The rate of hydrolysis of methyl parathion is considerably higher than that of parathion; at pH 1–5, 50% is hydrolyzed in 175 days at $20°$ C., and in 11 hours at $70°$ C.; the rate of hydrolysis in alkaline medium is still greater.

Methyl parathion is relatively unstable thermally and when it is heated to $140°-160°$ C. it is almost completely converted to the corresponding thiolo isomer; this reaction sometimes takes place explosively and a porous mass is formed that contains a large amount of carbon:

$$(CH_3O)_2\overset{\underset{\parallel}{S}}{P}-O-\underset{}{\langle\bigcirc\rangle}-NO_2 \xrightarrow{140°-160°C} \underset{CH_3S}{\overset{CH_3O}{>}}\overset{\underset{\parallel}{O}}{P}-O-\underset{}{\langle\bigcirc\rangle}-NO_2$$

This reaction also occurs on prolonged heating of an alcoholic solution of methyl parathion at $100°$ C. Among the products of the thermal decomposition of methyl parathion are found trimethylsulfonium salts that are formed as a result of methylation of dimethyl sulfide. The formation of this compound also is possible from O-methyl S-methyl O-4-nitrophenyl thiophosphate. Methyl parathion is a powerful alkylating agent and can methylate sulfides, amines, thiourea, and many other compounds:

$$(CH_3O)_2\overset{\underset{\parallel}{S}}{P}-O-\underset{}{\langle\bigcirc\rangle}-NO_2 + (C_2H_5)_3N \longrightarrow$$

$$[(C_2H_5)_3\overset{+}{N}CH_3]\left[\underset{O}{\overset{CH_3S}{>}}\overset{\underset{\parallel}{O}}{P}-O-\underset{}{\langle\bigcirc\rangle}-NO_2\right]^-$$

It is possible that the high methylating ability of methyl parathion is the cause of its lower toxicity for mammals, because part of the compound is decomposed as a result of the demethylation reaction before it reaches the reactive centers of the cholinesterase.

The other chemical properties of methyl parathion are similar to those of parathion. The toxicity of methyl parathion for mammals is considerably lower (LD_{50} 25–50 mg./kg.); it also penetrates the skin with greater difficulty than parathion, and this facilitates working with it. The maximum permissible concentration in air is 0.1 mg./m.³

Methyl parathion is marketed in the form of emulsions, wettable powders, and dusts. In selecting diluents for dusts and wettable powders note that even weakly alkaline diluents are unsuitable, since methyl parathion (and also parathion) is comparatively quickly decomposed in these diluents and loses its insecticidal properties. Like parathion the compound is used for the control of a wide variety of plant pests. Because of its lower toxicity for mammals, methyl parathion gradually is displacing parathion. The world production of methyl parathion is higher than that of parathion and the trend is toward further growth.

It can be synthesized by all the previously described methods of preparing mixed aliphatic-aromatic esters of thiophosphoric acid. The most important industrial method is the reaction of dimethyl chlorothiophosphate with *p*-nitrophenol in the presence of HCl acceptors or with sodium *p*-nitrophenolate; the reaction is carried out in water in the presence of emulsifiers (e. g., ammonium or amine naphthenates), and also in organic solvents. Chlorobenzene, xylene, or aliphatic ketones (acetone and methyl ethyl ketone) are most often used. When conducting the reaction in organic solvents it is necessary to distill off the solvent at the lowest possible temperature in continuously operating film evaporators; otherwise decomposition of the methyl parathion is possible, sometimes occurring with an explosion.

A product containing up to 96–98% O,O-dimethyl O-4-nitrophenyl thiophosphate is obtained when the reaction of dimethyl chlorothiophosphate with *p*-nitrophenol is conducted in the presence of anhydrous potassium carbonate in acetone with the use of sufficiently pure starting materials. This product for convenience of handling and transportation is diluted with 10–15% xylene.

By the aqueous method the compound is obtained with 85–90% of the principal product. The yield of methyl parathion by the different methods ranges from 75 to 90%.

Impurities in the technical grade product are small amounts of *p*-nitrophenol, O-methyl O,O-bis(4-nitrophenyl) thiophosphate, and trimethyl thiophosphate. Moreover, in the product obtained in organic solvents O,S-dimethyl O-4-nitrophenyl thiophosphate is an impurity.

The insecticidal properties of all the isomers of methyl parathion have been studied, including the *d*- and *l*-forms of O,S-dimethyl O-4-nitrophenyl thiophosphate, of which the *levo* isomer proved to be the more toxic for animals (LD_{50} for rats 25 mg./kg., toxicity of *dextro* isomer 135 mg./kg.). Of the three thiophosphate with different positions of the sulfur, the most toxic was O,O-dimethyl S-4-nitrophenyl thiophosphate (LD_{50} for mice 7.5 mg./kg.).

O-Methyl O-ethyl O-4-nitrophenyl thiophosphate (Thiophos-ME, Methylethylthiophos, methylethylparathion) is a clear oily liquid, b.p. 116° C. at 0.12 mm. of Hg, vapor pressure at 20° C. 7.6×10^{-6} mm. of Hg, volatility 0.11 mg./m.³. Its solubility in water is about 40 mg./liter; it is soluble in most organic solvents except the paraffinic hydrocarbons, d_4^{20} 1.3182, n_D^{20} 1.5460. The LD_{50} for rats is 6–8 mg./kg. The resorption toxicity is somewhat lower than that of parathion, and the maximum permissible concentration in air is 0.03 mg./m.³.

The chemical properties of methylethylparathion are similar to those of parathion. Thus, its rate of hydrolysis occupies an intermediate position between the hydrolysis rates of parathion and methyl parathion. A similar situation is also observed in the methylation of organic compounds and in thermal decomposition. However, because of the presence of a methoxyl group in methylethylparathion its chemical properties are closer to those of methyl parathion.

The Pishchemuka isomerization proceeds mainly with the formation of O-ethyl S-methyl O-4-nitrophenyl thiophosphate:

$$\begin{array}{c} CH_3O \\ C_2H_5O \end{array} \!\!> \!\!\overset{S}{\underset{\|}{P}}\!\!-\!\!O\!\!-\!\!\!\!\bigcirc\!\!\!\!-\!\!NO_2 \xrightarrow{t°} \begin{array}{c} CH_3S \\ C_2H_5O \end{array} \!\!> \!\!\overset{O}{\underset{\|}{P}}\!\!-\!\!O\!\!-\!\!\!\!\bigcirc\!\!\!\!-\!\!NO_2$$

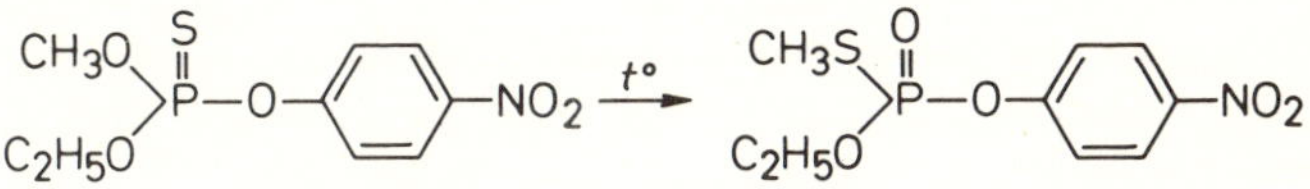

The presence of the isomeric O-methyl S-ethyl O-4-nitrophenyl thiophosphate in the reaction mixture is insignificant.

On plants and in other living organisms methylethylparathion breaks down somewhat more quickly than parathion. The paths of their conversions are identical. It is marketed in the form of a 20% emulsive concentrate that is used similarly to parathion.

It is produced by the condensation of methyl ethyl chlorothiophosphate with p-nitrophenol and an HCl acceptor in an organic solvent or with sodium p-nitrophenolate in aqueous medium in the presence of emulsifiers (yield 80–90%). Surface-active agents of the type of OP-7 and ammonium or amine naphthenates are used as emulsifiers:

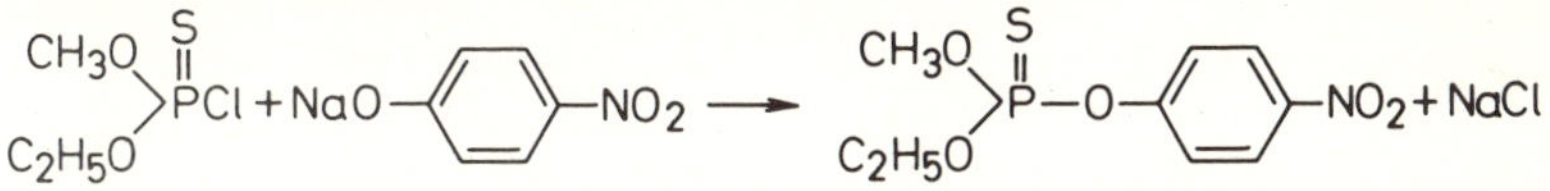

$$\begin{array}{c} CH_3O \\ C_2H_5O \end{array} \!\!> \!\!\overset{S}{\underset{\|}{P}}\!Cl + NaO\!\!-\!\!\!\!\bigcirc\!\!\!\!-\!\!NO_2 \longrightarrow \begin{array}{c} CH_3O \\ C_2H_5O \end{array} \!\!> \!\!\overset{S}{\underset{\|}{P}}\!\!-\!\!O\!\!-\!\!\!\!\bigcirc\!\!\!\!-\!\!NO_2 + NaCl$$

The compound that is produced in this reaction contains no less than 85% of the active ingredient. Impurities are small amounts of p-nitrophenol, ethyl dimethyl thiophosphate, and some other compounds. The compound was first synthesized and its production started in the Soviet Union.

The methyl ethyl chlorothiophosphate necessary for the production of methylethylparathion is obtained by the following reaction:

$$C_2H_5OH + PSCl_3 \rightarrow C_2H_5O\overset{S}{\underset{\|}{P}}Cl_2 + HCl$$

$$C_2H_5O\overset{S}{\underset{\|}{P}}Cl_2 + CH_3OH + NaOH \rightarrow NaCl + H_2O + \begin{array}{c} CH_3O \\ C_2H_5O \end{array}\!\!>\!\!\overset{S}{\underset{\|}{P}}Cl$$

The yield of methyl ethyl chlorothiophosphate is more than 85% calculated on phosphorus thiotrichloride. It is possible to use for the reaction a 40% aqueous solution of NaOH.

Because of the ready availability of methyl ethyl chlorothiophosphate, methylethylparathion is somewhat cheaper than parathion.

O,O-Dimethyl O-4-nitro-3-methylphenyl thiophosphate (methylnitrophos, metathion, Folithion, Sumithion, fenitrothion) is a clear liquid with an unpleasant odor, b.p. 95° C. at 0.01 mm. of Hg, vapor pressure at 20° C. 6.0×10^{-6} mm. of Hg, volatility 0.09 mg./m.^{3}, viscosity at 30° C. 20.8 centipoises, d_4^{20} 1.308, n_D^{20} 1.5505. Its solubility in water is about 30 mg./liter; it is highly soluble in most organic solvents and is miscible in all proportions with methyl and ethyl alcohols, alkyl acetates, ketones, and aromatic hydrocarbons. Its solubility in kerosine is about 4% and in petroleum ether about 7%. The LD_{50} for various experimental animals ranges from 142 to 1,000 mg./kg. The nature of its action on animals is similar to that of methyl parathion.

The chemical properties of fenitrothion practically do not differ from those of methyl parathion, but its rate of hydrolysis by water and alkalies is somewhat lower. Thus, in 0.1 N NaOH solution at 30° C., 50% of methyl parathion is hydrolyzed in 5 minutes, and 50% of fenitrothion in 12 minutes.

The thermal stability of this compound also is low, and when it is heated above 100° C. it undergoes Pishchemuka isomerization and may decompose explosively. Because of this, overheating of the compound must be avoided both in its production and in storage. The compound must be stored in enameled, aluminum, or glass containers. Iron promotes decomposition of the compound like that of most other organophosphorus compounds.

O,O-Dimethyl O-4-nitro-3-methylphenyl thiophosphate is obtained by the condensation of dimethyl chlorothiophosphate with sodium 4-nitro-3-methylphenolate in aqueous medium or of free nitrocresol in the presence of anhydrous potassium carbonate in acetone or methyl ethyl ketone:

$$(CH_3O)_2\overset{S}{\overset{\|}{P}}Cl + NO_2-\underset{}{\bigcirc}\!\!-ONa \longrightarrow (CH_3O)_2\overset{S}{\overset{\|}{P}}-O-\underset{}{\bigcirc}\!\!-NO_2 + NaCl$$

The synthesis of 4-nitro-3-methylphenol presents appreciable difficulties, since on direct nitration of *m*-cresol a mixture of the 4-nitro- and 6-nitro-isomers is formed that contains not more than 60% of the desired product. It is possible to produce quite pure 4-nitro-3-methylphenol either by nitration of esters of *m*-cresol with carboxylic acids or by oxidation of the corresponding nitrosophenol:

$$H_3C-\underset{}{\bigcirc}\!\!-OH + HNO_2 \longrightarrow H_3C-\underset{ON}{\bigcirc}\!\!-OH \xrightarrow{[O]} H_3C-\underset{O_2N}{\bigcirc}\!\!-OH$$

However, by both the first and second of these methods up to 30% of the valuable *m*-cresol is lost.

Systematic study of the insecticidal properties of mixtures of various compounds established that O,O-dimethyl O-6-nitro-3-methylphenyl thiophosphate is an active synergist for many organophosphorus insecticides: parathion, methyl parathion, methylethylparathion, and fenitrothion.

Each of the organophosphorus compounds listed, when mixed with O,O-dimethyl O-6-nitro-3-methylphenyl thiophosphate in a 1:1 ratio, is effective at the same concentrations as the pure individual compound. For use in agriculture a mixture of O,O-dimethyl O-4-nitro-3-methylphenyl thiophosphate with O,O-dimethyl O-6-nitro-3-methylphenyl thiophosphate has been recommended, which is obtained by the reaction of dimethyl chlorothiophosphate with isomeric nitrocresols that are a product of the direct nitration of *m*-cresol by nitric acid. This mixture practically does not differ in insecticidal and acaricidal activity from pure O,O-dimethyl O-4-nitro-3-methylphenyl thiophosphate. In the Soviet Union this preparation is called methylnitrophos. The compound is marketed in the form of a 30% emulsive concentrate.

O,O-Dimethyl O-4-nitro-3-chlorophenyl thiophosphate (Chlorthion) is a yellow crystalline substance, m.p. 21° C., d_4^{20} 1.437. It is highly soluble in aromatic hydrocarbons and their halogen derivatives, but slightly soluble in paraffinic hydrocarbons. Its solubility in water at 20° C. is 40 mg./liter.

The chemical properties of Chlorthion are similar to those of methyl parathion. Thus, 50% is hydrolyzed at 20° C. and pH 1–5 in 138 days, and at 70° C. and pH 5 in 5.3 hours. When acted on by oxidizing agents it is converted to the corresponding ester of phosphoric acid. The LD_{50} for rats is 880–980 mg./kg. It is somewhat less active insecticidally than methyl parathion and is marketed in the form of 10 and 50% emulsive concentrates and a 20% wettable powder.

Chlorthion is synthesized by the reaction of dimethyl chlorothiophosphate with 3-chloro-4-nitrophenol in methyl ethyl ketone in the presence of anhydrous potassium carbonate:

$$(CH_3O)_2\overset{S}{\overset{\|}{P}}Cl + HO-\underset{}{\bigcirc}\overset{Cl}{\underset{}{}}-NO_2 + K_2CO_3 \longrightarrow$$

$$(CH_3O)_2\overset{S}{\overset{\|}{P}}-O-\underset{}{\bigcirc}\overset{Cl}{\underset{}{}}-NO_2 + KHCO_3 + KCl$$

The yield is about 90% and the compound obtained is 97% pure. It is very important for good yields that the potassium carbonate be pulverized; the finer the grinding of the potassium carbonate, the more quickly the reaction takes place and the higher the yield of the final compound.

O,O-Dimethyl 2-chloro-4-nitrophenyl thiophosphate (dicapthon) is a white crystalline powder, m.p. 52°–53° C., highly soluble in aromatic hydrocarbons, halogenated hydrocarbons of the aliphatic and aromatic series, esters of carboxylic acids, and ketones, and slightly soluble in paraffinic hydrocarbons. Its solubility in water is about 35 mg./liter. The LD_{50} for rats is 330–400 mg./kg.

Its chemical properties are completely analogous to those of Chlorthion. Dicapthon is produced from dimethyl chlorothiophosphate and 2-chloro-

4-nitrophenol in toluene in the presence of potassium carbonate; the yield is 90%. It is used mainly in the form of emulsions or suspensions to control flies. Both Chlorthion and dicapthon have not yet been widely used.

O,O-Diethyl O-2,4-dichlorophenyl thiophosphate (Nemacide, V-C 13 Nemacide) is a clear oily liquid, b.p. 108° C. at 0.01 mm. of Hg, d_4^{20} 1.313. It is highly soluble in most organic solvents; its solubility in water is 245 mg./liter. The LD_{50} for rats is 270 mg./kg. Nemacide is more resistant to heating than the mixed esters that contain a nitro group in the aromatic radical; thus, it withstands heating at 120° C. for 24 hours.

Oxidizing agents convert it to the corresponding ester of phosphoric acid, like other mixed esters of thiophosphoric acid. It is similar to parathion in rate of hydrolysis.

Nemacide can be made by the reaction of diethyl chlorothiophosphate with sodium 2,4-dichlorophenolate in aqueous medium (yield about 80%) or by the reaction of diethyl chlorothiophosphate with 2,4-dichlorophenol in acetonitrile or other suitable organic solvent in the presence of potassium carbonate (yield 94%).

To control nematodes the compound is introduced into the soil at dosages on the order of 200 kg./ha. Since the compound does not have a systemic effect, nematodes that have penetrated into the roots are not destroyed. Nemacide is still used as yet on a limited scale, but it is an interesting example of organophosphorus nematicides.

O,O-Dimethyl O-2,4,5-trichlorophenyl thiophosphate (ronnel, Trolene, Etrolene, Dow ET-14, Dow ET-57, Korlan, Nankor, trichlorometafos) is a white crystalline substance, m.p. 41° C., b.p. 97° C. at 0.01 mm. of Hg, d_4^{25} 1.4850. The vapor pressure at 25° C. is 0.8×10^{-3} mm. of Hg. It is highly soluble in most organic solvents; the solubility in water is 44 mg./liter and it is stable at temperatures up to 80° C. In weakly alkaline medium it is hydrolyzed with the formation mainly of O-methyl O-2,4,5-trichlorophenyl thiophosphoric acid, while in strongly alkaline medium it is mainly O,O-dimethylthiophosphoric acid that is formed:

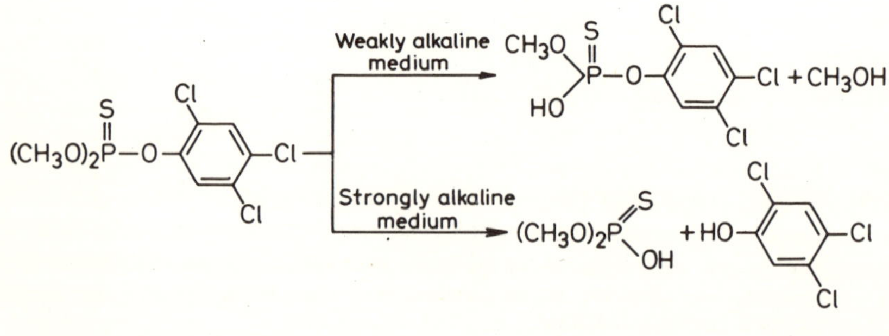

In the bovine organism and in rats about half of the compound introduced is excreted with the urine in the form of O-methyl O-2,4,5-trichlorophenylthiophosphoric acid, while in insects it breaks down predominantly to dimethylthiophosphoric acid and 2,4,5-trichlorophenol.

Ronnel is one of the compounds that are low in toxicity to warm-blooded animals, and it is used to control ectoparasites of domestic cattle both by application to the skin in the form of emulsions and by introduction into the stomach through the feed. The LD_{50} for animals is in the range 400–3,000 mg./kg.

Several methods have been described for making it, the most important of which are the following:

1. Reaction of sodium 2,4,5-trichlorophenolate with dimethyl chlorothiophosphate in aqueous medium in the presence of emulsifiers:

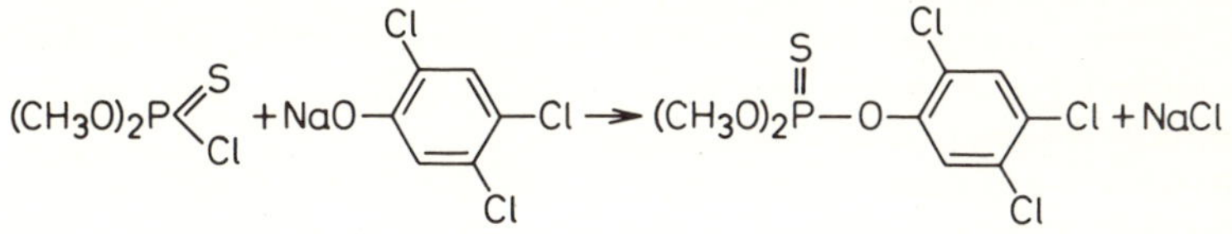

$$(CH_3O)_2P(S)Cl + NaO\text{-}C_6H_2Cl_3 \rightarrow (CH_3O)_2P(=S)\text{-}O\text{-}C_6H_2Cl_3 + NaCl$$

The yield of ronnel by this reaction is 85–90%; it contains as an impurity some 2,4,5-trichlorophenol, which can be washed out with a small amount of NaOH.

2. Reaction of dimethyl chlorothiophosphate with 2,4,5-trichlorophenol in methyl ethyl ketone in the presence of finely ground potassium carbonate with a yield of up to 80%.

3. Methanolysis of O-2,4,5-trichlorophenyl dichlorothiophosphate in the presence of NaOH:

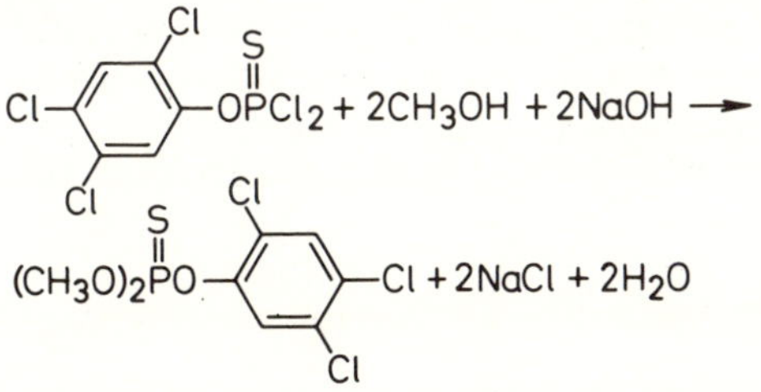

$$Cl_3C_6H_2\text{-}OP(=S)Cl_2 + 2CH_3OH + 2NaOH \rightarrow$$
$$(CH_3O)_2P(=S)O\text{-}C_6H_2Cl_3 + 2NaCl + 2H_2O$$

The O-2,4,5-trichlorophenyl dichlorothiophosphate necessary for this synthesis is obtained by the following reaction:

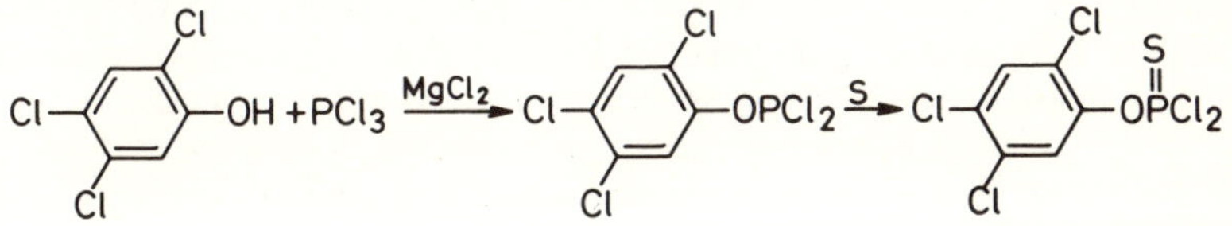

$$Cl_3C_6H_2\text{-}OH + PCl_3 \xrightarrow{MgCl_2} Cl_3C_6H_2\text{-}OPCl_2 \xrightarrow{S} Cl_3C_6H_2\text{-}OP(=S)Cl_2$$

O-2,4,5-Trichlorophenyl dichlorophosphite is produced in about 90% yield by prolonged heating of 2,4,5-trichlorophenol and phosphorus trichloride (used in large excess) with a catalytic quantity of anhydrous magnesium chloride. The addition of sulfur is accomplished without isolation of the O-2,4,5-trichlorophenyl dichlorophosphite in the pure form, and only after the second reaction has been carried out is the phosphorus trichloride distilled off. The distilled phosphorus trichloride is returned to the process for the production of O-2,4,5-trichlorophenyl dichlorophosphite.

In the United States the compound is marketed in two degrees of purity: about 98–99% and about 95%.

O-Methyl O-ethyl O-2,4,5-trichlorophenyl thiophosphate (trichlormetafos-3). The closest homolog of ronnel is trichlormetafos-3, which was first synthesized and proposed for use in agriculture in the Soviet Union. Pure trichlormetafos-3 is a colorless oily liquid, b.p. 127° C. at 0.15 mm. of Hg, vapor pressure at 20° C. about 0.6×10^{-3} mm. of Hg, volatility about 8 mg./m.^{3} It is highly soluble in most organic solvents; its solubility in water is less than 40 mg./liter, d_4^{20} 1.4345, n_D^{20} 1.5520. The LD_{50} for various experimental animals is 330–800 mg./kg.

It is used to control warble flies of cattle both by external application and by introduction into the stomach at dosages of the order of 35–40 mg./kg. live weight of the animal. At these dosages the compound is eliminated relatively rapidly from the animal's body and does not remain in the milk and meat. It is recommended also for the control of plant pests and is effective against many sucking insects and mites at a concentration of 0.1%. At these concentrations it does not damage plants (in contrast to ronnel, which is strongly phytocidal). Trichlormetafos-3 is used to control insect pests and flies harmful to public health. It is marketed in the form of 30 and 50% emulsive concentrates containing emulsifiers in addition to the active ingredient.

The principal method for producing trichlormetafos-3 with a yield of 85% is the reaction of O-methyl O-ethyl chlorothiophosphate with sodium 2,4,5-trichlorophenolate in aqueous medium in the presence of emulsifiers:

$$\begin{array}{c}CH_3O \\ C_2H_5O\end{array}\!\!>\!\!\overset{\overset{S}{\|}}{P}Cl + NaO\!-\!\!\underset{Cl}{\overset{Cl}{\diagdown}}\!\!-\!Cl \longrightarrow \begin{array}{c}CH_3O \\ C_2H_5O\end{array}\!\!>\!\!\overset{\overset{S}{\|}}{P}O\!-\!\!\underset{Cl}{\overset{Cl}{\diagdown}}\!\!-\!Cl + NaCl$$

It contains as impurities a small amount of 2,4,5-trichlorophenol, dimethyl ethyl thiophosphate, and some other compounds. To obtain trichlormetafos-3 of high quality it is very important that the starting materials, primarily the methyl ethyl chlorothiophosphate and 2,4,5-trichlorophenol, be pure. The former should not contain diethyl chlorothiophosphate as an impurity, and the latter should not contain dichlorophenol and isomeric trichlorophenols.

The chemical properties of trichlormetafos-3 are similar to those of ronnel, but it differs from ronnel in being somewhat more resistant to hydrolysis and reacting more slowly with ammonia. The hydrolysis of trichlormetafos-3 proceeds in a manner completely analogous to the hydrolysis of ronnel.

O,O-Diethyl O-2,4,5-trichlorophenyl thiophosphate has considerably higher toxicity to mammals than ronnel and trichlormetafos-3.

O,O-Dimethyl O-2,5-dichloro-4-bromophenyl thiophosphate (bromophos) is a white crystalline substance, m.p. 54° C., vapor pressure at 20° C. 1.3×10^{-4} mm. of Hg, solubility in water at 20° C. about 40 mg./liter, solubility in organic solvents at 20° C. in g./100 g.: acetone 109, diesel fuel 21, isopropyl alcohol 8, methyl alcohol 10, methyl ethyl ketone 120, xylene 90, chlorobenzene 98, and methylene chloride 112. The LD_{50} for various species of animals is 2,000–4,000 mg./kg.

In alkaline medium bromophos is hydrolyzed like ronnel with the formation of O,O-dimethylthiophosphoric acid, 2,5-dichloro-4-bromophenol, and O-methyl O-2,5-dichloro-4-bromophenyl phosphoric acid:

$$(CH_3O)_2\overset{S}{\overset{\|}{P}}-O-\underset{Cl}{\overset{Cl}{\bigcirc}}-Br + H_2O \longrightarrow \begin{cases} (CH_3O)_2P\overset{S}{\underset{OH}{\diagdown}} + HO-\underset{Cl}{\overset{Cl}{\bigcirc}}-Br \\ \underset{HO}{\overset{CH_3O}{\diagdown}}\overset{S}{\overset{\|}{P}}-O-\underset{Cl}{\overset{Cl}{\bigcirc}}-Br + CH_3OH \end{cases}$$

At 22° C., 50% is hydrolyzed at pH 13 in 3.5 hours. On plants, bromophos is practically completely broken down in 13–20 days, depending on the species.

Bromophos can be synthesized by all the methods that have been described for ronnel. The greatest difficulty is the preparation of the 2,5-dichloro-4-bromophenol, which apparently is best carried out by bromination of 2,5-dichlorophenol. In turn the 2,5-dichlorophenol is made by hydrolysis of 1,2,4-trichlorobenzene with NaOH in methanol at 160°–190° C., under pressure. In this case a mixture of 2,4- and 2,5-dichlorophenols is formed that is separated by fractional precipitation from alkaline solutions, since the isomers have different dissociation constants.

Bromophos has been proposed for the control of a wide variety of species of insect pests, including flies and other insects affecting public health. It is used in the form of 20 and 25% emulsive concentrates, 5% granulated preparations, and aerosols.

The closest homolog of bromophos is bromophos-ethyl, which is an oily liquid, b.p. 122°–123° C. at 10^{-3} mm. of Hg, vapor pressure at 30° C. 4.2×10^{-5} mm. of Hg. The LD_{50} of bromophos-ethyl for rats is 238 mg./kg. It is similar to bromophos in insecticidal activity.

O,O-Dimethyl O-2,5-dichloro-4-iodophenyl thiophosphate (Compound C-9491, iodophos) and O,O-diethyl O-2,5-dichloro-4-iodophenyl thiophosphate (Compound C-8874, iodophos-ethyl) have been proposed for use as insecticides. The first of these is a crystalline substance, m.p. 73° C., LD_{50}

for rats about 2,000 mg./kg.; the second melts at 47°–48° C., and its LD_{50} for rats is about 140 mg./kg. Both compounds have a broad spectrum of action on insects and mites and are being studied in detail.

O,O-Dimethyl O-4-cyanophenyl thiophosphate (cyanox) (m.p. 14° C., LD_{50} for rats about 995 mg./kg.), *P-chloro-2,4-dioxa-5-methyl-P-thiono-3-phosphabicyclo[4.4.0]-decane* (UC 8305) (b.p. 78° C. at 0.2 mm. of Hg, LD_{50} 120 mg./kg.), and *2-methoxy-4H-1,2,3-benzodioxaphosphorine-2-sulfide (salithion)* (m.p. 54°–55° C., LD_{50} for mice 91 mg./kg.) have been proposed for use as insecticides and acaricides:

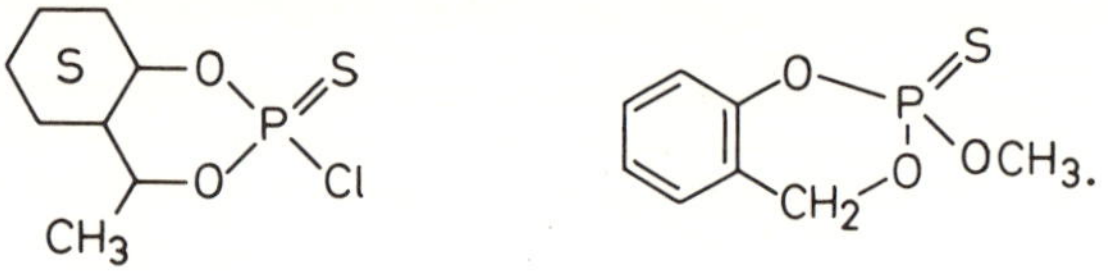

Compound UC 8305   Salithion

O,O-Dimethyl O-(4-methylmercapto-3-methylphenyl) thiophosphate (fenthion, Lebaycid, Baytex) is a colorless oil, b.p. 87° C. at 0.01 mm. of Hg, vapor pressure at 20° C. 3×10^{-5} mm. of Hg, volatility 0.46 mg./m.³ Its solubility in water at 20° C. is about 54 mg./liter; it is highly soluble in most organic solvents, including alcohols, ethers and esters, halogenated aromatic and aliphatic hydrocarbons, and is poorly soluble in petroleum ether. The LD_{50} for rats is 215–245 mg./kg.

Fenthion is more resistant to hydrolysis and heating than methyl parathion. At 80° C., 50% is hydrolyzed in acid medium in 36 hours, and in alkaline medium in 95 minutes. When acted on by oxidizing agents fenthion is oxidized first to the sulfoxide, and then to the sulfone:

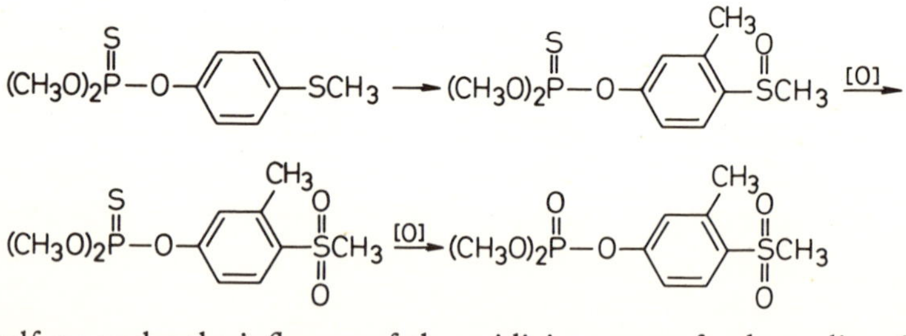

The sulfone under the influence of the oxidizing agents further splits off the thiono sulfur and goes to the corresponding ester of phosphoric acid.

Under natural conditions fenthion is stable and can be stored for a long time. In toxicity to insects fenthion approaches Chlorphos but the selectivity of its action is less, since its toxicity for mammals is higher. In animals and plants fenthion is oxidized to the sulfone, which then is hydrolyzed and breaks down to products that are harmless to animals. It is marketed in the form of a 50% emulsive concentrate, a 25% wettable powder, and a 3% dust.

Fenthion is obtained in 90% yield by the reaction of 4-methylmercapto-3-methylphenol with dimethyl chlorothiophosphate in methyl ethyl ketone at 60° C., in the presence of finely ground potassium carbonate. The 4-methylmercapto-3-methylphenol necessary for this is synthesized from dimethyl sulfoxide and *m*-cresol:

$$(CH_3)_2SO + \underset{CH_3}{\underset{|}{C_6H_3(OH)}} + HCl \longrightarrow \underset{\underset{S^+(CH_3)_2Cl^-}{|}}{\underset{CH_3}{\underset{|}{C_6H_2(OH)}}} + H_2O \longrightarrow \underset{\underset{SCH_3}{|}}{\underset{CH_3}{\underset{|}{C_6H_2(OH)}}} + CH_3Cl$$

4-Methylmercapto-3-methylphenol also can be obtained by the action of dimethyl sulfide and sulfuryl chloride on *m*-cresol:

$$(CH_3S)_2 + SO_2Cl_2 + 2\,\underset{CH_3}{\underset{|}{C_6H_3(OH)}} \longrightarrow 2\,\underset{\underset{SCH_3}{|}}{\underset{CH_3}{\underset{|}{C_6H_2(OH)}}} + SO_2 + 2HCl$$

A large number of analogs and homologs of fenthion have been synthesized and studied, but most of these compounds have a higher toxicity than fenthion for animals. In particular, both O,O-dimethyl O-4-methylmercaptophenyl thiophosphate and its oxygen analog are more toxic. In many mixed esters of thiophosphoric acid the introduction of a methyl group or halogen in the *meta*-position to the ester group lowers the toxicity to vertebrate animals and makes comparatively little change in the insecticidal properties (see fenitrothion, fenthion, Chlorthion, ronnel, trichlormetafos-3, and bromophos).

O,O-Diethyl O-p-(methylsulfinyl)phenyl thiophosphate (fensulfothion, Terracur-p, dasanit) is an oily liquid, b.p. 138°–141° C. at 0.01 mm. of Hg, soluble in most organic solvents, slightly soluble in water. The LD_{50} for rats is 10.5 mg./kg. It is used in the form of emulsions, wettable powders, dusts, and granules to control nematodes and soil insects.

Fensulfothion is produced by condensation of O,O-diethyl chlorothiophosphate with 4-methylsulphinyl-1-hydroxybenzene in the presence of an acid combining reagent:

$$(C_2H_5O)_2PSCl + HO\text{—}C_6H_4\text{—}\overset{O}{\underset{\|}{S}}\text{—}CH_3 \longrightarrow (C_2H_5O)_2\overset{S}{\underset{\|}{P}}\text{—}O\text{—}C_6H_4\text{—}\overset{O}{\underset{\|}{S}}\text{—}CH_3$$

or by reaction of O,O-diethyl chlorothiophosphate with 4-methylthiohydroxybenzene and oxidation of the resulting ester with hydrogen peroxide:

$$(C_2H_5O)_2PSCl + NaO-\langle C_6H_4 \rangle-SCH_3 \rightarrow (C_2H_5O)_2\overset{S}{\underset{\|}{P}}-O-\langle C_6H_4 \rangle-SCH_3$$

$$(C_2H_5O)_2\overset{S}{\underset{\|}{P}}-O-\langle C_6H_4 \rangle-SCH_3 + H_2O_2 \rightarrow (C_2H_5O)_2\overset{S}{\underset{\|}{P}}-O-\langle C_6H_4 \rangle-\overset{O}{\underset{\|}{S}}-CH_3$$

O,O,O',O'-*Tetramethyl* O,O'-*thiodi-p-phenylene phosphorothioate* (*Abate*) is a white crystalline substance, m.p. 30°–30.5° C.; the technical grade (90–95%) is a brown viscous liquid insoluble in water, hexane, methyl cyclohexane and other hydrocarbons, soluble in carbon tetrachloride, ethylene dichloride, lower alkyl ketones, and toluene. In aqueous solution no hydrolysis is observed at pH 8 at room temperature for several weeks or pH 11 at 40° C. for several hours. The LD_{50} is 4,000 mg./kg. It is used to control mosquitoes and agricultural pests and is synthesized by the reaction of dimethyl chlorothiophosphate with 4,4'-dihydroxyphenyl sulfide in 10% NaOH to pH 10–11 at 25°–60° C.:

$$2(CH_3O)_2PSCl + NaO-\langle C_6H_4 \rangle-S-\langle C_6H_4 \rangle-ONa \rightarrow 2NaCl +$$

$$+ (CH_3O)_2\overset{S}{\underset{\|}{P}}-O-\langle C_6H_4 \rangle-S-\langle C_6H_4 \rangle-O-\overset{S}{\underset{\|}{P}}(OCH_3)_2$$

In addition to the mixed esters of thiophosphoric acid that contain aromatic radicals, some use has been made of mixed esters of thiophosphoric acid of the heterocyclic series as insecticides and acaricides. Individual representatives of this series are effective and are widely used.

O,O-*Diethyl* O-*(2-isopropyl-4-methylpyrimidyl-6) thiophosphate* (*diazinon*) is a colorless oil, b.p. 89° C. at 0.1 mm. of Hg, vapor pressure at 20° C. 8.4×10^{-5} mm. of Hg, volatility 1.39 mg./m.^{3}, d_4^{20} 1.115, solubility in water at 20° C. 40 mg./liter. It is highly soluble in most organic solvents. The LD_{50} for various experimental animals is 76–320 mg./kg.

Diazinon is not as resistant to hydrolysis as parathion. In acid medium it is hydrolyzed 12 times as rapidly as parathion, and in alkaline medium the hydrolysis goes on at practically the same rate. In an excess of water the principal products of hydrolysis of diazinon are diethylthiophosphoric acid and 2-isopropyl-4-methyl-6-hydroxypyrimidine, but with insufficient water in acid medium a small amount of tetraethyl dithio- and thiopyrophosphates is formed.

It is marketed in the form of emulsive concentrates, wettable powders, dusts, and granulated formulations. It is used to control various plant pests and animal parasites. In spite of its relatively high toxicity to mammals, diazinon is considered in most countries as a safe compound, and it usually is assigned to the third grouping of toxic compounds.

It is produced in 85% yield by the reaction of 2-isopropyl-4-methyl-6-hydroxypyrimidine with diethyl chlorothiophosphate in the presence of potassium carbonate:

$$(C_2H_5O)_2\overset{\underset{\|}{S}}{P}Cl + \underset{iso\text{-}C_3H_7}{\text{[4-methyl-6-hydroxypyrimidine]}} \xrightarrow{K_2CO_3} (C_2H_5O)_2\overset{\underset{\|}{S}}{P}\text{-O-[pyrimidyl]-}C_3H_7\text{-}iso$$

This reaction goes well in dimethylformamide solution at 50°–60° C. 2-Isopropyl-4-methyl-6-hydroxypyrimidine can be synthesized in satisfactory yields by the following route:

$$(CH_3)_2CHCN + HCl + CH_3OH \longrightarrow (CH_3)_2CHC\begin{matrix}NH\cdot HCl\\OCH_3\end{matrix} \xrightarrow{+NH_3}$$

$$(CH_3)_2CHC\begin{matrix}NH\cdot HCl\\NH_2\end{matrix} \xrightarrow{CH_3COCH_2COOCH_3,\, NaOH} \text{[2-isopropyl-4-methyl-6-hydroxypyrimidine]}$$

O,O-Diethyl O-3,5,6-trichloropyridyl thiophosphate (Dursban) is a white crystalline substance, m.p. 41.5°–43° C., vapor pressure at 25° C. 1.9×10^{-5} mm. of Hg; it is highly soluble in most organic solvents, but almost insoluble in water. The LD_{50} is 150 mg./kg. for rats, and 500 mg./kg. for guinea pigs. In acid or alkaline media the compound is slowly hydrolyzed by water, forming diethylthiophosphoric and ethylthiophosphoric acids and trichlorohydroxypyridine. By oxidation it may be converted to the corresponding phosphate.

Dursban is an active insecticide for the control of sucking and chewing plant pests as well as household parasites. The duration of action of the compound on different surfaces is 6–11 weeks, but on the leaves of plants it is short-acting. This apparently is due to its rapid hydrolysis under the influence of enzymes; the compound is active on grain plants, however, for several weeks. Dursban is especially effective in controlling mosquito larvae. When swamps are treated with a dosage of 0.5 kg./ha., it gives 100% mortality of mosquitoes for two months. There are reports that it is effective against soil-inhabiting plant pests.

Dursban is synthesized by the reaction of the sodium salt of 3,5,6-trichloro-2-hydroxypyridine with diethyl chlorothiophosphate in dimethylformamide:

$$(C_2H_5O)_2\overset{\underset{\|}{S}}{P}Cl + \text{[3,5,6-trichloro-2-NaO-pyridine]} \longrightarrow (C_2H_5O)_2\overset{\underset{\|}{S}}{P}\text{-O-[3,5,6-trichloropyridyl]}$$

The yield from this reaction is 80–85%.

3,5,6-Trichloro-2-hydroxypyridine is produced by the hydrolysis of 2,3,5,6-tetrachloropyridine with NaOH at an elevated temperature:

O,O-Diethyl O-pyrazinyl thiophosphate (thionazin, Zinophos) is a clear liquid, vapor pressure at 30° C. 3×10^{-3} mm. of Hg. It is highly soluble in organic solvents; in water at 25° C. the solubility is 1.14 g./liter. The LD_{50} for rats is 10.7 mg./kg. Its chemical properties recall those of other thiophosphates.

Thionazin is a systemic nematicide, acaricide, and insecticide. At a dosage of 5–10 kg./ha. it gives a satisfactory effect in the eradication of plant-feeding nematodes. The insecticidal and acaricidal properties of the compound are evident at dosages of the order of 1 kg./ha.

It is produced by the reaction of diethyl chlorothiophosphate with the sodium derivative of 2-hydroxypyrazine:

O,O-Diethyl O-(4-methylcoumarinyl-7) thiophosphate (Potasan, E-838) is a white crystalline substance, m.p. 38° C. It is practically insoluble in water, but highly soluble in most organic solvents. At pH 5–8 it is resistant to the action of water, and only when the pH is increased is its hydrolysis substantially accelerated. The LD_{50} for rats is 19–42 mg./kg. It is a selective insecticide and has been proposed for the control of the Colorado potato beetle, but it has only limited use in combination with lindane.

The hydrolysis of Potasan takes place at the ethyl group, and the first product of hydrolysis that is formed is O-ethyl O-(4-methylcoumarinyl-7) thiophosphoric acid, which has low toxicity for both insects and animals. It is synthesized in >95% yield by prolonged heating at 80°–90° C. of diethyl chlorothiophosphate with methylumbelliferone in chlorobenzene in the presence of potassium carbonate:

Methylumbelliferone can be obtained in good yield by the condensation of resorcinol with acetoacetic ester:

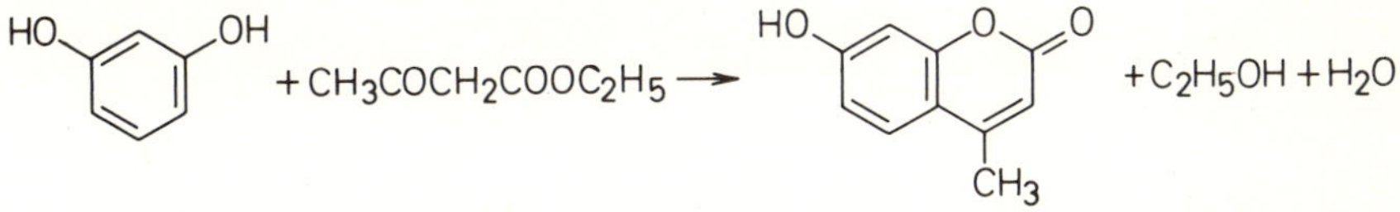

HO—C₆H₄—OH + CH₃COCH₂COOC₂H₅ → [7-hydroxy-4-methylcoumarin] + C₂H₅OH + H₂O

The closest analog of Potasan is Dition, which is less toxic to mammals. It has been produced on a small scale in Italy.

O,O-Diethyl O-(3,4-tetramethylenecoumarinyl-7) thiophosphate (Dition) is a white crystalline substance with a weak odor, m.p. 88°–89° C.; it is insoluble in water, but highly soluble in most organic solvents. The LD_{50} of the technical grade product for rats is about 150 mg./kg. Dition is similar to Potasan in its chemical properties. It has been recommended for use as an insecticide of low toxicity for the control of some insect pests affecting public health. It is produced by the reaction of diethyl chlorothiophosphate with 3,4-tetramethyleneumbelliferone in the presence of one mole of sodium ethylate:

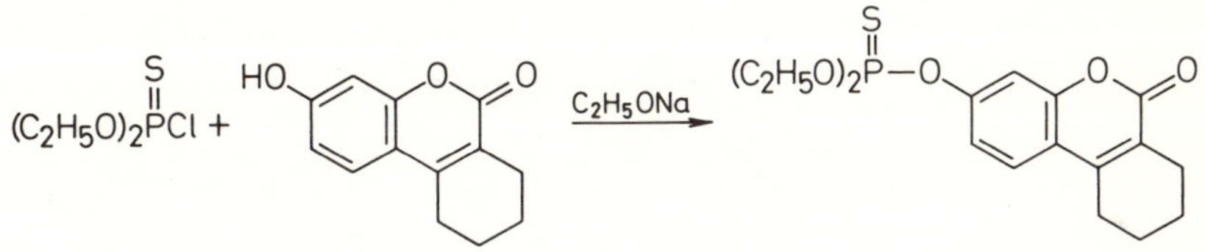

$(C_2H_5O)_2\overset{S}{\underset{\|}{P}}Cl$ + [3,4-tetramethyleneumbelliferone] $\xrightarrow{C_2H_5ONa}$ $(C_2H_5O)_2\overset{S}{\underset{\|}{P}}-O-$[coumarin ring system]

The corresponding thiophosphates obtained from 3,4-trimethyleneumbelliferone, which have approximately the same toxicity to animals and insects, also have been patented. However, practical use of these compounds still is small. A thiophosphate based on chloroumbelliferone is used in animal husbandry under the names of coumaphos (see below).

O,O-Diethyl O-(3-chloro-4-methylcoumarinyl-7) thiophosphate (coumaphos, Co-Ral, Asunthol, Resitox, Muscatox) is a white crystalline substance, m.p. 95° C., d_4^{20} 1.474, solubility in water at 20° C. about 1.5 mg./liter. It is slightly soluble in most organic solvents, but rather highly soluble in the lower esters of carboxylic acids and ketones. The LD_{50} for various experimental animals is 55–200 mg./kg. It is resistant to hydrolysis in alkaline and acid media and withstands boiling with 8% sodium carbonate solution for two hours. Its reaction with dilute and concentrated alkali follows different routes. The pyrone ring is opened by the action of dilute KOH. After subsequent acidification the ring again closes and the starting product is recovered in unchanged form. However, when coumaphos is subjected to the prolonged action of dilute alkali, opening of the ring occurs with partial splitting off of an ethoxy group. This process is irreversible.

Upon heating with concentrated alkali coumaphos is completely decomposed:

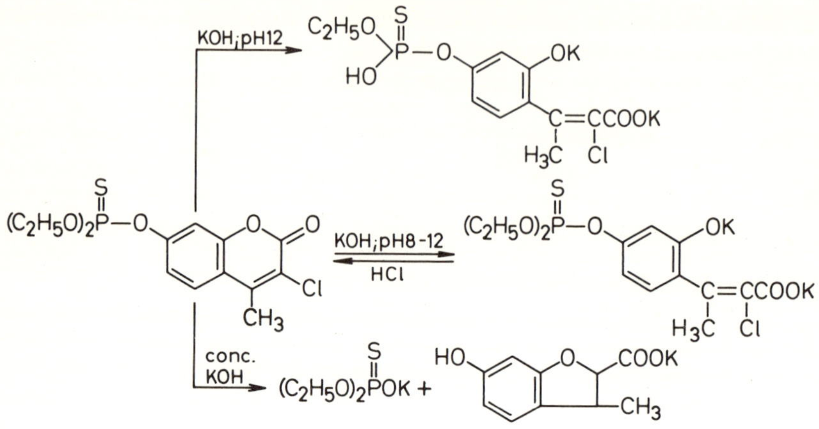

When coumaphos is oxidized with nitric acid or other oxidizing agents, O,O-diethyl O-(3-chloro-4-methylcoumarinyl-7) phosphate is formed, which is known as Coroxon:

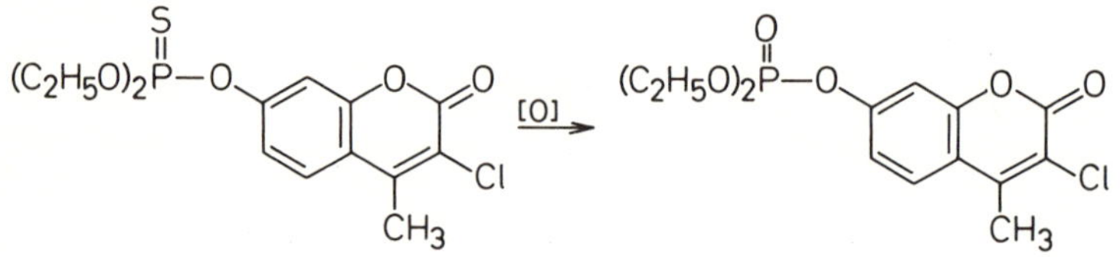

In the United States, Coroxon is used as an agent to control intestinal worms in domestic animals, in a mixture with phenothiazine. Coroxon is a white crystalline substance, m.p. 65° C.

Coumaphos is a valuable insecticide for the control of ectoparasites of domestic animals. In the animal body it is quickly broken down and converted to diethyl- and ethyl thiophosphoric and phosphoric acids, and also to H_3PO_4. Because of the low toxicity of coumaphos for fish, it is of interest also as an agent for the control of mosquito larvae.

It is produced in more than 90% yield by the reaction of diethyl chlorothiophosphate with 3-chloro-7-hydroxy-4-methylcoumarin in the presence of potassium carbonate:

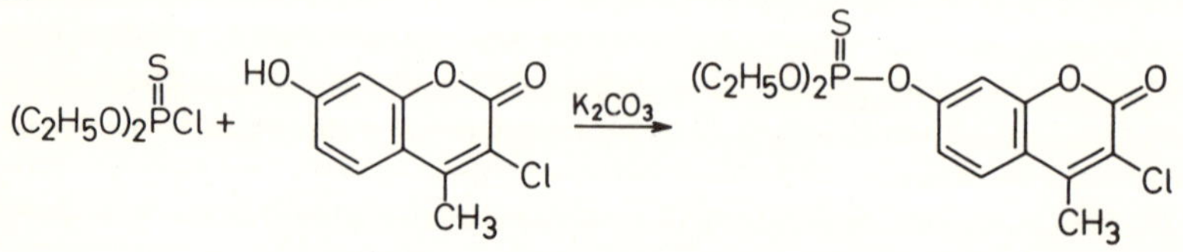

The reaction proceeds readily in methyl ethyl ketone at 60°–70° C., with thorough stirring.

Of the mixed esters of thiophosphoric acid containing heterocyclic radicals, mention should be made of endothion.

O,O-Dimethyl S-(5-methoxypyronyl-2-methyl) thiophosphate (endothion, Phosphopyrone) is a white crystalline substance, m.p. 90°–91° C. It is highly soluble in water (150 g. in 100 ml. of water), but only slightly soluble in hydrocarbons and oils. The LD_{50} for rats is 30–50 mg./kg. When acted on by alkalies, endothion is hydrolyzed to form as the first product O-methyl S-(5-methoxypyronyl-2-methyl) thiophosphate, which has low toxicity for animals and insects. This same product is obtained as a result of the metabolism of endothion in plants.

Endothion is a systemic insecticide with an effective duration of 8–14 days. It is marketed in the form of a 20% concentrate and a 50% wettable powder which are used to protect plants at a concentration of 0.05–0.075% active ingredient.

It is produced by the reaction of sodium O,O-dimethylthiophosphate with 5-methoxy-2-chloromethylpyrone in acetone or other suitable solvent:

$$(CH_3O)_2P\overset{O}{\underset{SNa}{\diagup}} + ClCH_2\text{-pyrone-}OCH_3 \longrightarrow$$

$$(CH_3O)_2\overset{O}{\overset{\|}{P}}SCH_2\text{-pyrone-}OCH_3 + NaCl$$

Of considerably greater interest as systemic insecticides and acaricides are the mixed esters of thiophosphoric acid that contain a sulfide or sulfoxyl group in one of the ester radicals. At present more than five compounds of this type are used in agriculture.

Demeton (Systox, mercaptophos). The technical grade product is a mixture of the thiono (70%) and thiolo (30%) isomers of O,O-diethyl 2-ethylmercaptoethyl thiophosphate. The thiono isomer is an oily liquid, b.p. 106° C. at 0.4 mm. of Hg, vapor pressure at 20° C. 2.48×10^{-4} mm. of Hg, volatility 3.5 mg./m.³, d_4^{20} 1.1193. At 20° C., its solubility in water is 60 mg./liter; it is highly soluble in organic solvents, including the petroleum hydrocarbons. The thiolo isomer has b.p. 100° C. at 0.25 mm. of Hg, vapor pressure at 20° C. 2.6×10^{-4} mm. of Hg, volatility 3.67 mg./m.³, d_4^{20} 1.1325. Its solubility in water at 20° C. is about 2,000 mg./liter; it is highly soluble in most organic solvents. The LD_{50} of the technical grade mixture of isomers for rats is 7–10 mg./kg. The LD_{50} of the thiono isomer is 30 mg./kg., and that of the thiolo isomer is 1.5 mg./kg. The maximum permissible concentration in air is 0.02 mg./m.³[3]

The thiono isomer of demeton is the more resistant to hydrolysis in alkaline medium (time for 50% hydrolysis of thiono isomer at 20° C. and pH 13 is 75 minutes and that of the thiolo isomer 0.85 minute); in acid medium the thiolo isomer is hydrolyzed half as fast as the thiono isomer.

One of the interesting reactions of demeton is the ability of the thiono isomer to rearrange to the thiolo isomer; this reaction takes place more rapidly at an elevated temperature and in polar solvents:

$$(C_2H_5O)_2\overset{\underset{\|}{S}}{P}OCH_2CH_2SC_2H_5 \rightarrow (C_2H_5O)_2\overset{\underset{\|}{O}}{P}SCH_2CH_2SC_2H_5$$

It is assumed that the rearrangement proceeds through the intermediate formation of a sulfonium ion:

$$\left[(C_2H_5O)_2P\overset{S}{\underset{O}{\diagdown}}\right]^{-}\left[\begin{array}{c}H_2C\\|\\H_2C\end{array}\!\!\!\diagdown SC_2H_5\right]^{+}$$

This assumption is supported by the fact that O,O-diethyl O-2-ethylsulfonylethyl thiophosphate does not rearrange to the corresponding thiol. The rearrangement of the thiono isomer of demeton also takes place readily in plant tissues. The metabolism of demeton in plants proceeds according to the following general scheme:

$$\begin{array}{c}
(C_2H_5O)_2\overset{\underset{\|}{S}}{P}OCH_2CH_2SC_2H_5 \rightarrow (C_2H_5O)_2\overset{\underset{\|}{S}}{P}OCH_2CH_2\overset{\underset{\|}{O}}{S}C_2H_5 \rightarrow (C_2H_5O)_2\overset{\underset{\|}{S}}{P}OCH_2CH_2\overset{\underset{\|}{\underset{\|}{O}}{\overset{O}{}}}{S}C_2H_5\\
\downarrow \qquad\qquad\qquad \downarrow \qquad\qquad\qquad \downarrow\\
(C_2H_5O)_2\overset{\underset{\|}{O}}{P}OCH_2CH_2SC_2H_5 \rightarrow (C_2H_5O)_2\overset{\underset{\|}{O}}{P}OCH_2CH_2\overset{\underset{\|}{O}}{S}C_2H_5 \rightarrow (C_2H_5O)_2\overset{\underset{\|}{O}}{P}OCH_2CH_2\overset{\underset{\|}{\underset{\|}{O}}{\overset{O}{}}}{S}C_2H_5\\
\downarrow \qquad\qquad\qquad \qquad\qquad\qquad \\
(C_2H_5O)_2\overset{\underset{\|}{O}}{P}SCH_2CH_2SC_2H_5 \rightarrow (C_2H_5O)_2\overset{\underset{\|}{O}}{P}SCH_2CH_2\overset{\underset{\|}{O}}{S}C_2H_5 \rightarrow (C_2H_5O)_2\overset{\underset{\|}{O}}{P}SCH_2CH_2\overset{\underset{\|}{\underset{\|}{O}}{\overset{O}{}}}{S}C_2H_5
\end{array}$$

Then hydrolysis occurs and further oxidation of the organic part of the demeton molecule with the formation of the simplest decomposition products. All the products shown in this diagram have been isolated and identified. Their properties are give in Table LII.

Oxidation of demeton with the formation of the listed compounds takes place also with such oxidizing agents as hydrogen peroxide, bromine water, hypochlorites, etc. The reaction with bromine is utilized for the quantitative determination of the thiolo isomer content in the mixture, since the isomers react differently with bromine.

Demeton with dimethyl sulfate and other alkylating agents yields sulfonium compounds of the type:

$$\left[(C_2H_5O)_2\overset{\underset{\|}{S}}{P}OCH_2CH_2\overset{\underset{|}{CH_3}}{S}C_2H_5\right]^{+}A^{-}$$

having a toxicity for animals considerably higher than that of the starting demeton.

Demeton has a powerful systemic effect and is used to control many sucking plant pests. The duration of action of the compound under field conditions depends on the crop that is treated and is from 4–6 weeks. It is

Table LII. *Properties of oxidation and metabolism products of demeton*

Compound	B. P. (°C./mm. Hg)	LD_{50} for rats (mg./kg.)
$(C_2H_5O)_2\overset{\overset{S}{\|}}{P}OCH_2CH_2SOC_2H_5$	160/0.3	100
$(C_2H_5O)_2\overset{\overset{S}{\|}}{P}OCH_2CH_2SO_2C_2H_5$	145/0.05	90
$(C_2H_5O)_2\overset{\overset{O}{\|}}{P}SCH_2CH_2SOC_2H_5$	160/0.3	2
$(C_2H_5O)_2\overset{\overset{O}{\|}}{P}SCH_2CH_2SO_2C_2H_5$	160/0.2	2
$(C_2H_5O)_2\overset{\overset{O}{\|}}{P}OCH_2CH_2SC_2H_5$	115–116/0.3	175
$(C_2H_5O)_2\overset{\overset{O}{\|}}{P}OCH_2CH_2SOC_2H_5$	170/1	10
$(C_2H_5O)_2\overset{\overset{O}{\|}}{P}OCH_2CH_2SO_2C_2H_5$	155–159/0.3	75

marketed in the form of 30 and 50% emulsive concentrates. Nonionic compounds of the type of polyethylene glycol esters are used as emulsifiers with the addition of small amounts of a calcium alkylarylsulfonate. However, because of the high toxicity of demeton its use in the Soviet Union is forbidden.

In industry demeton is produced by the reaction of diethyl chlorothiophosphate with 2-hydroxydiethyl sulfide in the presence of an HCl acceptor (NaOH or carbonates of the alkali metals):

$$(C_2H_5O)_2\overset{\overset{S}{\|}}{P}Cl + HOCH_2CH_2SC_2H_5 + NaOH \rightarrow$$
$$(C_2H_5O)_2\overset{\overset{S}{\|}}{P}OCH_2CH_2SC_2H_5 + H_2O + NaCl$$

The compound is partially isomerized directly in the process of preparation and the higher the reaction temperature the more isomerization occurs. The pure thiolo isomer of demeton can be obtained in good yield from salts of diethylthiophosphoric acid and 2-chlorodiethyl sulfide:

$$(C_2H_5O)_2P\begin{smallmatrix}O\\SK\end{smallmatrix} + ClCH_2CH_2SC_2H_5 \rightarrow (C_2H_5O)_2\overset{O}{\overset{\|}{P}}SCH_2CH_2SC_2H_5 + KCl$$

Methylmercaptophos (methyl demeton, Metasystox) is the methyl homolog of demeton. Technical grade methyl demeton is a mixture of thiolo and thiono (70%) isomers of O,O-dimethyl 2-ethylmercaptoethyl thiophosphate. The thiono isomer is a liquid with a characteristic unpleasant odor, b.p. 93° C. at 0.5 mm. of Hg, vapor pressure at 20° C. 1.85×10^{-5} mm. of Hg, volatility 23.3 mg./m.³, d_4^{20} 1.1904. About 330 mg. of the compound dissolves in 1 liter of water at 20° C. It is highly soluble in most organic solvents. The thiolo isomer is a liquid, b.p. 102° C. at 0.4 mm. of Hg, vapor pressure at 20° C. 3.6×10^{-4} mm. of Hg, volatility 4.5 mg./m.³, d_4^{20} 1.207. Its solubility in water at 20° C. is 3,300 mg./liter; it is highly soluble in organic solvents.

The chemical stability of methyl demeton is lower than that of demeton. For example, 50% of methyl demeton is hydrolyzed at pH 3 and 70° C. in 4.9 hours, and 50% of demeton in 10 hours; in alkaline medium the difference is still more appreciable. Rearrangement of the thiono to the thiolo isomer also takes place more rapidly. While the thiono isomer of demeton at 40° C. is converted to a mixture containing 90% thiono and 10% thiolo isomers in 91 days, the thiono isomer of methyl demeton under the same conditions is 10% isomerized in 8 days.

In the presence of traces of water methyl demeton can enter into an intermolecular methylation reaction:

$$(CH_3O)_2\overset{O}{\overset{\|}{P}}SCH_2CH_2SC_2H_5 \rightarrow$$

$$\left[(CH_3O)_2\overset{O}{\overset{\|}{P}}SCH_2CH_2\overset{CH_3}{\overset{|}{S}}C_2H_5\right]^+ \left[\begin{smallmatrix}CH_3O\\O\end{smallmatrix}\!\!\!\overset{O}{\overset{\|}{P}}SCH_2CH_2SC_2H_5\right]^-$$

which in a number of instances is the cause of spoilage of the product on prolonged storage. The stability of methyl demeton is negatively affected also by metallic iron and some iron compounds, and it should not therefore be kept in iron containers but rather in enameled or pure aluminum ones.

Methyl demeton is less toxic than demeton. The LD_{50} of the technical product is in the range 80–100 mg./kg. The LD_{50} of the thiono isomer is 180 mg./kg., and that of the thiolo isomer is 40 mg./kg. The maximum permissible concentration in air is 0.1 mg./m.³

The metabolism of methyl demeton in plants and insects proceeds similarly to that of demeton, but when methyl demeton enters the plant it is rapidly isomerized to the thiolo isomer, which is further metabolized by the same route as demeton. Oxidization products of methyl demeton have not been detected, but they have been obtained synthetically. The conversions and decomposition of methyl demeton in plants take place relatively quickly, and as a result the duration of its insecticidal action usually does not exceed three weeks.

Table LIII gives the products of oxidation of methyl demeton and their toxicity for rats.

Table LIII. *Oxidation products of methyl demeton and their toxicity for rats*

Compound	B. P. (°C./mm. Hg)	LD_{50} for rats (mg./kg.)
$(CH_3O)_2\overset{\overset{S}{\|}}{P}OCH_2CH_2SOC_2H_5$	95–96/0.01	600
$(CH_3O)_2\overset{\overset{S}{\|}}{P}OCH_2CH_2SO_2C_2H_5$	101/0.01	500
$(CH_3O)_2\overset{\overset{O}{\|}}{P}SCH_2CH_2SOC_2H_5$	105/0.01;	40
$(CH_3O)_2\overset{\overset{O}{\|}}{P}SCH_2CH_2SO_2C_2H_5$	115/0.01 m.p. 52° C.	40

Methyl demeton is used in the form of aqueous emulsions and is marketed as concentrates containing 30 and 50% of the mixture of isomers. In the Soviet Union a synergist has been found for methyl demeton, the addition of which permits lowering the dosage of the compound by 20–40% and decreases the toxicity almost three times. This compound is known as sinerfos.

Methyl demeton is produced by the reaction of 2-hydroxydiethyl sulfide with dimethyl chlorothiophosphate in the presence of an HCl acceptor:

$$(CH_3O)_2\overset{\overset{S}{\|}}{P}Cl + HOCH_2CH_2SC_2H_5 + NaOH \rightarrow$$
$$(CH_3O)_2\overset{\overset{S}{\|}}{P}OCH_2CH_2SC_2H_5 + NaCl + H_2O$$

The technical grade product contains as an impurity a small amount of trimethyl thiophosphate, 2-hydroxydiethyl sulfide, and some other esters of thiophosphoric acid. However, the total content of the thiono and thiolo isomers usually is not $<90\%$.

Tinox. Similar in properties to methyl demeton is a compound produced in the German Democratic Republic under the name of Tinox. The technical grade product is a mixture of the thiolo (60–70%) and thiono (30–40%) isomers of O,O-dimethyl 2-methylmercaptoethyl thiophosphate. The solubility of Tinox in water at 20° C. is about 500 mg./liter; it is highly soluble in most organic solvents.

The thiono isomer is a liquid, b.p. 115° C. at 2 mm. of Hg, vapor pressure at 20° C. 4×10^{-4} mm. of Hg, volatility about 35 mg./m.³, d_4^{20} 1.154. The LD_{50} for rats is more than 50 mg./kg.

The thiolo isomer is a liquid, b.p. 109° C. at 2 mm. of Hg; its solubility in water is about 2,000 mg./liter; it is highly soluble in most organic solvents. The LD_{50} for rats is about 40 mg./kg.

The chemical properties, insecticidal activity, and duration of action of Tinox are close to those of methyl demeton. It is marketed in the form of a 50% emulsive concentrate.

The methods of making Tinox are entirely analogous to the methods of producing demeton and methyl demeton.

O,O-Dimethyl S-ethylmercaptoethyl thiophosphate (Metasystox-I) is a rapidly acting systemic and contact compound. It is the thiolo isomer of methyl demeton, and there is no point in repeating the description of the properties of this compound. It is marketed in the form of a 50% emulsive concentrate. The best method of synthesizing the compound is the reaction of 2-chlorodiethyl sulfide with salts of dimethylthiophosphoric acid:

$$(CH_3O)_2P\begin{smallmatrix}O\\SK\end{smallmatrix} + ClCH_2CH_2SC_2H_5 \rightarrow (CH_3O)_2\overset{O}{\underset{\|}{P}}SCH_2CH_2SC_2H_5 + KCl$$

By this method the thiolo isomer of methyl demeton is obtained in 85–90% yields. The reaction can be carried out both in aqueous medium and in organic solvents. It is produced also when trimethyl thiophosphate reacts with 2-chlorodiethyl sulfide with heating somewhat above 100° C.:

$$(CH_3O)_3P = S + ClCH_2CH_2SC_2H_5 \rightarrow (CH_3O)_2\overset{O}{\underset{\|}{P}}SCH_2CH_2SC_2H_5 + CH_3Cl$$

Trimethyl thiophosphate is synthesized with very good yields from methyl dichlorothiophosphate and methyl alcohol in the presence of NaOH:

$$CH_3O\overset{S}{\underset{\|}{P}}Cl_2 + 2\,CH_3OH + 2\,NaOH \rightarrow (CH_3O)_3\overset{S}{\underset{\|}{P}} + 2\,NaCl + 2\,H_2O$$

O,O-Dimethyl S-(2-ethylsulfinyl)ethyl thiophosphate (oxydemetonmethyl, Metasystox-R) is a yellow liquid with a weak odor, b.p. 106° C. at 0.01 mm. of Hg, volatility 0.09 mg./m.³ at 20° C. It is highly soluble in water, but slightly soluble in hydrocarbons of the aliphatic series, highly soluble in methylene chloride and halogenated aromatic hydrocarbons. The LD_{50} for rats is 65–75 mg./kg. In acid medium it is more resistant to hydrolysis than methyl demeton, but in alkaline medium it is hydrolyzed more rapidly than the latter. Oxydemetonmethyl is close to methyl demeton in insecticidal activity; it has a less unpleasant odor. It is marketed in the form of a 25% concentrate and is used at 0.025% concentration of the active ingredient. In the United States the residue tolerance for oxydemetonmethyl is 0.75 mg./kg.

It is produced either by the oxidation of the thiolo isomer of methyl demeton with various oxidizing agents (hydrogen peroxide, hypochlorites, bromine, etc.):

$$(CH_3O)_2\overset{\overset{O}{\|}}{P}SCH_2CH_2SC_2H_5 \xrightarrow{[O]} (CH_3O)_2\overset{\overset{O}{\|}}{P}SCH_2CH_2SOC_2H_5$$

or by the reaction of salts of dimethylthiolophosphoric acid with 2-(ethylsulfinyl)ethyl bromide:

$$(CH_3O)_2\overset{\overset{O}{\|}}{P}SK + BrCH_2CH_2SOC_2H_5 \rightarrow (CH_3O)_2\overset{\overset{O}{\|}}{P}SCH_2CH_2SOC_2H_5 + KBr$$

O,O-Dimethyl S-[2-(ethylsulfinyl)isopropyl] thiophosphate (Metasystox-S) is an oily liquid, b.p. 115° C. at 0.02 mm. of Hg, vapor pressure at 20° C. 3.5×10^{-6} mm. of Hg, volatility 0.05–0.07 mg./m.^{3}, d_4^{20} 1.257. It is rather highly soluble in water, alcohols, ketones, and chlorinated derivatives of hydrocarbons; it is slightly soluble in hydrocarbons of the aliphatic series. The compound is resistant to hydrolysis in acid medium, but is hydrolyzed several times faster in alkaline medium. The LD_{50} for rats is 105 mg./kg.

Its insecticidal activity is almost the same as that of oxydemetonmethyl. It is marketed in the form of a 50% concentrate, which is used at a concentration of 0.05% active ingredient.

It is produced by the following scheme:

$$(CH_3O)_2\overset{\overset{O}{\|}}{P}SNa + Cl\overset{\overset{CH_3}{|}}{C}HCH_2SC_2H_5 \rightarrow (CH_3O)_2\overset{\overset{O}{\|}}{P}S\overset{\overset{CH_3}{|}}{C}HCH_2SC_2H_5 \xrightarrow{H_2O_2}$$

$$(CH_3O)_2\overset{\overset{O}{\|}}{P}S\overset{\overset{CH_3}{|}}{C}HCH_2\overset{\overset{O}{\|}}{S}C_2H_5$$

The total yield of the compound is 75–85%.

The hydrolytic stability of the sulfoxide is almost 100 times greater than that of the sulfide.

O,O-Dimethyl S-2-(1-N-methylcarbamoylethylmercapto)ethyl thiophosphate (vamidothion) is a white crystalline substance, m.p. 46°–48° C. The technical grade product, m.p. 33°–38° C., is highly soluble in water (4 kg./liter), acetone, methyl ethyl ketone, ethyl acetate, and acetonitrile, difficulty soluble in hexane, cyclohexane, and other paraffinic and cycloparaffinic hydrocarbons; in xylene the solubility is 125 g./liter. The LD_{50} of vamidothion (in mg./kg.) is 64–100 for rats, 43–68 for mice, and 85 for guinea pigs. The insecticidal activity of vamidothion approaches that of methyl demeton, but considerably exceeds it in duration of effect. It is somewhat more resistant to hydrolysis than methyl demeton and demeton. Oxidizing agents convert it first to the sulfoxide and then further to the sulfone:

$$(CH_3O)_2\overset{O}{\overset{\|}{P}}SCH_2CH_2S\overset{CH_3}{\overset{|}{C}H}CONHCH_3 \xrightarrow{[O]}$$

$$(CH_3O)_2\overset{O}{\overset{\|}{P}}SCH_2CH_2\overset{OCH_3}{\overset{\|}{\underset{\|}{S}}}CHCONHCH_3 \xrightarrow{[O]}$$

$$(CH_3O)_2\overset{O}{\overset{\|}{P}}SCH_2CH_2\overset{O}{\overset{\|}{\underset{\underset{O}{\|}}{S}}}-\overset{CH_3}{\overset{|}{C}H}CONHCH_3$$

The first product of the metabolism of vamidothion in plants also is the sulfoxide. Simultaneously with oxidation in plants hydrolysis of the compound also occurs with the formation of dimethylphosphoric acid and phosphoric acid.

Vamidothion is produced by the reaction of ammonium dimethylthiophosphate or an alkali dimethylthiophosphate with an α-(2-haloethylmercapto)propiomethylamide:

$$(CH_3O)_2\overset{O}{\overset{\|}{P}}SNH_4 + ClCH_2CH_2S\underset{\underset{CH_3}{|}}{C}HCONHCH_3 \rightarrow$$

$$(CH_3O)_2\overset{O}{\overset{\|}{P}}SCH_2CH_2S\underset{\underset{CH_3}{|}}{C}HCONHCH_3 + NH_4Cl$$

Great difficulties are presented in the preparation of α-(2-haloethylmercapto)propiomethylamide, which is synthesized by the following route:

$$HOCH_2CH_2SH + KOH + CH_3CHClCONHCH_3 \rightarrow$$

$$HOCH_2CH_2S\underset{\underset{CH_3}{|}}{C}HCONHCH_3 + H_2O + KCl$$

$$HOCH_2CH_2S\underset{\underset{CH_3}{|}}{C}HCONHCH_3 + SOCl_2 \rightarrow$$

$$ClCH_2CH_2S\underset{\underset{CH_3}{|}}{C}HCONHCH_3 + HCl + SO_2$$

Other routes also are possible for the synthesis of this compound.

Compounds that contain an ester group in one of the ester radicals also have been proposed for use as insecticides. Examples of these compounds are acetofos and methylacetofos.

O,O-Diethyl S-carboethoxymethyl thiophosphate (acetofos) is a colorless liquid with a specific unpleasant odor, b.p. 120° C. at 0.15 mm. of Hg, vapor pressure at 20° C. about 7.5×10^{-4} mm. of Hg. It is highly soluble in water and most organic solvents, d_4^{20} 1.1840. The LD_{50} for various animals is 300–700 mg./kg.

When acetofos reacts with water in acid or alkaline medium, hydrolysis of the compound occurs with the splitting off of the ester groups. Breakdown of this type of compound proceeds differently in warm-blooded animals and in insects, and acetofos consequently has little toxicity for mammals but is highly insecticidal. Hydrolysis in warm-blooded animals, according to O'Brien, takes place by the following route:

$$(C_2H_5O)_2\overset{\overset{O}{\|}}{P}SCH_2COOC_2H_5 + H_2O \rightarrow (C_2H_5O)_2\overset{\overset{O}{\|}}{P}SCH_2COOH + C_2H_5OH$$

while in insects splitting off of the acetic acid group and inhibition of vitally important enzymes take place.

Acetofos is a contact insecticide, but its activity approaches that of trichlorphon. It is produced by the reaction of the ethyl ester of monochloroacetic acid with the ammonium salt of diethylthiophosphoric acid in a suitable organic solvent (e. g., benzene):

$$(C_2H_5O)_2\overset{\overset{O}{\|}}{P}SNH_4 + ClCH_2COOC_2H_5 \rightarrow$$
$$(C_2H_5O)_2\overset{\overset{O}{\|}}{P}SCH_2COOC_2H_5 + NH_4Cl$$

The yield by this reaction is close to 90%.

O,O-Dimethyl S-carboethoxymethyl thiophosphate (methylacetofos) is a clear liquid with an unpleasant odor, b.p. 116°–120° C. at 0.35 mm. of Hg, d_4^{20} 1.250. It is highly soluble in water and in most organic solvents. It is similar to acetofos in chemical properties, but is less stable and in storage may undergo conversion as a result of its high alkylating ability. The main decomposition products are compounds of the sulfonium type. The LD_{50} of pure methylacetofos for rats is about 1,000 mg./kg. Methylacetofos is similar to acetofos in insecticidal activity.

It is synthesized similarly to acetofos, but because of side reactions the yield does not exceed 70%. The main impurity in the technical grade product is trimethyl thiophosphate, which greatly increases the toxicity of methylacetofos to vertebrate animals.

In addition to insecticides, among the derivatives of thiophosphoric acid are found fungicides, the simplest representative being kitazin.

O,O-Diethyl S-benzyl thiophosphate (kitazin) is a colorless liquid, b.p. 115° C. at 0.02 mm. of Hg. It is poorly soluble in water, but highly soluble in most organic solvents. The LD_{50} for rats is 238 mg./kg. Kitazin is used in Japan as a curative fungicide for combating piricularia of rice at dosages up to 1 kg./ha. as an aqueous emulsion or a 1.5% dust. It is marketed in the form of a 48% emulsive concentrate and a 1.5% dust.

In alkaline medium the compound is rapidly hydrolyzed, forming products of low toxicity to the plant disease agents. Hydrolysis takes place somewhat more slowly in acid medium. At high temperature O,O-diethyl S-benzyl thiophosphate also is unstable and breaks down comparatively quickly with the formation of dibenzyl sulfide.

It can be obtained in good yield by the reaction of benzyl chloride with salts of diethylthiophosphoric acid:

$$(C_2H_5O)_2\overset{\overset{O}{\|}}{P}SK + ClCH_2C_6H_5 \rightarrow (C_2H_5O)_2\overset{\overset{O}{\|}}{P}SCH_2C_6H_5 + KCl$$

This reaction is best carried out in some organic solvent. The yield of the desired product is about 85%.

O,O-Diisopropyl S-benzyl thiophosphate (Kitazin P) is a colorless liquid, b.p. 120–130° C. at 0.1 mm. of Hg. The LD_{50} is 660 mg./kg. (mice).

Mixed esters of thiophosphoric acid that contain a dialkylamino group in one of the ester radicals have been proposed for use as systemic insecticides.

O,O-Diethyl S-(2-diethylaminoethyl) thiophosphate (Amiton, Tetram, Inferno, Metramak) is used in the form of the acid oxalate (m.p. 100° to 101° C.) as a systemic insecticide to control aphids and mites on various species of plants. The free base is a colorless liquid, b.p. 80° C. at 0.01 mm. of Hg, highly soluble in water and most organic solvents. This ester forms salts with acids; with alkyl halides it yields quaternary ammonium salts that are toxic to vertebrate animals. The LD_{50} of the free base for different animals is 0.5–7 mg./kg. It hydrolyzes comparatively slowly in water.

It is produced by the reaction of sodium diethylaminoethylate with diethyl chlorothiophosphate with subsequent rearrangement of the thiono ester to the thiolo form:

$$(C_2H_5O)_2\overset{\overset{S}{\|}}{P}Cl + NaOCH_2CH_2N(C_2H_5)_2 \rightarrow$$

$$NaCl + (C_2H_5O)_2\overset{\overset{S}{\|}}{P}OCH_2CH_2N(C_2H_5)_2$$

$$(C_2H_5O)_2\overset{\overset{S}{\|}}{P}OCH_2CH_2N(C_2H_5)_2 \rightarrow (C_2H_5O)_2\overset{\overset{O}{\|}}{P}SCH_2CH_2N(C_2H_5)_2$$

This rearrangement takes place readily at a comparatively low temperature. The mechanism of the rearrangement apparently is entirely analogous to that of the thiono isomer of demeton to the thiolo isomer.

An example of the organophosphorus herbicides that are derived from thiophosphoric acid is Zytron.

O-Methyl O-2,4-dichlorophenyl N-isopropylamidothiophosphate (Zytron) is a colorless crystalline substance, m.p. 51.4° C., vapor pressure at 150° C. 2 mm. of Hg. One liter of water dissolves 5 mg. of the compound; it is highly soluble in most organic solvents. Its LD_{50} for guinea pigs is 210 mg./kg., for dogs 1,000 mg./kg. Zytron has been recommended for the control of crabgrass in lawns and some weed grasses in peanuts.

It is synthesized by the following scheme:

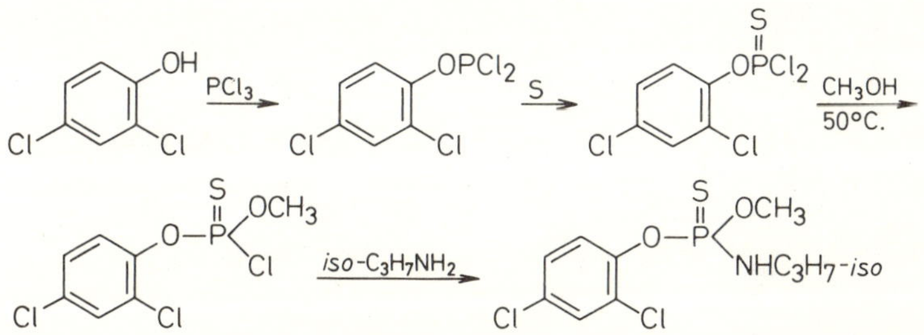

[reaction scheme: 2,4-dichlorophenol + PCl_3 → aryl-$OPCl_2$ → (S) → aryl-$OP(S)Cl_2$ → CH_3OH, 50°C → aryl-O-P(S)(OCH$_3$)Cl → iso-$C_3H_7NH_2$ → aryl-O-P(S)(OCH$_3$)(NHC$_3H_7$-iso)]

In addition to insecticidal properties, some individual amidoesters of thiophosphoric acid have a fungicidal effect.

O,O-Diethyl phthalimidothiophosphate is a white crystalline substance, m.p. 83°–84° C. It is highly soluble in benzene (130 g. in 100 g. of solvent), xylene, and ethyl acetate, and only slightly soluble in hexane and cyclohexane. Its LD_{50} for rats is about 5,000 mg./kg. At 0.03–0.06% it is effective in controlling diseases of apples, cherries, and other crops.

The method used in the preparation of this compound is:

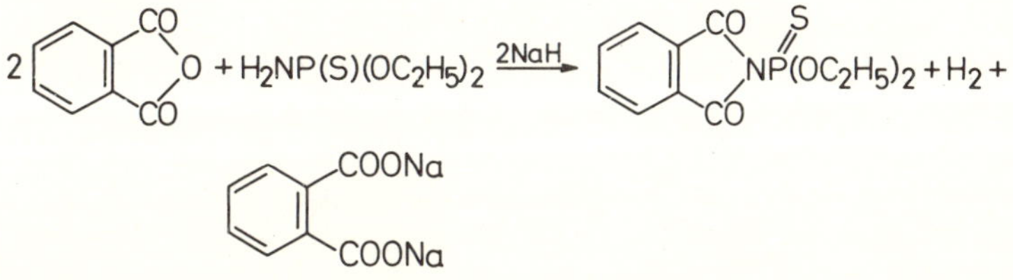

[reaction scheme: phthalimide-NK + $ClP(S)(OC_2H_5)_2$ → phthalimide-N–P(S)$(OC_2H_5)_2$ + KCl

phthaloyl-(COCl)$_2$ + $H_2NP(S)(OC_2H_5)_2$ $\xrightarrow{(C_2H_5)_3N}$ phthalimide-N–P(S)$(OC_2H_5)_2$

2 phthalic anhydride + $H_2NP(S)(OC_2H_5)_2$ $\xrightarrow{2NaH}$ phthalimide-NP$(OC_2H_5)_2$ + H_2 + disodium phthalate (COONa)$_2$]

Derivatives of dithiophosphoric acid

In addition to the derivatives of thiophosphoric acid, various derivatives of dithiophosphoric acid and trithiophosphoric acid are widely used as pesticides. Of the derivatives of dithiophosphoric acid the most important are the compounds used for the control of insect plant pests, and it is only recently that fungicides and herbicides have been found among them.

The general formulas of well-known pesticidal derivatives of dithiophosphoric acid are:

$$\begin{array}{c}RO\\R'O\end{array}\!\!\!>\!\!P\!\!<\!\!\begin{array}{c}S\\SCHCOOR''\\||\\R'''\end{array}\quad(IX) \qquad \begin{array}{c}RO\\R'O\end{array}\!\!\!>\!\!P\!\!<\!\!\begin{array}{c}S\\SCH_2CON\!<\!\!\begin{array}{c}R''\\R'''\end{array}\end{array}\quad(X)$$

$$\begin{array}{c}RO\\R'O\end{array}\!\!\!>\!\!P\!\!<\!\!\begin{array}{c}S\\SCH_2SR''''\end{array}\quad(XI) \qquad \begin{array}{c}RO\\R'O\end{array}\!\!\!>\!\!P\!\!<\!\!\begin{array}{c}S\\SCH_2CH_2SR''\end{array}\quad(XII)$$

$$\begin{array}{c}RO\\R'O\end{array}\!\!\!>\!\!P\!\!<\!\!\begin{array}{c}S\\SCH_2NHAr\end{array}\quad(XIII) \qquad \begin{array}{c}RO\\R'O\end{array}\!\!\!>\!\!P\!\!<\!\!\begin{array}{c}S\\SCH_2Ar\end{array}\quad(XIV)$$

$$\begin{array}{c}RO\\ArS\end{array}\!\!\!>\!\!P\!\!<\!\!\begin{array}{c}S\\NHR'\end{array}\quad(XV)$$

R, R', and R'' are lower aliphatic radicals, R''' is hydrogen or any radical, Ar is an aromatic or heterocyclic radical, and R'''' is an aliphatic or aromatic radical.

Going from the derivatives of thiophosphoric acid to the corresponding derivatives of dithiophosphoric acid in most cases the toxicity of the compound decreases, and its chemical stability increases. As a result of this, the duration of its action under field conditions becomes greater. Furthermore, the spectrum of action of the compounds is changed. Many derivatives of dithiophosphoric acid, especially those containing heterocyclic radicals, have high activity not only toward sucking plant pests, but also toward chewing insects.

For the series of compounds of general formula (IX) the following relationships can be noted between biological activity and structure:

1. All derivatives of dithiophosphoric acid with such a structure are less toxic to vertebrates than the corresponding derivatives of thiophosphoric and phosphoric acids.

2. Mixed esters in which R and R' are methyl groups are the least toxic. An increase in the number of carbon atoms in these radicals increases the toxicity to vertebrates without increasing the insecticidal activity.

3. A change in R''' has less effect on the toxicity of the mixed dithiophosphates to vertebrates, but significantly influences the insecticidal and acaricidal properties. The highest activity is shown by compounds in which R''' is the simplest aromatic radical. Introduction into R''' of a carboalkoxy

group strongly decreases the toxicity of the compound to vertebrates and has almost no effect on the insecticidal and acaricidal properties. This is true for aliphatic derivatives, but not for aromatic ones.

The great diversity in toxicity of this group of mixed esters of dithiophosphoric acid for insects and vertebrates is explained by the different routes of metabolism of the compounds in living organisms. Thus, for example, malathion in insects goes over to the more toxic ester of thiophosphoric acid, O,O-dimethyl S-1,2-dicarboethoxyethyl thiophosphate, while in warm-blooded animals hydrolysis of the group in the side chain takes place and a product is obtained that is practically nontoxic to animals:

$$(CH_3O)_2\overset{S}{\overset{\|}{P}}SCHCOOC_2H_5 \xrightarrow{[O]} (CH_3O)_2\overset{O}{\overset{\|}{P}}SCHCOOC_2H_5$$
$$\underset{CH_2COOC_2H_5}{|} \qquad \underset{CH_2COOC_2H_5}{|}$$

$$\downarrow H_2O$$

$$(CH_3O)_2\overset{S}{\overset{\|}{P}}SCHCOOH \xrightarrow{H_2O} (CH_3O)_2\overset{S}{\overset{\|}{P}}SCHCOOH$$
$$\underset{CH_2COOC_2H_5}{|} \qquad \underset{CH_2COOH}{|}$$

Compounds of general formula (X) usually have not only contact but also systemic insecticidal action.

High insecticidal activity and moderate toxicity to mammals are shown by compounds in which $R = R' = CH_3$. Replacement of even one methyl group by an ethyl or other group containing a larger number of carbon atoms increases the compound's toxicity to mammals without substantially changing the insecticidal activity. Thus, for example, O,O-dimethyl S-(N-methylcarbamoylmethyl) dithiophosphate has an LD_{50} for rats of 250 mg./kg., but O-methyl O-ethyl S-(N-methylcarbamoylmethyl) dithiophosphate has an LD_{50} of 12 mg./kg.

Replacement of a methyl radical in the amide group by ethyl does not substantially alter the toxicity of the compound to warm-blooded animals, but an increase in the number of carbon atoms in the hydrocarbon radical on the nitrogen decreases the insecticidal activity. Replacement of the second hydrogen atom on the nitrogen by a hydrocarbon radical also leads to a lowering of the insecticidal activity. When this hydrogen is replaced by a carboalkoxy or formyl group, the activity against insects is not decreased, but the toxicity of the compound to mammals is significantly increased. However, there are exceptions to this general rule. Compounds of general formula (X) are widely used in agriculture as insecticides and acaricides.

Compounds of formulas (XI) and (XII) also have strong systemic insecticidal and acaricidal effects. The general situation of the relationship of insecticidal activity of these two groups of compounds to their structure can be formulated thus:

1. In compounds of formula (XI) an increase in the number of carbon atoms by more than two in all the hydrocarbon radicals lowers the toxicity to animals and to plant pests. The most toxic compound is the ester in which $R = R' = R''' = C_2H_5$. Replacement of R'''' by an aromatic radical leads to some lowering of the toxicity to mammals while preserving the acaricidal and insecticidal properties.

2. Of the compounds of general formula (XII) the least toxicity to mammals is shown by the compounds with $R = R' = CH_3$. Replacement of one methyl radical by an ethyl or higher hydrocarbon radical leads to a sharp increase in toxicity, and also to some strengthening of the insecticidal properties. When the total number of carbon atoms in R and R' is more than four, compounds are produced that have relatively little insecticidal activity.

3. When various functional groups are introduced into R'', compounds also are obtained with a high insecticidal activity, but because of the complexity of preparing these compounds they are not yet in use.

4. Compounds in which R'''' is a dialkyldithiophosphoric acid group also have insecticidal properties.

Compounds of formulas (XIII) and (XIV) are extremely varied and it is very difficult to draw conclusions as to the relationship of their insecticidal activity to structure. Compounds with formula (XV) show mainly fungicidal properties. Their insecticidal activity is somewhat weaker.

The principal methods for preparing the mixed esters of dithiophosphoric acid are the following:

1. Reaction of salts of a dialkyldithiophosphoric acid with the appropriate halogen derivatives:

$$(RO)_2\overset{\underset{\|}{S}}{P}Na + ClCH_2R'' \rightarrow (RO)_2\overset{\underset{\|}{S}}{P}CH_2R'' + NaCl$$

In some cases it is possible to use not only the salts, but also the free acids. In this case the process goes more slowly and at a higher temperature:

$$(RO)_2\overset{\underset{\|}{S}}{P}H + ClCH_2R'' \rightarrow (RO)_2\overset{\underset{\|}{S}}{P}CH_2R'' + HCl$$

The first reaction usually is carried out in aqueous medium or in organic solvents and the second reaction without a solvent.

2. Addition of dialkyldithiophosphoric acids at the double bond of unsaturated compounds:

$$(RO)_2\overset{\underset{\|}{S}}{P}H + \underset{\underset{CHR'}{\|}}{CHR'} \rightarrow (RO)_2\overset{\underset{\|}{S}}{P}\underset{\underset{CH_2R'}{|}}{CHR'}$$

By this reaction compounds of types (IX) and (XII) can be obtained in good yields.

3. Reaction of dialkylchlorothiophosphates with the appropriate thiols:

$$(RO)_2\overset{\underset{\|}{S}}{P}Cl + NaSR' \rightarrow (RO)_2\overset{\underset{\|}{S}}{P}SR' + NaCl$$

The reaction goes well in the presence of HCl acceptors. Either caustic alkalies or various organic bases can be used.

The dialkyldithiophosphoric acids necessary for the synthesis of the mixed esters of dithiophosphoric acid are obtained in good yields by the reaction of phosphorus pentasulfide with alcohols:

$$P_2S_5 + 4\ ROH \rightarrow H_2S + 2\ (RO)_2\overset{\underset{\|}{S}}{P}SH$$

Trialkyl dithiophosphate is formed in small amounts as a by-product.

The purity of the starting phosphorus pentasulfide is very important in the synthesis of the dialkyldithiophosphoric acids, since impurities increase the number of by-products that are difficult to remove. The phosphorus pentasulfide is most often obtained by direct reaction of equimolecular quantities of sulfur and yellow phosphorus at an elevated temperature. When sufficiently pure starting materials are used, the phosphorus pentasulfide does not require further purification. In the opposite case it is necessary to carry out its purification either by crystallization from carbon disulfide or by vacuum distillation.

At the present time more than 25 different derivatives of dithiophosphoric acid are used in agriculture in various countries and their total production is more than 20,000 tons a year.

O,O-Dimethyl S-1,2-dicarboethoxyethyl dithiophosphate (malathion, carbophos) is a colorless liquid, b.p. 120° C. at 0.2 mm. of Hg, d_4^{25} 1.23, n_d^{25} 1.4985, vapor pressure at 20° C. 1.25×10^{-4} mm. of Hg, volatility 2.26 mg./m.³ At 20° C., 145 mg. dissolves in 1 liter of water; it is highly soluble in most organic solvents, with the exception of saturated hydrocarbons. The LD_{50} for various experimental animals is 500–1,500 mg./kg. The maximum permissible concentration in air is 0.5 mg./m.³

Malathion on prolonged heating at 150° C. is isomerized and goes over to the corresponding thiolo isomer:

$$(CH_3O)_2\overset{\underset{\|}{S}}{P}SCHCOOC_2H_5 \rightarrow \overset{CH_3O}{\underset{CH_3S}{\diagdown\diagup}}\overset{\underset{\|}{O}}{P}SCHCOOC_2H_5$$
$$|\qquad\qquad\qquad\qquad\qquad\qquad\qquad |$$
$$CH_2COOC_2H_5 \qquad\qquad\qquad\quad CH_2COOC_2H_5$$

At a higher temperature this reaction proceeds violently and a considerable part of the product is decomposed, sometimes even explosively. When the product contains such impurities as trimethyl dithiophosphate, dimethyldithiophosphoric acid, and some others, intensive decomposition of malathion may take place even at a lower temperature.

Hydrolysis of malathion follows different paths in acid and alkaline media. In acid medium the main products of hydrolysis are dimethyldithiophosphoric acid and the ester of mercaptosuccinic acid, while in alkaline medium the salt of dimethyldithiophosphoric acid and the ester of fumaric acid are formed:

$$(CH_3O)_2\overset{\overset{S}{\|}}{P}SCHCOOC_2H_5 \underset{CH_2COOC_2H_5}{|} \begin{array}{l} \xrightarrow{H_2O,\ H^+} (CH_3O)_2\overset{\overset{S}{\|}}{P}OH + \underset{HSCHCOOC_2H_5}{\overset{CH_2COOC_2H_5}{|}} \\ \xrightarrow{NaOH} (CH_3O)_2\overset{\overset{S}{\|}}{P}SNa + \underset{\overset{\|}{CHCOOC_2H_5}}{CHCOOC_2H_5} \end{array}$$

This reaction is used for the quantitative polarographic determination of malathion.

When malathion is oxidized by nitric acid or other strong oxidizing agents, the thiono sulfur atom is split off and the corresponding ester of thiolophosphoric acid is formed:

$$(CH_3O)_2\overset{\overset{S}{\|}}{P}SCHCOOC_2H_5 + 8\ HNO_3 \rightarrow$$
$$\underset{CH_2COOC_2H_5}{|}$$

$$(CH_3O)_2\overset{\overset{O}{\|}}{P}SCHCOOC_2H_5 + 8\ NO_2 + H_2SO_4 + 3\ H_2O$$
$$\underset{CH_2COOC_2H_5}{|}$$

Malathion on prolonged contact with iron or materials containing iron breaks down and completely loses its insecticidal properties. It therefore cannot be kept in iron containers; it is best kept in glass ones.

It is marketed in the form of an emulsive concentrate containing 30 to 60% active ingredient, an emulsifier, and a solvent (e.g., xylene). It also is possible to use malathion in the form of dusts or of suspensions prepared from a wettable powder. It usually is used in agriculture at 0.2–0.25% active ingredient. In the Soviet Union a limit of 8 mg./kg. has been established for residues of malathion.

The principal method of producing malathion is the addition of dimethyldithiophosphoric acid to maleic acid ester. The reaction takes place very readily in the presence of basic catalysts in various organic solvents, or without a solvent:

$$(CH_3O)_2\overset{\overset{S}{\|}}{P}SH + \underset{\overset{\|}{CHCOOC_2H_5}}{CHCOOC_2H_5} \rightarrow (CH_3O)_2\overset{\overset{S}{\|}}{P}SCHCOOC_2H_5$$
$$\underset{CH_2COOC_2H_5}{|}$$

It is possible to combine this reaction with the preparation of the dimethyldithiophosphoric acid from methyl alcohol and phosphorus pentasulfide. In that case the reaction is carried out in diethyl maleate. However, the reaction sometimes take place so violently that spontaneous decomposition of the product occurs. The technical grade compound obtained by the method described contains small amounts of trimethyl dithiophosphate, dimethyl maleate, and solvent. Xylene is most often used as the solvent for the malathion synthesis. The yield of malathion by this reaction is more than 80%, but when the condensation process and the preparation of the dimethyldithiophosphoric acid are carried out simultaneously the yield decreases somewhat.

Malathion also is produced by prolonged boiling of a solution of the sodium or potassium salt of dimethyldithiophosphoric acid with a diethyl halosuccinate in alcohol or other polar solvent, but this method is more complicated and is rarely used:

$$(CH_3O)_2\overset{\overset{S}{\|}}{P}SK + BrCHCOOC_2H_5 \rightarrow (CH_3O)_2\overset{\overset{S}{\|}}{P}SCHCOOC_2H_5 + KBr$$
$$\phantom{(CH_3O)_2\overset{\overset{S}{\|}}{P}SK + Br}\underset{CH_2COOC_2H_5}{|} \phantom{\rightarrow (CH_3O)_2\overset{\overset{S}{\|}}{P}SC}\underset{CH_2COOC_2H_5}{|}$$

A large number of analogs and homologs of malathion have been synthesized and studied, but most of them have higher toxicity to animals and therefore are not used. Of interest for the control of flies is O,O-diethyl S-carboethoxymethyl dithiophosphate, which is known as acethion (b.p. 92° C. at 0.01 mm. of Hg, d_4^{20} 1.176, LD_{50} for rats 1,050–1,110 mg./kg.). It is synthesized by the reaction of ethyl monochloroacetate with diethyl dithiophosphate in aqueous medium with prolonged heating:

$$(C_2H_5O)_2\overset{\overset{S}{\|}}{P}SNa + ClCH_2COOC_2H_5 \rightarrow (C_2H_5O)_2\overset{\overset{S}{\|}}{P}SCH_2COOC_2H_5 + NaCl$$

This compound, although it is not used, is of great theoretical interest, because its low toxicity for animals was predicted on the basis of a study of the metabolism of organophosphorus compounds.

When an ester group on the acetic acid radical in acethion or its homologs is replaced by an amide group, the insecticidal activity of the compound is sharply increased with simultaneous increase also in its toxicity to mammals. However, the increase in toxicity to mammals is less appreciable; consequently many such amides are of considerable practical interest and some of them are used in agriculture. At present seven compounds of this group are in use, and a number of compounds are being investigated.

O,O-Dimethyl S-(N-methylcarbamoylmethyl) dithiophosphate (dimethoate, phosphamide, Rogor, Bi-58) is a snow-white crystalline substance, m.p. 51°–52° C., b.p. 107° C. at 0.05 mm. of Hg, vapor pressure at 20° C. 8.5×10^{-6} mm. of Hg, volatility 0.107 mg./m.³ It is highly soluble in water (39 g./liter) and most organic solvents, but is slightly soluble in paraffinic hydrocarbons. The LD_{50} of the pure compound for rats is 250–265 mg./kg.; the technical grade compound is more toxic (LD_{50} 150–160 mg./kg.) due to the presence of impurities, the most important of which is O,O-dimethyl S-(N-methylcarbamoylmethyl) thiophosphate (LD_{50} for rats 55 mg./kg.).

Dimethoate is thermally unstable and on heating it decomposes, first undergoing Pishchemuka isomerization:

$$(CH_3O)_2\overset{S}{\overset{\|}{P}}SCH_2CONHCH_3 \rightarrow \overset{CH_3O}{\underset{CH_3S}{>}}\overset{O}{\overset{\|}{P}}SCH_2CONHCH_3$$

The O,S-dimethyl S-(N-methylcarbamoylmethyl) dithiophosphate obtained is more toxic to mammals than the starting dimethoate (LD_{50} for rats 100 mg./kg.).

When dimethoate is oxidized by various oxidizing agents or by the oxygen of the air (oxidation by the oxygen of the air proceeds especially well on the green leaves of plants), it is converted to O,O-dimethyl S-(N-methylcarbamoylmethyl) thiophosphate:

$$(CH_3O)_2\overset{S}{\overset{\|}{P}}SCH_2CONHCH_3 + [O] \xrightarrow{H_2O}$$
$$(CH_3O)_2\overset{O}{\overset{\|}{P}}SCH_2CONHCH_3 + H_2SO_4$$

The hydrolysis of dimethoate takes place most readily in alkaline medium; in acid medium it is more stable (50% at pH 9 is hydrolyzed at 70° C. in 0.8 hour, but at pH 2 it requires 21 hours).

The metabolism of dimethoate in plants is apparently different from that in animals. In a study of its metabolism in cows and rats the following products were isolated:

$$(CH_3O)_2\overset{S}{\overset{\|}{P}}SCH_2CONHCH_3$$

$(CH_3O)_2\overset{S}{\overset{\|}{P}}SCH_2COOH \qquad \overset{CH_3O}{\underset{HO}{>}}\overset{S}{\overset{\|}{P}}SCH_2CONHCH_3 \qquad (CH_3O)_2\overset{S}{\overset{\|}{P}}SH$

$\downarrow$
$(CH_3O)_2\overset{S}{\overset{\|}{P}}OH$
$\downarrow$
$(CH_3O)_2\overset{O}{\overset{\|}{P}}OH$
$\downarrow$
$\rightarrow H_3PO_4$

while the following substances were isolated from plants:

$$(CH_3O)_2\overset{\overset{S}{\|}}{P}SCH_2CONHCH_3$$

$$(CH_3O)_2\overset{\overset{O}{\|}}{P}SCH_2CONHCH_3 \qquad (CH_3O)_2\overset{\overset{S}{\|}}{P}OH \qquad \overset{CH_3O}{\underset{HO}{\diagup}}\overset{\overset{S}{\|}}{P}SCH_2CONHCH_3$$

$$(CH_3O)_2\overset{\overset{O}{\|}}{P}OH$$

$$\longrightarrow H_3PO_4 \longleftarrow$$

Depending on the dosages of dimethoate applied, it usually is broken down in plants in 15–20 days. Of course, this also depends on the nature of the plant and on the weather conditions.

In storage dimethoate is relatively unstable and rapidly breaks down, especially at an elevated temperature. Impurities present in dimethoate catalyze its decomposition. It decomposes very quickly when it contains traces of organic basis, and when it breaks down, intermolecular alkylation occurs. Solutions of dimethoate in organic solvents are more stable.

Usually dimethoate is marketed in the form of a 40% emulsive concentrate that contains an emulsifier and an organic solvent in addition to the active ingredient. It is used to control plant-feeding mites, aphids, and other sucking plant pests at dosages of 700–1,000 g./hg. (active ingredient). At higher concentrations dimethoate is phytocidal. The duration of action of dimethoate is about 15 days. In many countries the minimum interval has been set as 10–15 days.

Dimethoate can be produced by the following two methods:

1. Reaction of salts of dimethyldithiophosphoric acid with *N*-methylchloroacetamide:

$$(CH_3O)_2\overset{\overset{S}{\|}}{P}SK + ClCH_2CONHCH_3 \rightarrow (CH_3O)_2\overset{\overset{S}{\|}}{P}SCH_2CONHCH_3 + KCl$$

This reaction is most often carried out in a medium consisting of water and some organic solvent. The yield is 80–95%; to obtain a good yield of dimethoate by this reaction, it is very important to have pure starting materials.

2. Reaction of O,O-dimethyl S-(carboaroxymethyl)- or S-(carboalkoxymethyl)dithiophosphate at low temperature with aqueous methylamine:

$$(CH_3O)_2\overset{\overset{S}{\|}}{P}SCH_2COOC_6H_5 + CH_3NH_2 \rightarrow$$

$$(CH_3O)_2\overset{\overset{S}{\|}}{P}SCH_2CONHCH_3 + C_6H_5OH$$

Both the phenyl and ethyl esters are used. The yield is more than 90%. However, the dimethoate obtained by this method is less stable. For stabilization of the compound very careful purification from all impurities is necessary and primarily from excess methylamine, traces of which can cause decomposition of dimethoate.

As with most other organophosphorus compounds, dimethoate should not be kept in an iron container. Even stainless steel on standing promotes a more rapid decomposition of the compound.

O,O-Dimethyl S-(N-ethylcarbamoylmethyl) dithiophosphate (Fitios, B/77) is the closest homolog of dimethoate. It is a white crystalline substance, m.p. 67°–68° C. (the technical grade product melts at 65°–66.7° C.), d_4^{70} 1.1670. At 25° C., 8.5 g. of the compound dissolves in 1 liter of water; it is highly soluble in alcohols, acetone, ether, and chloroform, moderately soluble in benzene, but poorly soluble in xylene, methylnaphthalene, and other hydrocarbons. The LD_{50} for mice is 350 mg./kg.

The chemical properties of Fitios are reminiscent of those of dimethoate, but the compound is somewhat more stable than dimethoate in storage. Its metabolism in plants and animals is completely analogous to that of dimethoate.

The best method of producing Fitios is the reaction of ammonium or potassium dimethyldithiophosphate with N-ethylmonochloroacetamide, which takes place either in aqueous medium or in organic solvents:

$$(CH_3O)_2\overset{\overset{S}{\|}}{P}SK + ClCH_2CONHC_2H_5 \rightarrow (CH_3O)_2\overset{\overset{S}{\|}}{P}SCH_2CONHC_2H_5 + KCl$$

The yield of technical grade Fitios by this reaction is close to 90%. To purify the product it is recrystallized from an organic solvent. Either halogen derivatives of aliphatic hydrocarbons or aromatic hydrocarbons are used as the solvent.

Fitios approaches dimethoate in insecticidal activity, but it is longer acting. It is marketed in the form of a 20% emulsive concentrate.

O,O-Diethyl S-(N-isopropylcarbamoylmethyl) dithiophosphate (FAC 20) is a second homolog of dimethoate. The active ingredient of FAC 20 is a white crystalline substance, m.p. 24°–25° C. It is practically insoluble in water, but soluble in most organic solvents with the exception of the alkanes. It is entirely analogous to dimethoate in chemical properties, but it is more stable in storage. Its LD_{50} for rats is 8–20 mg./kg. It is marketed in the form of a 20% emulsive concentrate. However, because of its relatively high toxicity its use is limited. Its methyl homolog, O,O-dimethyl S-(N-isopropylcarbamoylmethyl) dithiophosphate, has been synthesized and studied; it melts at 77°–78° C., LD_{50} for rats 250 mg./kg.

FAC 20 is produced by the same methods as dimethoate. The yields by the first and second methods are close to 90%.

In addition to the compounds described, practical use has been made in agriculture of some other compounds that contain more complex radicals on the nitrogen. The simplest compound of this kind is amidithion.

O,O-Dimethyl S-(N-2-methoxyethylcarbamoylmethyl) dithiophosphate (amidithion, Thiocron) is a crystalline substance, m.p. 46° C. The technical grade product is an unpleasant smelling liquid that does not distill in high vacuum. It is moderately soluble in water, more so in alcohols and esters, but poorly soluble in aliphatic hydrocarbons. Its LD_{50} for rats is 600 to 660 mg./kg.

The chemical properties of amidithion are similar to those of dimethoate. Amidithion is synthesized by the reaction of alkali or ammonium dimethyldithiophosphates with the methoxyethylamide of monochloroacetic acid:

$$(CH_3O)_2\overset{\overset{S}{\|}}{P}SK + ClCH_2CONHCH_2CH_2OCH_3 \rightarrow$$

$$(CH_3O)_2\overset{\overset{S}{\|}}{P}SCH_2CONHCH_2CH_2OCH_3 + KCl$$

It is a good contact acaricide and insecticide with prolonged systemic action for 15–20 days. It is marketed in the form of emulsive concentrates. The compound is not yet used on a large scale, but it, undoubtedly, is interesting because of its low toxicity to animals and man. Its metabolism in plants apparently is analogous to that of dimethoate, but there are no accurate data yet.

O,O-Dimethyl S-(morpholinocarbamoylmethyl) dithiophosphate (morphothion, Ekatin-F, Ekatin-M) is a white crystalline substance, m.p. 63° to 64° C. It is poorly soluble in water, highly soluble in alcohols, ketones, and esters of acetic acid, and only slightly soluble in aliphatic saturated hydrocarbons. Its LD_{50} for rats is 190 mg./kg. The chemical properties of morphothion are similar to those of the compounds described. It is marketed in small quantities in Switzerland in the form of an emulsive concentrate.

Morphothion is produced by the reaction of salts of dimethyldithiophosphoric acid with monochloroacetomorpholide:

$$(CH_3O)_2\overset{\overset{S}{\|}}{P}SNa + ClCH_2CON\underset{}{\bigcirc}O \rightarrow (CH_3O)_2\overset{\overset{S}{\|}}{P}SCH_2CON\underset{}{\bigcirc}O + NaCl$$

O,O-Dimethyl S-(N-methyl-N-formylcarbamoylmethyl) dithiophosphate (formothion, Antio) is a clear liquid, b.p. 25°–26° C., d^{26} 1.361, poorly soluble in water but highly soluble in organic solvents. Its LD_{50} for rats is 330 mg./kg.

Its chemical properties are similar to those of dimethoate, but it is more stable in storage and on heating. One of its principal metabolites in animals is O,O-dimethyl S-(carboxymethyl) dithiophosphate, which is further hydrolyzed to dimethylthiophosphoric acid and phosphoric acid. Formothion is produced by the reaction of salts of dimethyldithiophosphoric acid with N-formyl-N-methylchloroacetamide:

$$(CH_3O)_2\overset{S}{\overset{\|}{P}}SK + ClCH_2CON(CH_3)CHO \rightarrow (CH_3O)_2\overset{S}{\overset{\|}{P}}SCH_2CON(CH_3)CHO + KCl$$

It is marketed in the form of a 25% emulsive concentrate and is used to control various sucking and some chewing plant pests at dosages of 0.5–1 kg./ha.

O,O-Diethyl S-(N-methyl-N-carboethoxycarbamoylmethyl) dithiophosphate (mecarbam) is a clear liquid, b.p. 144° C. at 0.02 mm. of Hg, m.p. 9° C., d^{20} 1.222, volatility at 40° C. 6×10^{-6} mg./m.^{3} About 0.1% of the compound dissolves in water at room temperature, 4% in hexane, and 2% in kerosine; it is highly soluble in acetone, acetonitrile, and many other organic solvents. The LD_{50} for rats is 36–39 mg./kg. Under normal conditions of storage it is rather stable. At pH 3 it is hydrolyzed according to the following scheme:

$$(C_2H_5O)_2\overset{S}{\overset{\|}{P}}SCH_2CON(CH_3)COOC_2H_5 + 2\,H_2O \rightarrow$$
$$(C_2H_5O)_2\overset{S}{\overset{\|}{P}}SCH_2COOH + CO_2 + C_2H_5OH + CH_3NH_2$$

The metabolism of mecarbam in plants apparently proceeds in a manner similar to that of dimethoate. It is produced by the reaction of sodium (or potassium) diethyldithiophosphate with N-carboethoxy-N-methylchloroacetamide. To control sucking and chewing plant pests it is used at dosages on the order of 100–200 g./ha. It is most often marketed in the form of a 40% emulsive concentrate.

The next group of insecticidal derivatives of dithiophosphoric acid are compounds of the general formula (XI). The most important representatives of this group are phorate (Thimet), carbophenothion (Trithion), phenkapton, ethion, etc.

O,O-Diethyl S-(ethylthiomethyl) dithiophosphate (phorate, Thimet) is a clear liquid, b.p. 100° C. at 0.4 mm. of Hg, vapor pressure at 28° C. 8.4×10^{-4} mm. of Hg, volatility 12.4 mg./m.^{3} It is highly soluble in most organic solvents; its solubility in water is about 70 mg./liter. Phorate is relatively unstable to hydrolysis and at 70° C. and pH 8 50% is saponified in 2 hours; in acid medium it is more stable. It belongs in the category of very toxic compounds; the LD_{50} for rats is 1.1–2.3 mg./kg.

It is easily oxidized to the corresponding sulfoxide, which is more resistant to hydrolysis than the starting compound. This usually explains the comparatively long insecticidal action of phorate in plants. The metabolism of phorate in plants can be represented as follows:

$$
\begin{array}{c}
\overset{S}{\underset{\|}{}} \\
\longmapsto (C_2H_5O)_2PSCH_2SC_2H_5 \longmapsto
\end{array}
$$

$$
\begin{array}{ccc}
\overset{S}{\underset{\|}{}}\;\overset{O}{\underset{\|}{}} & & \overset{O}{\underset{\|}{}}\;\overset{O}{\underset{\|}{}} \\
(C_2H_5O)_2PSCH_2SC_2H_5 & \longrightarrow & (C_2H_5O)_2PSCH_2SC_2H_5 \\
\downarrow & & \downarrow \\
\overset{S}{\underset{\|}{}}\;\overset{O}{\underset{\|}{}} & & \overset{O}{\underset{\|}{}}\;\overset{O}{\underset{\|}{}} \\
(C_2H_5O)_2PSCH_2SC_2H_5 & \longrightarrow & (C_2H_5O)_2PSCH_2SC_2H_5 \\
\underset{\|}{}\;\underset{O}{} & & \underset{\|}{}\;\underset{O}{} \\
\downarrow & & \downarrow \\
\overset{S}{\underset{\|}{}} & & \overset{S}{\underset{\|}{}} \\
(C_2H_5O)_2PSH & \longrightarrow & (C_2H_5O)_2POH \\
& \searrow \; H_3PO_4 \; \swarrow &
\end{array}
$$

The metabolites of phorate in most cases are as highly toxic to animals as the starting compound. The final products of metabolism are practically inactive biologically.

Phorate is marketed in various formulations for application to the soil. The most frequently used is a powder on activated carbon that contains 25 or 44% active ingredient. It is used mainly to protect cotton seedlings from infestation by aphids and mites. Its effect is so long lasting that it provides complete protection to the plants for 20–25 days.

Two methods are described in the literature for its synthesis:

1. Reaction of diethyldithiophosphoric acid with formaldehyde and ethyl mercaptan:

$$
\overset{S}{\underset{\|}{}}\qquad\qquad\qquad\overset{S}{\underset{\|}{}}
$$
$$(C_2H_5O)_2PSH + CH_2O + C_2H_5SH \rightarrow (C_2H_5O)_2PSCH_2SC_2H_5 + H_2O$$

Phorate is obtained by this reaction in about 70% yield. The process is very convenient and takes place at room temperature in aqueous medium. The reaction product is easily isolated by extraction with hydrocarbons or their chlorine derivatives.

2. Reaction of salts of diethyldithiophosphoric acid with chloromethyl ethyl sulfide:

$$
\overset{S}{\underset{\|}{}}\qquad\qquad\qquad\overset{S}{\underset{\|}{}}
$$
$$(C_2H_5O)_2PSNa + ClCH_2SC_2H_5 \rightarrow (C_2H_5O)_2PSCH_2SC_2H_5 + NaCl$$

The process is carried out in acetonitrile or other suitable solvent. The final product is obtained in more than 80% yield. The second method is somewhat more complicated, since it requires preliminary preparation of chloromethyl ethyl sulfide.

O,O-Diisopropyl S-(ethylsulfonylmethyl) dithiophosphate (Aphidan) is the closest analog of phorate. It is produced by oxidation of the appropriate homolog of phorate with hydrogen peroxide. It is marketed in the form of a 5% granulated formulation and is used to control aphids by application to the soil. The LD_{50} for rats is 84 mg./kg.

As a result of the ready availability of this group of compounds, a large number of analogs and homologs of phorate have been synthesized, some of them with high insecticidal activity and moderate toxicity to mammals.

O,O-Diethyl S-(4-chlorophenylthiomethyl) dithiophosphate (carbophenothion, Trithion) is a light-yellow oil, b.p. 130° C. at 1 mm. of Hg, vapor pressure at 20° C. 3.05×10^{-7} mm. of Hg, volatility 0.0057 mg./m.^{3} It is almost insoluble in water but highly soluble in organic solvents. The LD_{50} for rats is 10–30 mg./kg. Carbophenothion is a rather powerful contact acaricide and insecticide; it is used to control various sucking plant pests. In the United States the following residues of carbophenothion are permitted (in p.p.m.): in quinces and apricots 0.8, in citrus fruits 2, and in sugar beets 5. It is marketed in the form of a 25% wettable powder or emulsive concentrate.

Methyl Trithion is the methyl homolog of carbophenothion (b.p. 125° C. at 0.01 mm. of Hg). It is considerably less toxic to mammals and to insect plant pests. The LD_{50} for rats is 180 mg./kg. It has to be used at higher dosages, and its use, therefore, is not economical.

Carbophenothion and Methyl Trithion are produced by the reaction of salts of the dialkyldithiophosphoric acid with *p*-chlorophenyl chloromethyl sulfide; the latter compound can be synthesized in satisfactory yields from *p*-chlorothiophenol, formalin, and concentrated HCl:

O,O-Diethyl S-(2,5-dichlorophenylthiomethyl) dithiophosphate (phenkapton) is a clear oil, b.p. 120° C. at 0.001 mm. of Hg, d_4^{21} 1.3507, vapor pressure at 20° C. 4.1×10^{-8} mm. of Hg, volatility 0.00085 mg./m.^{3} It is practically insoluble in water but highly soluble in most organic solvents. The LD_{50} is 200–250 mg./kg. It is a selective acaricide of prolonged action. It is marketed in the form of a 25% wettable powder and an emulsive concentrate and is used at 0.02% active ingredient.

Phenkapton is produced similarly to carbophenothion. The 2,5-dichlorothiophenol necessary for the reaction is synthesized as follows:

Somewhat similar to the compounds described is ethion.

O,O,O,O-Tetraethyl S,S'-methylenebis(dithiophosphate) (ethion) is a yellow oil, b.p. 164°–165° C. at 0.3 mm. of Hg, d_4^{20} 1.2277. It is poorly soluble in aliphatic hydrocarbons, highly soluble in aromatic hydrocarbons and their halogen derivatives and practically insoluble in water. The LD_{50} for rats is 55 mg./kg. It is used to control plant-feeding mites and aphids, and marketed in the form of a 25% wettable powder and an emulsive concentrate.

Ethion is produced by the condensation of salts of diethyldithiophosphoric acid with chlorobromomethane or dibromomethane with heating in an organic solvent:

$$2(C_2H_5O)_2\overset{S}{\underset{\|}{P}}SNH_4 + CH_2Br_2 \rightarrow (C_2H_5O)_2\overset{S}{\underset{\|}{P}}SCH_2\overset{S}{\underset{\|}{P}}(OC_2H_5)_2 + 2\,NH_4Br$$

When dibromomethane is used, the yield does not exceed 50%, but with chlorobromomethane it may reach 80%.

A large number of different analogs and homologs of ethion have been studied and among them have been found active insecticides and acaricides, but they have not yet been used practically. Of the compounds of general formula (XII), thiometon, disulfoton, and others are used in agriculture.

O,O-Dimethyl S-(2-ethylmercaptoethyl) dithiophosphate (thiometon, Compound M-81, Intrathion, Ekatin) is a colorless oil with a strong, unpleasant odor, b.p. 104° C. at 0.3 mm. of Hg, vapor pressure at 20° C. 3×10^{-4} mm. of Hg, volatility 4.0 mg./m.³, d_4^{20} 1.209. At 25° C., its solubility in water is 200 mg./liter; it is highly soluble in most organic solvents with the exception of aliphatic and some alicyclic hydrocarbons. At normal temperatures thiometon is stable, but it may break down when heated. The first step in its thermal conversion is a Pishchemuka rearrangement. It is hydrolyzed by water similarly to methyl demeton.

The metabolic path of thiometon in plants is represented as follows:

$$(CH_3O)_2\overset{S}{\underset{\|}{P}}SCH_2CH_2SC_2H_5$$

$$(CH_3O)_2\overset{S}{\underset{\|}{P}}SCH_2\overset{O}{\underset{\|}{S}}C_2H_5 \longrightarrow (CH_3O)_2\overset{O}{\underset{\|}{P}}SCH_2\overset{O}{\underset{\|}{S}}C_2H_5$$

$$(CH_3O)_2\overset{S}{\underset{\|}{P}}SCH_2\overset{O}{\underset{\|}{S}}C_2H_5 \longrightarrow (CH_3O)_2\overset{O}{\underset{\|}{P}}SCH_2\overset{O}{\underset{\|}{S}}C_2H_5$$

$$\longrightarrow (CH_3O)_2\overset{O}{\underset{\|}{P}}OH \longleftarrow$$

$$H_3PO_4$$

The LD_{50} of thiometon for different animals is 70–120 mg/kg. The maximum permissible concentration in air is 0.1 mg./m.³

Thiometon is a good systemic insecticide with about the same duration of action as methyl demeton, but its initial insecticidal activity is somewhat weaker. It is assumed that its contact insecticidal properties develop after conversion of part of the compound to the thiolo isomer of methyl demeton or its metabolites. Thiometon is marketed in the form of emulsive concentrates containing from 20–50% active ingredient, and also in granulated formulations.

The principal method of producing it in a yield of about 90% is by reacting salts of dimethyldithiophosphoric acid with 2-chlorodiethyl sulfide:

$$(CH_3O)_2\overset{S}{\underset{\|}{P}}SNa + ClCH_2CH_2SC_2H_5 \rightarrow (CH_3O)_2\overset{S}{\underset{\|}{P}}SCH_2CH_2SC_2H_5 + NaCl$$

The process is carried out either in aqueous medium with good stirring or in organic solvents. Thiometon can be synthesized also by the reaction of 2-hydroxydiethyl sulfide with dimethyldithiophosphoric acid in the presence of sodium hydroxide and *p*-toluenesulfonyl chloride:

$$HOCH_2CH_2SC_2H_5 + (CH_3O)_2\overset{S}{\underset{\|}{P}}SH + CH_3\text{-}\langle\text{-}\rangle\text{-}SO_2Cl + 2NaOH \longrightarrow$$

$$(CH_3O)_2\overset{S}{\underset{\|}{P}}SCH_2CH_2SC_2H_5 + CH_3\text{-}\langle\text{-}\rangle\text{-}SO_2ONa + NaCl + 2H_2O$$

However, the yield in this case does not exceed 75–78%.

It has been surmised that an intermediate product in this process is the ester of *p*-toluenesulfonic acid with 2-hydroxydiethyl sulfide, but it has not been possible to isolate this compound in the reaction process.

O,O-Diethyl S-(2-ethylmercaptoethyl) dithiophosphate (disulfoton, Compound M-74, Di-Syston, Dithiosystox) is a colorless liquid with an unpleasant odor, b.p. 113° C. at 0.4 mm. of Hg, vapor pressure at 20° C. 1.8×10^{-4} mm. of Hg, volatility 2.7 mg./m.³, d_4^{20} 1.144. Its solubility in water at 20° C. is about 25 mg./liter; it is highly soluble in most organic solvents. In acid medium disulfoton is resistant to hydrolysis, but in alkaline medium it is hydrolyzed more rapidly. At 70° C., 50% is hydrolyzed in 60 hours at pH 5 and in 7.2 hours at pH 9.

The chemical properties of disulfoton are similar to those of thiometon, but disulfoton is considerably more stable. Its metabolism in plants follows an identical path. Like thiometon it is capable of adding various alkyl halides and dimethyl sulfate at the sulfur of the ester group and yielding sulfonium compounds. These sulfonium compounds are toxic to mammals but do not show contact insecticidal properties. This apparently is due to the low solubility of the sulfonium compounds in lipoids, and their low penetrative ability through the cutaneous covering of insects. The anticholinesterase activity of this group of compounds is very great.

Disulfoton belongs to the toxic group of compounds (LD_{50} for rats 2.5–12.5 mg./kg.) and its use is not permitted in the Soviet Union. In other countries this compound is widely used by its application to the soil to control sucking pests of cotton and other crops. However, for the control of vectors of potato virus diseases its use is permitted on only seed potatoes. In countries other than the Soviet Union disulfoton is marketed in the form of a 50% mixture with activated carbon and as 2.5, 5, and 10% granulated formulations.

Several methods are known for the preparation of disulfoton; the two most important are:

1. Reaction of salts of diethyldithiophosphoric acid with 2-chlorodiethyl sulfide:

$$(C_2H_5O)_2\overset{S}{\underset{\|}{P}}SK + ClCH_2CH_2SC_2H_5 \rightarrow (C_2H_5O)_2\overset{S}{\underset{\|}{P}}SCH_2CH_2SC_2H_5 + KCl$$

2. Heating diethyldithiophosphoric acid with 2-chlorodiethyl sulfide at about 100 C.:

$$(C_2H_5O)_2\overset{S}{\underset{\|}{P}}SH + ClCH_2CH_2SC_2H_5 \rightarrow (C_2H_5O)_2\overset{S}{\underset{\|}{P}}SCH_2CH_2SC_2H_5 + HCl$$

Disulfoton can be synthesized by the first method with yields of more than 90%, but the product obtained by the second method is impure and the yield does not exceed 80%.

In Czechoslovakia a new compound of this series, Tetrathion, has been produced which has mixed ester radicals.

O-Methyl O-ethyl S-(2-ethylthioethyl) dithiophosphate (Tetrathion) is a thick, oily liquid with an unpleasant odor. Its LD_{50} for experimental animals is 1.2–22.3 mg./kg. The chemical properties of this insecticide are similar to those of thiometon and disulfoton. It is marketed in the form of a 50% emulsive concentrate and is intended for application to the soil by irrigation of the plants with emulsions of the compound.

It is produced in the same manner as the compounds described but the starting material is O-methyl O-ethyldithiophosphoric acid. This acid can be synthesized from phosphorus pentasulfide and a mixture of methyl and ethyl alcohols, used in equimolecular quantities:

$$P_2S_5 + 2\,CH_3OH + 2\,C_2H_5OH \rightarrow 2\;\begin{matrix}CH_3O\\C_2H_5O\end{matrix}\!\!>\!\!P\!\!<\!\!\begin{matrix}S\\SH\end{matrix} + H_2S$$

Since small amounts of dimethyl- and diethyldithiophosphoric acids are obtained in this reaction, the final product contains thiometon and disulfoton as impurities.

O,O-Dimethyl S-(2-ethylsulfinyl)ethyl dithiophosphate (LD_{50} 60–80 mg./kg.) and *O,O-diethyl S-(2-ethylsulfinyl) dithiophosphate* (LD_{50} 3.6 mg. per kg.) have been proposed for use as insecticides and acaricides. A large number of analogs and homologs of the previously described compounds have been synthesized, but no active insecticides with low toxicity for mammals have yet been found.

Compounds of the general formula (XIII) have a strong insecticidal effect not only on sucking plant pests, but also on various leaf-eating insects. In many cases these compounds are very promising as substitutes for the chlorine-containing insecticides such as DDT and its analogs. The properties of the most important compounds of this type that are used in agriculture are described in detail below.

O,O-Dimethyl S-(N-phthalimidomethyl) dithiophosphate (Imidan, Phthalophos) is a white crystalline substance, m.p. 72°–72.7° C. At 25° C., about 25 mg. of the compound dissolves in 1 liter of water; it is highly soluble in acetone, methyl ethyl ketone, cyclohexanone, methylene chloride, xylene, and other organic solvents, with the exception of the aliphatic hydrocarbons, in which it is slightly soluble. The LD_{50} for experimental animals is 40–200 mg./kg.

In acid medium Imidan is resistant to hydrolysis and at pH 4.5 in aqueous solution it is 50% hydrolyzed in 15 days, while at pH 7 50% is hydrolyzed in 12 hours, and at pH 8.3 in four hours. The final products of the hydrolysis are phthalimide, dimethylthiophosphoric acid, and formaldehyde:

$$(CH_3O)_2\overset{S}{\overset{\|}{P}}SCH_2N\underset{CO}{\overset{CO}{\diagup}}\!\!\!\!\bigcirc + 2H_2O \longrightarrow$$

$$(CH_3O)_2\overset{S}{\overset{\|}{P}}OH + \underset{CO}{\overset{CO}{\diagup}}\!\!\!\!\bigcirc NH + CH_2O + H_2S$$

However, these compounds are capable of further hydrolysis to phosphoric and phthalic acids. Oxidizing agents also break down Imidan and the oxidation is directed primarily at the sulfur atom.

Imidan is marketed in the form of a 20% emulsive concentrate, a 50% wettable powder, and a 10% granulated formulation. It is recommended for the control of various pests of cotton, fruits, and other crops. Its use in place of DDT to control the codling moth is of interest, since it breaks down relatively easily and does not leave toxic residues on the fruits.

It is produced by the following method:

$$\bigcirc\!\!\!\!\underset{CO}{\overset{CO}{\diagup}}NH + CH_2O \longrightarrow \bigcirc\!\!\!\!\underset{CO}{\overset{CO}{\diagup}}NCH_2OH \xrightarrow{HCl}$$

$$\bigcirc\!\!\!\!\underset{CO}{\overset{CO}{\diagup}}NCH_2Cl \xrightarrow{(CH_3O)_2\overset{S}{\overset{\|}{P}}SNa} (CH_3O)_2\overset{S}{\overset{\|}{P}}SCH_2N\underset{CO}{\overset{CO}{\diagup}}\!\!\!\!\bigcirc$$

In the first step hydroxymethylphthalimide is obtained from formalin and phthalimide; the hydroxymethylphthalimide is converted to chloromethylphthalimide by reaction with HCl; by reaction of the chloromethylphthalimide with salts of dimethyldithiophosphoric acid Imidan is obtained. All the steps in this process give good yields.

O,O-Diethyl S-(6-chlorobenzoxazolinyl-3-methyl) dithiophosphate (phosalone) is a white crystalline substance with a garlicky odor, m.p. 45°–47° C. It is practically insoluble in water, but highly soluble in many organic solvents. Its LD_{50} for rats is 135 mg./kg., and for mice 180 mg./kg. It is relatively stable in acid medium; in alkaline medium it is rapidly hydrolyzed. The principal products of the hydrolysis are 6-chlorobenzoxazolone, diethylthiophosphoric acid, and formaldehyde.

When oxidizing agents act on phosalone, the primary step is splitting off of the thiono sulfur and conversion of the compound to O,O-diethyl S-(6-chlorobenzoxazolinyl-3-methyl) thiophosphate, which is relatively unstable and quickly breaks down. It is assumed that this compound is the first metabolite of phosalone in plants.

Phosalone has been recommended for the control of pests of various crops and it is very promising as a substitute for DDT and other persistent chlorine-containing insecticides. It is used also to treat seeds in order to protect the seedlings from damage by insect pests and mites.

Phosalone is produced by the condensation of sodium or ammonium diethyldithiophosphate with 6-chloro-3-(chloromethyl)benzoxazolone:

O,O-Dimethyl S-(3,4-dihydro-4-keto-1,2,3-benzotriazinyl-3-methyl) dithiophosphate (azinphos methyl, Guthion, Gusathion) is a white crystalline substance, m.p. 73°–74° C., vapor pressure 2.2×10^{-7} mm. of Hg at 20° C., volatility 0.004 mg./m.^{3} When heated in a high vacuum it breaks down with the evolution of gaseous decomposition products. At 20° C., about 30 mg. of azinphos methyl dissolves in 1 liter of water.

Azinphos methyl belongs to the group of compounds with high toxicity to mammals; its LD_{50} for rats is 15–17 mg./kg. It is chemically stable and can be stored under ordinary conditions for an unlimited length of time. At pH 5 50% is hydrolyzed at 70° C. in 8.9 hours, and at 20° C. in 240 days. In alkaline medium azinphos methyl breaks down several times faster. When it is carefully hydrolyzed in acid medium, azinphos methyl yields 3,4-dihydro-4-keto-1,2,3-benzotriazine, but in alkaline medium it yields a hydroxymethyl derivative:

[Reaction scheme: benzotriazinone-NCH₂SP(OCH₃)₂ with S and O + H₂O → benzotriazinone-NCH₂OH + (CH₃O)₂PSH (with S); then → benzotriazinone-NH + CH₂O]

The action of oxidizing agents on azinphosmethyl causes splitting off of the thiono sulfur and formation of *P-O*-azinphos methyl.

The ethyl homolog of azinphos methyl (azinphos ethyl, Gusathion-A, m.p. 53° C.) has similar properties.

Azinphos methyl is recommended for the control of various plant pests and gives good results on cotton, fruits, and many other crops. It is marketed in the form of emulsive concentrates and wettable powders for spraying in aqueous suspensions.

It is produced by the reaction of the *N*-halomethyl derivatives of azimidobenzoyl with salts of dimethyldithiophosphoric acid:

[Reaction: benzotriazinone-NCH₂Br + (CH₃O)₂PSNa → benzotriazinone-NCH₂SP(OCH₃)₂ + NaBr]

When the methylbromo derivative is used, azinphos methyl is obtained in practically quantitative yield. The 4-oxo-3,4-dihydro-1,2,3-benzotriazine necessary for the synthesis of azinphos methyl can be prepared from the amide of anthranilic acid by the action of sodium nitrite in acid medium:

[Reaction: anthranilamide + HNO₂ → benzotriazinone + 2H₂O]

The broad spectrum of action of the heterocyclic derivatives of dithiophosphoric acid has attracted the attention of investigators to this class of compounds, and new experimental compounds are continuously appearing and being studied for use as pesticides. Tests are going on in different countries of *O,O*-dimethyl *S*-(5-methoxy-1,3,4-thiadiazolinyl-3-methyl) dithiophosphate (XVI) (m.p. 39°–40° C., LD_{50} 25–48 mg./kg.) and its 5-ethoxy homolog (XVII) (m.p. 49°–50° C., LD_{50} 225 mg./kg.):

[Structures: (CH₃O)₂PSCH₂N-thiadiazoline-OCH₃ (XVI); (CH₃O)₂PSCH₂N-thiadiazoline-OC₂H₅ (XVII)]

In England and some other countries a compound known as menazon is used to control aphids.

O,O-Dimethyl S-(4,6-diamino-1,3,5-triazinyl-2-methyl) dithiophosphate (menazon, Sayfos) is a white crystalline substance, m.p. 160°–162° C. Its solubility in water is about 0.1%; it is slightly soluble in most organic solvents. It is a systemic insecticide with low toxicity to mammals. Its LD_{50} is 900 mg./kg. It is capable of penetrating plants through the root system and imparting an insecticidal effect of long duration.

Its chemical properties are similar to those of the other dithiophosphates of this type. It is marketed in the form of a 70% wettable powder as well as granules which are applied to the soil. It has given especially good results in the control of vectors of the viruses that cause potato diseases.

Menazon is produced by the reaction of alkali or ammonium dimethyldithiophosphates with 4,6-diamino-2-chloromethyl-1,3,5-triazine:

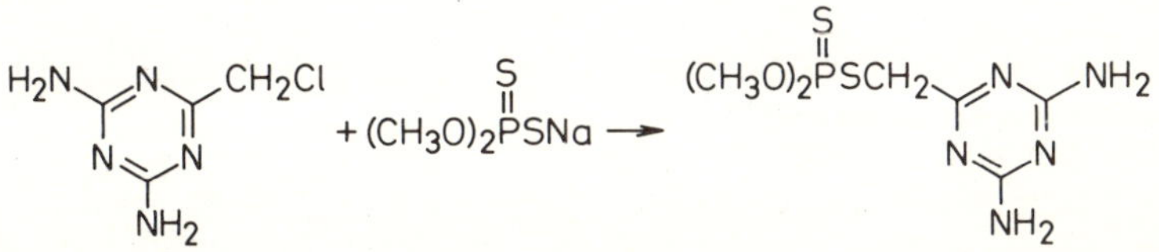

1,4-Dioxane-2,3-S,S'-bis(O,O-diethyl dithiophosphate) (dioxathion, Delnav) is a dark-yellow oil with an unpleasant odor which does not distill without decomposition in vacuum. The technical grade product is a mixture of 70% *cis-* and *trans-*isomers of 1,4-dioxane-2,3-S,S'-bis(diethyldithiophosphate) with 30% tetraethyltrithiopyrophosphate and O,O-diethyl S-(1,4-dioxenyl) dithiophosphate. The LD_{50} of the *cis*-isomer is 65 mg./kg., and that of the *trans*-isomer is 240 mg./kg. Dioxathion is marketed in the form of a 25% wettable powder and an emulsive concentrate.

The pure *trans*-compound is a crystalline substance, m.p. 80°–81° C., while the *cis-* compound is a liquid. The *cis*-isomer is an effective insecticide, but the toxicity of the *trans*-isomer to insects is only about one-third as great.

Dioxathion is thermally unstable and even at 130°–140° C. it decomposes, giving off one molecule of diethyldithiophosphoric acid. The *trans*-isomer breaks down especially easily:

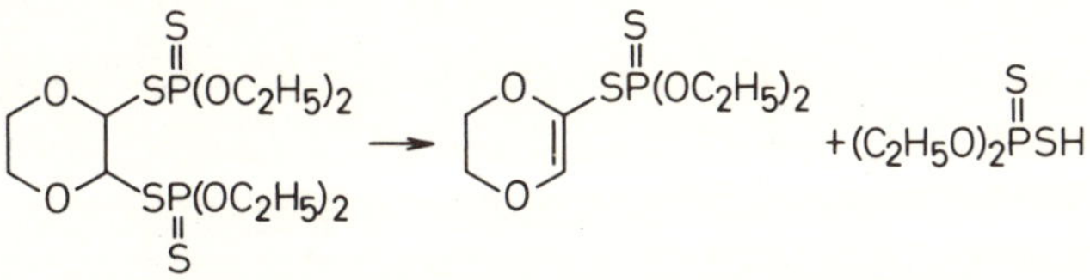

Upon oxidation of dioxathion the corresponding P-O-compound is formed. The best method of synthesizing dioxathion apparently is the addition of bis(diethoxythiophosphone) disulfide to 1,4-dioxene:

$$\text{\Large O} \diagdown\!\!\diagup \text{\Large O} \quad + \quad \left[(C_2H_5O)_2\overset{\overset{S}{\|}}{P}S \right]_2 \quad \longrightarrow \quad \text{\Large O} \diagdown\!\!\diagup \overset{\overset{\displaystyle O \diagdown SP(OC_2H_5)_2 \; \overset{S}{\|}}{}}{\underset{\displaystyle O \diagup SP(OC_2H_5)_2 \; \underset{S}{\|}}{}}$$

It also is possible to prepare it from diethyldithiophosphoric acid or its salts and 2,3-dichloro-1,4-dioxane:

$$\begin{array}{c}\text{O}\diagdown\!\!\diagup\text{Cl} + (C_2H_5O)_2\overset{\overset{S}{\|}}{P}SH \xrightarrow{C_5H_5N} \\ \text{O}\diagup\!\!\diagdown\text{Cl} + (C_2H_5O)_2\overset{\overset{S}{\|}}{P}SH \xrightarrow{ZnCl_2} \end{array} \quad \text{O}\diagdown\!\!\diagup\overset{O\diagdown SP(OC_2H_5)_2}{\underset{O\diagup SP(OC_2H_5)_2}{}}$$

In the first case pyridine or another tertiary amine is used to bind the HCl; the latter reaction takes place at 70° C. in the presence of catalytic amounts of anhydrous zinc chloride.

As yet, of the mixed esters of dithiophosphoric acid with the general formula (XIV) a use has been found for only a few compounds. First mention should be made of Cidial.

O,O-Dimethyl S-(1-carboethoxybenzyl) dithiophosphate (Cidial) is an oily liquid with a characteristic unpleasant odor, not distillable in vacuum without decomposition, highly soluble in most organic solvents and insoluble in water, d_4^{20} 1.226. Its ignition temperature is 168°–172° C. It is thermally unstable and when heated to 120° C. for 110 hours almost 50% decomposes; it withstands 30 days heating at 50° C. without noticeable decomposition.

Cidial may show insecticidal effects for 12–25 days on plants. It is an insecticide with a broad range of action and is used to control plant-feeding mites and sucking insects and also leaf-eating plant pests. It gives good results in the control of the codling moth. Its LD_{50} for rats is 250–300 mg./kg.

Cidial is produced by the reaction of alkali dimethyldithiophosphates with esters of halophenylacetic acids:

$$C_6H_5CHXCOOC_2H_5 + (CH_3O)_2\overset{\overset{S}{\|}}{P}SNa \rightarrow (CH_3O)_2\overset{\overset{S}{\|}}{P}S\overset{\overset{C_6H_5}{|}}{C}HCOOC_2H_5 + NaX$$

In addition to the insecticides and acaricides, a number of compounds having a fungicidal effect have been found among the derivatives of dithiophosphoric acid. An example of these compounds is phosbutyl, which displays an intraplant fungicidal effect.

O-Ethyl S-phenyl N-butylamidodithiophosphate (phosbutyl) is a colorless liquid with a specific unpleasant odor, b.p. 150°–151° C. at 0.3 mm. of Hg. It is poorly soluble in water but highly soluble in most organic solvents,

d_4^{20} 1.1445. Its LD_{50} for rats is about 300 mg./kg. It is used in the form of an emulsive concentrate and has curative fungicidal properties for many plants that are infected with fungus diseases. It is rapidly absorbed by plants and is not washed off by rain even the day after application. The dosage of the compound is 1–2 kg./ha. When it is stored under ordinary conditions it is stable, but when it is heated with alkalies it is hydrolyzed with the formation of phosphoric acid, alcohol, thiophenol, and butylamine. In plants in addition to hydrolysis it undergoes oxidation.

Phosbutyl is produced by the reaction of butylamine with O-ethyl S-phenyl chlorodithiophosphate in the presence of an HCl acceptor:

$$\begin{matrix} C_2H_5O \\ C_6H_5S \end{matrix} \!\!\! \overset{S}{\underset{}{\text{PCl}}} + C_4H_9NH_2 + NaOH \rightarrow \begin{matrix} C_2H_5O \\ C_6H_5S \end{matrix} \!\!\! \overset{S}{\underset{}{\text{PNHC}_4H_9}} + NaCl + H_2O$$

O,O-Diisopropyl S-(2-benzenesulfamidoethyl) dithiophosphate (Compound R-4461) is a new experimental herbicide for the control of weeds in cotton, alfalfa, rice, cabbage, and other crops. The technical grade product is a brown liquid that cannot be distilled, d^{22} 1.25. It is poorly soluble in water and organic solvents of petroleum origin but highly soluble in aromatic hydrocarbons and their halogen derivatives. Its LD_{50} for rats is 770 mg./kg. It is marketed in the form of emulsive concentrates and granules for application to the soil and is effective at dosages of 6–7 kg./ha.

Compound R-4461 is produced by the following reaction:

$$(C_3H_7O)_2\overset{S}{\underset{\|}{P}}SNa + ClCH_2CH_2NHSO_2C_6H_5 \rightarrow$$
$$NaCl + (C_3H_7O)_2\overset{S}{\underset{\|}{P}}SCH_2CH_2NHSO_2C_6H_5$$

O-Ethyl S,S-diphenyl dithiophosphate (hinosan) is a yellow oil, insoluble in water, soluble in organic solvents. The LD_{50} for rats is 212 mg./kg., for mice 218 mg./kg. It is marketed in the form of 40% emulsive concentrate or 3% dusts. The compound is a selective fungicide used for the control of *Piricularia oryzae*.

Hinosan is synthesized by reaction of O-ethyl dichlorophosphate with thiophenol in the presence of an acid-combining reagent:

$$C_2H_5OPOCl_2 + 2\,C_6H_5SH \rightarrow C_2H_5O\overset{O}{\underset{\|}{P}}(SC_6H_5)_2$$

Recently a number of new organophosphorus insecticides have been under investigation, for example, *O,O-dimethyl S-2-(acetylamido)ethyl dithiophosphate (amophos)*, m.p. 22°–23° C. It is highly soluble in most organic solvents. It has a systemic effect and is used in the form of aqueous emulsions. Its LD_{50} for experimental animals is 200–400 mg./kg.

The compound butifos, S,S,S-tributyl trithiophosphate, is used as a defoliant for cotton. It boils at 150° C. at 0.3 mm. of Hg, LD_{50} 170 to 250 mg./kg. It is prepared by the oxidation of tributyl trithiophosphite or by the reaction of butyl mercaptan with phosphorus oxychloride. It is used in the form of solutions in oil or of emulsive concentrates at dosages of 1–1.5 kg./ha.

Derivatives of pyrophosphoric acid

The first organophosphorus insecticide, which was manufactured in Germany in 1943, was tetraethyl pyrophosphate, known as Bladan or tepp. With the emergence of information on the insecticidal properties of tetraethyl pyrophosphate and related compounds, a systematic investigation into pesticidal properties of the organic compounds of phosphorus was started. Among the derivatives of pyrophosphoric acid that have been studied a number of compounds have been found that still retain their importance.

The following rules have been determined for the relationship of the insecticidal activity to the structure of esters of pyrophosphoric acid. The insecticidal activity of tetraalkyl pyrophosphates decreases with an increase in size of the ester radical. The most active compound is tetraethyl pyrophosphate; the activity of tetramethyl pyrophosphate is somewhat lower.

The tetraalkyl monothiopyrophosphates show greater insecticidal activity than the tetraalkyl pyrophosphates, but their toxicity to mammals also is increased. Introduction of a second sulfur atom on the phosphorus lowers the toxicity of the compound to mammals without appreciable lowering of the insecticidal activity. Introduction of a third sulfur on the phosphorus decreases the insecticidal activity of the compound. Thus, O,O,O,O-tetraalkyl dithiopyrophosphates have greater insecticidal effect than O,O,O,O-tetraalkyl trithiopyrophosphates with the same hydrocarbon radicals.

Tetraalkyl dithiopyrophosphates containing different hydrocarbon radicals have almost the same insecticidal activity as the corresponding symmetrical tetraalkyl dithiopyrophosphates of the same molecular weight.

All the pyrophosphates have strong contact insecticidal effects and show practically no systemic insecticidal effect, a fact which is associated with their low hydrolytic stability and their rapid decomposition on the plant with the formation of nontoxic products.

Systemic activity appears going from the esters of pyrophosphoric acid to the amides with a decrease in contact insecticidal activity. Octamethylpyrophosphoramide has been proposed as a systemic insecticide. It does not show contact insecticidal activity and, therefore, is used to protect mulberry trees from sucking insect pests.

In the literature many general methods are described for making esters and amides of pyrophosphoric, thio-, dithio-, and trithiopyrophosphoric acids. Only the most important methods that have importance in the production of these compounds will be shown.

The principal methods of syntheses of the tetraalkyl pyrophosphates are:

1. Reaction of trialkyl phosphates with phosphorus oxychloride or other acid chlorides:

$$5\,(RO)_3PO + POCl_3 \rightarrow 3\,(RO)_2\overset{\underset{\|}{O}}{P} - O - \overset{\underset{\|}{O}}{P}(OR)_2 + 3\,RCl$$

$$2\,(RO)_3PO + SOCl_2 \rightarrow (RO)_2\overset{\underset{\|}{O}}{P} - O - \overset{\underset{\|}{O}}{P}(OR)_2 + 2\,RCl + SO_2$$

These reactions take place at a comparatively high temperature and the tetraalkyl pyrophosphates are produced in satisfactory yields, but the technical grade products are somewhat contaminated with other derivatives of phosphoric acid. A dialkyl chlorophosphate also can be used as the acid chloride:

$$(RO)_2\overset{\underset{\|}{O}}{P}Cl + (RO)_3\overset{\underset{\|}{O}}{P} \rightarrow (RO)_2\overset{\underset{\|}{O}}{P} - O - \overset{\underset{\|}{O}}{P}(OR)_2 + RCl$$

2. Reaction of dialkyl chlorophosphates with water in the presence of bases:

$$2\,(RO)_2\overset{\underset{\|}{O}}{P}Cl + H_2O \rightarrow (RO)_2\overset{\underset{\|}{O}}{P} - O - \overset{\underset{\|}{O}}{P}(OR)_2$$

Either organic tertiary amines or alkali carbonates can be used as the bases. The purest product is obtained by this reaction in good yield. This method is suitable also for the synthesis of symmetrical tetraalkyl dithiopyrophosphates from dialkyl chlorothiophosphates:

$$2\,(RO)_2\overset{\underset{\|}{S}}{P}Cl + H_2O \rightarrow (RO)_2\overset{\underset{\|}{S}}{P} - O - \overset{\underset{\|}{S}}{P}(OR)_2$$

When the water is replaced by hydrogen sulfide, tetraalkyl thio- and trithiopyrophosphates can be prepared in satisfactory yields:

$$2\,(RO)_2\overset{\underset{\|}{X}}{P}Cl + H_2S \rightarrow (RO)_2\overset{\underset{\|}{X}}{P} - S - \overset{\underset{\|}{X}}{P}(OR)_2 + 2\,HCl \quad (X = O \text{ or } S)$$

The unsymmetrical tetraalkyl pyrophosphates are produced when salts of dialkylphosphoric acids react with the appropriate dialkyl chlorophosphates:

$$(RO)_2\overset{\overset{X}{\|}}{P}Cl + NaS\overset{\overset{X}{\|}}{P}(OR')_2 \rightarrow NaCl + (RO)_2\overset{\overset{X}{\|}}{P}-S-\overset{\overset{X}{\|}}{P}(OR')_2$$

This is a convenient method of preparing thio- and dithiopyrophosphates.

Tetraalkyl dithiopyrophosphates are synthesized in good yield by the reaction of trialkyl phosphites with *bis*(dialkoxythiophosphone) disulfides:

$$(RO)_3P + [(RO)_2\overset{\overset{S}{\|}}{P}S]_2 \rightarrow (RO)_2\overset{\overset{S}{\|}}{P}-S-\overset{\overset{O}{\|}}{P}(OR)_2 + RS\overset{\overset{S}{\|}}{P}(OR)_2$$

For the industrial production of octamethylpyrophosphoramide and its analogs use is made of the reaction of tetramethyl diamidochlorophosphate with water in the presence of organic or inorganic bases:

$$[(CH_3)_2N]_2\overset{\overset{O}{\|}}{P}Cl + H_2O \rightarrow [(CH_3)_2N]_2\overset{\overset{O}{\|}}{P}-O-\overset{\overset{O}{\|}}{P}[N(CH_3)_2]_2$$

Tetraethyl pyrophosphate (tepp, TEPP) is a colorless liquid with an unpleasant odor, b.p. 82° C. at 0.05 mm. of Hg, d_4^{20} 1.185, vapor pressure at 20° C. 1.55×10^{-4} mm. of Hg, volatility 2.5 mg./m.³ It is miscible with water in all proportions and is highly soluble in alcohol, acetone, aromatic hydrocarbons, and carbon tetrachloride, but poorly soluble in petroleum ether and ligroin. When tepp is heated to 170° C. it breaks down with the formation of ethylene; at 208–215° C. the decomposition takes place violently. Tepp was originally known as Bladan.

Tepp is rapidly hydrolyzed by water even at room temperature, forming diethylphosphoric acid, which is nontoxic for both insects and animals:

$$(C_2H_5O)_2\overset{\overset{O}{\|}}{P}-O-\overset{\overset{O}{\|}}{P}(OC_2H_5)_2 + H_2O \rightarrow 2\,(C_2H_5O)_2\overset{\overset{O}{\|}}{P}OH$$

Under the influence of sunlight the diethylphosphoric acid breaks down completely to phosphoric acid, methane, and carbon monoxide. The formation of the last two products goes through a stage of acetaldehyde formation:

$$(C_2H_5O)_2\overset{\overset{O}{\|}}{P}OH \rightarrow [H_3PO_4 + 2\,CH_3CHO] \rightarrow H_3PO_4 + 2\,CH_4 + 2\,CO$$

Tepp easily reacts with alcohols and amines. In the first case the trialkyl phosphate and diethylphosphoric acid are obtained, and in the second case the amide and the salt of diethylphosphoric acid are obtained. The reaction of tepp with anhydrous HF or potassium bifluoride is interesting;

in this reaction diethyl fluorophosphate is formed, which is toxic to mammals.

Tepp belongs to the category of compounds that are extremely toxic to animals and man, therefore in working with it great care must be exercised. The LD_{50} for rats is 1.12 mg./kg. The insecticidal activity of tepp is only about one-third that of parathion, and its use is continuously decreasing. It is used in the form of a 0.2% aqueous solution to control plant-feeding mites and aphids.

Tepp is synthesized by one of the general methods already described.

Tetraethyl dithiopyrophosphate (sulfotepp, pirofos, Bladafum) is a colorless liquid, b.p. 92° C. at 1 mm. of Hg, vapor pressure at 20° C. 1.7×10^{-4} mm. of Hg, volatility 9 mg./m.^{3}, solubility in water 25 mg./liter; it is highly soluble in most organic solvents except aliphatic hydrocarbons. The LD_{50} for rats is about 5 mg./kg.

Sulfotepp is similar to parathion in insecticidal activity. In the German Federal Republic it is used in small amounts as a fumigant to control mites in greenhouses. It is effective against mollusks and soft scales, and is also produced by one of the general methods described. A large number of homologs and analogs of sulfotepp have been synthesized. The LD_{50} of tetraethyl monothiopyrophosphate is 0.5 mg./kg.

Tetra-n-propyl dithiopyrophosphate (propyl thiopyrophosphate, NPD) is a liquid, b.p. 104° C. at 0.01 mm. of Hg. It is practically insoluble in water, highly soluble in most organic solvents, but slightly soluble in petroleum ether and ligroin. The LD_{50} for rats is 100 mg./kg. It has moderate contact insecticidal activity and is marketed in the form of a 25% wettable powder and an 85% emulsive concentrate. In the United States it is produced on on experimental basis.

Like sulfotepp, it is resistant to hydrolysis and its chemical properties are reminiscent of those of the other pyrophosphates.

Octamethylpyrophosphoramide (schradan, octamethyltetraamide of pyrophosphoric acid, octamethyl, OMPA, Pestox-III, Systam) is a colorless liquid, b.p. 126° C. at 1 mm. of Hg, vapor pressure at 20° C. 6.5×10^{-4} mm. of Hg, volatility 9.5 mg./m.^{3}, crystallizes on cooling, m.p. about 20° C. Schradan is miscible with water in all proportions; it is highly soluble in alcohols, ketones, and halogen derivatives of hydrocarbons, difficultly soluble in petroleum ether and ligroin. The LD_{50} for rats is about 9 mg./kg. The maximum permissible concentration in air is 0.02 mg./m.^{3}

An aqueous solution of the pure compound is neutral and can be stored for an unlimited time. In acid medium schradan breaks down rapidly, but in alkaline medium it is more stable. Thus, 50% of the compound is hydrolyzed in 100 years in neutral aqueous solution, in 200 min. in $1 N$ HCl, and in 70 days in $1 N$ NaOH. Schradan has almost no contact insecticidal activity and acts only as a systemic insecticide through the plant, because in plants and animals it is converted to a more toxic metabolite. The metabolic path of schradan in plants and animals is shown below:

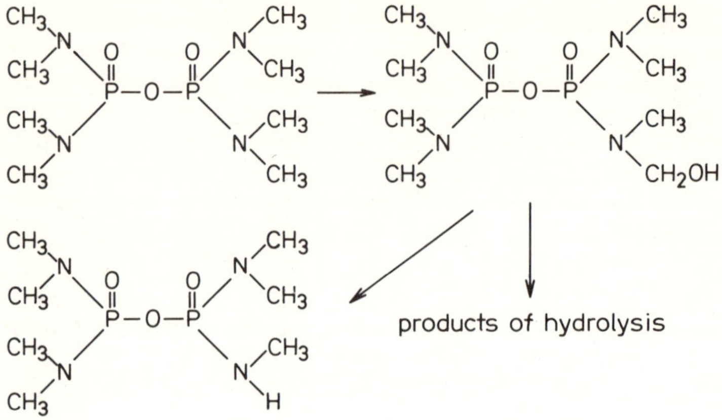

CH₃ structures (schradan oxidation scheme):

$$[(CH_3)_2N]_2P(O)-O-P(O)[N(CH_3)_2]_2 \longrightarrow [(CH_3)_2N]_2P(O)-O-P(O)[N(CH_3)_2][N(CH_3)(CH_2OH)]$$

$$\downarrow$$

$$[(CH_3)_2N]_2P(O)-O-P(O)[N(CH_3)_2][NH(CH_3)]$$

↙   ↓

products of hydrolysis

Schradan can be oxidized to the hydroxymethyl compound also by potassium permanganate or other oxidizing agents. This compound is toxic to warm-blooded animals.

When schradan is chlorinated, it forms as the first step of the process the corresponding chloromethyl derivative; 4 atoms of chlorine can be introduced in all.

Schradan is used on a small scale as a systemic insecticide of prolonged action. In the Soviet Union it is used to control sucking pests of the mulberry tree, since it is harmless to the silkworms that feed on the mulberry leaves.

The principal industrial method for the production of schradan is the reaction of tetramethyl diamidochlorophosphate with water in the presence of tertiary amines, for example, potassium carbonate, pyridine or, still better, triethylamine. The synthesis of schradan is also accomplished by the reaction of ethyl tetramethyldiamidophosphate with tetramethylamidochlorophosphate at 40° C.:

$$[(CH_3)_2N]_2\overset{O}{\overset{\|}{P}}Cl + C_2H_5O\overset{O}{\overset{\|}{P}}[N(CH_3)_2]_2 \rightarrow$$

$$[(CH_3)_2N]_2\overset{O}{\overset{\|}{P}}-O-\overset{O}{\overset{\|}{P}}[N(CH_3)_2]_2 + C_2H_5Cl$$

Derivatives of phosphonic acids

Among the large number of derivatives of alkyl- and arylphosphonic acids, alkyl- and arylthio- and dithiophosphonic acids, and dialkyl- and diarylphosphinic acids that have been studied, active insecticides, acaricides, fungicides, and herbicides have been found, but only a few of the compounds have been used in agriculture; most of the organophosphorus derivatives of this type are in the investigational stage.

The toxicity of the esters of methyl- and ethylphosphonic acids to warm-blooded animals in most cases is higher than that of the analogous mixed esters of phosphoric acid. The corresponding derivatives of thiophosphonic acids are less toxic than the analogous derivatives of thiophosphoric acid. However, it is more complicated to prepare them and as a result their cost is higher. The toxicity of the mixed esters of benzenethiophosphonic acid to mammals as a rule is lower than that of the corresponding mixed esters of thiophosphoric acid. Some compounds of this type are used in agriculture.

The introduction of a second hydrocarbon radical on the phosphorus atom often lowers the insecticidal activity of the compound without lowering the toxicity to mammals.

Introduction of various substituents into a hydrocarbon radical that is connected to phosphorus changes the toxicity of the compound in one direction or the other depending on the nature of the substituent. For example, a hydroxyl group or a halogen atom usually lowers the insecticidal activity of the compound and the toxicity to mammals, although there are exceptions to this rule, an example of which is trichlorfon.

An increase in size of the hydrocarbon radical connected with phosphorus lowers the toxicity and the insecticidal activity of the compound. For example, the LD_{50} for rats (in mg./kg.) of O-ethyl O-(4-nitrophenyl) methylthiophosphonate is 2.5; that of O-ethyl O-(4-nitrophenyl) isopropylthiophosphonate is 25; that of O-ethyl O-(4-nitrophenyl) n-hexylthiophosphonate is 250.

The toxicity to mammals is changed depending on the structure of the ester radicals in the same order as in the esters of phosphoric and thiophosphoric acids. Thus, O-alkyl O-(4-nitrophenyl) esters of alkylphosphonic acids are more toxic than O-alkyl O-(2,4,5-trichlorophenyl) esters of these acids.

Derivatives of the alkylphosphonic and -thiophosphonic acids are prepared by the same methods as the corresponding derivatives of their phosphoric and thiophosphoric analogs. The greatest difficulty is experienced in producing a carbon-phosphorus bond. Usually for the synthesis of esters and amides of phosphonic, phosphinic, thiophosphonic, and thiophosphinic acids use is made of the Arbuzov reaction, the Mikhaelis-Bekker reaction, and other methods that are considered in the description of individual compounds.

The most important compound among the phosphonic acid derivatives is trichlorfon (chlorophos), with a world-wide production of several tens of thousands of tons a year.

O,O-Dimethyl (1-hydroxy-2,2,2-trichloroethyl)phosphonate (trichlorfon, chlorophos, Dipterex, Dylox) is a white crystalline powder, m.p. 73° to 74° C., b.p. 100° C. at 0.1 mm. of Hg, solubility of the compound (in g./100 g.) in water 12.3, benzene 15.2, chloroform 75. It is slightly soluble in paraffinic hydrocarbons. Determination of the molecular weight shows that it is a bimolecular compound. The vapor pressure of trichlorfon at

20° C. is 7.8×10^{-6} mm. of Hg, volatility 0.11 mg./m.^{3} Its LD_{50} for rats is 630 mg./kg. The maximum permissible concentration in air is 0.5 mg./m.^{3} The compound is widely used to control various plant pests (sucking and chewing), including *Eurygaster integriceps*, the codling moth, and the European cornborer, as well as various species of flies, and warble flies of cattle.

It is stable in acid medium and is rapidly hydrolyzed in alkaline medium. The hydrolysis under different conditions follows two paths. In acid medium the first product of hydrolysis is O-methyl-(1-hydroxy-2,2,2-trichloroethyl)phosphoric acid, which is further hydrolyzed with mineralization of the molecule:

$$(CH_3O)_2\overset{O}{\underset{\|}{P}}-\overset{OH}{\underset{|}{C}}HCCl_3 + H_2O \rightarrow CH_3OH + \underset{CH_3O}{\overset{HO}{>}}\overset{O}{\underset{\|}{P}}-\overset{OH}{\underset{|}{C}}HCCl_3$$

Trichlorfon breaks down especially rapidly in the light in dilute solutions. In alkaline medium trichlorfon is dehydrochlorinated and simultaneously rearranges. The principal product of the reaction is O,O-dimethyl O-(2,2-dichlorovinyl) phosphate (dichlorvos, DDVP):

$$(CH_3O)_2\overset{O}{\underset{\|}{P}}-\overset{OH}{\underset{|}{C}}HCCl_3 + KOH \rightarrow (CH_3O)_2\overset{O}{\underset{\|}{P}}OCH=CCl_2 + KCl + H_2O$$

Dichloroacetaldehyde, dimethylphosphoric acid, and some other compounds are formed as products of side reactions. Trichlorfon is a good methylating agent; for example, by the action of potassium iodide it is possible to obtain methyl iodide from it.

Upon prolonged storage of aqueous solutions of trichlorfon partial hydrolysis of the compound occurs, as a result of which the solutions become acid and contain phosphoric and dimethylphosphoric acids, and HCl. Trichlorfon should not be stored in containers of unprotected iron. Reducing agents also decompose trichlorfon.

Two methods may be used for its production on an industrial scale:
1. Condensation of dimethyl phosphite with chloral:

$$(CH_3O)_2PHO + CCl_3CHO \rightarrow (CH_3O)_2\overset{O}{\underset{\|}{P}}-\overset{OH}{\underset{|}{C}}H-CCl_3$$

The reaction takes place at room temperature with the evolution of heat. Trichlorfon can be obtained in practically quantitative yield. In this method it is very important that the starting dimethyl phosphite be pure; the purer the dimethyl phosphite, the greater the yield and the better the quality of the trichlorfon that is obtained.

When technical grade chloral and dimethyl phosphite react, the yield of trichlorfon is about 90%. The product contains as impurities some of the acid ester and other substances that are formed from methyl phosphite and chloral.

2. In the second method the production of dimethyl phosphite and its reaction with chloral are combined in one step. The process is carried out in an organic solvent with the removal of heat:

$$3\ CH_3OH + PCl_3 + CCl_3CHO \rightarrow (CH_3O)_2\overset{O}{\underset{\|}{P}}-\overset{OH}{\underset{|}{C}}HCCl_3 + CH_3Cl + 2\ HCl$$

The yield of trichlorfon by the combined process is >80–85%. To purify the trichlorfon it is recrystallized from water or an organic solvent. The mother liquors from the recrystallization of trichlorfon usually are used for the production of dichlorvos, which is receiving more and more agricultural and household use. Many modification of the second method of preparing trichlorfon exist; various solvents are used (carbon tetrachloride, methyl chloride, chlorobenzene, etc.), and the process is carried out over a wide range of temperatures.

A large number of analogs and homologs of trichlorfon have been synthesized, but their insecticidal properties are considerably weaker than those of trichlorfon. For example, O,O-diethyl (1-hydroxy-2,2,2-trichloroethyl)phosphonate is almost 15 times less active toward flies than trichlorfon, and is more toxic to warm-blooded animals. Among the derivatives of trichlorfon some of its esters with different acids are of interest and have been used as insecticides.

O,O-Dimethyl (1-butyryl-2,2,2-trichloroethyl)phosphonate (butonate) is a colorless liquid, almost odorless, b.p. 112°–114° C. at 0.03 mm. of Hg, d_4^{20} 1.3998, n_D^{20} 1.4740. It is moderately soluble in water but highly soluble in most organic solvents. The LD_{50} for rats is 700 mg./kg. Butonate has a strong contact action. It is used mainly to control ectoparasites of domestic animals and is marketed in the form of an emulsive concentrate. The compound is more resistant to hydrolysis than trichlorfon. Its hydrolysis goes rapidly in alkaline medium; the end products are phosphoric, hydrochloric, and butyric acids.

Butonate is produced by the action of butyric anhydride or acid chloride on trichlorfon:

$$(CH_3O)_2\overset{O}{\underset{\|}{P}}\underset{|}{\underset{OH}{C}}HCCl_3 + (CH_3CH_2CH_2CO)_2O \rightarrow$$

$$(CH_3O)_2\overset{O}{\underset{\|}{P}}\underset{|}{\underset{OCOCH_2CH_2CH_3}{C}}HCCl_3 + C_3H_7COOH$$

O-(Trichlorophenyl) diethylthiophosphinate (Agvitor) is a clear liquid with a weak unpleasant odor, practically insoluble in water, highly soluble in organic solvents. It is used to control carrot and onion flies by application to the soil. Its LD_{50} for rats is 17 mg./kg. It is produced by the reaction of diethylchlorothiophosphinate with sodium or other alkali 2,4,5-trichlorophenolate:

$$(C_2H_5)_2\overset{\overset{S}{\|}}{P}Cl + NaO-\underset{Cl}{\overset{Cl}{\bigcirc}}-Cl \longrightarrow (C_2H_5)_2\overset{\overset{S}{\|}}{P}-O-\underset{Cl}{\overset{Cl}{\bigcirc}}-Cl$$

In addition to the esters of aliphatic phosphonic acids, use is made in agriculture of derivatives of the simplest aromatic phosphonic acids; one of these is EPN.

O-Ethyl O-(4-nitrophenyl) benzenethiophosphonate (EPN) is a white crystalline substance, m.p. 36° C. The technical grade product is a liquid with an unpleasant odor, d_4^{25} 1.268, n_D^{20} 1.5978. It is practically insoluble in water, highly soluble in most organic solvents. In alkaline medium it is relatively rapidly hydrolyzed with the formation of benzenethiophosphonic acid, alcohol, and *p*-nitrophenol. In acid and neutral media EPN is more stable. Its LD_{50} for rats is 8–36 mg./kg. It is a powerful acaricide and the strength and duration of its insecticidal and acaricidal action are only a little less than those of parathion and methyl parathion. In the United States the tolerance for residues of EPN on fruits is 3 p.p.m. It is marketed in the form of wettable powders, dusts, and emulsive concentrates. However, it is more expensive than methyl parathion.

For the production of EPN the following series of reactions can be used:

$$PCl_3 + \bigcirc \xrightarrow{AlCl_3} \bigcirc-PCl_2 \xrightarrow{PSCl_3} C_6H_5\overset{\overset{S}{\|}}{P}Cl_2 \xrightarrow{C_2H_5ONa}$$

$$\underset{C_2H_5O}{\overset{C_6H_5}{>}}\overset{\overset{S}{\|}}{P}Cl \xrightarrow{NaO-\bigcirc-NO_2} \underset{C_2H_5O}{\overset{C_6H_5}{>}}\overset{\overset{S}{\|}}{P}-O-\bigcirc-NO_2$$

Of the analogs of EPN mention should be made of O-ethyl O-2,4-dichlorophenyl benzenethiophosphonate, which has recently been produced in Japan for the control of the onion fly. It is prepared entirely similarly to EPN, but it is considerably less toxic to warm-blooded animals.

2-Chloroethylphosphonic acid (ethrel). This acid is a very hygroscopic white crystalline substance, m.p. 74°–76° C. It is very soluble in water, alcohol, and other hydrophilic solvents. It is stable in aqueous solutions below a pH of 3.5; as the pH rises above 3.5, disintegration of the molecule takes place and it releases free ethylene gas and chloride and phosphate ions. The LD_{50} is 4,200 mg./kg. It is an active plant-growth regulator. Its mode of action appears to be related to its ability to release ethylene to plant tissues. The acid undergoes a chemical decomposition which can best be described as a base-catalyzed elimination reaction, as illustrated:

$$ClCH_2CH_2\overset{O}{\underset{O^-}{\overset{\|}{P}}}-OH + OH^- \rightarrow CH_2=CH_2 + \overset{O}{\underset{O^-}{\overset{\|}{P}}}-(OH)_2 + Cl^-$$

O-Ethyl O-2,4,5-trichlorophenyl ethylthiophosphonate (trichloronate) (LD_{50} 35 mg./kg.) and O-ethyl S-phenyl ethyldithiophosphonate (difonate) (LD_{50} 16.5 mg./kg.) have been used as insecticides for control of pests in soil.

General references

ADDOR, R. W.: U.S. Patent 3,364,230.
ARBUZOV, A. E.: Izbrannye trudy [Selected works]. Publishing House of the Academy of Sciences of the USSR (1952).
ARBUZOV, B. A.: In book Reaktsii i metody issledovaniya organicheskikh soedinenii [Reactions and methods of investigation of organic compounds], vol. 3, pp. 7–72. State Scientific and Technical Publishing House of Chemical Literature [USSR] (1954).
BEYNON, K. I., and A. N. WRIGHT: J. Sci. Food Agr. 20, 250 (1969).
DERKACH, G. I., I. N. ZHMUROVA, A. V. KIRSANOV, V. E. SHEVCHENKO, and A. S. SHTEPANEK: Fosfazosoedineniya [Phosphazo compound]. Publishing House "Naukova dumka" Kiev (1965).
GOLIKOV, S. N., and V. I. ROZENGART: Kholinesterazy i antikholinesteraznye veshchestva [Cholinesterases and anticholinesterase agents]. Publishing House "Meditsina" (1964).
GRAPOV, A. F.: In book Reaktsii i metody issledovaniya organicheskikh soedinenii [Reactions and methods of investigation of organic compounds], vol. 15, pp. 43–231. Publishing House "Khimiya" (1966).
HORN, H. G.: Chem. Ztg. 93, 241–251 (1969).
KABACHNIK, M. I., A. P. BRESTKIN, and M. YA. MIKHELSON: O mekhanizme fiziologicheskogo deistviya fosfororganicheskikh soedinenii [The mechanism of the physiological action of organophosphorus compounds]. Publishing House "Nauka" (1965).
KAGAN, YA. S.: Toksikologiya fosfororganicheskikh insektitsidov i gigiena truda pri ikh primenenii [Toxicology of organophosphorus insecticides and the hygiene of working with them]. State Publishing House of Medical Literatur [USSR] (1963).
Khimiya i primenenie FOS [Chemistry and use of organophosphorus compound]. Publishing House of Academy of Science of USSR (1957) and (1962).
LOVELL, J. B., and R. W. BAER: U.S. Patent 3,390,209.
MEL'NIKOV, N. N.: Uspekhi Khim. 22, 253 (1953); Zhur. Vsesoyuz. Khim. Obshchestva im. D. I. Mendeleeva 5, 275 (1960); 9, 524 (1964); 13, 248 (1968); Khim. v Sel'skom Khozyaistve, No. 2, 20 (1967); Arch. Pflanzensch. 5, 3–24 (1969).
—, A. F. GRAPOV: Zhur. Vsesoyua. Khim. Obshchestva im. D. I. Mendeleeva 13, 291 (1968).
—, YA. A. MANDEL'BAUM, K. D. SHVETSOVA-SHILOVSKAYA: Khim. Sredstva Zashchity Rastenii No. 3, 5–57 (1957).
MENN, J. J., and K. SZABO: J. Econ. Entomol. 58, 734 (1965).
METCALF, R. L.: Organic insecticides. New York-London: Interscience (1955).
Novye pestitsidy [New pesticides]. Publishing House "Mir" (1964).
O'BRIEN, R. D.: Toxic phosphorus esters. New York: Academic Press (1960).
— Insecticides. New York-London: Academic Press (1967).
—, and E. Y. SPENCER: J. Agr. Food Chem. 1, 946 (1953).
— — J. Agr. Food Chem. 3, 56 (1955).

Pianka, M.: J. Sci. Food Agr. **19**, 403, 475 (1968).
—, and J. D. Edwards: J. Sci. Food Agr. **19**, 399 (1968).
Plets, V. M.: Organicheskie soedineniya fosfora [Organophosphorus compounds] (1940).
Pudovik, A. N., I. V. Gur'anova, and A. Istmaeva: In book Reaktsii i metody issledovaniya organicheskikh soedinenii [Reactions and methods of investigation of organic compounds], vol. 19, pp. 9–848. Publishing House "Khimiya" (1968).
Saunders, B. C.: Some aspects of the chemistry and toxic action of organic compounds containing phosphorus and fluorine. London and New York: Cambridge Univ. Press (1957).
Schrader, G.: Die Entwicklung neuer insektizider Phosphorsäure-Ester. Weinheim/Bergstraße: Verlag Chemie GmbH. (1963).
Sittig, M.: Pesticide production processes. New Jersey: Noyes Development Corporation (1967).
Spencer, E. Y., and R. D. O'Brien: J. Agr. Food Chem. **1**, 716 (1953).
— —, and R. W. White: J. Agr. Food Chem. **5**, 123 (1957).
Tempel, A., J. Meltzer, and B. G. van den Bos: Netherlands J. Plant Pathol. **74**, 133 (1968).
Tolkmith, H.: Nature **211**, 522 (1966).
—, P. B. Budde, D. R. Mussell, and R. A. Nyquist: J. Med. Chem. **10**, 1074 (1967).
—, and D. R. Mussell: World Rev. Pest Control **6**, 74–79 (1967).
—, J. N. Selber, P. B. Budde, and D. R. Mussell: Science **158**, 1462 (1967).
—, H. O. Senkbeil, and D. R. Mussell: Science **155**, 85 (1967).
Tsvetkov, E. N., and M. I. Kabachnik: In book Reaktsii i metody issledovaniya organicheskikh soedinenii [Reactions and methods of investigation of organic compounds], vol. 13, pp. 267–427. Publishing House "Khimiya" (1965).
Uspekhi v oblasti bor'by s vreditelyami rastenii [Progress in the field of plant pest control]. Foreign Literature Publishing House [USSR] (1960).
Uspekhi v oblasti izucheniya pestitsidov [Progress in the field of investigation of pesticides]. Foreign Literatur Publishing House [USSR] (1962).
Vladimirova, I. L., A. F. Grapov, and V. I. Lomakina: In book Reaktsii i metody issledovaniya organicheskikh soedinenii [Reactions and methods of investigation of organic compounds], vol. 16. Publishing House "Khimiya" (1966).

XXVII. Arsenic compounds

General characteristics of pesticidal properties

The compounds of arsenic have high biological activity. Among the numerous arsenic compounds are known substances that show powerful insecticidal, fungicidal, herbicidal, and zoocidal effects. A common disadvantage of all the compounds of arsenic without exception is their high toxicity to man and domestic animals. Even some organoarsenic compounds that are relatively low in toxicity to warm-blooded animals are converted to toxic inorganic compounds when they undergo mineralization in animal organisms, in the soil, or in plants. When arsenic compounds are used systematically they can accumulate in the soil and then enter plants. For example, in the United States arsenic has been found in tobacco decades after use had been discontinued on the plots where it was grown. When arsenic compounds have been used for plant protection, massive poisonings of domestic animals have been observed, especially in places where forage crops have been treated with such preparations.

Cases have been recorded of cancer of the lungs, liver, and other organs in people who have worked with arsenic compounds for ten years or more.

Because of these facts, the use of arsenic compounds in agriculture is continually decreasing, and in some countries it is completely forbidden (German Democratic Republic, German Federal Republic, and others).

Only organic compounds of arsenic that can be used at low dosages are promising as seed disinfectants and for the treatment of growing plants.

Inorganic compounds of arsenic

Of the inorganic compounds of arsenic, the salts of arsenic and arsenous acids with various bases and metals are used as pesticides.

Sodium arsenite is a white crystalline substance, highly soluble in water and practically insoluble in organic solvents. The technical grade compound is a mixture of the ortho- and metaarsenites in which the latter predominates. Sodium arsenite is produced by the reaction of white arsenic (arsenous oxide) with sodium carbonate or NaOH:

$$As_2O_3 + Na_2CO_3 \rightarrow 2\ NaAsO_2 + CO_2$$

The technical grade product usually has the appearance of large black lumps or powder. It is used to control locusts and scabies mites of domestic animals, especially of sheep.

Calcium arsenite is a white solid substance, very slightly soluble in water and practically insoluble in organic solvents. It is decomposed by acids with the formation of white arsenic and the corresponding calcium salt. It was used previously as a seed disinfectant and also in controlling rodents and locusts. It is produced by the reaction of lime with white arsenic:

$$As_2O_3 + Ca(OH)_2 \rightarrow Ca(AsO_2)_2 + H_2O$$

The technical grade product consists principally of calcium metaarsenite with a small trace of calcium orthoarsenite [$Ca_2(AsO_3)_2$]. At the present time there is practically no use made of calcium arsenite.

Some use is made in the United States of *zinc arsenite* and *arsenate* for wood impregnation. For this purpose zinc arsenite or arsenate is dissolved in an aqueous ammonia solution and the wood is saturated with the solution. After evaporation of the water and ammonia, the water-insoluble zinc arsenite or arsenate is deposited on the wood and preserves it from damage by microorganisms and insects. Because of the poor solubility of the zinc salts of the arsenic acids in water, they are not leached out of the wood for a long time.

Paris green (the double salt of copper acetate and copper arsenite), a green powder, is poorly soluble in water. Its LD_{50} for animals is about 18–25 mg./kg. Chronic poisoning by the vapors of Paris green has been observed when it has been used as a dye. It previously was used to control chewing plant pests but it is phytotoxic.

The compound can be produced by several methods; the most important is the reaction of sodium arsenite with sodium acetate and copper sulfate:

$$6\ NaAsO_2 + 2\ CH_3COONa + 4\ CuSO_4 \rightarrow$$
$$(CH_3COO)_2Cu \cdot 3\ Cu(AsO_2)_2 + 4\ Na_2SO_4$$

or the reaction of white arsenic with copper carbonate and acetic acid:

$$3\ As_2O_3 + 4\ CuCO_3 + 2\ CH_3COOH \rightarrow$$
$$(CH_3COO)_2Cu \cdot 3\ Cu(AsO_2)_2 + 4\ CO_2 + H_2O$$

The latter method is convenient because when it is used there is no wastewater contaminated with arsenic compounds.

Of the compounds of pentavalent arsenic, lead arsenate, calcium arsenate, zinc arsenate, and others are used as pesticides. In the Soviet Union the agricultural use of lead arsenate is forbidden because of the high chronic toxicity of lead.

Calcium arsenate is a white, poorly water-soluble substance that is used to control codling moths that are resistant to DDT and other pesticides, and also to control various chewing plant pests. The technical grade product contains a mixture of salts of arsenic acid, in which the basic salt of the composition $Ca_3(AsO_4)_2 \cdot Ca(OH)_2$ predominates. The As_2O_5 content in it is about 40% and the arsenous acid content is not more than 1%. It should not contain more than 0.6% of water-soluble arsenic compounds, which increase the phytocidal effect of the preparation. In order to diminish the phytocidal effect a small amount of zinc sulfate or oxide is sometimes added. Calcium arsenate can be used both for dusting plants and for spraying in aqueous suspensions.

Many methods are known for the synthesis of calcium arsenate; the two most important are the following:

1. Reaction of water-soluble salts of arsenic acid with lime:

$$2\ Na_3AsO_4 + 4\ Ca(OH)_2 \rightarrow Ca_3(AsO_4)_2 \cdot Ca(OH)_2 + 6\ NaOH$$

The sodium arsenate necessary for this process is obtained by oxidation of sodium arsenite in the presence of copper sulfate as a catalyst:

$$As_2O_3 + 6\ NaOH \rightarrow 2\ Na_3AsO_3 + 3\ H_2O$$
$$2\ Na_3AsO_3 + O_2 \rightarrow 2\ Na_3AsO_4$$

2. Reaction of arsenic acid with lime:

$$2\ H_3AsO_4 + 4\ Ca(OH)_2 \rightarrow Ca_3(AsO_4)_2 \cdot Ca(OH)_2 + 6\ H_2O$$

This reaction can be carried out either in aqueous medium or by a semi-dry method (mixing concentrated solutions of arsenic acid with lime). The arsenic acid necessary for this process is obtained in good yields by oxidation of white arsenic with nitric acid:

$$3\ As_2O_3 + 4\ HNO_3 + 7\ H_2O \rightarrow 6\ H_3AsO_4 + 4\ NO$$

Iodine is used as a catalyst. The nitric oxide evolved in this reaction is oxidized by the air to nitrogen dioxide which then with water yields nitric acid. Thus, considerable quantities of arsenous anhydride can be oxidized with a small amount of nitric acid.

Zinc and copper arsenates are used as disinfectants for wood. Furthermore various mixtures of salts containing arsenic acid or its salts are used. The compositions of some of the disinfectants used in countries other than the Soviet Union are given below:

AsCu compound (in %): $Na_2HAsO_4 \cdot 7\ H_2O$ 9.3, $Na_2Cr_2O_7 \cdot 2\ H_2O$ 48.0, and $CuSO_4 \cdot 5\ H_2O$ 39.

Boliden-bis (in %): H_3AsO_4 33–37, Na_2HAsO_4 34–38, and $Na_2Cr_2O_7 \cdot 2\ H_2O$ 27–31.

Volman's salt (in %): NaF 25, Na_2HAsO_4 25, $Na_2Cr_2O_7$ 37.5, and dinitrophenol 12.50. The disinfectant Tanalite also has a similar composition.

Chemonite, which is used to impregnate wood, has the following composition (in %): $CuSO_4$ 41, NaOH 12.8, arsenous anhydride 5.4, and NH_4OH 30.8. After evaporation of the ammoniacal solution water-insoluble copper arsenite is formed in the wood.

Organic compounds of arsenic

The biological activity of the organic compounds of arsenic has been studied in ample detail; among them have been found a large number of active substances. Comparatively few of them, however, have been used as pesticides. The insecticidal activity of arsenic compounds of the aliphatic series has no practical importance, since most of them do not exceed the inorganic arsenates and arsenites in toxicity to insects. Furthermore, all the organic arsenic compounds are more expensive. The simplest arsenic compounds are of interest only as fungicides or herbicides. The fungicidal activity of the derivatives of trivalent arsenic is higher than that of the compounds of pentavalent arsenic. The most active compounds are those containing one hydrocarbon radical connected with the arsenic. The derivatives of dialkyl- and diarylarsine occupy an intermediate position. The trialkyl- and triarylarsines have comparatively low fungicidal and herbicidal activity.

The toxicity of most derivatives of trivalent arsenic is proportional to their content of arsenic, which in animals and the soil changes to inorganic compounds. Such compounds as the aryldichloroarsines and the corresponding oxides cause a strong irritating effect on the skin. The irritating properties of the alkyldichloroarsine oxides are weaker than those of the corresponding alkyldichloroarsines. The irritating properties of the alkylarsine sulfides are still less, and some of them are used in agriculture to control plant diseases.

The fungicidal activity of the alkylarsines and their derivatives decreases with an increase in molecular weight of the compound; for example, the fungicidal activity of methylarsine oxide is higher thant that of amylarsine oxide. A similar relationship is observed in the sulfide series.

The fungicidal activity of the aromatic compounds of arsenic is greatly influenced by various substituents in the aromatic radical. However, compounds of this kind have not yet been used in agriculture.

Some heterocyclic compounds of arsenic containing an arsenic atom in the ring have high fungicidal activity. Such compounds include derivatives of phenarsazine (I) and phenoxarsine (II):

Some of them approach the organomercury compounds in fungicidal activity.

A disadvantage common to this group of compounds is their ability to irritate the upper respiratory passages. The irritating properties are clearly expressed in the phenarsazine derivatives. 9-Chlorodihydrophenarsazine is listed among the poisonous warfare agents under the name of adamsite. However, in countries outside the Soviet Union some derivatives of adamsite are used to control growths on marine vessels.

Alkyl-, alkylaryl-, and dialkylarsenic acids are used in the form of salts with the alkali metals or amines as herbicides. The simplest compound of this series is cacodylic acid, which is used to control undesirable vegetation.

The great hindrance to the wide use of organoarsenic compounds is the high toxicity and the persistence of arsenic, since all the organic compounds of arsenic are finally converted in the environment to inorganic compounds. The latter may accumulate in the soil and enter into plants and food products.

The most important method for the synthesis of derivatives of methylarsine is the reaction of salts of arsenous acid with alkyl halides or other alkylating agents:

$$Na_3AsO_3 + CH_3X \rightarrow CH_3\overset{\overset{O}{\|}}{As}(ONa)_2 + NaX$$

The salt obtained is converted to a trivalent arsenic compound by reduction with sulfurous acid in the presence of small amounts of iodine. This reduction takes place most readily in acid medium. When excess HCl is used, methyldichloroarsine is formed:

$$CH_3\overset{\overset{O}{\|}}{As}(ONa)_2 + 2\,HCl \rightarrow CH_3\overset{\overset{O}{\|}}{As}(OH)_2 + 2\,NaCl$$

$$CH_3\overset{\overset{O}{\|}}{As}(OH)_2 + SO_2 \rightarrow CH_3AsO + H_2SO_4$$

Methylarsine sulfide (MAS, Rhizoctol, Urbasulf) is a white crystalline substance, m.p. 110° C. It is practically insoluble in water but highly soluble in carbon disulfide. In 100 g. of solvent the following amounts of methylarsine sulfide dissolve (in g.): in acetone 0.66, benzene 3.3, methyl alcohol 0.096, ethyl alcohol 0.095.

The compound withstands heating up to 150° C. without decomposition, is resistant to the action of acids, but on prolonged heating with alkalies it is gradually hydrolyzed with the formation of methylarsine oxide and the sulfide of the corresponding metal. Oxidizing agents of the type of nitric acid and potassium permanganate oxidize methylarsine sulfide to methylarsenic acid and sulfuric acid:

$$CH_3AsS + 2\,H_2O + 5O \rightarrow CH_3\overset{\overset{O}{\|}}{As}(OH)_2 + H_2SO_4$$

This reaction can be used for destruction of the compound with the formation of highly water-soluble compounds.

Methylarsine sulfide is used as a disinfectant or fungicide for the control of the fungi *Rhizoctonia, Pythium,* and *Piricularia* in the form of dusts and wettable powders containing about 10% active ingredient. Other formulations for dusting contain 0.15% active ingredient. The LD_{50} for rats is about

100 mg./kg. On long contact with the human skin the compound causes irritation and hyperemia.

Methylarsine sulfide is obtained by the reaction of hydrogen sulfide with methylarsine oxide in aqueous medium:

$$CH_3AsO + H_2S \rightarrow CH_3AsS + H_2O$$

Methylarsine bis(lauryl sulfide) (MALS) is a thick, clear oil with unpleasant odor, d^{20} 1.04. It is practically insoluble in water, miscible with benzene and acetone, 3.2 g. of it dissolves in 100 ml. of methyl alcohol, and 24.9 g. in 100 ml. of ethyl alcohol. The LD_{50} is about 1,000 mg./kg.

The chemical properties of MALS are close to those of methylarsine sulfide. On prolonged heating with aqueous or alcoholic solutions of alkalies, it is hydrolyzed with the formation of methylarsine oxide and lauryl mercaptide. Oxidizing agents yield dodecanesulfonic acid and methylarsenic acid:

$$CH_3As(SC_{12}H_{25})_2 + 2\ H_2O + 7O \rightarrow CH_3\overset{O}{\overset{\|}{As}}(OH)_2 + 2\ C_{12}H_{25}SO_3H$$

MALS is produced by the action of lauryl mercaptan on methylarsine oxide or by the reaction of alkali mercaptides with methyldichloroarsine:

$$CH_3AsCl_2 + 2\ C_{12}H_{25}SNa \rightarrow CH_3As(SC_{12}H_{25})_2 + 2\ NaCl$$

In the soil under the influence of microorganisms it is oxidized in a similar manner with complete mineralization of the arsenic. The compound is used in the form of a 0.4% dust to control *Piricularia* of rice at a dosage of 90 kg./ha.

Methylarsine bis(dimethyldithiocarbamate) (Urbazid, Tuzet) is a white crystalline substance, m.p. 144° C., practically insoluble in water but highly soluble in carbon disulfide and some other organic solvents. The LD_{50} for rats is 175 mg./kg. It is used in the form of aqueous suspensions at 0.025 to 0.03% active ingredient to control diseases of the cacao tree. It is marketed in the form of an 80% wettable powder. Under the name Tuzet a preparation of the following composition (in %) is used to control apple scab: methylarsine bis(dimethyldithiocarbamate) 20, zinc dimethyldithiocarbamate 20, tetramethylthiuram disulfide 40, diluent and auxiliary materials 20.

When it is heated with alkali hydroxides, methylarsine bis(dimethyldithiocarbamate) is hydrolyzed with the formation of methylarsine oxide and the corresponding salts of dimethyldithiocarbamic acid. Oxidizing agents act on it in the same way as on methylarsine sulfide.

To produce it, use is made of the reaction of methyldichloroarsine with salts of dimethyldithiocarbamic acid:

$$2(CH_3)_2N\overset{S}{\overset{\|}{C}}SNa + CH_3AsCl_2 \rightarrow \left[(CH_3)_2N\overset{S}{\overset{\|}{C}}S \right]_2 AsCH_3 + 2\ NaCl$$

This reaction can be carried out either in aqueous medium or in organic solvents. The target product is obtained in practically quantitative yield.

Tris(dimethyldithiocarbamoyl)arsine (Azomat) is a yellow-green crystalline substance, m.p. 224°–226° C. It is insoluble in water, slightly soluble in most organic solvents. The LD_{50} for mice is 1,300 mg./kg. It is used to control powdery mildew fungi.

When it is heated with alkali hydroxides Azomat is decomposed with the formation of salts of arsenous and dimethyldithiocarbamic acids:

$$As\left[\overset{\overset{O}{\|}}{S}CN(CH_3)_2\right]_3 + 6\ KOH \rightarrow 3(CH_3)_2N\overset{\overset{S}{\|}}{C}SK + K_3AsO_3 + 3\ H_2O$$

Oxidizing agents also break this compound down with the formation of arsenic acid. It is produced by the reaction of salts of dimethyldithiocarbamic acid with arsenic trichloride:

$$AsCl_3 + 3(CH_3)_2N\overset{\overset{S}{\|}}{C}SNa \rightarrow As\left[\overset{\overset{S}{\|}}{S}CN(CH_3)_2\right]_3 + 3\ NaCl$$

This compound is marketed in Japan in the form of a 40% wettable powder for use in aqueous suspensions.

Methylarsenic acid and its salts. Various salts of methylarsenic acid are used as herbicides. Most of its salts with the alkali metals and with organic bases are highly soluble in water and poorly soluble in organic solvents. As herbicides they give satisfactory results at dosages of 2.5–10 kg./ha. The use of arsenic compounds at these dosages, however, highly contaminates the soil with arsenic.

The salts of methylarsenic acid are obtained by the previously described reaction of the salts of arsenous acid with methyl halides or methyl sulfate (see earlier). The LD_{50} of the sodium salt is 700 mg./kg.

The ferric ammonium salt under the name of Neo-azazin is used in Japan to control *Piricularia* of rice.

Cacodylic acid in the pure form is a white crystalline substance, m.p. 200° C. It is highly soluble in water and alcohols, considerably less soluble in other organic solvents. The LD_{50} for mice is 185 mg./kg. It has been proposed for pre-emergence destruction of weeds on pastures and as a desiccant. It is used mainly in the form of its water-soluble salts.

Cacodylic acid is produced by the oxidation of cacodyl oxide:

$$[(CH_3)_2As]_2O + H_2O + 2O \rightarrow 2(CH_3)_2\overset{\overset{O}{\|}}{A}sOH$$

The synthesis of cacodyl oxide is accomplished by dry distillation of a mixture of arsenous anhydride with anhydrous potassium acetate:

$$4\ CH_3COOK + As_2O_3 \rightarrow [(CH_3)_2As]_2O + 2\ K_2CO_3 + 2\ CO_2$$

When sodium acetate is heated with arsenous anhydride, almost no cacodyl oxide is formed.

S-Phenoxyarsinyl O,O-diethyl dithiophosphate (Thiarsine) is a white crystalline substance, m.p. 64°–65° C. It is poorly soluble in water, and moderately soluble in organic solvents. The LD_{50} for experimental animals is 50–300 mg./kg.

Thiarsine has been proposed for use as a seed disinfectant instead of Granosan and other mercury compounds. When it is finely ground it is somewhat irritant to the mucous membranes of the nasopharynx. In working with Thiarsine, therefore, it is necessary to protect the respiratory passages with a gas mask. A disinfectant based on Thiarsine contains 10–15% phenanthrenequinone, 2.5–5% Thiarsine, and an insecticide. This preparation for the disinfection of grain seeds is used at dosages of 1–2 kg./ton of seed.

Thiarsine is produced by the reaction of chlorophenoxyarsine with sodium diethyldithiophosphate in aqueous-organic medium:

$$\text{[phenoxarsine-Cl]} + (C_2H_5O)_2\overset{S}{\underset{}{P}}SNa \longrightarrow \text{[phenoxarsine-SP(OC}_2\text{H}_5)_2\text{=S]} + NaCl$$

The chlorophenoxyarsine necessary for this reaction is made by the condensation of diphenyl ether with arsenic trichloride in the presence of anhydrous aluminium chloride with prolonged heating of the mixture:

$$AsCl_3 + (C_6H_5)_2O \xrightarrow{AlCl_3} \text{[phenoxarsine-Cl]} + 2HCl$$

Technical grade Thiarsine contains a small trace of phenoxyarsine oxide.

A large number of other arsenic compounds have been studied and some of them are active fungicides, but they have not yet been used in practice.

General references

BABIN, V. V., L. I. ISAKOVA, G. S. LEVSKAYA, E. P. UGRYUMOV, and L. N. SHIRYAEVA: Khim. v Sel'skom khozyaistve 4, 923 (1966).

GABRIELOVA, M. G., and M. A. MOROZOVA: Proizvodstvo yadokhimikatov [Production of pesticides]. State Scientific and Technical Publishing House of Chemical Literature [USSR] (1953).

FREIDLINA, R. KH.: Sinteticheskie metody v oblasti metalloorganicheskikh soedinenii mysh'yaka [Synthetic methods in the field of organometallic compounds of arsenic]. Publishing House of the Academy of Sciences of the U.S.S.R. (1945).

GUTHRIE, F. E., and T. G. BOWERY: Residue Reviews **19**, 31 (1967).
KAMAI, G. KH., and B. D. CHERNOKAL'SKII: In book "Reaktsii i metody issledovaniya organicheskikh soedinenii" [Reactions and methods of investigation of organic compounds], vol. 13, pp. 7–126. Publishing House "Khimiya" (1964).
LAZAREV, N. V., ed.: Vrednye veshchestva v promyshlennosti [Harmful substances in industry]. Handbook, part II, pp. 163–185. Publishing House "Khimiya" (1965).
Neorganicheskie insektitsidy, fungitsidy i zootsidy [Inorganic insecticides, fungicides, and zoocides]. Coll. vol. State Scientific and Technical Publishing House of Chemical Literature [USSR] (1960).

XXVIII. Heterocyclic compounds with one heteroatom in the ring

General characteristics of pesticidal properties

Heterocyclic compounds occupy nearly the first place among the other classes of organic compounds with respect to variety of biological activity. Within the ranks of the heterocyclic compounds with one heteroatom in the ring are found substances with various pesticidal properties (insecticides, fungicides, herbicides, growth stimulators and inhibitors, nematicides, insect repellents, etc.). Among the heterocyclic compounds also are some alkaloids that are used in agriculture as pesticides. These include nicotine and anabasine, which show strong insecticidal properties.

Various heterocycles containing three or more atoms in the ring have pesticidal properties. The first representative of the three-membered heterocyclic compounds, ethylene oxide, is used as a fumigant. Because of its high explosive hazard it can be used only in a mixture with carbon dioxide [or other flammability suppressant. R.L.B.]. The homologs of ethylene oxide, for example propylene oxide, also show an insecticidal effect. The α-oxides of polycyclic compounds, such as dieldrin and endrin, are effective insecticides (see chapter IV). Ethylenimine and ethylene sulfide do not have independent importance as chemical agents for plant protection, but they are used for the synthesis of pesticidal compounds. Many ethylenimides of carboxylic and phosphoric acids and cyclotriphosphazenes are being studied as insect chemosterilants. Compounds of this group are highly toxic to vertebrates.

The five- and six-membered heterocyclic compounds have the widest use as pesticides.

Five-membered heterocyclic compounds

Furan and thiophene and their dihydro and tetrahydro derivatives have only slight pesticidal activity, but some of their derivatives are highly active.

Bis(3,4-dichlorofuranon-5-yl-2) ether (I), the simplest derivative of the dihydrofuranone series, is a white crystalline substance, m.p. 141°–145° C. It is slightly soluble in water but highly soluble in aromatic hydrocarbons, alcohols, and ketones. The LD_{50} for rats is about 2,000 mg./kg. It has been proposed as an experimental fungicide for the control of apple scab. It strongly irritates the skin and, in spite of its relatively high fungicidal activity, it has not been used practically in agriculture. It is obtained by dehydration of mucochloric acid.

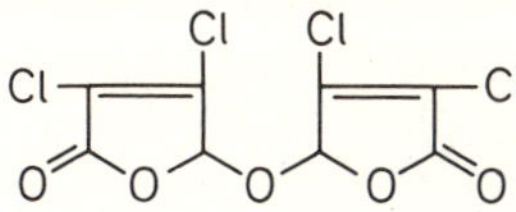

Insecticidal activity is also shown by furfuralacetone, which is easily obtained by condensation of acetone with furfural. Many derivatives of furfural are used in animal and poultry husbandry as curative and prophylactic agents for various diseases, such as coccidiosis.

5-Nitrofurfural semicarbazone is a light-yellow crystalline substance, m.p. 268°–269° C. (dec.). It shows rather strong bactericidal and fungicidal properties and is used in medicine. It is synthesized by the reaction of 5-nitrofurfural with semicarbazide:

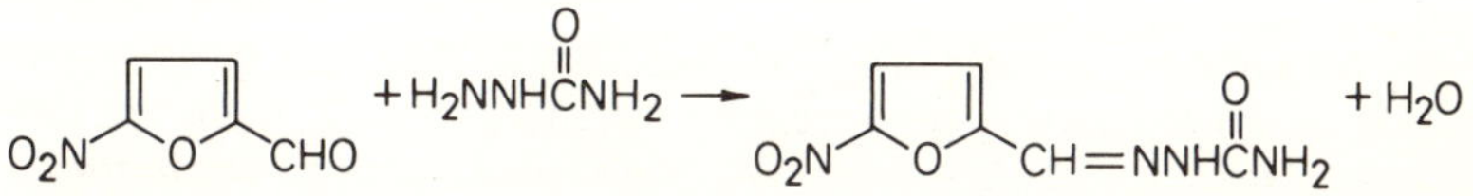

4a-Formyl-1,4,4a,5a,6,9,9a,9b-octahydrodibenzofuran(II) is a clear oily liquid, b.p. 307° C. at 760 mm. of Hg, d_{20}^{20} 1.120, n_D^{20} 1.5240. It is practically insoluble in water and highly soluble in most organic solvents. The LD_{50} for rats is about 2,500 mg./kg. It is used as an agent to repel flies, stable flies, and cockroaches, and also for the treatment of milk cows in combination with the pyrethrins and some other insecticides. It is produced by the condensation of furfural with divinyl.

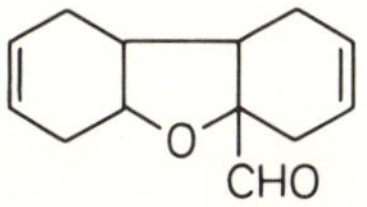

1,3,4,5,6,7,8,8-Octachloro-3a,4,7,7a-tetrahydro-4,7-endomethylenenaphthalene (isobenzan, Telodrin) is a white crystalline substance, m.p. 122° to 123° C. It is the most active chlorine-containing insecticide. Its LD_{50} for rats is 5–10 mg./kg. It is produced by chlorination of the adduct of hexachlorocyclopentadiene with dihydrofuran:

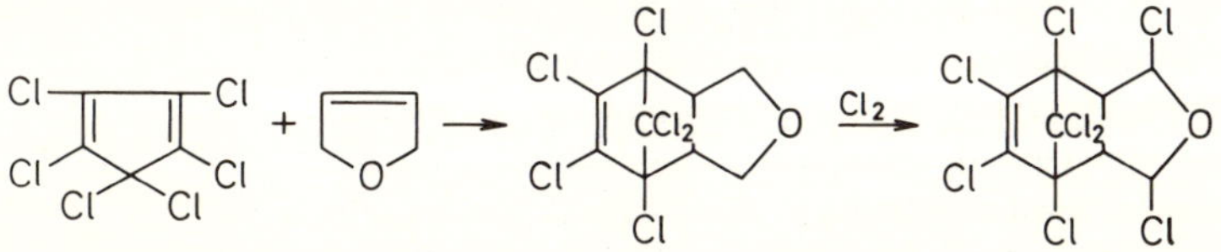

In spite of the good insecticidal activity of isobenzan it is used only to a limited extent in agriculture because of its high toxicity to warm-blooded animals and man. An advantage of this compound in addition to its effectiveness it its relatively slight stability in alkaline medium. It is used in the form of emulsions or suspensions at a dosage of 100 to 300 g./ha. In the Soviet Union the use of isobenzan is not permitted.

The pesticidal activity of the five-membered heterocyclic compounds with a sulfur atom in the ring is low. 3,4-Dichlorotetrahydrothiophene dioxide finds practical use as an experimental nematicide.

3,4-Dichlorotetrahydrothiophene dioxide is a white crystalline substance, m.p. 129°–131° C., highly soluble in organic solvents; 0.22 g. of the compound dissolves in 100 g. of water at 20° C. It is produced by chlorinating the adduct of sulfurous anhydride with divinyl:

$$SO_2 + CH_2=CHCH=CH_2 \longrightarrow \underset{\underset{O}{\overset{\|}{S}}\overset{\|}{O}}{\bigcirc} \xrightarrow{Cl_2} \underset{\underset{O}{\overset{\|}{S}}\overset{\|}{O}}{\overset{ClCl}{\bigcirc}}$$

It is unstable toward the action of alkalies. It is used to control nematodes by application to the soil.

Of considerably greater interest for use as pesticides are the heterocyclic compounds with an atom of nitrogen in the ring; among them are known substances with a strong insecticidal effect, plant growth regulators, and fungicides.

Nirosan was used in Germany during World War II as a stomach insecticide:

Nirosan

Its active principal is 1,3,6,8-tetranitrocarbazole (m.p. 285° C., LD_{50} for rats about 250 mg./kg.), which is prepared by direct nitration of carbazole by nitric acid in the presence of sulfuric acid.

1,8-Dichloro-3,6-dinitrocarbazole (m.p. 235°–237° C.) also has insecticidal properties.

Of the heterocyclic compounds containing nitrogen, derivatives of indole, the 3-indolylalkylcarboxylic acids, which are plant growth regulators, are considerably more interesting. Of this group of substances 3-indolylacetic and 3-indolyl-γ-butyric acids are used to accelerate root formation in the vegetative propagation of plants by cuttings.

3-Indolylacetic acid (heteroauxin) is a colorless crystalline substance, m.p. 168°–169° C. (dec.). It is moderately soluble in water, highly soluble in alcohol and benzene; in the light it quickly turns dark. It is stable in alkaline medium in the form of its salts, but is decomposed by acids.

Heteroauxin is the sole growth stimulator that has been isolated from plants (except for the group of gibberellins), from products of the vital activity of various microorganisms, etc. In plants it apparently is formed by the oxidative deamination of tryptophane:

[Indole-CH₂CHCOOH-NH₂ → Indole-CH₂COCOOH and Indole-CH₂CH₂NH₂ → Indole-CH₂CHO → Indole-CH₂CH₂OH → Indole-CH₂COOH]

In addition to heteroauxin, 3-indolylacetaldehyde, 3-indolylacetonitrile, and some esters have been isolated from plants.

Heteroauxin can be synthesized both by the reaction of indole with monochloroacetic acid, which takes place at an elevated temperature, and by the reaction of indole with formaldehyde and HCN or its salts:

Indole + ClCH₂COOH → Indole-CH₂COOH + HCl

Indole + CH₂O + HCN → Indole-CH₂CN → Indole-CH₂COOH

Esters of heteroauxin are formed when indole reacts with diazoacetic ester:

Indole + N₂CHCOOR → Indole-CH₂COOR + N₂

Other methods also are known for the preparation of this compound, but the first of the methods described is of the most practical interest.

Heteroauxin in the form of aqueous solutions containing from 50 to 200 mg. of the compound/liter of water is used for rooting cuttings.

3-Indolyl-γ-butyric acid, a more effective rooting stimulator, is a white crystalline substance, m.p. 124° C. It is less soluble in water than heteroauxin and is highly soluble in most organic solvents. It is obtained in satisfactory yields by a Fischer cyclization reaction of the phenylhydrazone of the ethyl ester of δ-formylvalerianic acid:

$C_6H_5NHN=CHCH_2CH_2CH_2CH_2COOC_2H_5 \longrightarrow$

Indole-CH₂CH₂CH₂COOC₂H₅

Another route for the synthesis of indolylbutyric acid is the reaction of indolylmagnesium halides with butyrolactone:

$$\text{Indole-N-MgX} + \text{butyrolactone} \rightarrow \text{Indole-CH}_2\text{CH}_2\text{CH}_2\text{COOMgX} \xrightarrow{HCl} \text{Indole-CH}_2\text{CH}_2\text{CH}_2\text{COOH} + \text{MgClX}$$

A large number of other indolylalkylcarboxylic acids have been synthesized and studied, but they all are less active than those described.

Six-membered heterocyclic compounds

Among the group of compounds with six atoms in the ring a considerably larger number of practically important pesticides have been discovered in the coumarin, pyran, pyridine, quinoline, dipyridyl, and other series.

Some of the classes of organic compounds enumerated are characterized by specific biological activity. Thus, derivatives of coumarin have a strong zoocidal effect as a result of the presence of blood anticoagulant properties; derivatives of pyran show fungicidal and bactericidal properties with relatively low toxicity to warm-blooded animals; fungicidal and herbicidal action is characteristic of many derivatives of pyridine and quinoline. When pyridine and its analogs are hydrogenated, the fungicidal and herbicidal activity of the compounds is diminished and insecticidal properties appear. Most nitrogen-containing heterocyclic compounds are highly toxic to warm-blooded animals.

3-Acetyl-6-methylpyrandione-2,4 (dehydroacetic acid) is a white crystalline substance, odorless and tasteless, m.p. 269.9° C.; it sublimes, however, at a temperature of about 110° C. It is almost insoluble in water and moderately soluble in most organic solvents. It is used as a fungicide for the treatment of fresh vegetables and fruits, and also for impregnation of packaging materials.

It is characterized by low toxicity to warm-blooded animals. The LD_{50} for rats is more than 1,000 mg./kg. Feeding rats 0.3 g./kg. daily for five weeks did not cause any symptoms of poisoning, although a decrease in weight of the animals was noted. People receiving a dose of 10 mg./kg. of the compound for five months showed no objective or subjective symptoms of poisoning.

Dehydroacetic acid, because of the presence of three carbonyl groups, forms highly water-soluble salts with alkalies. It is produced by condensation of two molecules of acetoacetic ester on prolonged heating with sodium bicarbonate:

$$2CH_3COCH_2COOC_2H_5 \xrightarrow{NaHCO_3} \underset{CH_3}{\underset{|}{\text{[2,2-dimethyl-6-carbobutoxy-2,3-dihydropyranone-4]}}} + 2C_2H_5OH$$

2,2-Dimethyl-6-carbobutoxy-2,3-dihydropyranone-4 (Indalone) is a colorless liquid, b.p. 258° C. at 760 mm. of Hg, d_{25}^{25} 1.054, n_D^{25} 1.4750. It is insoluble in water, but highly soluble in most organic solvents, with which it is miscible in all proportions. The LD_{50} for rats is 7,400 mg./kg. Prolonged application to the skin does not cause irritation.

Indalone is used as a repellent for blood-sucking diptera by application to the skin. It is most often used not alone, but mixed with various other repellents. Thus, for example, the preparation Rutgers-622 contains 60% dimethyl phthalate, 20% Indalone, and 20% 2-ethylhexanediol-1,3, and the preparation DID contains 75% dimethyl phthalate, 20% Indalone, and 5% dimethyl carbate. These preparations have a long protective action of five to eight hours.

Indalone is produced by condensation of mesityl oxide with dibutyl oxalate. This reaction occurs easily in the presence of sodium butylate or, better still, magnesium butylate. Sodium butylate can be made easily by the reaction of NaOH and butyl alcohol, with the water being distilled off as an azeotropic mixture with the alcohol:

$$(CH_3)_2C=CHCOCH_3 + (COOC_4H_9)_2 \longrightarrow \text{[pyranone product]} + C_4H_9OH$$

As already indicated, various derivatives of coumarin have anticoagulant properties that disrupt the normal coagulability of the blood, making it possible to use them as agents for rodent control. When coumarin is systematically introduced into their food, internal bleeding and death of the animals results.

3-(α-Acetonylbenzyl)-4-hydroxycoumarin (warfarin) in the pure form is a white crystalline substance, m.p. 159°–161° C., tasteless and odorless. It does not dissolve in water, is highly soluble in acetone and dioxane, and dissolves readily in alkalies yielding the corresponding derivative of the enol form. The salts of the enol with acid anhydrides or acid chlorides easily form the corresponding esters; m.p. of enol acetate 117°–118° C.

Warfarin readily forms an oxime, m.p. 182°–183° C. and a dinitrophenylhydrazone, m.p. 205°–210° C.

The compound is used for rodent control in the form of food baits containing 0.5–1% active ingredient. To obtain the maximum lethal effect, the animal should receive the compound with the food not less than 4–5 times in the course of several days.

It is obtained in good yields by the condensation of 4-hydroxycoumarin with benzalacetone:

$$\text{4-hydroxycoumarin} + C_6H_5CH=CHCOCH_3 \longrightarrow \text{3-(}\alpha\text{-acetonylbenzyl)-4-hydroxycoumarin}$$

The 4-hydroxycoumarin necessary for this synthesis is made by condensation of the methyl ester of acetylsalicylic acid under the influence of metallic sodium:

$$\text{methyl acetylsalicylate} \xrightarrow{Na} \text{4-hydroxycoumarin} + CH_3OH$$

An advantage of warfarin and similar compounds is their relatively low toxicity from a single use, making it possible to avoid accidents from the chance introduction of the compound into human beings and beneficial animals.

3-(α-Acetonyl-4'-chlorobenzyl)-4-hydroxycoumarin (coumachlor) is a colorless crystalline substance, m.p. 169°–171° C. It is practically insoluble in water, poorly soluble in hydrocarbons, more soluble in alcohols, acetone, and chloroform. It is the closest derivative of warfarin and is similar to warfarin in both chemical and toxicological properties. It is produced by condensation of 4-hydroxycoumarin with 4-chlorobenzalacetone:

$$\text{4-hydroxycoumarin} + Cl\text{-}C_6H_4\text{-}CH=CHCOCH_3 \longrightarrow \text{coumachlor}$$

3-(α-Tetralyl)-4-hydroxycoumarin (coumatetralyl) is a white crystalline substance, m.p. 172°–176° C. It is practically insoluble in water but soluble in aqueous solutions of alkali hydroxides and in most organic solvents. It is somewhat more toxic to rodents than warfarin and coumachlor. To prepare coumatetralyl, 4-hydroxycoumarin is condensed with α-tetralol:

$$\text{4-hydroxycoumarin} + \alpha\text{-tetralol} \longrightarrow \text{coumatetralyl} + H_2O$$

The LD_{50} for rodents is 0.3 mg./kg. for introduction five times (total 1.5 mg. per kg.). The LD_{50} for fish with 96-hour exposure is 1,000 mg./kg. Vitamin K is an antidote for this compound. It is used in food baits containing 0.038–0.044% active ingredient.

3,3'-Methylene-bis(4-hydroxycoumarin) (dicoumarol) is a white crystalline substance, m.p. 285°–293° C. It is almost insoluble in water but highly soluble in most organic solvents and in aqueous solutions of alkali hydroxides. It is one of the least toxic compounds of this series for rodents. Its scale of use is considerably less than that of the coumarin derivatives already described. It is produced by the condensation of 4-hydroxycoumarin with formaldehyde:

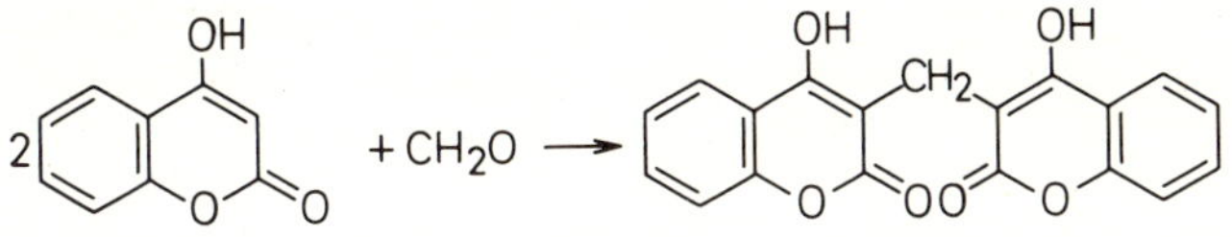

The six-membered heterocyclic compounds with a sulfur atom in the ring are not yet used as pesticides, while among their nitrogen analogs compounds are found that have varied biological activity.

Dipropyl ester of 2,5-pyridinedicarboxylic acid (Repellent MGK-326, dipropyl isocinchomeronate) is a colorless liquid, b.p. 186°–187° C. at 760 mm. of Hg, d^{20} 1.082, n_D^{25} 1.4979. It is employed as a repellent for house flies. It is used in a mixture with methylumbelliferone, since without the addition of the latter it is rapidly hydrolyzed under conditions of high humidity. It is made by the oxidation of 2,5-dialkylpyridines (most often 2-methyl-5-ethylpyridine) by various oxidizing agents, with subsequent esterification of the pyridine-2,5-dicarboxylic acid with *n*-propyl alcohol:

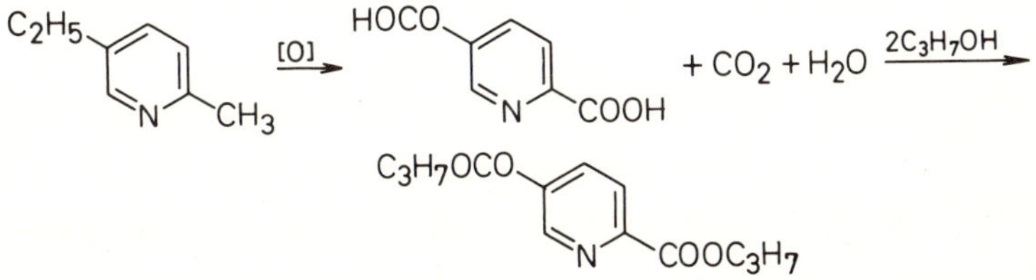

3,5,6-Trichloro-4-aminopicolinic acid (picloram, Tordon) is one of the most important derivatives of the pyridinecarboxylic acid series. This is a colorless crystalline substance, m.p. 215° C. (dec.); solubility in water 0.043%, acetone 2%, ethyl alcohol 1%. The amine and potassium salts of this acid are highly soluble in water. Its vapor pressure at 35° C. is 6×10^{-7} mm. of Hg. Its LD_{50} for various experimental animals is 2–8 g./kg. Picloram is stable at room temperature and is stored without change for a practically unlimited time.

At dosages up to 2 kg./ha. it destroys perennial and annual weeds, including ones that are very difficult to control with other compounds. For example, it is an active herbicide for the control of pink smartweed, field bindweed, amaranthus, and others. Picloram is produced from α-picoline by chlorination in the presence of 0.5–1% anhydrous aluminium chloride:

[Chlorination scheme: 2-picoline + Cl₂ at 95°–105°C gives 3,4,5,6-tetrachloro-2-trichloromethylpyridine (30%) + 3,4,5-trichloro-2-trichloromethylpyridine (32%) + 3,4,5-trichloro-2-dichloromethylpyridine (16%).]

The mixture of chlorination products obtained is further treated with ammonia at 100° C. under pressure and after complete chlorination the trichloromethyl group is hydrolzyed to carboxyl with 80% sulfuric acid at 120°–160° C.:

[Reaction scheme showing: 3,4,5-trichloro-2-CCl₃-pyridine + NH₃ → 3,5-dichloro-4-amino-2-CCl₃-pyridine; then + Cl₂ → 3,5,6-trichloro-4-amino-2-CCl₃-pyridine; also 3,4,5-trichloro-2-CHCl₂-pyridine + NH₃ → 3,5-dichloro-4-amino-2-CHCl₂-pyridine, then + Cl₂ → same intermediate; and 3,4,5,6-tetrachloro-2-CCl₃-pyridine + NH₃ → same intermediate; final H₂SO₄ hydrolysis → 3,5,6-trichloro-4-amino-2-COOH-pyridine (picloram).]

The chlorination of picoline and its conversion products goes better when there is strong illumination.

2,3,5-Trichloro-4-hydroxypyridine (Daxtron) is a second herbicide of the pyridine series. It is a crystalline substance, m.p. about 216° C.; at 25° C. its solubility in water is 0.057%, acetone 4%, 1% aqueous NaOH solution 5.35%. The vapor pressure at 26° C. is 5.5×10^{-6} mm. of Hg. Its LD_{50} for rats is 80 mg./kg. It gives good results in the nonselective eradication of vegetation at dosages of 10–20 kg./ha. This compound has considerably less general herbicidal activity than picloram.

Daxtron is made by the action of a 10% solution of NaOH on 2,3,4,5-tetrachloropyridine at 160°–190° C., under pressures of the order of 10 to 15 atm.:

[2,3,4,5-tetrachloropyridine + NaOH → 2,3,5-trichloro-4-hydroxypyridine]

It has as yet been used only under experimental conditions.

Besides herbicidal properties, the pyridine derivatives, particularly the N-oxides of substituted pyridines, show high fungicidal activity.

Pyridine-thiol-2 oxide-1 is an example. It is put out in the form of the iron (Omadine OM-1565), manganese (Omadine OM-1564), and zinc (OM-1563) salts and the disulfide (OM-1456) for the control of apple scab, peach leaf curl, and other diseases of fruit crops.

Pyridinethiol-2 oxide-1 exists in two tautomeric forms, of which the thiolo form predominates under ordinary conditions:

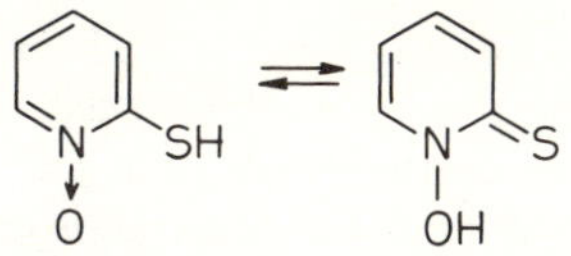

With salts of trivalent iron it forms a chelate complex which has a deep blue color; similar complexes are obtained with other metals, e.g., zinc. It is synthesized by the reaction of 2-bromopyridine oxide-1 with sodium hydrosulfide or thiourea:

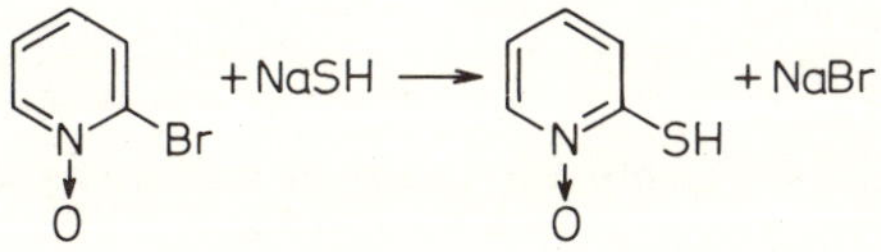

2,6-Dichloro-4-phenylpyridinedicarbonitrile-3,5 (pyridinitrile) is a novel fungicide which shows a broad spectrum of activity against various harmful fungi in fruit-growing, viniculture, and horticulture. It is a white crystalline substance, m.p. 208°–210° C., b.p. 218° C. at 0.1 mm. of Hg. The LD_{50} is 5,000 mg./kg. It is synthesized by the following scheme:

$$C_6H_5CHO + NH_3 + 2\ NCCH_2COOR \rightarrow$$

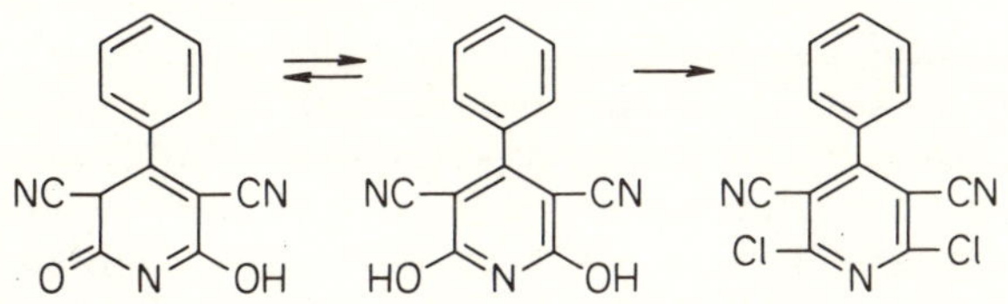

Although the homologs of pyridine have appreciable insecticidal activity, it is insufficient for their use in agriculture. The insecticidal activity of compounds increases when one of the hydrogen atoms in the pyridine ring is replaced by a nitrogen-containing heterocyclic radical. Examples of such compounds are nicotine, nornicotine, and anabasine.

1,3-(1-Methylpyrrolidyl)pyridine (nicotine) is a clear, oily liquid, b.p. 247.6° C. at 760 mm. of Hg, d_4^{20} 1.0092, n_D^{20} 1.5239, vapor pressure at 25° C. 0.0455 mm. of Hg. It is highly soluble in water, miscible with methyl and ethyl alcohols in all proportions, and highly soluble in most other organic solvents. The use of nicotine in agriculture is very limited. It has been almost entirely displaced by organophosphorus compounds that act on a wider circle of pests, are cheaper, and in many cases are less toxic to man and animals. The LD_{50} orally is 50–60 mg./kg., and intravenously 8–10 mg./kg. When it comes in contact with uncovered areas of the skin nicotine is rapidly absorbed, causing serious poisoning.

Nicotine and its analogs are mainly for the preparation of nicotinic acid which is formed in good yields when nicotine, nornicotine, and anabasine are oxidized by various oxidizing agents. The salts of nicotine with citric and malic acids occur in all species of tobacco. To isolate the free base, tobacco wastes usually are treated with lime and then extracted with dichloroethane or trichloroethylene. The nicotine is extracted from solution in the organic solvent with an aqueous solution of sulfuric acid with which it forms a stable sulfate. A 40% aqueous solution of nicotine sulfate is sold. Nicotine sulfate does not have contact insecticidal activity, and therefore when plants are sprayed with it soap or other alkaline substances are added to liberate the free base from the salt. To control plant pests 0.05–0.3% solutions of nicotine sulfate are used with the addition of 0.3% soap.

3-(2-Pyrrolidyl)pyridine (nornicotine) is a colorless, viscous liquid, b.p. 270°–271° C. at 760 mm. of Hg. Nornicotine was discovered in tobacco along with nicotine. Like nicotine it occurs in tobacco as salts of citric and malic acid.

Nicotine   Nornicotine

Nornicotine is close to nicotine in toxicity to insects and warm-blooded animals.

1-3-(2-Piperidyl) pyridine (anabasine) is a colorless liquid, b.p. 280.9° C., d^{20} 1.0481, n_D^{20} 1.5443. The vapor pressure at 80° C. is about 2.55 mm. of Hg. It is miscible with water in all proportions and is highly soluble in most organic solvents. Anabasine was first discovered by A. P. OREKHOV in 1929 in the plant *Anabasis aphylla,* which grows in large quantities in Central Asia. [This compound had previously been synthesized by scientists in the United States. R.L.B.] It is extracted from plants in a manner similar to nicotine. For use in agriculture a 40% aqueous solution of the sulfates of the total alkaloids of *Anabasis* is put out; the total alkaloids contain 70% of anabasine and 30% other alkaloids (aphylline, aphyllidine, lupinine, methylanabasine). Anabasine is practically equal to nicotine in its insecticidal activity and toxicity to warm-blooded animals.

Going from anabasine to the dipyridyls, one of the isomers of which is a dehydrogenation product of anabasine, the insecticidal activity decreases appreciably. However, all the isomers of dipyridyl have a rather strong contact insecticidal effect. When the dipyridyls are alkylated on the nitrogen, compounds are obtained with a strong contact herbicidal effect; some of them are used in agriculture.

Derivatives of dipyridyls. Derivatives of 2,2'-dipyridyl and 4,4'-dipyridyl are used as herbicides. The most widely used method of producing the dipyridyls is the dehydrogenation of pyridine:

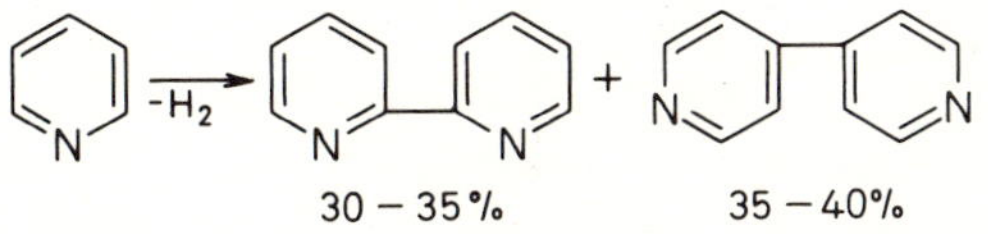

This reaction takes place in the presence of various metals (sodium, potassium, aluminum, magnesium) or of ferric chloride at a temperature from 120°–350° C. The yield of the mixed dipyridyls (containing 35–40% 4,4'-dipyridyl, 30–35% 2,2'-dipyridyl, and other isomers) is 70%. 4,4'-Dipyridyl can be synthesized also by a Dimroth reaction:

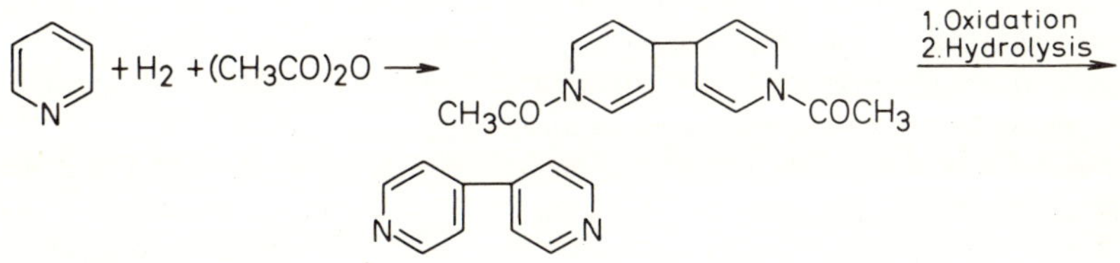

The reduction of pyridine is carried out in acetic anhydride with zinc dust and the amide obtained is oxidized and hydrolyzed. The yield of dipyridyl does not exceed 60%, calculated on the starting pyridine.

When the 4,4'-dipyridyl is alkylated with methyl chloride, dimethyl sulfate, trimethyl phosphate, trimethyl thiophosphate, or trimethyl dithiophosphate, the corresponding quaternary ammonium salts are formed, which are strong contact herbicides of nonselective action. All the salts are hygroscopic crystalline substances that rapidly deliquesce in the air. They are not usually isolated in the crystalline form and are marketed in the form of aqueous solutions containing 20% active ingredient (calculated on the basis of the quaternary ammonium cation).

1,1'-Dimethyl-4,4'-dipyridylium chloride (paraquat) is a typical representative of the dipyridyls. It is used in agriculture for the nonselective eradication of plants prior to planting of a crop, and also as a desiccant. Paraquat is synthesized by the reaction of 4,4'-dipyridyl with methyl chloride, which takes place at high temperature and pressure:

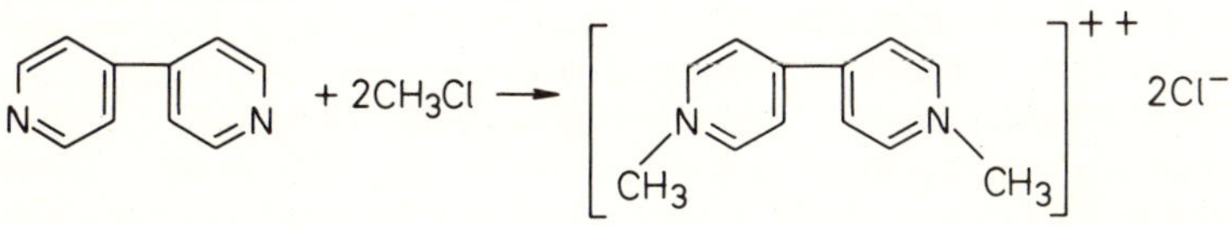

By the reaction of 4,4'-dipyridyl with esters of the phosphorus acids, it is possible to obtain products of methylation on one or two nitrogen atoms depending on the conditions:

$$\text{4,4'-dipyridyl} \xrightarrow[2(CH_3O)_3P=X]{(CH_3O)_3P=X} \left[\text{mono-methylated}\right]^+ (CH_3O)_2\overset{X}{\overset{\|}{P}}O^- \;/\; \left[\text{di-methylated}\right]^{++} 2(CH_3O)_2\overset{X}{\overset{\|}{P}}O^-$$

Both groups of compounds have a herbicidal effect and have been proposed as selective herbicides at dosages of 0.6–1 kg./ha.

A disadvantage common to this group of herbicides is their relatively high toxicity to animals, which ranges within the limits 30–70 mg./kg., calculated as the cation. Because they are only slightly volatile and are not absorbed through the skin, working with these compounds under field conditions does not present a great hazard if they do not enter the stomach. The advantages of the compounds of the paraquat group include their rapid breakdown in the soil and in plants which takes place in the first few hours after application. The derivatives of dipyridyl are used in farming without tillage by eradicating the weedy vegetation before sowing the grain crops. Paraquat also has fungicidal activity and has a favorable action in the control of root rots.

New compounds obtained by the condensation of 4,4'-dipyridyl with amides of monochloroacetic acid have been proposed as selective herbicides for the control of some weeds that are resistant to 2,4-D.

1,1'-Di(pentamethylenecarbamoylmethyl)-4,4'-dipyridylium chloride (PP-407) (III), the first of these herbicides, is obtained by the condensation of monochloropentamethylenacetamide with 4,4'-dipyridyl at elevated temperature:

$$\left[\bigcirc\!\!\text{NCOCH}_2\text{N}\!\!=\!\!\bigcirc\!\!-\!\!\bigcirc\!\!=\!\!\text{NCH}_2\text{CON}\!\!\bigcirc\right]^{++} 2Cl^- \qquad (III)$$

This is a white crystalline substance, poorly soluble in hydrophobic organic solvents, highly soluble in water (about 20%); with water it yields a monohydrate that is dehydrated at 100° C., m.p. 303° C. (dec.). It is unstable in alkaline solutions. The LD_{50} orally is 800–1,600 mg./kg., and intraperitoneally it is 12–25 mg./kg. PP-407 at dosages of 1–1.5 kg./ha. is effective in controlling such weeds as snakeweed, chickweed, field spurrey, scentless mayweed, etc.

1,1'-Bis(3,5-dimethylmorpholinecarbamoylmethyl)-4,4'-dipyridylium chloride (PP-745), the second compound of this group, is prepared similarly to PP-407 by the reaction of 4,4'-dipyridyl with the 3,5-dimethylmorpholide of monochloroacetic acid at elevated temperature:

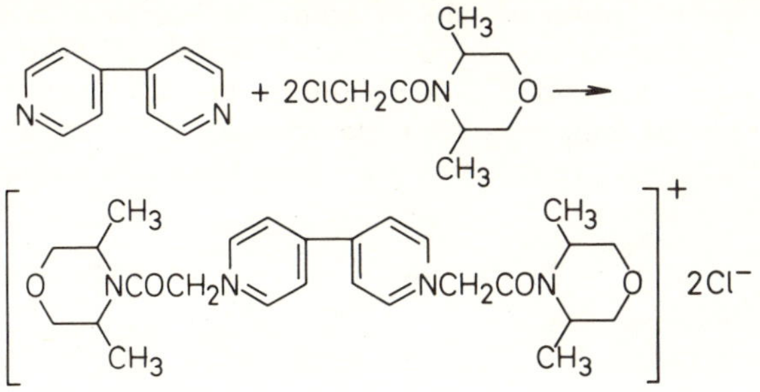

The compound is a white crystalline substance that decomposes at about 300° C. It is highly soluble in water, slightly soluble in hydrophobic organic solvents, and moderately soluble in alcohols and ketones. Its LD_{50} orally is 400–800 mg./kg., and intraperitoneally it is 25–50 mg./kg.

Both compounds are used in the form of aqueous solutions for spraying.

1,1'-Ethylene-2,2-dipyridylium bromide (diquat) is a white crystalline substance, m.p. 335°–340° C. At 20° C. 70 g. of the compound dissolves in 100 g. of water. It is usually marketed in the form of an aqueous solution containing 20% of the quaternary cation. The LD_{50} is about 400 mg./kg. It is produced by the reaction of 2,2'-dipyridyl with 1,2-dibromoethane at elevated temperature and pressure.

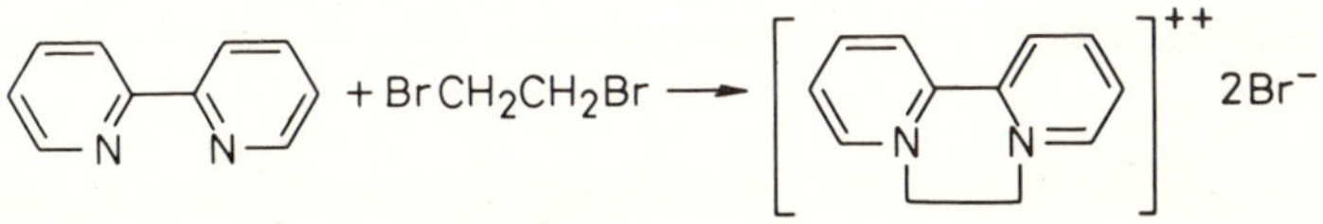

It is similar to paraquat in chemical and herbicidal properties, but is somewhat less active. In soil and plants diquat breaks down rapidly, but its metabolism in plants has not been studied a great deal as yet. A great disadvantage of herbicides of the dipyridyl group is their comparatively high cost, which is due to the complexity of their production.

8-Hydroxyquinoline (8-quinolinol, oxine) is a white crystalline substance, m.p. 76° C., poorly soluble in water, highly soluble in acids and alkali hydroxides with which it forms the corresponding salts. Because of the hydroxy group in the aromatic nucleus, 8-quinolinol is close to phenol and naphthol in its properties. Thus, it is easily halogenated, thiocyanated, and nitrated and the substituents are directed primarily into the 5- and 7-positions. When 8-quinolinol is halogenated or thiocyanated the bactericidal and fungicidal properties are strengthened, but at the same time the protein index increases and consequently the use of the substituted compounds is less advantageous. The salts of 8-quinolinol with metals and acids are employed to protect plants from diseases and also for disinfection of nonmetallic materials. With heavy metals 8-quinolinol forms stable chelates.

8-Quinolinol is produced by three methods:

1. By alkaline fusion of salts of quinoline-8-sulfonic acid with alkali hydroxides (similar to the preparation of naphthols). The reaction goes rather easily and the yield of 8-quinolinol reaches 80%. Isomeric quinolinols that are formed as impurities are removed either by fractional crystallization or by fractional precipitation with acids from alkaline solutions:

$$\text{(quinoline-8-SO}_3\text{Na)} + 2\text{NaOH} \longrightarrow \text{(8-quinolinol-ONa)} + \text{Na}_2\text{SO}_3 + \text{H}_2\text{O}$$

The quinoline-8-sulfonic acid necessary for the reaction is prepared by direct sulfonation of quinoline. Separation of the sulfonic acid isomers is carried out in a manner similar to the separation of the isomeric naphthalenesulfonic acids.

2. By a Skraup reaction from o-aminophenol:

$$\text{(o-aminophenol)} + \begin{array}{c}\text{CH}_2\text{OH}\\ \text{CHOH}\\ \text{CH}_2\text{OH}\end{array} \xrightarrow{[O]} \text{(8-quinolinol)}$$

The yields of 8-quinolinol do not exceed 70%.

3. By diazotization of 8-aminoquinoline with subsequent decomposition of the diazo compound in aqueous medium:

$$\text{(8-NH}_2\text{-quinoline)} + \text{HNO}_2 + \text{HCl} \longrightarrow \text{(8-N}_2^+\text{Cl}^-\text{-quinoline)} \xrightarrow{\text{H}_2\text{O}} \text{(8-OH-quinoline)}$$

This last method is seldom used because it is more difficult to carry out and involves many steps; furthermore, the 8-quinolinol that is produced contains many tarry products.

The LD_{50} of pure 8-quinolinol for warm-blooded animals lies between 1,000–1,200 mg./kg.

For the control of plant diseases 8-quinolinol is most often used in the form of copper 8-quinolinolate (copper oxinate), 8-quinolinol sulfate (quinosol), or benzoate.

Copper 8-quinolinolate is a greenish-yellow crystalline powder, insoluble in water, alcohol, and most other ordinary organic solvents. It is stable in storage and withstands heating to 150°–160° C. without decomposition. Caustic alkalies and strong mineral acids decompose copper 8-quinolinolate. In the first case cupric oxide and the quinolate of the alkali metal are formed, and in the second case the copper salt of the acid used and 8-quinolinol or its salt with the acid are obtained.

It is produced by precipitation from solutions of copper salts with 8-quinolinol. Copper 8-quinolinolate has a chelate structure (IV):

(IV)

It is used both for the control of plant diseases and as a disinfectant for nonmetallic materials to protect them from damage by microorganisms. Copper 8-quinolinolate (15%) makes up the composition of seed disinfectants for grains and other crops, and also is used in the form of a wettable powder for spraying of green plants and for the treatment of nonmetallic materials, e. g., textiles.

Quinosol is a yellow crystalline substance, m.p. 175°–178° C. It is highly soluble in water and alcohol, poorly soluble in hydrophobic organic solvents. Its LD_{50} for experimental animals is 800–1,000 mg./kg. It is synthesized by the reaction of 8-quinolinol with sulfuric acid. Quinosol has a systemic fungicidal effect and is used on a large scale for the control of some diseases of grain crops. Simultaneously with quinosol, 8-quinolinol benzoate is employed, which is close to quinosol in fungicidal activity but is somewhat less phytotoxic.

Many derivatives of the hydrogenation products of quinoline and isoquinoline have the ability to repel blood-sucking insects. The simplest derivatives of this series are formyl- and acetyl-1,2,3,4-tetrahydroquinolines, which have been widely tested in the Soviet Union under the names kyuzol and kyuzol-A.

1-Formyl-1,2,3,4-tetrahydroquinoline (kyuzol) is a colorless crystalline substance, m.p. 38.5° C., practically insoluble in water but highly soluble in most organic solvents. It is low in toxicity to animals and man, but when it is applied to the skin in many cases it causes irritation and hyperemia. This fact hinders the use of the compound for the treatment of exposed parts of the skin. Less irritating to the skin is 1-acetyl-1,2,3,4-tetrahydroquinoline (liquid, b.p. 143°–153° C. at 6–8 mm. of Hg, d^{20} 1.11, n_D^{20} 1.5750). Kyuzol is produced in 85% yield when quinoline is heated with formic acid in the presence of sodium formate:

$$\text{quinoline} + 3HCOOH \longrightarrow \text{1-formyl-1,2,3,4-tetrahydroquinoline} + 2CO_2 + H_2O$$

1-Acetyl-1,2,3,4-tetrahydroquinoline (kyuzol-A) is obtained in practically quantitative yield by the acetylation of 1,2,3,4-tetrahydroquinoline with acetic anhydride or acetic acid (in the latter case with removal of water by azeotropic distillation with toluene or other hydrocarbon). The 1,2,3,4-tetrahydroquinoline necessary for the reaction is synthesized by hydrogenation of quinoline with hydrogen in the presence of Raney catalyst at elevated temperature and a pressure on the order of 100 atm.:

$$\text{quinoline} + 2H_2 \longrightarrow \text{1,2,3,4-tetrahydroquinoline}$$

General references

ELDERFIELD, R.: Geterotsiklicheskie soedineniya [Heterocyclic compounds]. Foreign Literature Publishing House [USSR], vol. 1 (1953), vol. 2 (1954), vol. 3 (1954), vol. 4 (1955) *.

KOST, A. S., and L. G. YUDIN: Zhur. Obshchei Khim. 25, 1947 (1955).

MEL'NIKOV, N. N., and YU. S. BASKAKOV: Khimiya gerbitsidov i regulyatorov rosta rastenii [Chemistry of herbicides and plant growth regulators]. State Scientific and Technical Publishing House of Chemical Literature [USSR] (1962).

MOHR, G., D. ERDMANN, S. LUST, and G. SCHNEIDER: 20th Internat. Symposium Ghent, p. 1293 (1968).

---

* Cited here are dates of presumed Russian translations from English. — R.L.B.

XXIX. Heterocyclic compounds with two heteroatoms in the ring

General characteristics of pesticidal properties

The pesticidal properties of a large number of heterocyclic systems containing two heteroatoms in the ring have been studied, but only a limited number of the compounds is used in agriculture and industry.

Heterocyclic compounds with two atoms of oxygen in a five-membered ring, derivatives of methylenedioxybenzene, are widely used as synergists for the pyrethrins and their synthetic analogs. Among them are butylcarbitol piperonate (I) (*n*-butoxyethoxyethyl-3,4-methylenedioxybenzoate, b.p. 176°–178° C. at 0.5 mm. of Hg), piperonyl butoxide (II) (3,4-methylenedioxy-6-propylbenzyl-6-butyldiethyleneglycol ether, b.p. 180° C. at 1 mm. of Hg), piperonyl cyclonene (III) (m.p. 50° C.), and sesamex (Sesoxane) (IV) (b.p. 137°–141° C. at 0.08 mm. of Hg). To this group also should be assigned propyl isome (V) and sesamin (VI) (m.p. 122.5° C.); the first of these compounds is prepared by the condensation of dipropyl maleate with isosafrole, and the second is isolated from sesame oil.

[Structure V: bicyclic compound with methylenedioxy, COOC₃H₇ groups and CH₃] (V)

[Structure VI: bis-methylenedioxybenzene dimer with dioxolane bridge] (VI)

The heterocyclic oxygen compounds also include some insecticides of plant origin, for example, rotenone (m.p. 163° C.; a dimorphous form melts at 181° C.) (VII), which is a selective contact insecticide obtained by extraction from the roots of a plant species.

[Structure VII: rotenone] (VII)

An advantage of this compound is its comparatively short-lived action and its low toxicity to vertebrates. In spite of the existence of a large number of effective synthetic compounds, rotenone still is used at the present time, although on a limited scale.

Most of the five-membered heterocyclic compounds that contain even one nitrogen atom have a fungicidal effect. An example is benzoxazolinone (VIII), which is found in a bound form in some plants. It is easily prepared by heating 2-aminophenol with urea:

$$\text{2-aminophenol} + NH_2CONH_2 \rightarrow \text{benzoxazolinone} + NH_3 \quad \text{(VIII)}$$

Its *N*-trichloromethylthio derivative has been patented as an active fungicide useful for the control of diseases of growing plants. The fungicidal action of benzoxazolinone is increased when halogen is introduced in the aromatic nucleus.

The corresponding thio derivatives also have fungicidal properties, and 4-chloro-2-hydroxybenzothiazolinyl-3-acetic acid (IX) (m.p. 193° C., LD_{50} 1,000 mg./kg.) has been proposed as a plant growth regulator for the control of chickweed and other weeds.

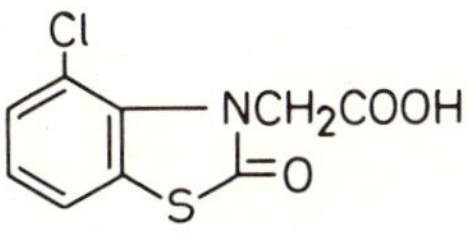

$$\text{(IX)} \quad \underset{S}{\underset{\|}{\bigcirc}}\text{-NCH}_2\text{COOH, Cl substituent}$$

(IX)

Salts of imidazoline, which are powerful acaricides, have been used for a number of years to control plant diseases.

A sexual sterilizing action is exerted on house flies by imidazolidinone-2 (X) (m.p. 131°–132° C., LD_{50} 5,000 mg./kg.), which is obtained by the reaction of ethylenediamine with urea:

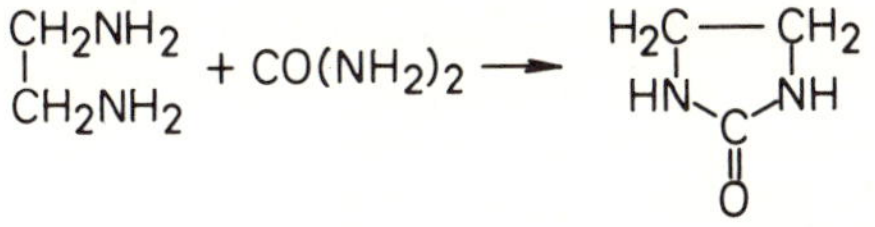

$$\begin{array}{c} CH_2NH_2 \\ | \\ CH_2NH_2 \end{array} + CO(NH_2)_2 \longrightarrow \begin{array}{c} H_2C-CH_2 \\ | \quad\quad | \\ HN\diagdown_{\underset{\|}{C}}\diagup NH \\ O \end{array} \quad \text{(X)}$$

When house flies are fed a diet containing 0.05% imidazolidinone-2, egg laying ceases almost completely. However, its action on other species of insects is considerably weaker and not of practical importance.

It has been proposed that the o-chlorophenylhydrazone of 3-methylisoxazoledione-4,5 (XI) (m.p. 167° C., LD_{50} for rats 126 mg./kg.) be used as an active fungicide with a wide range of action both for the control of diseases of growing plants and as a seed disinfectant for various crops.

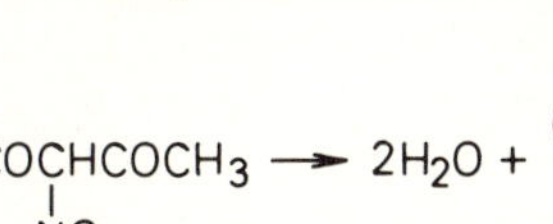

(XI)

1-Aryl-3,5-dimethyl-4-nitrosopyrazoles substituted in the aromatic nucleus (XII) are used as fungicides; these compounds are prepared by the condensation of isonitrosoacetylacetone with the appropriate arylhydrazines:

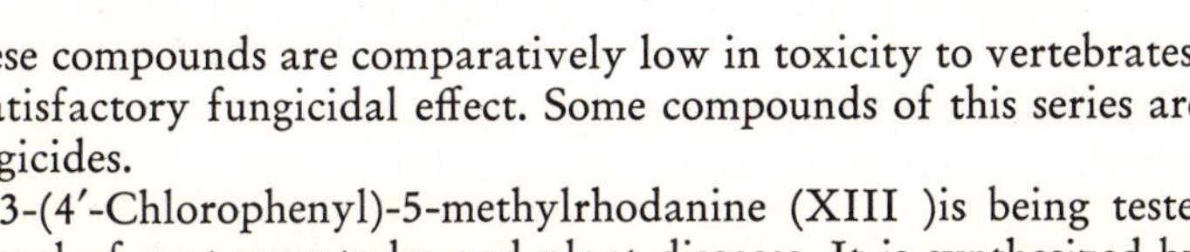

$$ArNH-NH_2 + CH_3COCHCOCH_3 \longrightarrow 2H_2O + \underset{NO}{|}$$

(XII)

These compounds are comparatively low in toxicity to vertebrates and have a satisfactory fungicidal effect. Some compounds of this series are systemic fungicides.

3-(4'-Chlorophenyl)-5-methylrhodanine (XIII) is being tested for the control of root nematodes and plant diseases. It is synthesized by the condensation of ammonium 4-chlorophenyldithiocarbamate with alkali or ammonium α-chloropropionates:

$$\text{Cl-C}_6\text{H}_4\text{-NHCSNH}_4 + \text{CH}_3\text{CHClCOONa} \longrightarrow \underset{\substack{|\\ \text{C}_6\text{H}_4\text{Cl}}}{\text{[5-methyl-3-(4-chlorophenyl)-2-thiohydantoin]}} + \text{NaCl} + \text{H}_2\text{O} + \text{NH}_3 \qquad (\text{XIII})$$

In the series of six-membered heterocyclic compounds, derivatives of uracil are used to control undesirable vegetation; derivatives of pyrazole are used as selective herbicides, and derivatives of pyrimidine are used to control rodents.

Piperazine is used in animal husbandry as an anthelminthic; similar properties are shown by phenothiazine (thiodiphenylamine) which has been proposed also for the control of chewing insects.

Five-membered heterocycles

Of the large number of five-membered heterocyclic compounds with two heteroatoms in the ring that have been investigated, in agriculture the derivatives of imidazoline containing in the 2-position a hydrocarbon radical with a large number of carbon atoms have been widely used. In the United States 2-heptadecylimidazoline-2 acetate as well as its derivatives are used.

2-Heptadecylimidazoline acetate (glyodin) is a bright-orange crystalline substance, m.p. 62°–68° C. It is slightly soluble in water, highly soluble in alcohols; in isopropyl alcohol at 25° C. its solubility is about 39%. The action of alkalies converts it to the free base, which has a wax-like consistency, m.p. 94° C. When glyodin is heated or stands for a prolonged period in a solution of alkali hydroxides it is hydrolyzed, forming the 2-aminoethylamide of stearic acid. The LD_{50} of glyodin for experimental animals is about 1,000 mg./kg.

Glyodin is produced by the condensation of ethylenediamine with stearic acid at an elevated temperature. The amide of stearic acid is first formed and this is further dehydrated on heating in vacuum:

$$\text{H}_2\text{NCH}_2\text{CH}_2\text{NH}_2 + \text{C}_{17}\text{H}_{35}\text{COOH} \longrightarrow \text{H}_2\text{NCH}_2\text{CH}_2\text{NHCOC}_{17}\text{H}_{35} + \text{H}_2\text{O}$$

$$\text{H}_2\text{NCH}_2\text{CH}_2\text{NHCOC}_{17}\text{H}_{35} \longrightarrow \underset{\substack{|\\ \text{C}_{17}\text{H}_{35}}}{\text{[2-heptadecylimidazoline]}} + \text{H}_2\text{O}$$

The yield of the compound by this reaction is almost quantitative. The presence of oleic acid as an impurity in the stearic acid is not permissible, because in that case a phytocidal product is obtained. The base obtained is treated with an equimolecular quantity of acetic acid and the 2-heptadecylimidazoline acetate is isolated:

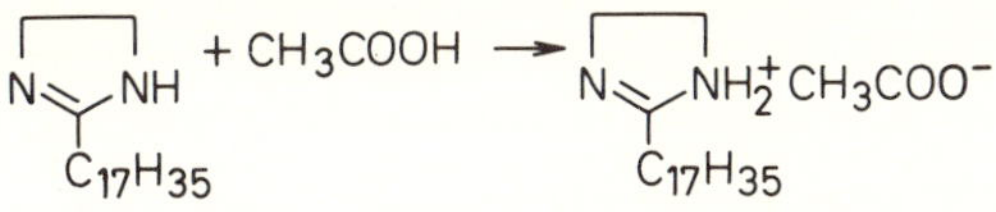

$$\underset{C_{17}H_{35}}{\underset{|}{N\diagdown\diagup NH}} + CH_3COOH \longrightarrow \underset{C_{17}H_{35}}{\underset{|}{N\diagdown\diagup NH_2^+}} CH_3COO^-$$

Glyodin is sold as a 34% solution in isopropyl alcohol. It also is sometimes marketed in a mixture with 1-hydroxyethyl-2-heptadecylimidazoline (XIV). The latter compound is obtained by the condensation of 2-hydroxyethylethylenediamine with ethyl stearate at an elevated temperature:

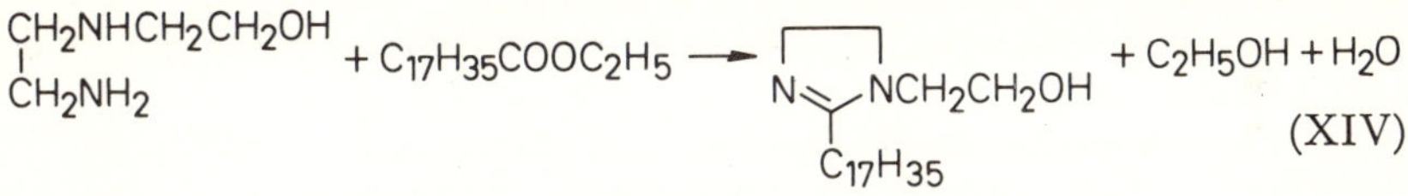

$$\underset{CH_2NH_2}{\overset{CH_2NHCH_2CH_2OH}{|}} + C_{17}H_{35}COOC_2H_5 \longrightarrow \underset{C_{17}H_{35}}{\underset{|}{N\diagdown\diagup NCH_2CH_2OH}} + C_2H_5OH + H_2O \qquad (XIV)$$

The base that is formed is a soap-like substance with m.p. about 50° C.; its solubility in water is about 0.01% at 75° C.

The toxicity and other properties of this compound are entirely analogous to those of glyodin. It is marketed as a formulation containing 30% glyodin and 4% of 1-hydroxyethyl-2-heptadecylimidazoline on kaolin. The kaolin is added to improve the emulsibility of the preparation.

1-Methyl-2-phenyl-3-dodecylbenzimidazolinium ferrocyanide (Fungilon, Bayer-32394, B-169-ferrocyanide) is a yellow-green crystalline substance, m.p. 178°–180° C., practically insoluble in water, soluble in acetone, ethyl alcohol, chloroform, and methylene chloride. It decomposes in alkaline medium. Its LD_{50} for rats by oral administration is 500 mg./kg.; intraperitoneally it is 10 mg./kg. Upon prolonged contact it irritates the skin of rabbits.

Fungilon is produced by the reaction of dodecyl halides with 1-methyl-2-phenylbenzimidazole and subsequent conversion of the halogen salt to the ferrocyanide:

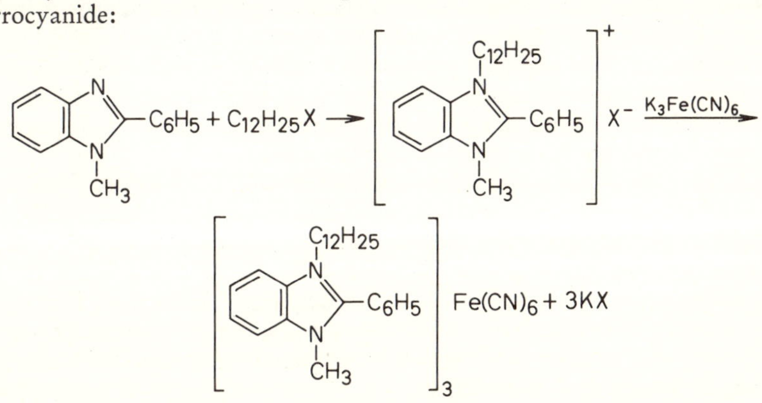

[benzimidazole-CH$_3$]—C$_6$H$_5$ + C$_{12}$H$_{25}$X ⟶ [benzimidazolinium with C$_{12}$H$_{25}$, CH$_3$, C$_6$H$_5$]$^+$ X$^-$ $\xrightarrow{K_3Fe(CN)_6}$

[benzimidazolinium with C$_{12}$H$_{25}$, CH$_3$, C$_6$H$_5$]$_3$ Fe(CN)$_6$ + 3KX

It is marketed in the form of a 20% wettable powder. It has a fungicidal effect not only on apple and pear scab, but also on powdery mildew fungi; however, it is phytocidal to some varieties of pears.

Methyl-1-(butylcarbamoyl)-2-benzimidazolecarbamate (benlate, duPont fungicide 1991) is a novel protective and eradicant fungicide with systemic activity, effective against a broad range of fungi and against mites. It is a

white crystalline solid with a faint acrid odor, is decomposed with heat, and is insoluble in water and hydrocarbons. The LD_{50} is 9,590 mg./kg. It is used in the form of a 50% wettable powder to control different plant diseases on several crops.

2-(2-Furyl)benzimidazole (furidazol, Voronit) is a crystalline solid, m.p. 284°–288° C., practically insoluble in water, soluble in acetone, ethanol, and methanol. The LD_{50} is 1,000 mg./kg. It is used as a fungicide suitable for seed treatment against *Fusarium nivale* and *F. culmorum*, in the form of dusts mixed with hexachlorobenzene.

2-(Thiazolyl)benzimidazole also is used as a fungicide for seed treatment.

5,6-Dichloro-1-phenoxycarbonyl-2-trifluoromethylbenzimidazole (fenazaflor) is a greyish-yellow crystalline powder, m.p. 103° C., vapor pressure 4×10^{-5} mm. of Hg at 25° C. The LD_{50} is 283 mg./kg. It is a nonsystemic acaricide showing considerable activity against mites, including organophosphate-resistant strains and is recommended for use at 0.03–0.04% active ingredient.

4,5-Dichloro-2-trifluoromethylbenzimidazole (chlorflurazole) is a crystalline substance, m.p. 213° C., solubility in water 60 p.p.m. The LD_{50} is 300–400 mg./kg. This herbicide has a contact effect on seedlings of a range of broad-leaved weeds, while grasses have a high degree of resistance. Hence it has been investigated for the control of annual weeds in cereals. Effective dosages are in the range of 0.25–1 kg./ha.

4,5,7-Trichloro-2,1,3-benzothiadiazole (PH 40-21, TH 052 H) is a crystalline substance, m.p. 131°–132° C., solubility in water 2.5 p.p.m. at 20° C., in acetone 3%, benzene 10%, chloroform 20%, xylene 10% at 20° C. The LD_{50} is 1,620 mg./kg. It is a soil-applied herbicide affecting weeds in the early stages of germination. PH 40–21 has been suggested for post-emergence use at dosages of 5–9 kg./ha. in rice, cotton, peas, and other crops. It is synthesized by the reaction of 2,4,5-trichlorophenylenediamine with thionyl chloride in the presence of pyridine:

$$\underset{\underset{Cl}{\bigvee}}{\overset{Cl}{\bigvee}}\overset{Cl}{\underset{-NH_2}{\overset{-NH_2}{\bigvee}}} + SOCl_2 \xrightarrow{+C_5H_5N} \underset{\underset{Cl}{\bigvee}}{\overset{Cl}{\bigvee}}\overset{Cl}{\underset{=N}{\overset{=N}{\bigvee}}}S + 2HCl + H_2O$$

2-Amino-4-methyl-5-carboxanilidothiazole (Seedvax) is a white crystalline substance, m.p. 220°–222° C., solubility in water 0.1%, methanol 15.5%, acetone 15% at 25° C. The LD_{50} is 141° mg./kg. It is an effective systemic fungicide that controls by seed treatment such organisms as *Rhizoctonia solani* on cotton, beans, sugar beets, and potatoes. It is synthesized by the condensation of α-chloroacetoacetanilide with thiourea:

$$CH_3COCHClCONHC_6H_5 + NH_2CSNH_2 \longrightarrow \underset{H_2N-C\diagdown_S\diagup C-CONHC_6H_5}{\overset{N\!=\!\!=\!\!C-CH_3}{}}$$

2,4-Dimethyl-5-carboxanilidothiazole (G-696) is a white solid, m.p. 140°–142° C., solubility in water 0.246% at 25° C. The LD_{50} is 5,600 mg. per kg. It is a systemic fungicide used in seed treatment. It is synthesized by the reaction of α-chloroacetoacetanilide with thioacetamide:

$$CH_3COCHClCONHC_6H_5 + CH_3CSNH_2 \longrightarrow \underset{CH_3-C\diagdown_S\diagup C-CONHC_6H_5}{\overset{N\!=\!\!=\!\!C-CH_3}{}}$$

A large number of other five-membered heterocyclic compounds also has been studied, including derivatives of hydantoin, thiazole, benzothiazole, oxazole, and others, but these compounds have not yet been practically used in agriculture.

Six-membered heterocycles

A considerably larger number of heterocyclic compounds with a six-membered ring is used in agriculture and industry. Of the six-membered heterocycles with two heteroatoms the following have been studied as pesticides: derivatives of pyridazine, pyrimidine, uracil, thiouracil, piperazine, morpholine, thiomorpholine, and many others.

3-Hydroxypyridazone (maleic hydrazide) is a white crystalline substance, dec. temperature 300°–320° C. It is poorly soluble in water (0.6% at 25° C.), alcohol, and acetone, more soluble in dimethylformamide (2.4%). In 0.5% solution it has a pH of 4. This compound was one of the first to be used as a herbicide and plant growth inhibitor.

Maleic hydrazide can exist in three isomeric forms, but in the crystalline material the pyridazone structure (XV) predominates:

$$\text{(XV)}$$

It dissolves readily in aqueous solutions of alkali hydroxides and amines with which it forms highly water-soluble salts. Its salts with the alkaline earths and heavy metals are practically insoluble in water. Its LD_{50} for mice is more than 2,000 mg./kg. Rats withstand up to 5% of the compound in their food without symptoms of poisoning. It is marketed in the form of the sodium or diethanolamine salts. The former contains 50% of the hydrazide, and the latter 30%.

Maleic hydrazide is produced by heating equimolecular quantities of hydrazine salts and maleic acid or maleic anhydride in aqueous solution:

$$NH_2NH_2 + \text{(maleic anhydride)} \rightarrow \text{(maleic hydrazide)} + H_2O$$

1-Phenyl-4-amino-5-chloropyridazone-6 (chlorazine, Pyramin, phenazon, pyrazon) is a white crystalline substance, m.p. 205°–206° C., vapor pressure at 40° C. 0.074 atm. Its solubility in water at 20° C. is 0.03%, in methyl alcohol 3.5%, and in acetone 2.8%. Its LD_{50} for rats is about 3,500 mg./kg. It is used at dosages of 2–4 kg./ha. to control weeds in sugar beets. When there is sufficient moisture, it is highly effective on dicotyledonous and partially so on monocotyledonous weeds and does not harm sugar beets. When the weather is dry, its action is weaker.

Chlorazine is produced in the following manner:

$$HOCCCl=CClCOOH + C_6H_5NHNH_2 \rightarrow$$

$$\xrightarrow{NH_3}$$

The condensation of mucochloric acid with phenylhydrazine is carried out in aqueous medium by heating either in the presence of some HCl or without it. Phenylhydrazine hydrochloride may be used for the condensation. The amination process takes place at an elevated temperature in an autoclave. The mucochloric acid necessary for this synthesis can be obtained by two methods:

1. Chlorination of furfural in HCl medium:

[furan-CHO] + $3H_2O + 5Cl_2 \rightarrow$ HOCCCl=CClCOOH + CO_2 + 8HCl

2. Oxidative chlorination of butynediol:

$HOCH_2C \equiv CCH_2OH + 4 Cl_2 + H_2O \rightarrow$ HOCCCl = CClCOOH + 6 HCl

The yield of mucochloric acid by the procedures indicated, depending on the conditions, varies from 50–80%. When mucochloric acid comes in contact with uncovered parts of the skin, it causes severe irritation; blisters appear that are reminiscent of those caused by a thermal burn. Mucochloric acid has some fungicidal action but because of its high phytocidal effect it is not possible to use it for this purpose.

Chlorazine is marketed as an 80% wettable powder and is used for soil treatment. A large number of different analogs of chlorazine are described in the literature and some of them have herbicidal properties, but they are not yet in use.

An interesting group of herbicides comprises the derivatives of uracil, a number of which are used in agriculture.

5-Bromo-3-sec-butyl-6-methyluracil (bromacil, Hyvar) is a white crystalline substance, m.p. 158°–159° C. At 25° C., 815 mg. of the compound dissolves in 1 liter of water. It is moderately soluble in acetone, acetonitrile, and ethyl alcohol and slightly soluble in hydrocarbons. Bromacil is stable at room temperature in aqueous solutions and in the presence of bases, but is decomposed by acids; it withstands heating to its melting point without decomposition. Its LD_{50} for rats is 5,200 mg./kg. It is marketed as an 80% wettable powder for use in aqueous suspensions. It is a nonselective herbicide and when it is applied to the soil it destroys most annual plants.

Bromacil is synthesized by the condensation of bromoacetoacetic acid with *sec*-butylurea. This reaction takes place in two stages:

$CH_3COCHBrCOOC_2H_5 + H_2NCONHCH(CH_3)C_2H_5 \rightarrow$

$C_2H_5CH(CH_3)NHCONHC(CH_3)=CBrCOOC_2H_5 \rightarrow$ [uracil ring structure with Br, CH$_3$, CHC$_2$H$_5$, CH$_3$ substituents]

A large series of homologs and analogs of bromacil has been studied; the most effective are compounds 732 (terbacil), 733, and 634 (lenacil).

5-Chloro-3-tert-butyl-6-methyluracil (terbacil, Compound 732) melts at 184° C. Its solubility in water at 25° C. is 710 mg./liter; it is highly soluble in dimethylformamide, moderately soluble in xylene and some other organic solvents. The LD_{50} is about 5,000 mg./kg.

5-Bromo-3-tert-butyl-6-methyluracil (Compound 733) is a white crystalline substance, m.p. 188° C. Its solubility in water at 25° C. is about

410 mg./liter; it is entirely similar to terbacil with respect to its solubility in organic solvents. The LD_{50} is about 5,000 mg./kg.

Both compounds are used as 80% wettable powders and at dosages of 1–4 kg./ha. They give satisfactory results in the eradication of weeds in plantings of citrus, pineapple (only terbacil), asparagus, alfalfa, etc. They are produced in a manner similar to bromacil.

3-Cyclohexyl-5,6-trimethyleneuracil (lenacil) is a white crystalline substance, m.p. 315.6°–316.8° C. At 25° C. its solubility in water is only 6 mg./liter; it is slightly soluble in most organic solvents (<1%), but soluble in pyridine. It is decomposed by the action of concentrated alkalies. It is rather stable to heating. The LD_{50} for experimental animals is more than 10,000 mg./kg. This compound gives interesting results as an agent for the control of weeds in sugar beet plantings. At dosages of 0.5 to 6 kg./ha. it gives satisfactory eradication of annual weeds.

Lenacil is produced by the condensation of cyclohexylurea with an ester of cyclopentanone-2-carboxylic acid:

$$\text{Cyclohexyl-NHCONH}_2 + \text{cyclopentanone-COOC}_2\text{H}_5 \rightarrow \text{lenacil} + C_2H_5OH + H_2O$$

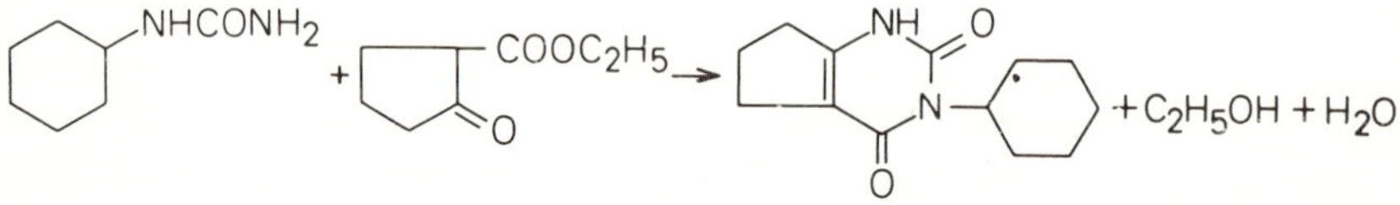

The ester of cyclopentanone-2-carboxylic acid necessary for this synthesis is formed in good yield from the ester of adipic acid by a Claisen reaction:

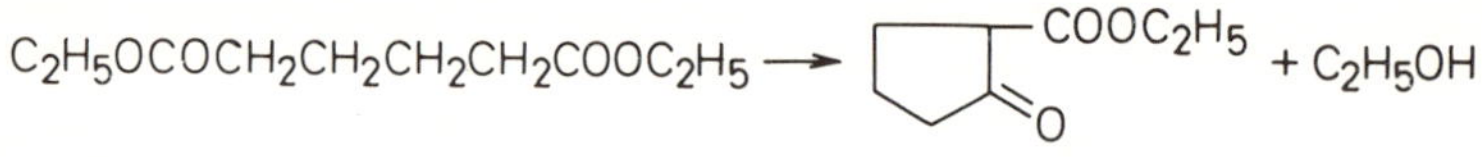

$$C_2H_5OCOCH_2CH_2CH_2CH_2COOC_2H_5 \rightarrow \text{cyclopentanone-COOC}_2H_5 + C_2H_5OH$$

In addition to the uracil derivatives, a large number of derivatives of pyrimidine have been studied, and active herbicides and zoocides have been found among them.

2-Chloro-4-dimethylamino-6-methylpyrimidine (Castrix) is a white crystalline substance, m.p. 87° C. The technical grade product is a brown waxy material, poorly soluble in water but highly soluble in most organic solvents. Castrix is very toxic to most animals; its LD_{50} for rats is 1.25 mg. per kg. and for rabbits it is 5 mg./kg. It has been used in grain baits to control rodents; the content of active ingredient in the bait is 0.1%. It is produced by the following scheme:

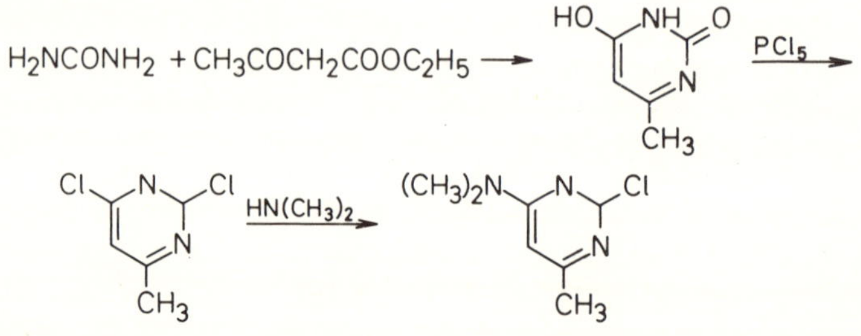

$$H_2NCONH_2 + CH_3COCH_2COOC_2H_5 \rightarrow \text{(HO, NH, O, CH}_3\text{ pyrimidine)} \xrightarrow{PCl_5}$$

$$\text{(Cl, N, Cl, CH}_3\text{ pyrimidine)} \xrightarrow{HN(CH_3)_2} \text{((CH}_3)_2N, N, Cl, CH_3 \text{ pyrimidine)}$$

2-Dimethylamino-4-hydroxy-5-n-butyl-6-methylpyrimidine (dimethirimol) is a white crystalline solid with no odor, m.p. 102° C., solubility in water 0.12 g./100 ml., acetone 4.5 g./100 ml., chloroform 120 g./100 ml., ethanol 6.5 g./100 ml., and xylene 36 g./100 ml., all at 25° C. It is stable to heat and in alkaline and acid solutions. The LD_{50} is 4,000 mg./kg. Dimethirimol is a new systemic fungicide. It has proved particularly effective, both as a soil and spray treatment, for the control of powdery mildew affecting cucumbers, melons, and certain ornamentals.

2,3-Dihydro-5-carboxanilido-6-methyl-1,4-oxathiin (Vitavax) is a solid, m.p. 91.5°–92.5° C.; a dimorphic form has m.p. 98°–100° C. Solubility in water 170 p.p.m. at 25° C., in methanol 21 g./100 g., in ethanol 11 g./100 g., in benzene 15 g./100 g. The LD_{50} is 3,200 mg./kg. It is a systemic fungicide of promise for the seed treatment of cereals against loose smuts and of cotton, sugar beet, and vegetables against *Rhizoctonia*.

Two methods are used for the production of Vitavax:

1. Reaction of α-chloroacetoacetanilide with 2-mercaptoethanol:

$$(CH_2CO)_2 + C_6H_5NH_2 \rightarrow CH_3COCH_2CONHC_6H_5$$

$$CH_3COCH_2CONHC_6H_5 + SO_2Cl_2 \rightarrow CH_3COCHClCONHC_6H_5 + SO_2 + HCl$$

$$CH_3COCHClCONHC_6H_5 + HOCH_2CH_2SH \xrightarrow{KOH} \text{[oxathiin ring: } CH_2\text{–O–C(CH}_3\text{)=C(CONHC}_6H_5\text{)–S–CH}_2\text{]}$$

2. From α-chloroacetoacetate and 2-mercaptoethanol:

$$CH_3COCH_2COOR + SO_2Cl_2 \rightarrow CH_3COCHClCOOR + HCl + SO_2$$

$$CH_3COCHClCOOR + HOCH_2CH_2SH \xrightarrow{KOH} \text{[oxathiin ring with COOR]} + HCl$$

[oxathiin-COOR] $\xrightarrow{H_2O}$ [oxathiin-COOH] $\xrightarrow{SOCl_2}$ [oxathiin-COCl] $\xrightarrow{C_6H_5NH_2}$ [oxathiin-CONHC_6H_5]

2,3-Dihydro-5-carboxanilido-6-methyl-1,4-oxathiin-4,4-dioxide (Plantvax) is a white crystalline substance, m.p. 127.5°–130° C., solubility in water 1,000 p.p.m., in methanol 7 g./100 g., in ethanol 3 g./100 g., in acetone 36 g./100 g., in benzene 3.4 g./100 g., all at 25° C. The LD_{50} is 2,000 mg./kg. It is a systemic fungicide of promise for the treatment of cereals and vegetables agains rusts. It is synthesized by the oxidation of 2,3-dihydro-5-carboxanilido-6-methyl-1,4-oxathiine with hydrogen peroxide:

$$\begin{array}{c}\text{CH}_2\text{-O-C-CH}_3\\\text{CH}_2\text{-S-C-CONHC}_6\text{H}_5\end{array} \xrightarrow{H_2O_2} \begin{array}{c}\text{CH}_2\text{-O-C-CH}_3\\\text{CH}_2\text{-S(O}_2\text{)-C-CONHC}_6\text{H}_5\end{array}$$

N-Tridecyl-2,6-dimethylmorpholine (E-236) is a liquid, soluble in most organic solvents, practically insoluble in water. The LD_{50} is 650 mg./kg. It is a fungicide used for the control of *Erysiphe graminis*.

N-Tritylmorpholine (Frescon) is a colorless crystalline solid, m.p. 174° to 176° C., soluble in most organic solvents, insoluble in water. The LD_{50} is 1,000 mg./kg. (percutaneous). It is hydrolyzed in mild acids to triphenylcarbinol. It is a molluscicide. The LD_{50} for 24 hours' exposure, followed by a recovery period of 48 hours, for the snails *Australorbis grabatrus*, *Bulimus truncatus*, *B. globosus*, and *Biomphalaria pfeifferi* was 0.025, 0.053, 0.050 and 0.014 p.p.m., respectively.

Dibenzo-1,4-thiazine (phenothiazine, thiodiphenylamine) is a light-yellow crystalline substance, m.p. 185° C. It is practically insoluble in water, poorly soluble in most organic solvents. In the air it is slowly oxidized, yielding phenothiazone (XVI) and thinol (XVII):

(XVI)   (XVII)

It is assumed that the mechanism of action of phenothiazine on insects is associated with the formation of a leucothionol compound in the haemolymph, which suppresses the activity of the respiratory enzymes. The oxidation products of phenothiazine also have a slight fungicidal effect.

Phenothiazine is one of the oldest pesticides with low toxicity to mammals. It is used mainly as an anthelminthic to control intestinal worms of domestic animals, and also in small amount to control larvae of the malaria mosquito. It is produced by the reaction of diphenylamine with sulfur in the presence of iodine at 180°–220° C., in practically quantitative yield:

$$(C_6H_5)_2NH + 2S \longrightarrow H_2S + \text{phenothiazine}$$

To purify phenothiazine it is recrystallized from benzene or acetone, and also is sublimed. The freshly sublimed compound is almost colorless while the technical product has a gray-green color and is used in the form of a finely dispersed powder.

1,4-Dithiaanthraquinone-2,3-dinitrile (XVIII) (dithianon) is a brownish powder, m.p. 225° C. It is insoluble in water but soluble in dioxane, chlorobenzene, and chloroform. It was produced in 1962 in the form of a 75% wettable powder as an experimental fungicide for the control of diseases of growing plants. Its LD_{50} for rats is 1,015 mg./kg. Lengthy investigation of this compound has shown that it is low in toxicity and does not have a cancerogenic effect.

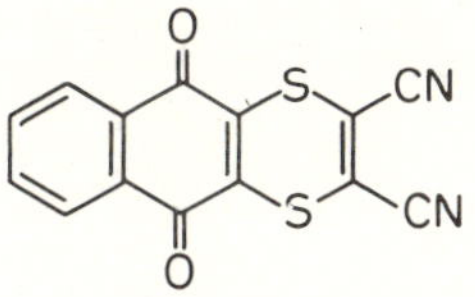

(XVIII)

It is incompatible with materials of an alkaline nature, since in alkaline medium it breaks down relatively rapidly.

The compounds of formulas XIX, XX, and XXI also have herbicidal properties.

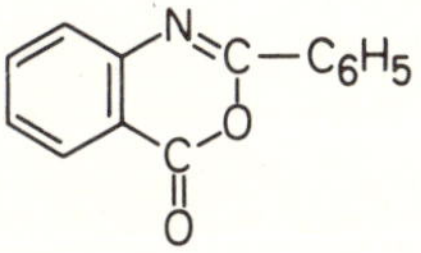

(XIX)

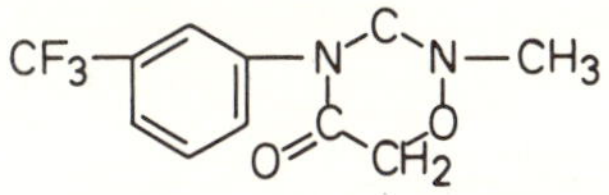

(XX)

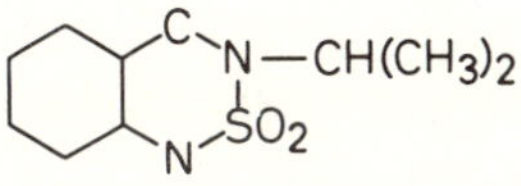

(XXI)

General references

BYRDE, R. J., and C. W. HARPER: Rep. Agr. Hort. Research Sta. Univ. Bristol, p. 151 (1955).
DAAMS, J.: Proc. 7th Brit. Weed Control Conf., p. 1091 (1964).
DELORAINE, J., P. DESAYMARD, J. DESMORAS, J. B. DODEL, P. JAKQUET, J. METTIVIER, and A. PELANARD: Proc. 2nd Symposium New Herbicides, Paris (1965).
DELP, C. J., and H. L. KLOPPING: Plant Disease Reporter 52, 95 (1968).
DYSON, G., and P. MAY: Khimiya sinteticheskikh lekarstvennykh veshchestv [Chemistry of synthetic drugs]. Publishing House "Mir" (1964).
ELDERFIELD, R.: Geterotsiklicheskie soedineniya [Heterocyclic compounds], vol. 6. Foreign Literature Publishing House [USSR] (1960).
ELIAS, I., R. S. SHEPHARD, M. C. SNELL, and J. STUBS: Nature 219, 1160 (1968).
ERWIN, D. C.: World Review of Pest Control 8, 6 (1969).
FISCHER, A.: Abstr. 6th Internat. Congress Plant Prot. Vienna, p. 446 (1967).

MEL'NIKOV, N. N., and YU. A. BASAKOV: Khimiya gerbitsidov i regulyatorov rosta rastenii [Chemistry of herbicides and plant growth regulators]. State Scientific and Technical Publishing House of Chemical Literature [USSR] (1962).
PFEIFFER, R. K.: Proc. 8th Brit. Weed Control Conf., p. 394 (1966).
POMMER, E. H., and J. KRADEL: 19th Internat. Symposium, Ghent, p. 735 (1967).
POPA, C., and R. DRIMUS: Chimia Produselor Fitofarmaceutica. Bucharest (1965).
REINBERGS, E.: Canad. J. Plant Sci. **48,** 31 (1968).
SCHMELING, B., and M. KULKA: Science **152,** 659 (1966).
SCHUHMANN, G.: Nachrichtenpflanzenschutz **20,** 3 (1968).
SPENCER, E. Y.: Guide to the chemicals used in crop protection, Canada (1968).
SUMMERS, L. A., R. J. BYRDE, and C. W. HARPER: Ann. Applied Biol. **62,** 45 (1968).

XXX. Heterocyclic compounds with three heteroatoms in the ring

General characteristics of pesticidal properties

Among heterocyclic compounds with three atoms of nitrogen, nitrogen and sulfur, or nitrogen and oxygen in the ring a number of effective materials have been found, and some of them are used in agriculture.

The five-membered heterocyclic compounds (derivatives of triazole, thiadiazole, and tetrazole, and also the corresponding benzo derivatives) are more active toward the lower and higher plants than toward insects and mites, and they, therefore, are used mainly as herbicides or fungicides. Even some organic compounds of phosphorus that contain heterocyclic radicals have predominantly fungicidal, and not insecticidal, activity. An example of such compounds is 5-amino-3-phenyl-1,2,4-triazolyl-1-bis(dimethylamido)phosphate (Wepsin) (I) (m.p. 164°–168° C., LD_{50} for rats 10–20 mg. per kg.):

$$\begin{array}{c}\text{(CH}_3\text{)}_2\text{N}\\ \text{(CH}_3\text{)}_2\text{N}\end{array}\!\!\!\overset{\overset{\text{O}}{\|}}{\text{P}}\!-\!\text{N}\!-\!\!\!\begin{array}{c}\text{NH}_2\\ \text{N}\\ \text{N}\end{array}\!\!\!-\!\!\text{C}_6\text{H}_5 \qquad\qquad (I)$$

It has been proposed for use in the control of fungous diseases of roses and powdery mildew of apple. Wepsin also shows slight acaricidal effect, but its fungicidal activity is considerably weaker.

3-Amino-1,2,4-triazole (amitrole) shows high herbicidal activity and is used to control weeds either independently or mixed with various other compounds.

Benzotriazole has interesting physiological activity. When it enters a plant through the root system, it changes the shape of the leaves (formative effect). The quaternary dialkylbenzotriazolium salts are characterized by high physiological activity for plants, and also by fungicidal and bactericidal effects. The most effective in this respect is dioctylbenzotriazolium chloride (or bromide) (II):

$$\text{(II)}$$

Compounds of this type are easily produced by prolonged heating of benzotriazole with alkyl halides.

High fungicidal and bactericidal activities are characteristic of the quaternary salts of tetrazole and other similar compounds containing a hydrocarbon radical with 7–16 carbon atoms on a nitrogen. Higher molecular compounds are inactive. However, all these compounds are toxic to man and animals and show a curare-like effect.

Six-membered heterocyclic compounds are used as fungicides, herbicides, and soil sterilants. Of the group of six-membered heterocyclic compounds, derivatives of symmetrical triazine are used to control weeds in a great variety of agricultural crops. Since this group of compounds is very important, they are considered below in more detail.

Some use has been made also of derivatives of tetrahydrothiazine (III):

$$\text{(III)}$$

In the series of heterocyclic compounds with a seven-membered ring, endosulfan (Thiodan) (IV), a cyclic ester of sulfurous acid, is used in agriculture:

$$\text{(IV)}$$

3-Amino-1,2,4-triazole (aminotriazole) and its derivatives

3-Amino-1,2,4-triazole (amitrole, aminotriazole) is a white crystalline substance, m.p. 159° C. At 23° C. 28 g. of the compound dissolves in 100 ml. of water; at 75° C. 26 g. dissolves in 100 ml. of ethyl alcohol. It is almost insoluble in ether and acetone. It is physiologically active toward plants. It is used as a herbicide, defoliant, and plant growth regulator. Amitrole forms stable salts with acids. Compounds of it with some metals and their salts also have been described.

It is resistant to hydrolysis and the action of oxidizing agents, but enters readily into condensation reactions, for example with aldehydes, ketones, and other compounds. It is rapidly absorbed both by roots and by aboveground parts of plants and causes chlorosis that is not removed by the addition of iron salts to the soil or treatment of the plants with them. This apparently is associated with the displacement by amitrole of pyrrole rings from the chlorophyll molecule and the formation of chelates with ions of metals that are vitally important to plants. Amitrole has been proposed for use in a mixture with ammonium thiocyanate against undesirable vegetation; its effect is intensified by several times in the presence of ammonium thiocyanate. It gives good results in the control of couch grass, Johnson grass, aquatic vegetation, etc. To control couch grass it is used at dosages of 3–7 kg./ha. It also has been proposed as an additive to defoliants, for example to magnesium chlorate. In this case the dosage of the defoliant is decreased. Its LD_{50} for rats is 1,100 mg./kg.

3-Amino-1,2,4-triazole is produced in good yield by the condensation of aminoguanidine with formic acid:

$$H_2N\overset{NH}{\underset{\|}{C}}NHNH_2 + HCOOH \longrightarrow H_2N\overset{NH}{\underset{\|}{C}}NHNHCHO \longrightarrow \underset{NH_2}{\underset{|}{\underset{N\diagdown\diagup N}{\bigcap}}}^{NH}$$

The aminoguanidine necessary for this purpose is obtained by reduction of nitroguanidine:

$$H_2N\overset{NH}{\underset{\|}{C}}NHNO_2 \xrightarrow{2H_2} H_2N\overset{NH}{\underset{\|}{C}}NHNH_2 + 2H_2O$$

or by reaction of hydrazine with methylisothiourea:

$$H_2N\overset{NH}{\underset{\|}{C}}SCH_3 + H_2NNH_2 \longrightarrow H_2N\overset{NH}{\underset{\|}{C}}NHNH_2 + CH_3SH$$

It is possible to synthesize 3-amino-1,2,4-triazole with a yield of about 50% by heating triazine and aminoguanidine to 100°–200° C.

$$2 \underset{N}{\underset{\diagdown}{\bigcap}}\overset{N\diagup}{\overset{\diagdown}{N}} + 6H_2N\overset{NH}{\underset{\|}{C}}NHNH_2 \longrightarrow 6\underset{NH_2}{\underset{|}{\underset{N\diagdown\diagup N}{\bigcap}}}^{NH} + 6NH_3 + N_2$$

Substituted 3-amino-1,2,4-triazoles can be prepared in a similar manner by the use of substituted aminoguanidines; however, they have received practically no use as yet.

The herbicidal action of various derivatives of 3-amino-1,2,4-triazole that contain substituents on both the amino group and the carbon atoms is considerably weaker. Some acylated aminotriazoles are no less effective than amitrole. Among these compounds the highest herbicidal activity is shown by 1,2,4-triazolyl-3-urea, which has defoliant and desiccant properties.

Derivatives of tetrahydrothiadiazine

Derivatives of tetrahydrothiadiazine, which are obtained by the condensation of alkyldithiocarbamic acids or their salts with formaldehyde and primary amines, are soil sterilants and fungicides. The simplest representative of this series of compounds is 3,5-dimethyltetrahydro-1,3,5-thiadiazinethione-2 (V), known by the trade name of Mylone and in the Soviet Union as thiazone. Its carboxy derivative, 3-methyl-5-carboxymethyltetrahydro-1,3,5-thiadiazinethione-2 (VI) (thiadiazinethione) is an active nematicide, herbicide, and fungicide:

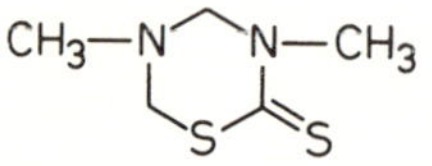
CH3—N
N—CH3
S
S

(V)

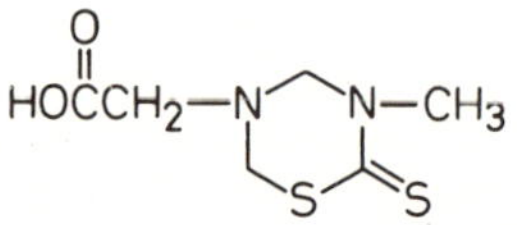
O
HOCCH2—N
N—CH3
S
S

(VI)

Japanese investigators also have proposed as a fungicide ethylene-bis-(3-methyltetrahydro-1,3,5-thiadiazinethion-2-yl-5) (VII), which is obtained by the condensation of sodium methyldithiocarbamate with formaldehyde and ethylenediamine in aqueous medium:

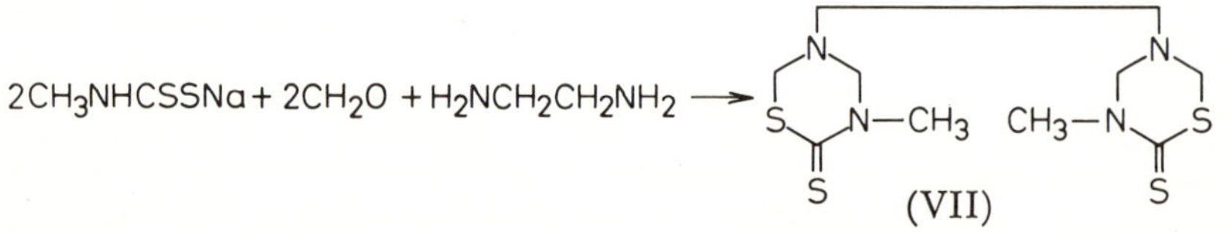
2CH3NHCSSNa + 2CH2O + H2NCH2CH2NH2
N
S
N—CH3
S
CH3—N
N
S
S

(VII)

It is supposed that the biocidal effect of this group of compounds is associated with their decomposition in the soil to form the corresponding isothiocyanates, which in most cases show nematicidal, fungicidal, and herbicidal properties. These compounds also have insecticidal activity, although not always.

3,5-Dimethyl-1,2,3,5-tetrahydro-1,3,5-thiadiazinethione-2 (thiazone, Mylone) is a white crystalline substance, m.p. 104° C. (dec.). Its solubility in water at 25° C. is about 0.12%; it is poorly soluble in ethyl alcohol, benzene and other hydrocarbons, ether, and carbon tetrachloride, highly soluble in chloroform and acetone. Its LD_{50} for white mice is 650 mg./kg. However, when dust of the compound is inhaled for a long time, slight irritation of the upper respiratory passages is observed. When heated to 100° C., thiazone gradually breaks down, forming methyl isothiocyanate and dimethylthiourea. In the presence of water, breakdown on heating occurs considerably more rapidly. It is assumed that thiazone in the soil under the influence of moisture is converted to formaldehyde and methylammonium methyldithiocarbamate, which is further converted to methyl isothiocyanate:

$$CH_3-N\underset{S}{\overset{}{\diagdown}}\underset{}{\overset{}{\diagup}}N-CH_3 + 2H_2O \longrightarrow 2CH_2O + \left[CH_3NH\overset{S}{\overset{\|}{C}}S\right]^- [NH_3CH_3]^+ \longrightarrow$$

$$CH_3NCS + H_2S + CH_3NH_2$$

Most investigators ascribe the nematicidal and other biocidal activity of this compound to the formation of methyl isothiocyanate, which in their opinion reacts with vitally important enzyme systems of nematodes, insects, fungi, and weeds.

Thiazone is marketed in the form of 85 or 50% wettable powders with added surface-active agents. The most commonly used diluents are organic materials of plant origin, for example dry wood powder, bran, finely ground straw, and the like. Inorganic diluents catalyze the decomposition of the compound in storage and they therefore are seldom used. The compound is used in the form of a suspension or an aqueous solution (0.1%) to control nematodes by introduction into the soil at dosages of 55–1,000 kg. per ha. At the last dosage complete sterilization of the soil occurs.

Thiazone is produced by the reaction of sodium methyldithiocarbamate with formaldehyde and methylamine in aqueous medium at room temperature. When the components are mixed, the thiazone gradually separates out in the form of a white precipitate, which is filtered off and dried at as low a temperature as possible:

$$CH_3NH\overset{S}{\overset{\|}{C}}SNa + 2CH_2O + CH_3NH_2 \longrightarrow CH_3-N\diagdown\diagup N-CH_3 + H_2O + NaOH$$

The yield is more than 90%. By-products obtained are small amounts of symmetrical dimethylthiourea and methylammonium sulfide, the formation of which can be represented as follows:

$$CH_3NHCSSH + 2\ CH_3NH_2 \rightarrow (CH_3NH)_2CS + CH_3NH_3SH$$

3-Methyl-5-carboxymethyltetrahydro-1,3,5-thiadiazinethione-2 (thiadiazinethione) is a white crystalline substance, m.p. 139°–140° C. It is poorly soluble in water, soluble in methyl and ethyl alcohols, highly soluble in dimethylformamide, acetone, and dioxane. The chemical properties of thiadiazinethione are similar to those of thiazone, but thiadiazinethione is more stable toward heating. It reacts with water similarly to thiazone, but under certain conditions it is hydrolyzed with the formation of carbon disulfide and this fact is utilized for its quantitative determination. It is used in the form of granules containing 75% active ingredient. At dosages of 100–200 kg./ha. it gives satisfactory suppression of nematodes, fungi, and weeds. The LD_{50} for rats is 1,000 mg./kg.; for intraperitoneal introduction it is 300–375 mg./kg.

Thiadiazinethione is produced by the reaction of formalin with sodium methyldithiocarbamate in the presence of aminoacetic acid in aqueous medium:

$$CH_3NH\overset{\overset{S}{\|}}{C}SNa + 2CH_2O + H_2NCH_2COOH \longrightarrow CH_3-N\underset{S}{\overset{\frown}{}}\overset{\frown}{\underset{\|}{C}}N-CH_2COOH + NaOH$$

3,3'-Ethylene-bis(tetrahydro-4,6-dimethyl-2H-1,3,5-thiadiazine-2-thione) (duPont fungicide 328) is a white crystalline substance, m.p. 140° to 141° C., slightly soluble in water. It slowly decomposes in the presence of air or water and is unstable in many organic solvents. The LD_{50} is 5,000 mg./kg. It is an effective fungicide for control of apple scab, cedarapple rust, sooty blotch, blue mold of tobacco, and late blight of potatoes. It is synthesized by the reaction of ethylenediamine with carbon disulfide, ammonia, and acetaldehyde:

$$\begin{array}{c} CH_2-NH_2 \\ | \\ CH_2-NH_2 \end{array} + 2CS_2 + 2NH_3 + 4CH_3CHO \longrightarrow$$

[structure of product] $+ 4H_2O$

s-Triazines

In spite of the fact that study of the pesticidal properties of derivatives of triazine was started comparatively recently, considerable success has been achieved in this area and a large number of these compounds have received practical use in agriculture and industry. Triazines of the following types (VIII, IX, X, XI, XII) are used as pesticides:

[Structures VIII, IX, X with Cl substituent; XI with OR''; XII with SR'']

Compounds of general formula (VIII) that contain an aliphatic radical on the nitrogen are not sufficiently stable for practical use, although they have herbicidal properties. When the aliphatic radical on the nitrogen is replaced by an aromatic one, the phytocidal properties are diminished and the fungicidal activity is increased. The most active compound is 2,4-dichloro-6-(o-chlorophenylamino)-s-triazine, which is used under the name Dyrene to control plant diseases. Its isomers that have the halogen in other positions are less active.

Compounds of type (IX) show high herbicidal activity if the hydrocarbon radicals on the nitrogen have no more than four carbon atoms. When the length of the carbon chain is increased the herbicidal activity is lowered. A similar rule is observed in the series of compounds of general formula (X). Replacement of one of the hydrocarbon radicals by an alkoxyl decreases the stability of the compound and also decreases the duration of its action, which is of great practical importance for some compounds.

When the halogen atom is replaced by a hydroxyl the herbicidal properties are completely lost, but its replacement by an alkoxyl (compounds of formula (XI)) or an alkylthio group (compounds of formula (XII)) changes the selective action of the compound, while preserving its herbicidal properties. The optimum activity in compounds of this type is observed when the alkoxyl or alkylthio group contains one carbon atom. An increase in the number of carbon atoms in this radical lowers the herbicidal activity.

2-Chloro-4,6-bis(alkylamino)-s-triazines can exist in three tautomeric forms, but the equilibrium in most cases is shifted in the direction of the amine form (IX):

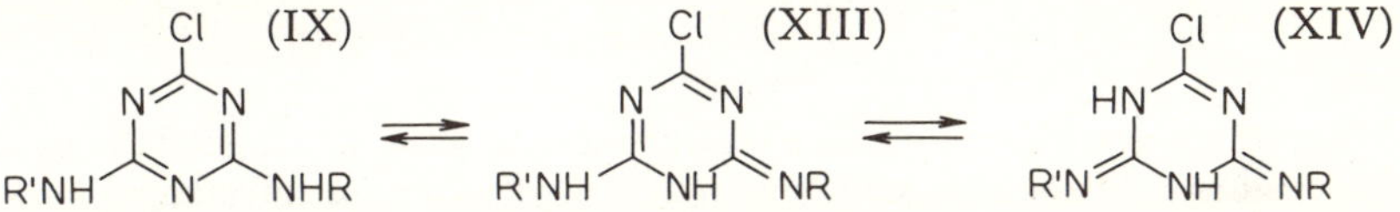

For compounds of formula (X) only two tautomeric forms are possible, and 2-chloro-4,6-bis(dialkylamino)-s-triazines (XVI) exist only in the amine

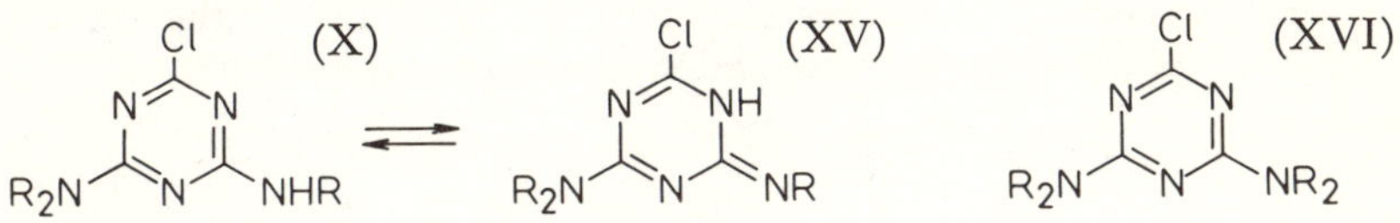

form. These two groups of compounds also differ in their physical properties. 2-Chloro-4,6-bis(alkylamino)-s-triazines are predominantly solid crystalline substances with high melting points, poorly soluble in water and organic solvents, while the 2-chloro-4,6-bis(dialkylamino)-s-triazines have low melting points and are soluble in many organic solvents.

The 2-chloro-4,6-bis(alkylamino)-s-triazines are stable in storage at room temperature and can be kept for a practically unlimited time without change. When they are heated with water, especially in the presence of organic or inorganic bases, they are hydrolyzed to inactive hydroxy compounds:

$$\underset{\underset{RNH}{\overset{Cl}{\bigtriangleup}}}{\text{triazine}} + H_2O \longrightarrow \underset{\underset{RNH}{\overset{OH}{\bigtriangleup}}}{\text{triazine}} + HCl$$

The halogen atom can be replaced by other groups; thus, by reaction with metal alcoholates the corresponding 2-alkoxy-4,6-bis(alkylamino)-s-triazines are formed, and by reaction with mercaptides the 2-alkylthio-4,6-bis(alkylamino)-s-triazines are formed:

$$\underset{\underset{RNH}{\overset{Cl}{\bigtriangleup}}}{\text{triazine}} + R'XNa \longrightarrow \underset{\underset{RNH}{\overset{XR'}{\bigtriangleup}}}{\text{triazine}} + NaCl$$

$$(X = O, S)$$

Both of these are valuable herbicides.

2-Chloro-4,6-bis(alkylamino)-s-triazines are produced in good yields by the reaction of cyanuric chloride with amines. If both amine groups contain the same hydrocarbon radicals, the reaction is carried out in one step:

$$\underset{\underset{Cl}{\overset{Cl}{\bigtriangleup}}}{\text{triazine}} + 2RNH_2 + 2NaOH \longrightarrow \underset{\underset{RNH}{\overset{Cl}{\bigtriangleup}}}{\text{triazine}} + 2NaCl + 2H_2O$$

When there are different radicals in the amino groups, a two-stage process is expedient. In the first stage one molecule of amine is introduced into the reaction; the 2,4-dichloro-6-alkylamino-s-triazine formed is further treated with one molecule of the second amine. Either organic or inorganic bases or an excess of amine are used as HCl acceptors.

The cyanuric chloride necessary for the synthesis of the substituted triazines is obtained by polymerization of cyanogen chloride:

$$3ClCN \longrightarrow \underset{\underset{Cl}{\overset{Cl}{\bigtriangleup}}}{\text{triazine}}$$

The reaction is carried out in the presence of activated carbon at 350° to 400° C., or in the liquid phase under pressure in various organic solvents with the use of anhydrous aluminum chloride, boron fluoride, HCl, etc., as catalyst.

Cyanuric chloride also can be obtained by the action of phosphorus pentachloride on cyanuric acid in a yield not exceeding 55–59%, while it is produced in more than 90% yield by the polymerization of cyanogen chloride:

$$\underset{\text{HO}}{\underset{\text{N}}{\text{OH}}}\underset{\text{N}}{\overset{\text{N}}{\bigotimes}}\text{OH} + 3\text{PCl}_5 \longrightarrow \underset{\text{Cl}}{\underset{\text{N}}{\text{N}}}\underset{\text{N}}{\overset{\text{Cl}}{\bigotimes}}\text{Cl} + 3\text{POCl}_3 + 3\text{HCl}$$

Cyanuric chloride is obtained in 85% yield by the chlorination of methyl thiocyanate, but purifying it of other products is difficult.

Pure cyanuric chloride is a white crystalline substance, m.p. 146° to 147° C., practically insoluble in water, highly soluble in most organic solvents. It is readily hydrolyzed by water, being converted to cyanuric acid. The hydrolysis takes place especially rapidly at an elevated temperature and in the presence of bases. This must be taken into consideration in using cyanuric chloride for the synthesis of substituted triazines in aqueous medium. In the presence of bases cyanuric chloride reacts with many compounds that contain an active hydrogen.

2-Chloro-4,6-bis(ethylamino)-s-triazine (simazine) is a white crystalline substance, m.p. 227°–228° C. The technical grade product contains more than 95% 2-chloro-4,6-bis(ethylamino)-s-triazine and melts at 224°–225° C. At 22° C., 1 liter of water dissolves about 5 mg. of the compound; the solubility in organic solvents (in g./100 g. of solvent) is methanol 0.4, petroleum ether 0.002, chloroform 0.9.

When simazine is heated with caustic alkalies the chlorine is replaced by hydroxyl and 2-hydroxy-4,6-bis(ethylamino)-s-triazine is formed, which does not have herbicidal properties. Simazine reacts similarly with sodium methylate and mercaptide, being converted to 2-methoxy-4,6-bis(ethylamino)-s-triazine and 2-methylthio-4,6-bis(ethylamino)-s-triazine, respectively, which are known as simetone and simetryne. However, these two compounds, in spite of their high herbicidal activity, are practically unused.

Simazine is thermally stable and withstands heating to a temperature higher than 150° C. It is not explosive under ordinary conditions and burns only when it is heated above its melting point. Its LD_{50} for rats is >5,000 mg./kg. It is used for application to the soil at dosages of 1–4 kg./ha., depending on the nature of the soil. Simazine is distinguished by the length of its residual action; therefore crops sensitive to simazine cannot be planted in plots treated with large dosages the previous year. It is used in mixtures with various other herbicides, which permits its use in small doses and avoids residual effects the following year. At high dosages simazine can be used as a nonselective herbicide on airports, railroads, and in some other cases.

It is produced in >90% yield by the reaction of cyanuric chloride with ethylamine and NaOH in aqueous medium, although this reaction also can be carried out in organic solvents. It is marketed as a 50% wettable powder with kaolin or chalk as the diluent.

2-Chloro-4-ethylamino-6-isopropylamino-s-triazine (atrazine) is a white crystalline substance, m.p. 173°–175° C. In water at 22° C. the solubility of atrazine is 70 mg./liter. Its solubility in organic solvents (in %) is ethyl ether 1.2, chloroform 5.2, methyl alcohol 1.8, and pentane 0.036. Under ordinary conditions of storage atrazine is stable and remains unchanged for a long time. When it is heated with caustic alkalies, or with alcoholates or mercaptides of the alkali metals, the chlorine is replaced by the corresponding groups. The product of the reaction of atrazine with sodium methylmercaptide is known as ametryne and is used as a selective herbicide. The chemical and biological properties of atrazine are similar to those of simazine, but its herbicidal effect differs with respect to duration and selectivity. The LD_{50} for rats is 3,080 mg./kg.

In contrast to simazine, atrazine is produced only in two stages. In the first stage, reaction of cyanuric chloride with the amine and NaOH produces 2,4-dichloro-6-ethylamino-s-triazine, which by further reaction with a molecule of isopropyl amine and NaOH gives atrazine:

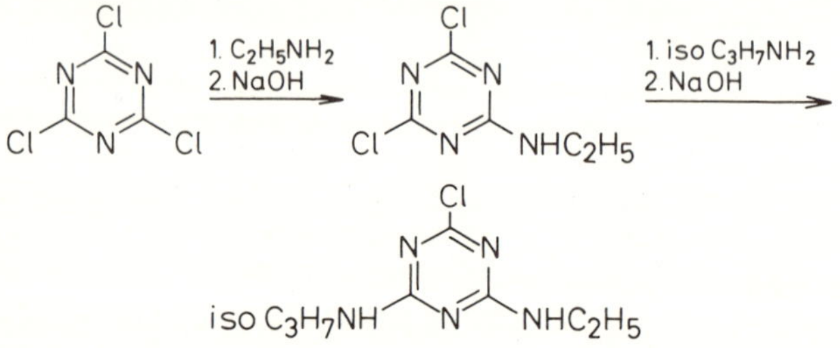

This reaction is easily carried out even in one apparatus, since the second reaction takes place at a higher temperature than the first. It also is possible to prepare atrazine by the use of a mixture of ethyl- and isopropylamines, but in this case the atrazine is contaminated with a large amount of simazine and to some extent propazine.

For the control of weeds in corn, atrazine is used at approximately the same dosages as simazine. It also is used as a nonselective herbicide on railroads. It is marketed in the form of a 50% wettable powder for the preparation of aqueous suspensions.

2-Chloro-4,6-bis(isopropylamino)-s-triazine (propazine) is a white crystalline substance, m.p. 210°–212° C. At 22° C., 8.6 mg. of the compound dissolves in 1 liter of water; it is poorly soluble in most organic solvents. At room temperature propazine is stable; heating with caustic alkalies converts it to the corresponding hydroxy derivative. The chemical properties of propazine are similar to those of simazine and atrazine. Because of its poor solubility in water it remains in the soil for a long time without decomposition. Propazine differs in selective action from simazine and has been proposed for the control of weeds in umbelliferous crops. However, at high dosages propazine damages carrots, especially when the humidity is high. The LD_{50} of propazine for rats is more than 5,000 mg./kg.

The synthesis of propazine is similar to that of simazine (from cyanuric chloride and isopropylamine); the reaction is carried out in either one or two steps. It is marketed as a 50% wettable powder.

A large number of other chlorotriazines have been studied [2-chloro-4-ethylamino-6-diethylamino-s-triazine (trietazine), m.p. 100°–102° C.; 2-chloro-4-isopropylamino-6-diethylamino-s-triazine (ipazine), m.p. 112° to 114° C.; 2-chloro-4,6-bis(diethylamino)-s-triazine (chlorazine), m.p. 27° to 29° C.; etc.], but they have received practically no use.

2-Methylthio-4,6-bis(isopropylamino)-s-triazine (prometryne) is a white crystalline substance, m.p. 118°–120° C. Its solubility in water at 20° C. is 0.048%; it is highly soluble in organic solvents. The LD_{50} for rats is 2,500–3,750 mg./kg. It is one of the important present-day herbicides that are used to control weeds in plantings of onions, cabbage, carrots, and grains. It is marketed in the form of a 50% wettable powder. The dosage used is 0.5–2 kg./ha. (active ingredient).

Prometryne is produced by the reaction of propazine with sodium mercaptide:

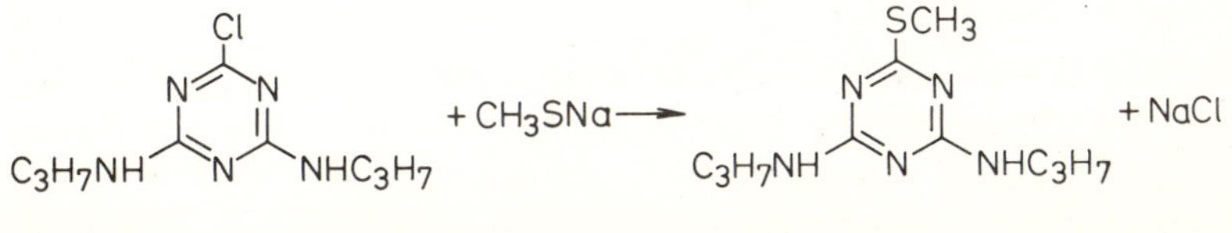

[Chemical reaction: propazine (Cl-triazine with two NHC3H7 groups) + CH3SNa → prometryne (SCH3-triazine with two NHC3H7 groups) + NaCl]

It also can be synthesized by another route. Reaction of propazine with thiourea leads to the corresponding thiuronium derivative, which on reaction with dimethyl sulfate or other methylating agent yields prometryne:

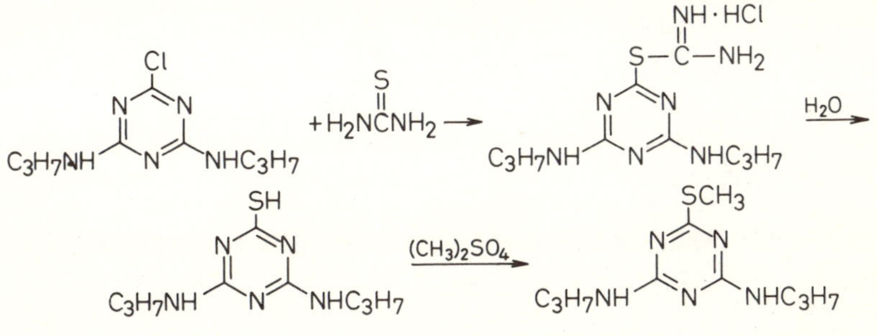

[Chemical reactions: propazine + H2NC(S)NH2 → thiuronium derivative (S-C(NH·HCl)=NH2 on triazine) → (H2O) → SH-triazine → (CH3)2SO4 → SCH3-triazine (prometryne)]

2-Methylthio-4-methylamino-6-isopropylamino-s-triazine (desmetryne) is a white crystalline substance, m.p. 54°–56° C. Its solubility in water is about 580 mg./liter; it is highly soluble in many organic solvents. The LD_{50} for rats is 1,390 mg./kg. It has been proposed for the control of weeds in cabbage (but not cauliflower) at dosages of 1.2–2 kg./ha. The duration of action of desmetryne in the soil is six to eight weeks. It is marketed in the form of a 25% wettable powder. Desmetryne is produced similarly to prometryne.

2-Methylthio-4-isopropylamino-6-(3-methoxypropylamino)-s-triazine (Gesaran) is a white crystalline substance, m.p. 68°–70° C. Its solubility in water is about 0.1%; it is highly soluble in a number of organic solvents. The LD_{50} for rats is more than 5,000 mg./kg. It has been proposed for the control of weeds in planting of winter rye, wheat, and barley at dosages of 4–4.5 kg./ha. Gesaran contains 3.3–4% simazine, 15% indicated triazine, and 24.3% MCPA. It is produced by the following route:

[Structure: 2,4-dichloro-6-ethylamino-s-triazine] + CH₃OCH₂CH₂CH₂NH₂ →(NaOH)

[Structure: 2-chloro-4-ethylamino-6-(3-methoxypropylamino)-s-triazine] →(CH₃SNa)

[Structure: 2-methylthio-4-ethylamino-6-(3-methoxypropylamino)-s-triazine]

For the control of weeds in grain plantings use also is made of 2-methylthio-4-ethylamino-6-*tert*-butylamino-*s*-triazine (HS-14260) (XVII) (m.p. 104° C., LD$_{50}$ 2,900 mg./kg.), which is prepared in an entirely analogous manner:

[Structure XVII: 2-methylthio-4-ethylamino-6-tert-butylamino-s-triazine]   (XVII)

2-Methoxy-4-ethylamino-6-isopropylamino-s-triazine (atratone, atraton) is a white crystalline substance, m.p. 94°–95° C. At 20° C., about 1,800 mg. of the compound dissolves in 1 liter of water. The LD$_{50}$ for rats is about 2,400 mg./kg. It is produced by the action of sodium methylate on atrazine:

[Structure: atrazine] + CH₃ONa → [Structure: atratone] + NaCl

It has been proposed for use on flax, cotton, small grains, and legumes.

2-Azido-4-isopropylamino-6-methylmercapto-s-triazine (C-7019) is a white crystalline substance, m.p. 91°–93° C., solubility in water 75 p.p.m. The LD$_{50}$ is 5,830 mg./kg. It is a selective herbicide for use in brassicaceous crops and sunflowers. This herbicide controls the most important annual weeds before emergence, at the time of emergence, or after emergence until the first true leaf.

2-Azido-4-ethylamino-4-t-butylamino-s-triazine (WL 9385) is a white crystalline substance, m.p. 101°–104° C., vapor pressure 7.4×10^{-7} mm. of Hg at 20° C., solubility in water 72 p.p.m. at 20° C. The LD_{50} is 460 mg./kg. It is a selective herbicide against annual grasses.

2-(4-Chloro-6-ethylamino-1,3,5-triazine-2-ylamino)-2-methylpropionitrile (WL 19805) is a white crystalline solid, m.p. 166.5°–167° C. (technical grade material). Its solubility in water is 171 p.p.m. at 25° C., in ethanol 4.5%, benzene 1.5%, chloroform 21.0%, carbon tetrachloride 1%, methylcyclohexane 21.0%, hexane 1.5%. Its vapor pressure is 1.6×10^{-9} mm. of Hg. The LD_{50} is 182 mg./kg. It is a selective herbicide for use on corn (maize).

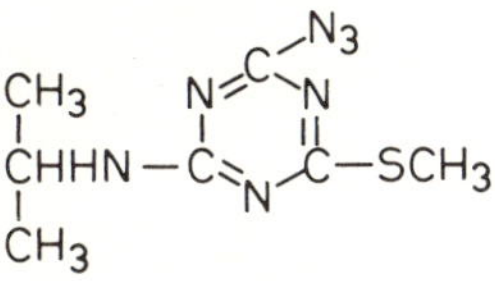

C-7019 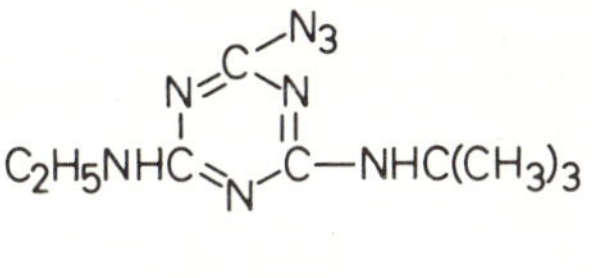   WL 9385 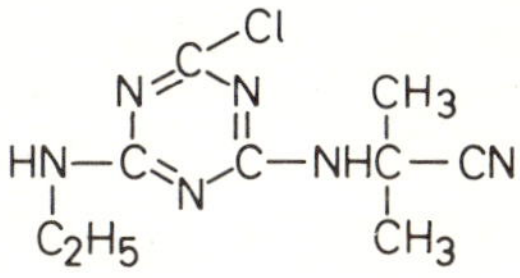   WL 19805

Other heterocyclic compounds

Besides the heterocyclic compounds described, a number of compounds are now used under working conditions as insect sex sterilants.

Among the compounds that have a sterilizing effect are 5-fluorouracil (XVIII), 5-fluorouracilcarboxylic acid (XIX), and some derivatives of folic acid (XX):

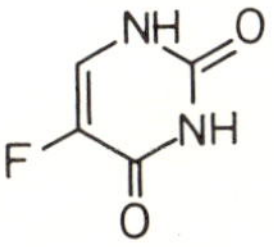

(XVIII) 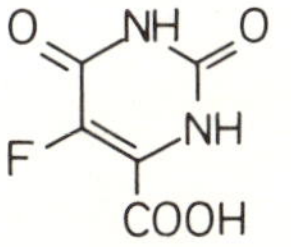   (XIX)

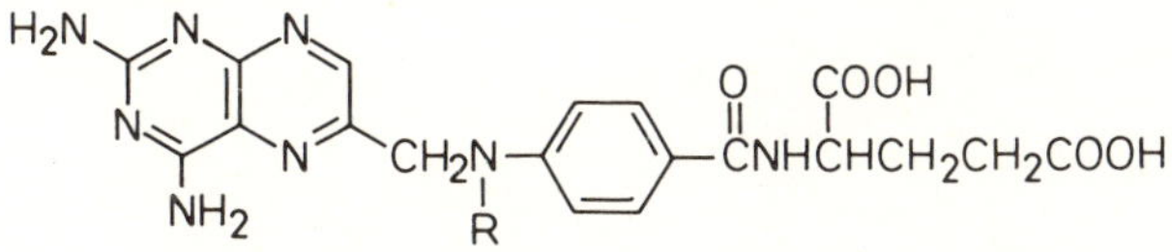

(XX)

The effect of these compounds apparently is based on replacement of metabolites. A sterilizing effect also is shown by 4-amino-1H-pyrazolo-3,4-pyrimidine (sulfate) (XXI), tris(ethyleneimino)-s-triazines, and tris(methylethyleneimino)-s-triazines (XXII):

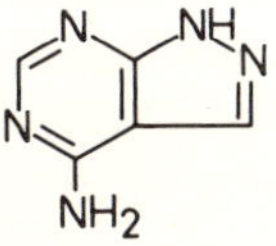

(XXI) 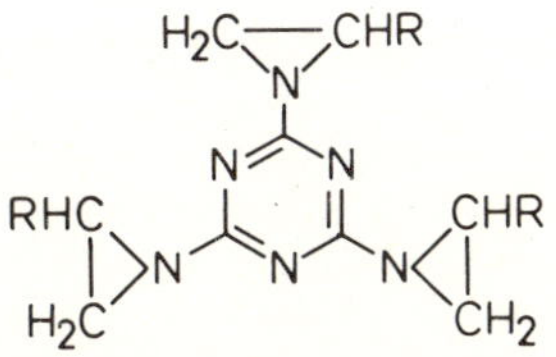   (XXII)

General references

ASCHER, K. R.: World Rev. of Pest Control 3, No. 1 (1964).
BARNSLEY, G. E., and P. A. GABBOTT: Proc. 8th Brit. Weed Control Conf., p. 372 (1966).
BASKAKOV, YU. A.: Zhurn. Vsesoyuz. Khim. Obshchestva im. D. I. Mendeleeva 9, 486 (1964).
CHAPMAN, T., D. JORDAN, D. H. PAYNE, W. J. HUGHES, and R. H. SCHIEFFERSTEIN: Proc. 9th Brit. Weed Control Conf. 2, 1018 (1968).
EBERT, E., and P. W. MÜLLER: Experientia 24, 1 (1968).
ELDERFIELD, R.: Geterotsiklicheskie soedineniya [Heterocyclic compounds], vol. 7. Publishing House "Mir" (1965).
GYSIN, H., and E. KNÜSLI: Adv. Pest Control Research 3, 280 (1961).
MEL'NIKOV, N. N., and YU. A. BASKAKOV: Khimiya gerbitsidov i regulyatorov rosta rastenii [Chemistry of herbicides and plant growth regulators]. State Scientific and Technical Publishing House of Chemical Literature [USSR] (1962).
MÜLLER, P.: Biochem. J. 101 (No. 2), 1 (1966).
SMITH, M., and T. G. MARKS: Proc. 9th Brit. Weed Control Conf. 1, 300 (1968).
VOL'FSON, L. G.: Zhurn. Vsesoyuz. Khim. Obshchestva im. D. I. Mendeleeva 5, 260 (1960).

XXXI. Inorganic pesticides

Sulfur and its compounds

One of the most important inorganic pesticides is sulfur, with its various compounds. Sulfur is practically nonpoisonous. In the finely ground condition or in the form of colloidal preparations it is widely used to control plant-feeding mites and powdery mildew fungi. To obtain a fine grind, sulfur is milled in special mills called micronizers. Often sulfur precipitated in the form of a colloid (from the purification of coke gas) is used after it is washed free of thiocyanates and other salts. In countries other than the Soviet Union colloidal sulfur for agricultural needs is obtained by precipitation from aqueous solutions of hyposulfite. Concentrated solutions of hyposulfite are carefully acidified with sulfuric or hydrochloric acid and the sulfur that separates out is freed of water-soluble salts by dialysis.

In agriculture sulfur is used also as a diluent for dusts of various compounds, for example DDT. Premises are fumigated with sulfur to control stored product pests and rodents. However, with the appearance of new effective preparations the use of sulfur for fumigation with sulfur dioxide is continuously decreasing. The dosage of colloidal sulfur or of sulfur in the form of a wettable powder for the control of mites or plant diseases is 4–15 kg./ha., depending on the crop and the objective.

For the control of plant-feeding mites and as a fumigant use is made of calcium polysulfide, which is known as lime-sulfur. This name arose from the method of preparation. Lime-sulfur usually is prepared on the site of use by boiling sulfur with slaked lime in water. Usually 1.8 parts of sulfur, 1 part of lime, and 10 parts of water are mixed for the preparation. Then the mixture is boiled for an hour (replacing water that evaporates to keep the level constant) and the solution obtained is filtered. After dilution to 40–50 times the volume with water it is used to treat plants. The dosage of lime-sulfur with a density of 1.161 g. cm.^{3} is 13–150 l./ha. For the treatment of plants the working solution should have a density of about 1.0037 g. per cm.^{3}

The effect of lime-sulfur is based on the fact that upon reaction with the carbon dioxide of the air or upon oxidation finely dispersed sulfur is deposited on the plant and has an acaricidal and fungicidal effect:

$$CaS_n + H_2O + CO_2 \rightarrow CaCO_3 + H_2S + S_{(n-1)}$$
$$2\,CaS_n + 3\,O_2 \rightarrow 2\,CaS_2O_3 + 2\,S_{(n-2)}$$

A similar effect is shown by barium polysulfide (solbar), which is prepared by mixing ground sulfur with barium sulfite powder. An advantage of this preparation is the compactness and convenience of transportation. It is highly soluble in water and is rather stable in storage where air and water do not have access to it. Solbar is more toxic to man than lime-sulfur.

A new fungicide for the control of powdery mildew fungi is calcium tetrathionate, a white crystalline substance poorly soluble in water and organic solvents. The LD_{50} for rats is about 600 mg./kg. It is used in the form of aqueous suspensions prepared from a 50% wettable powder. It is not inferior to colloidal sulfur in effectiveness. It is synthesized as follows:

$$2\ SO_2 + 2\ H_2O + S_2Cl_2 \rightarrow H_2S_4O_6 + 2\ HCl$$
$$H_2S_4O_6 + CaCl_2 \rightarrow CaS_4O_6 + 2\ HCl$$

Compounds of copper

One of the most important groups of fungicides is comprised of the various compounds of copper, which are widely used alone or mixed with organic compounds to control plant diseases. Copper sulfate has been used for more than 100 years as a fungicide on growing plants in the form of so-called Bordeaux mixture. This mixture is produced by precipitation of basic copper sulfate with lime from a 1% solution of copper sulfate. To prepare Bordeaux mixture, 1 kg. of copper sulfate is dissolved in 90 liters of water and to the solution obtained is added with thorough stirring 10 liters of freshly prepared 10% milk of lime:

$$4\ CuSO_4 + 3\ Ca(OH)_2 \rightarrow [Ca(OH)_2]_3 \cdot CuSO_4 + 3\ CaSO_4$$

Basic copper sulfate settles out in the form of a gelatinous precipitate that covers the foliage and fruit of plants well and is retained on their surface for a long time. With respect to retention on plants Bordeaux mixture occupies first place among fungicides. In preparing it there should not be an excess of copper sulfate; otherwise a strongly phytocidal preparation is obtained.

Basic copper sulfate also can be made under factory conditions. It is precipitated from concentrated solutions of copper sulfate, filtered off, dried, and ground with a diluent. However, the effectiveness of such a preparation is considerably less because of its poorer retention on plants.

A no less widely used copper fungicide is copper oxychloride. Its production is based on the following cycle of reactions:

$$4\ Cu + O_2 \rightarrow 2\ Cu_2O$$
$$Cu_2O + 2\ HCl \rightarrow 2\ CuCl + H_2O$$
$$2\ CuCl + O + 2\ HCl \rightarrow 2\ CuCl_2 + H_2O$$
$$CuCl_2 + Cu \rightarrow 2\ CuCl$$
$$4\ CuCl_2 + 3\ CaCO_3 + 3\ H_2O \rightarrow 3\ Cu(OH)_2 \cdot CuCl_2 + 3\ CaCl_2 + 3\ CO_2$$

Two other methods also are known for making this compound, but they are less economical and give a product of poorer quality. It usually is marketed in the form of a wettable powder containing 50 or 90% copper oxychloride. In addition to the active ingredient, the 50% formulation contains a diluent (most often kaolin) and a surface-active agent. The 90% formulation does not contain a diluent.

Other copper salts also have been proposed as fungicides, including the phosphate, arsenate, borate, etc., but for a number of reasons these compounds have received practically no use.

As indicated previously copper arsenate is used for the disinfection of wood. Basic copper carbonate is used as a seed disinfectant; this compound is obtained by the action of sodium carbonate on solutions of copper sulfate:

$$2\ CuSO_4 + 2\ Na_2CO_3 \xrightarrow{H_2O} 2\ Na_2SO_4 + CO_2 + CuCO_3 \cdot Cu(OH)_2$$

When copper sulfate reacts with chalk a preparation is formed that consists of the basic sulfate and basic carbonate of copper with gypsum as a contaminant. This product is known in the Soviet Union as Compound AB.

In England and the United States cuprous oxide is used and is a satisfactory protective fungicide and seed disinfectant. It is marketed in the form of preparations containing 20 and 80% copper (calculated as metal). It is produced by precipitation from water-soluble salts of copper in an alkaline reducing solution. The reducing agent usually is formaldehyde or glucose. The red or yellow cuprous oxide powder is practically insoluble in water and organic solvents. It is rather stable in storage. It is used in the form of aqueous suspensions. Wettable powders of cuprous oxide are prepared by drying in a spray dryer an aqueous suspension of cuprous oxide containing sulfite waste liquor as a surface-active agent. By this means a wettable powder preparation is obtained that suspends well in water and forms a fine dispersion.

Among the copper compounds mention should be made of the double salt of copper sulfate and dihydrazine sulfate, which is obtained by simple mixing of copper sulfate with hydrazine sulfate in equimolecular proportions. This double salt has been proposed for control of diseases of roses at a concentration of 0.2–0.9 g./liter; higher concentrations are phytocidal.

Also proposed as a fungicide is the quaternary salt cadmium-calcium-copper-zinc chromate-sulfate obtained by mixing the appropriate components. The composition of this salt in % is: CdO 4.5, CuO 11.7, ZnO 4.8, CaO 32.9, CrO_3 5.9, SO_3 11.7, and H_2O 26.5. This preparation is marketed in the form of a 90% wettable powder for the control of a number of plant diseases.

Other preparations based on copper also are known.

Halogen compounds

Salts of the hydrohalogen acids such as sodium chloride, for example, have fungicidal and bactericidal effects, but at relatively high concentrations. Concentrated solutions of sodium, potassium, magnesium, and other chlorides also show herbicidal properties. However, the fluorides have the highest biological activity. Sodium, potassium, and ammonium fluorides have been used as disinfectants for wood and also as agents for the control of moths by impregnation of materials.

The salts of hydrofluosilicic acid, such as sodium, potassium, ammonium, and magnesium fluosilicates, have similar properties. Magnesium fluosilicate, which is highly soluble in water, is convenient for the impregnation of wood and other fibrous materials. Sodium and barium fluosilicates were previously used as stomach insecticides, but they have now been displaced by more effective organic compounds. Sodium fluosilicate is produced by the use of fluorine-containing volatile substances that are evolved in the production of superphosphate. The fluorine-containing gases are trapped with water, forming hydrofluosilicic acid, which with sodium chloride yields the difficultly water-soluble (6.5 g./liter) sodium fluosilicate:

$$3\ SiF_4 + 3\ H_2O \rightarrow 2\ H_2SiF_6 + H_2SiO_3$$
$$H_2SiF_6 + 2\ NaCl \rightarrow Na_2SiF_6 + 2\ HCl$$

Magnesium fluosilicate and other highly water-soluble salts of hydrofluosilicic acid are produced by the reaction of hydrofluosilicic acid or silicon tetrafluoride with the appropriate oxides or salts:

$$H_2SiF_6 + MgO \rightarrow MgSiF_6 + H_2O$$

All the salts of hydrofluosilicic acid are incompatible with alkalies and especially with lime. When they react with lime or with calcium salts in the presence of bases, calcium fluoride is formed, which is insoluble in water and shows neither insecticidal nor fungicidal properties:

$$Na_2SiF_6 + 3\ Ca(OH)_2 \rightarrow 2\ NaOH + 3\ CaF_2 + H_2SiO_3$$

When most salts of hydrofluosilicic acid are heated they decompose into the fluoride salts and silicon fluoride:

$$Na_2SiF_6 \rightarrow 2\ NaF + SiF_4$$

Mention also must be made of cryolite, which is encountered as a mineral and can be prepared synthetically from sodium fluoride and aluminum fluoride or from aluminum fluoride, sodium chloride, and ammonium fluoride, and also by various other methods:

$$3\ NaF + AlF_3 \rightarrow Na_3AlF_6$$

Cryolite is a stomach insecticide for the control of chewing insects by spraying plants with 0.2% aqueous suspensions. However, the effectiveness of this preparation is inadequate. This compound has a low acute toxicity to animals, but its systematic consumption with food in a dose of 2–3 mg. per liter of water results in damage to the teeth. In spite of the cheapness and availability of cryolite, it has not been very widely used.

The oxygen compounds of the halogens, and especially of chlorine, have high biological activity. Thus, for example, powerful bactericidal and fungicidal effects are exerted by chlorine dioxide, the hypochlorites of the alkali metals, chloride of lime, and calcium hypochlorite, which are used for disinfection in veterinary practice. The chlorates of various metals show a herbicidal effect. Thus, magnesium and sodium chlorates and mixtures of them with some other compounds are used as nonselective herbicides and as defoliants for cotton and other crops. Potassium chlorate also is used as a herbicide. Sodium and potassium chlorates are most often produced by an electrochemical method from sodium or potassium chloride, respectively. They also are produced by the chlorination of the alkali hydroxides with elemental chlorine:

$$KOH + 3\,Cl_2 \rightarrow 5\,KCl + KClO_3 + 3\,H_2O$$

Magnesium chlorate is produced by melting sodium chlorate with magnesium chloride:

$$MgCl_2 + 2\,NaClO_3 \rightleftarrows Mg(ClO_3)_2 + 2\,NaCl$$

The melt is used as a defoliant for cotton without separation of the components. Isolation of pure magnesium chlorate presents considerable difficulties and in most cases is not carried out.

All the chlorates are strong oxidizing agents. The presence of easily oxidizable organic compounds or sulfur in chlorates is dangerous and often leads to fires and explosions. Somewhat of an exception is magnesium chlorate, which is hygroscopic and consequently the ignition of mixtures of it with organic substances takes place with more difficulty. To decrease the fire and explosion hazard of sodium chlorate, sodium borates are added to it, which also are physiologically active and in this case are not inert materials. Such mixtures contain 40–45% sodium chlorate and 50–60% sodium borate and are produced by mechanical mixing of the chlorate and borate.

A mixture of calcium chloride and chlorate (calcium chloridechlorate) obtained by chlorination of lime is used as a defoliant. This preparation is inconvenient to transport and use, since it is put out as an aqueous solution. An aqueous solution of magnesium chlorate gives satisfactory results in removing the leaves from cotton at dosages of 6–12 kg./ha. Also used as a herbicide is potassium chlorate, which is recommended for application to the soil six to ten months before planting of crops in amounts of 200 to 250 kg./ha. In this case the potassium chlorate is completely reduced by substances and microorganisms in the soil to potassium chloride, which is a satisfactory potassium fertilizer.

Compounds of phosphorus

Two compounds of phosphorus have found practical use as pesticides: zinc phosphide and aluminum phosphide. Zinc phosphide is used in baits to control murine species and other rodents, and aluminum phosphide is used to control stored-product pests. The effect of zinc phosphide on rodents apparently is based on formation in the animal's stomach of hydrogen phosphide (phosphine) which poisons the animal. The effect of aluminum phosphide on stored-product pests also is associated with the evolution of hydrogen phosphide as a result of reaction of the aluminum phosphide with moisture. Aluminum phosphide usually is distributed in the grain warehouse in special packets (Phostoxin) from which hydrogen phosphide is evolved under the influence of moisture:

$$AlP + 3 H_2O \rightarrow PH_3 + Al(OH)_3$$

Hydrogen phosphide, however, is evolved from zinc phosphide only by the action of HCl:

$$Zn_3P_2 + 6 HCl \rightarrow 2 PH_3 + 3 ZnCl_2$$

Zinc phosphide is produced by the reaction of fused zinc with phosphorus vapors at a temperature of about 500° C.:

$$2 P + 3 Zn \rightarrow Zn_3P_2$$

The zinc phosphide formed is further ground and used in baits.

A better method of synthesizing aluminum phosphide apparently is an alumothermic method. Aluminum powder is mixed with an equimolecular quantity of red phosphorus and the mixture is heated at any point until the reaction starts. The product thus obtained does not require grinding.

Other inorganic compounds

The pesticidal properties of a large number of other inorganic compounds also have been studied, but only a few are used in agriculture and industry. For example, barium carbonate has been used to control rodents and barium chloride to control the beet weevil, and inorganic compounds of mercury have been employed as seed disinfectants and insecticides, but now these compounds have been replaced by more effective and economically more advantageous organic preparations.

A new insecticide is finely pulverized silica gel, the action of which is based on obstruction of the respiratory organs [or on contact dehydration of the insect — Ed.].

For the control of murine rodents thallium sulfate is used. Food baits containing it as an additive are readily eaten by rodents. The sulfate and other salts of monovalent thallium are highly toxic to animals. The LD_{50} for various species of experimental animals is 12–18 mg./kg. The salts of trivalent thallium are considerably less toxic. A characteristic of compounds of this element is their capacity to cause balding in animals and man when high doses of the salts or organic compounds of thallium have been taken. When the hair has once fallen out, after some time it grows again and the coat of the animals (for example, rabbits, sheep, etc.) is completely restored. This property of thallium compounds has been suggested as a substitute for shearing sheep. However, because the toxic concentrations of the compounds and the concentrations that cause artificial shedding are very close, this method has not been practically used in agriculture.

Boric acid and its salts have a strong herbicidal effect on many plants and fungi. Compounds of boron have been proposed as disinfectants for the preservation of nonmetallic materials and as herbicides, but their use for these purposes is limited because of their relatively high cost and moderate effectiveness. Compounds of other elements are not used as pesticides in agriculture.

General references

FREAR, D.: Khimiya insektitsidov i fungitsidov [Chemistry of insecticides and fungicides]. Foreign Literature Publishing House [USSR] (1948) *.
GABRIELOVA, M. G., and M. A. MOROZOVA: Proizvodstvo yadokhimikatov [Production of pesticides]. State Scientific and Technical Publishing House of Chemical Literature [USSR] (1953).
VOL'FKOVICH, S. I., and R. E. REMEN: In book, Neorganicheskie insektitsidy, fungitsidy i zootsidy [Inorganic insecticides, fungicides, and zoocides], pp. 32–42. State Scientific and Technical Publishing House of Chemical Literature [USSR] (1960).
—, G. M. STRONGIN, R. E. REMEN, K. E. PISAREV, and A. I. SHISHKINA: *Ibid.*, pp. 5–31 (1960).

---

* Exact edition unknown; cited here is date of presumed Russian translation from English. — R.L.B.

Subject Index

Abate, properties and uses 338
— synthesis 338
Acaricides, definition 3
Acetaldehyde 111, 378
Acethion, properties and uses 359
— synthesis 359
Acetone oxime N-phenylcarbamate, see Proxypham
3-(α-Acetonylbenzyl)-4-hydroxycoumarin, see Warfarin
3-(α-Acetonyl-4′-chlorobenzyl)-4-hydroxycoumarin, see Coumachlor
Acetophos, properties and uses 351
— synthesis 351
3-Acetyl-6-methylpyrandione-2,4, see Dehydroacetic acid
1-Acetyl-1,2,3,4-tetrahydroquinoline, see Kyuzol-A
Acrex, see Dinobuton
Acrolein 111
Acrolein phenylhydrazone 279
Acrylonitrile 122, 128
Acylate, properties and uses 203
— synthesis 203
Adamsite 390
Aerogels, see Silica gel
Aerosol compositions 24
Aerosols 12, 20, 22 ff.
Agriculture, losses from pests 1, 5 ff.
Agronal 295
Agvitor, properties and uses 384
— synthesis 384
Alanap, properties and uses 152
— synthesis 152
Alcohols, pesticidal properties 89
Aldehydes, pesticidal properties 111 ff.
Aldicarb, properties and uses 197, 198
— synthesis 198
Aldrin, analytical method 61, 62
— conversion to dieldrin 61
— permissible concentration in air 62
— properties and uses 42, 53, 60 ff.
— synthesis 60, 61
— tolerance 62

Alfalfa 120, 157, 175, 209, 232, 276, 375, 422
Algae control (see also Slime control) 5, 115
Algalol, see Aretan
Algicides, definition 3
Alipur 229
Allethrin 131 ff.
Allyl alcohol 89
Allyl bromide 35
Allyl chloride 39
Allyl isothiocyanate 260
Alodan, permissible concentration in air 65
— properties and uses 52, 64 ff.
— synthesis 64, 268
Aluminium phosphide, properties and uses 445, 446
Ametryne 436
Amiben 148
Amides, pesticidal properties 124 ff.
Amidithion, properties and uses 363
— synthesis 363
Amines, pesticidal properties 83 ff.
Amibenzenesulfonyl carbamate, see Compound MB-9057
Aminocarb, see Matacil
Aminoguanidine, synthesis 429
2-Amino-4-methyl-5-carboxanilidothiazole, see Seedvax
Aminoparathion 324
5-Amino-3-phenyl-1,2,4-triazolyl-1-bis(dimethylamido)phosphate, see Wepsin
4-Amino-1H-pyrazolo-3,4-pyrimidine sulfate 439
3-Amino-1,2,4-triazole, see Amitrole
Amiton, see Tetram
Amitrole, mode of action 429
— properties and uses 427—430
— synthesis 429
Ammate, properties and uses 262, 263
— syntheses 262
Ammonium sulfamate, see Ammate

Ammonium thiocyanate, properties and uses 252, 259, 429
Amoben, properties and uses 217
— synthesis 217
Amophos 375
AMS, see Ammate
Amylarsine oxide 390
Anabasine 396, 405, 406
Aniline 83
Animal parasite control 50, 65, 76, 255, 257, 296, 299, 303 ff., 317, 332—334, 342, 383, 387, 397, 424
Animals, number of species 1
Anisic acid esters, pesticidal properties 141
Anthracene oil 21, 31, 32
Anthranilic acid esters, pesticidal properties 141
Anticoagulants 115
Antifeeding agents 299
Antio, see Formothion
Antiseptics, definition 3
Antu, properties and uses 227, 237, 238
— synthesis 238
APC 192
Aphicides, definition 3
Aphidan 366
Aphid reproductive rate 2
Apholate 318
Aphoxide 318
Apples 100, 102, 155, 188, 238, 264, 274, 283, 289, 315, 392, 396, 405, 418, 427, 432
— pest-caused losses 1, 5
Apricots 366
Aramite and cancer 266
— photodecomposition 266
— properties and uses 266, 267
— stabilization 266
— synthesis 266
Arboricides, definition 3, 5
Aresin, see Monolinuron
Aretan, properties and uses 292
— synthesis 292
Argox 295
Arsenic acid, synthesis 389
— and cancer 387
— compounds, irritant properties 390
— compounds, pesticidal properties 387 ff.
— in tobacco 387
N-Aryl-N',N'-dialkylureas, syntheses 230
Aryloxyacetic acid derivatives, pesticidal properties 159 ff.
— acids, reactions and syntheses 161 ff.
Aryloxyacetylamino acids, pesticidal properties 159

Aryloxyacetylhydroxamic acids, pesticidal properties 160
Aryloxyalkylcarboxylic acids, pesticidal properties 157 ff., 160, 175 ff.
Aryloxybutyric acids, pesticidal properties 173 ff.
Aryloxypropionic acids, pesticidal properties 172 ff.
Arylsulfonic acid esters, ovicidal properties 271, 272
AsCu compound 389
Asparagus 232, 422
Asulam 204
Asunthol, see Coumaphos
Atratone, properties and uses 438
— synthesis 438
Atrazine, properties and uses 435, 436, 438
— synthesis 436
Attractants 3, 10, 89, 139
— definition 3
Auxins 138
Avadex, see Di-allate
Avenin, properties and uses 317
— synthesis 317
2-Azido-4-ethylamino-4-*tert*-butylamino-*s*-triazine, see WL 9385
2-Azido-4-isopropylamino-6-methyl-mercapto-*s*-triazine, see C-7019
Azinphos ethyl 372
Azinphos methyl, oxygen analog 372
— methyl, properties and uses 371, 372
— methyl, synthesis 372
Azobenzene, properties and uses 280, 281
— synthesis 281
Azobenzide, see Azobenzene
Azo compounds, pesticidal properties 278 ff.
Azomat 393
Azosulfides 279
Azoxybenzene 278

B/77, see Fitios
B-169-ferrocyanide, see Fungilon
B-995, see DMSA
Bacteria control 30, 82, 84 ff., 91, 93, 102, 105, 107, 111, 118, 142, 148 ff., 213, 227, 275, 276, 280, 283 ff., 387, 397, 400, 409, 443, 445
Bactericides, definition 3
Baits (see also Rodent control) 12, 25
Balding 446
Banol, see Carbanolate
Banvel-D, see Dicamba
Banvel-T 147
Barban, analytical method 202
— decomposition 202

Subject Index

Barban, persistence 202
— properties and uses 184, 185, 202
— syntheses 202, 203
Barium carbonate 446
Barium chloride 446
Barium fluosilicate 444
Barium polysulfide, see Solbar
Barley 125, 155, 202, 210, 211, 438
Barthrin 131 ff.
Bayer-22,555, see Dexon
Bayer-30,686, see Eradex
Bayer-32,394, see Fungilon
Bayer-37,344, see Mesurol
Bayer-39,007, see Baygon
Baygon, properties and uses 193, 194
— synthesis 193, 194
Baytex, see Fenthion
Beans 1, 104, 113, 115, 159, 174, 206, 209, 210, 216, 419
— pest-caused losses 5
Bees and organophosphorus compounds 304 ff.
Beets 113, 155, 204, 209, 211, 292
Benachlor 68
Benefin 84
Benlate, properties and uses 418
Benzene hexabromides 44
Benzene hexachloride, see BHC
Benzenesulfonyl fluoride 276
Benzimine, synthesis and uses 152
Benzoic acids, pesticidal properties 142, 144 ff.
Benzopyrenes 31, 32
Benzoquinone 4-chlorobenzoylhydrazone 279
Benzoquinone monoguanidinehydrazone thiosemicarbazone, synthesis and uses 280
Benzoquinones 115
4-Benzothienyl N-methylcarbamate, see Mobam
Benzotriazole, formative effect 427
Benzoxazolinone, synthesis 414
Benzoylhydrazinoaniline 280
Benzoylhydrazone of quinone oxime, see Cerenox
Benzylamine 83
Benzyl benzoate 141
Benzylphenol 94
Betanal, properties and uses 204
— synthesis 204
Beta-oxidation 160
BHC, analytical method 46
— dehydrochlorination 45, 49
— formulations 14, 18, 21
— isomerization and isomers 43—46

BHC, oxidation 68
— permissible concentration in air 44
— properties and uses 32, 42 ff., 50, 68, 69, 104, 166, 216, 287, 295
— smokes, composition 23
— stereoisomers 42 ff.
— synthesis 46 ff.
Bi-58, see Dimethoate
Bicycloheptylcarbinol norcamphane-methanol 91
Bidisin 155
Bidrin, properties and uses 314
— synthesis 314
Binapacryl 100
Bi-PC, properties and uses 201, 229
— synthesis 201
Biphenyl, properties and uses 26, 27, 67, 82, 294
— tolerance 27
Bird repellents 117
Birlane, properties and uses 314, 315
— synthesis 314, 315
Bis(4-chlorophenoxy)methane, see Mikazin
Bis(3,4-dichlorofuranon-5-yl-2) ether 396, 397
Bis(2,4-dichlorophenoxyethyl) phosphite 308
1,1'-Bis(3,5-dimethylmorpholinecarbamoyl-methyl)-4,4'-dipyridylium chloride, see PP-745
Bis(ethylxanthogen) disulfide, properties and uses 177, 178, 180, 181
— disulfide, synthesis 181
Bis(ethylxanthogen) tetrasulfide 181
Bis(ethylxanthogen) trisulfide, properties and uses 177, 178, 181
— trisulfide, synthesis 181
Bis(methylxanthogen) disulfide, properties and uses 180
— disulfide, synthesis 180
Bis(pentachlorocyclopentadien-2,4-yl) 55
Bis(pentachlorophenyl) disulfide 49
Bis(tetrachloroethyl) disulfide 249
Bis(trichlorophenyl) disulfide 49
Biuret, properties and uses 225
Bladafume, see Sulfotepp
Bladan, see TEPP
Blastin, properties and uses 92
— synthesis 92
Blastogenic effect 6
Blood anticoagulants 400 ff.
Boats, marine growths 297
Boliden-bis 389
Bordeaux mixture, synthesis and uses 188, 442
Boric acid, pesticidal properties 447
Bornyl thiocyanoacetate, see Thanite

Bran 111
Brassisan, properties and uses 48, 49, 81
— synthesis 81
Bread 118
Brestan, properties and uses 297, 299, 301, 302
— stability 302
Broccoli 195
Bromacil, properties and uses 421
— synthesis 421
5-Bromo-3-*sec*-butyl-6-methyluracil, see Bromacil
5-Bromo-3-*tert*-butyl-6-methyluracil, see Compound 733
N-4-Bromo-3-chlorophenyl-*N*'-methyl-*N*'-methoxymethylurea 235
1-Bromo-3-chloropropene 40
Bromodiphenyl sulfones as acaricides 243
Bromonitrostyrene, properties 80
Bromophenylethane 67
N-4-Bromophenyl-*N*'-methyl-*N*'-methoxyurea, see Patoran
Bromophenyl sulfones 243
Bromophos, persistence 335
— properties and uses 320, 321, 335, 337
— synthesis 335
Bromophos-ethyl 335
Bromostyrene 67
Bromoxynil, properties and uses 150
— synthesis 150
Bromoxynil octanoate, properties and uses 150
— octanoate, synthesis 150
Bulan 80
Bulbosan 81
Butifos, properties and uses 259, 308, 376
— synthesis 376
Butiphos, see Butifos
Butonate, properties and uses 383
— synthesis 384
N-Butoxyphenyl-*N*',*N*'-dimethylurea 226
β-Butoxy-β'-thiocyanodiethyl ether, see Lethane-384
Butyl bromide 35
Butylcarbitol piperonate 413
N-Butyl-*N*',*N*'-dimethylurea 225
Butyl mercaptan 308
2-(4-*tert*-Butylphenoxy)propyl-2-chloroethyl sulfite, see Aramite
3-*sec*-Butylphenyl *N*-methylcarbamate, see H-28
3-*tert*-Butylphenyl *N*-methylcarbamate, see H-22
Butylphenyl *N*-methylcarbamates 184
Butynediol 89

C-2059, see Fluometuron
C-3126, see Patoran
C-3470, properties and uses 232
— synthesis 232
C-6989 110
C-7019 438, 439
C-8353 192
Cabbages 81, 109, 180, 209, 314, 315, 375, 437
— pest-caused losses 1, 5
Cacao trees 316, 392
Cacodylic acid, properties and uses 391, 393, 394
— acid, synthesis 393
Cacodyl oxide, synthesis 393
Cadmium-calcium-copper-zinc chromate-sulfate 443
Calcium arsenate, properties and uses 388, 389
— arsenate, syntheses 389
Calcium arsenite, properties and uses 388
— arsenite, synthesis 388
Calcium chloridechlorate 445
Calcium hypochlorite 445
Calcium polysulfide 188, 441
Calcium propionate 118
Calcium sulfamate 263
Calcium tetrathionate, properties and uses 246, 264, 442
— tetrathionate, synthesis 264, 442
Camphene, chlorination 50 ff.
Cancer (see also Blastogenic effect) 6, 32, 266, 387, 425
Captan and fermentation of grapes 247
— properties and uses 240, 247—250, 263
— synthesis 248
Captan analog No. 6, see Mesulfan
Carbamates, cholinesterase inhibition (see also specific compounds) 183
— structure-activity relationships 183 ff.
Carbamic acid derivatives, pesticidal properties 183 ff.
— acid esters, syntheses 186 ff., 198 ff.
Carbanolate 190
Carbathene, see Thioneb
Carbaryl, metabolites 189
— permissible concentration in air 188
— properties and uses 184, 188 ff.
— synthesis 188 ff.
— tolerance 189
Carbolic oil 31
Carbon disulfide, explosive concentrations 181
— disulfide, properties and uses 36, 37, 178—182, 211—222, 246, 247, 260, 431, 432

Carbon disulfide, synthesis 182
Carbonic acid derivatives, pesticidal
    properties 177 ff.
— acid esters, syntheses 178
Carbon sulfoxide 178
Carbon tetrachloride, permissible concentration in air 37
— tetrachloride, properties and uses
    35, 36, 38, 181
— tetrachloride, synthesis 36, 37, 55
Carbonyl sulfide, synthesis 208, 209
Carbophenothion, properties and uses
    364, 366
— synthesis 366
— tolerances 366
Carbophos, see Malathion
Carbothion, analytical method 213
— decomposition 212
— properties and uses 206, 211—213, 254
— stabilization 212
— synthesis 213
Carboxylic acid amides and imides,
    pesticidal properties 124 ff.
— acid esters, pesticidal properties 122 ff.
— acids, pesticidal properties 118 ff.,
    130 ff., 141 ff., 153 ff.
Carrots 109, 125, 147, 148, 155, 184, 200,
    209, 226, 232, 234, 235, 314, 384, 436,
    437
Castrix, properties and uses 422
— synthesis 422
Cattle 75, 255, 257, 333, 334, 382
Cauliflower 125, 437
CBP-55 40
CCC, properties and uses 88
— synthesis 88
CDEC, decomposition 223
— properties and uses 223, 224
— synthesis 223
Cereals (see also specific kinds) 1, 25, 423,
    424
Ceredon-Special 295
Cerenox, properties and uses 116, 117, 279,
    281, 282
— synthesis 116, 282
Ceresan, properties and uses 282, 294, 295,
    298
— synthesis 295
Ceresan-M, properties and uses 289
— synthesis 289, 290
Ceresan-Universal Nassbeize, see Aretan
Ceresan-Universal Trockenbeize, see
    Methoxyethylmercuric silicate
Cheese 118
Chemistry of pesticides (see also specific
    compounds) 1 ff.

Chemonite 389
Chemosterilants 3, 122, 318, 396, 415, 439
Chemotherapeutic index 283
Cherries, pest-caused losses 5
Chinomethionate, properties and uses 182
— synthesis 182
Chipcote, properties and uses 291, 292
— synthesis 291
Chloral, properties and uses 382, 383
— synthesis 71, 72
Chloralchloroacetamide 125, 128
Chloramines B and T 275
Chloranil, properties and uses 48, 49, 105,
    115
— synthesis 115
Chlorazine, properties and uses 420, 421,
    437
— synthesis 420, 421
Chlorbenside, properties and uses 241, 242
— synthesis 242
Chlorbenside sulfoxide and sulfone 242
Chlordane, analytical methods 57
— fluorine analogs 59
— properties and uses 42, 52, 55 ff., 59
— stabilization 57
— synthesis 56
— tolerance 57
Chlordene 56—59
Chlorendic acid 59, 289, 292
Chlorex 107
Chlorfenidim, see Monuron
Chlorfenson, see Ovex
Chlorflurazole, properties and uses 418
Chlorinat, see Barban
Chlorinated terpenes, stabilization 57
Chlorine dioxide 445
Chloroacetic acid, properties and uses
    120, 363
— acid, syntheses 120
Chloroacrylic acid 122
Chloroalkylphenols, pesticidal properties
    93, 94
2-Chloroallyl N,N-diethyldithiocarbamate,
    see CDEC
Chloroaniline, synthesis 201
Chlorobenzaldehydes, pesticidal properties
    112
4-Chlorobenzenediazothiourea,
    see Promurite
p-Chlorobenzenesulfonic acid, dimethyl
    amide 275
Chlorobenzilate, analytical method 142
— properties and uses 92, 141, 142
— synthesis 141
Chlorobenzoquinone benzoylhydrazone 279
6-Chlorobenzoxazolone 371

4-Chlorobenzyl phenyl ether 241
2-Chloro-4,6-bis(diethylamino)-s-triazine, see Chlorazine
2-Chloro-4,6-bis(ethylamino)s-triazine, see Simazine
2-Chloro-4,6-bis(isopropylamino)-s-triazine, see Propazine
Chlorobromomethane 367
5-Chloro-3-*tert*-butyl-6-methyluracil, see Terbacil
2-Chloro-4-*tert*-butylphenol 317, 318
4-Chlorobutyn-2-yl N-(m-chlorophenyl) carbamate, see Barban
Chlorocarbonic acids, pesticidal properties 177 ff.
2-Chloro-3-(4-chlorophenyl)propionic acid, methyl ester, see Bidisin
Chlorocholine chloride, see CCC
Chlorocresol 168
3-Chloro-6-cyanonorbornanone-2 oxime O,N-methylcarbamate, see Compound UC-20047A
1-Chloro-2,3-dibromopropane, see Nemagon
2-Chlorodiethyl sulfide 368, 369
9-Chlorodihydrophenarsazine, see Adamsite
2-Chloro-4-dimethylamino-6-methylpyrimidine, see Castrix
4-Chloro-3,5-dimethylphenoxyethanol 108
Chlorodimethylphenoxypropionic acid 173
Chlorodinitronaphthalene 81
4-Chlorodiphenyl sulfone, see Sulphenone
Chlorodiphenylsulfones, pesticidal properties 243
Chloroethanol, see Dicofol
2-Chloro-4-ethylamino-6-diethylamino-s-triazine, see Trietazine
2-Chloro-4-ethylamino-6-isopropylamino-s-triazine, see Atrazine
2-(4-Chloro-6-ethylamino-1,3,5-triazine-2-ylamino)-2-methylpropionitrile, see WL 19805
2-Chloroethyl m-chlorophenylcarbamate 185
α-(2-Chloroethylmercapto)propiomethylamide, synthesis 350
2-Chloroethyl phenylcarbamate 185
2-Chloroethylphosphonic acid, see Ethrel
Chloroform, properties and uses 35, 121
— synthesis 36
Chloroguaiacolacetic acids 159
4-Chloro-2-hydroxybenzothiazolinyl-3-acetic acid 414, 415
Chlorohydroxydiphenyls 95, 96
Chloroindan, see Chlordane

Chloro-IPC, see Chloropropham
2-Chloro-4-isopropylamino-6-diethylamino-s-triazine, see Ipazine
Chloromaleic acid anilides 128
2-Chloromethyl N-methylcarbamate, see CPMC
Chloromethylphenols, synthesis 169
Chloromethylphenoxyacetic acid, see MCPA
Chloromethylphenoxybutyric acid, see MCPB
4-Chloro-2-methylphenoxyethanol, synthesis 180
Chloromethylphenoxypropionic acid, see Mecoprop
Chloromethylphenyl dimethylformamidine 129
Chloromethylphthalimide, synthesis 371
1-Chloro-2-methylpropene-2, see Methallyl chloride
2-Chloro-6-methylquinone benzoylhydrazone 279
Chloronaphthalene 68
Chloronitrobenzene, synthesis 201
Chloronitroethanes 78, 79
2-Chloro-4-nitrophenol 331, 332
Chloronitropropanes 78, 79
Chloronitrostyrene 80
N-(Chloronorbornyl-2)-N',N'-dimethylurea 226, 230
Chloroparacide, see Chlorbenside
Chlorophene 51
4-Chlorophenol, properties and uses 274
— dissociation constant 309
— synthesis 272, 273
Chlorophenols, pesticidal properties 94, 318
4-Chlorophenoxyacetic acid 157, 161, 162, 172
4-Chlorophenoxyaniline 232
Chlorophenoxyarsine, synthesis 394
2,4-Chlorophenoxyethanol 307, 308
N-4-(4'-Chlorophenoxy)phenyl-N',N'-dimethylurea, see Tenoram
4-Chlorophenoxyphenyl isocyanate 232
4-Chlorophenyl benzenesulfonate, see Fenson
4-Chlorophenyl benzyl ether 241
N-(4-Chlorophenyl)carbamate 184
4-Chlorophenyl-4'-chlorobenzenesulfonate, see Ovex
4-Chlorophenyl-4'-chlorobenzyl ether 241
4-Chlorophenyl-4'-chlorobenzyl sulfide, see Chlorbenside
p-Chlorophenyl chloromethyl sulfide, synthesis 366

p-Chlorophenyl chloromethylsulfone, synthesis 240
N-4-Chlorophenyl-*N'*,*N'*-dimethylurea, see Monuron
4-Chlorophenylmethanesulfonamide, synthesis 250
N-4-Chlorophenyl-*N'*-methyl-*N'*-methoxyurea, see Monolinuron
3-(4'-Chlorophenyl)-5-methylrhodanine, synthesis and uses 415, 416
4-Chlorophenyl-2,4,5-trichlorophenylazo sulfide, see Mikazin
N'-4-Chlorophenyl-*O*,*N*,*N*-trimethylisourea, see Trimeturon
Chlorophos, see Trichlorfon
Chloropicrin and P compounds 78
— and SH compounds 78
— permissible concentration in air 79
— properties and uses 38, 78, 79, 276
— synthesis 79
Chloropinene 50
Chloropromurite, properties and uses 237
— synthesis 237
Chloropropham, properties and uses 185, 186, 200, 202
— synthesis 200, 201
4-Chloropyrocatechol from 2,4-D 165
Chloroguinones, pesticital properties 115
Chlorosulfacide, see Chlorbenside
Chlorothene, permissible concentration in air 52
— properties and uses 52
— synthesis 52
β-Chloro-β'-thiocyanodiethyl ether, see Compound 47
p-Chlorothiophenol 281, 366
2-Chloro-1-(2,4,5-trichlorophenyl)vinyl dimethyl phosphate, see Gardona
Chloroumbelliferone 341
6-Chloro-3,4-xylenyl *N*-methylcarbamate, see Carbanolate
Chlorphos 336
Chlorthion, properties and uses 331, 332, 337
— synthesis 331
Chlorthion oxon 331
Cholinesterase inhibition 183 ff., 303 ff., 309, 327, 368
Chrysanthemumic acid, synthesis 135, 136
Cidial, properties and uses 374
— stability 374
— synthesis 374
Cinerins, air inactivation 132
— properties and uses 130 ff.
Ciodrin, see Crotoxyphos
CIPC, see Chloropropham

Citrus fruits 95, 366, 422
CMPT, properties and uses 127
Coal-tar oils, composition 30, 31
— oils, frost protection 33
— oils, properties and uses 27, 30 ff., 33
Coke gas 441
Complexons (see also Trilon B) 163, 175
Compound 47, properties and uses 253, 258
— 47, synthesis 258
Compound 125 102
Compound 732, see Terbacil
Compound 733 421, 422
Compound 1068, see Chlordane
Compound AB 443
Compound C-8874, see Iodophos-ethyl
Compound C-9491, see Iodophos
Compound C-10015 196
Compound M-1 155
Compound M-74, see Disulfoton
Compound M-81, see Thiometon
Compound MB-8882 275
Compound MB-9057 275
Compound MB-9555 275
Compound No. 23 222, 223
Compound R-4461, properties and uses 375
— R-4461, synthesis 375
Compound UC-20047A, properties and uses 198
— UC-20047A, synthesis 198
Concentrate spraying 10, 19, 24, 28
Concentrated emulsions, see Emulsive concentrates
Contact herbicides, definition 4
Contact insecticides, definition 3
Copper acetate, see Paris green
Copper arsenate 389, 443
Copper arsenite, see Paris green
Copper borate 443
Copper carbonate 443
Copper compounds, pesticidal properties 442 ff.
Copper ethylenebis(dithiocarbamate) 221
Copper naphthenate 130
Copper oxinate, properties and uses 410, 411
— oxinate, synthesis 411
Copper oxychloride 18, 215, 221, 442, 443
Copper pentachlorophenolate 105
Copper phenylsalicylate, properties and uses 149
— phenylsalicylate, synthesis 149
Copper phosphate 443
Copper 8-quinolinolate, see Copper oxinate
Copper sulfate, pesticidal properties 6
Copper sulfate-dihydrazine sulfate 278, 443

Subject Index

Copper 2,4,5-trichlorophenolate 50, 103, 104, 216
Co-Ral, see Coumaphos
Corn (see also Maize) 1, 112, 125, 147, 176, 210, 216, 225, 234, 280, 282, 308, 436, 439
Coroxon 342
Cotoran, see Fluometuron
Cotton and cottonseed 1—3, 49, 62, 65, 82, 84, 89, 104, 113, 115, 122, 155, 157, 173, 179, 181, 184, 188, 193, 196, 225, 229, 231, 233, 234, 236, 254, 259, 280, 282, 290, 294, 308, 313, 318, 365, 369, 370, 372, 375, 376, 418, 419, 423, 438, 445
— goods, disinfectants 107
Coumachlor, properties and uses 402
— synthesis 402
Coumaphos, properties and uses 321, 341—343
— synthesis 342
Coumaphos oxon, see Coroxon
Coumatetralyl, properties and uses 402
— synthesis 402
CP 31675 126
CP 50144 126
CP 52223 126
CPB, see Fenson
CPMC 190
Creolin, see Coal-tar oils
Creosote oil 31
Cresol 93
Crotonic aldehyde 111
Crotoxyphos, properties and uses 312, 313
— syntheses 313
Crotylin 168
CRP-32 81
Cryolite, properties and uses 444
— tooth damage 444
Cucumbers 264, 423
Cuprocin 221
Cuprous oxide, synthesis and uses 18, 443
Curare 428
Currants 65, 175
Cyanamide 239
Cyano-2,4-dichlorophenylacrylic acid, properties and uses 154
— acid, synthesis 154
Cyanogen chloride 434
Cyanox 336
Cyanuric acid 434, 435
Cyanuric chloride, properties 435, 436
— chloride, synthesis 434
Cyclethrin 131 ff.
Cyclohexanecarboxylic acids 139

3-Cyclohexyl-5,6-trimethyleneuracil, see Lenacil
N-Cyclooctyl-N',N'-dimethylurea, see OMU
Cyclopropanecarboxylic acid anilides, pesticidal properties 139
Cyolane, see Dithiolane
Cypromid 127
Cystogon, properties and uses 222, 223
— synthesis 223

D-263, properties and uses 113
D-497, properties and uses 113
2,4-D, dichlorophenol in 163, 166
— from 2,4-DB 175
— from Falone 308
— impurities in 165, 166
— photodecomposition 163
— properties and uses 120, 157—159, 161—166, 171, 176, 408
— syntheses 165, 166
— tolerance 165
2,4-D esters, properties and uses 166—168
— esters, synthesis 167
2,4-D salts, properties 163, 164
Dacthal, properties and uses 144, 152, 153
— synthesis 153
Dalapon, properties and uses 121, 122
— synthesis 121
DAS-893, see Dacthal
Dasanit, see Fensulfothion
Daxtron, properties and uses 404
— synthesis 404
DB-905 82
2,4-DB oxidation to 2,4-D 175
— properties and uses 175
— synthesis 175
DDD, see TDE
DDE 73
DD Mixture 35, 38, 39
DDT 12, 20, 42, 50, 58, 63, 68, 76, 80, 91, 115, 240, 245, 256, 273, 275, 370, 371, 388, 441
— analogs and homologs (see also specific compounds) 74 ff.
— analytical methods 73, 74
— C-14 labelled, synthesis 72
— composition 69 ff.
DDT-dehydrochlorinase 73, 275
DDT dehydrochlorination 73, 76
— formulations 13, 14, 17—21
— hydrolysis 73
— in the environment 74
— in milk 74
— isomers 69 ff.
— permissible concentration in air 74

Subject Index

DDT persistence 74
— photodecomposition 73
— properties and uses 69 ff., 74
— residues, decomposition 73
— smokes, composition 23
— stabilization 73
— storage in fat 74
— syntheses 71 ff.
— telomerization 74
— thermal decomposition 72, 73
— tolerances 74
DDT-type compounds, insecticidal activities 74 ff.
DDVP, see Dichlorvos
Decachlor 55
Decotex, see MCPA
Decyl alcohol as pesticide 89
Decyl thiocyanate 253
Deet, properties and uses 142, 143, 151, 152
— syntheses 151
Defoliant 2929 RP, properties and uses 259, 260
— 2929 RP, synthesis 259, 260
Defoliants 3, 5, 89, 143, 178, 236, 252, 254, 259, 260, 303, 308, 376, 428, 429, 445
— definition 3
Dehydroacetic acid, properties and uses 400, 401
— acid, synthesis 400, 401
Delnav, see Dioxathion
Demeton, analytical method 344
— isomer rearrangement 344
— isomers 343
— metabolism 344, 345
— oxidation products 345
— permissible concentration in air 343
— persistence 345
— properties and uses 343—346, 348, 353
— synthesis 345, 346
Demeton sulfonium compounds 344
Dephosphorylation 304
Desiccants 3, 5, 95, 105, 107, 122, 178—181, 252, 393, 407
— definition 3
Desmethyl Gardona 315
Desmetryne, properties and uses 437
Dessin, see Dinobuton
DETA, see Deet
Dexon, properties and uses 281
— synthesis 281
DFDT, persistence 76
— properties and uses 76
— synthesis 76
Di-allate, decomposition 210
— properties and uses 210
— synthesis 211

4-Diallylamino-3,5-dimethylphenyl N-methylcarbamate, see APC
Diallylchloroacetamide, see Randox
Diammonium ethylenebis(dithiocarbamate), see Amoben
Diazinon, properties and uses 321, 338, 339
— synthesis 339
Dibasic acids, pesticidal properties 152 ff.
Dibenzo-1,4-thiazine, see Phenothiazine
Dibenzylamine 83
Dibenzyl sulfide 352
Dibrom, see Naled
Dibromobenzenes 67
Dibromobutene 40
Dibromobutyne 41
Dibromodichlorocyclohexane 44
1,2-Dibromoethane, properties and uses 35, 38, 39
— synthesis 38
3,5-Dibromo-4-hydroxybenzonitrile, see Bromoxynil
Dibromomethane 367
Dibromosalicylanilide 149
Dibromotetrachlorocyclohexane 44
Dibutyl adipate, properties and uses 123, 124
N,N-Dibutyl-4-chlorobenzenesulfonamide, see WARF antiresistant
O,O-Dibutyl O-2,2-dichlorovinyl phosphate 311
Dibutyl phthalate, properties and uses 142
Dibutyl succinate, properties and uses 123
— succinate, synthesis 123
Dibutyltin dilaurate and maleate 299
Dicamba, properties and uses 48, 49, 147
— synthesis 147
Dicapthon, properties and uses 331, 332
— synthesis 331, 332
Dichlone, properties and uses, 115, 116
— synthesis 116
Dichloralurea, properties and uses 18, 226, 228
— synthesis 228
Dichloran, properties and uses 87
— synthesis 87
Dichloroacetaldehyde 310, 382
Dichloroacetic acid 241
Dichloroacetoacetic acid, synthesis 314
S-2,3-Dichloroallyl N,N-diisopropylthio-carbamate, see Di-allate
2,5-Dichloro-3-aminobenzoic acid, see Amiben
Dichloroaniline 83, 124, 231
3,4-Dichlorobenzenediazothiourea, see Chloropromurite
Dichlorobenzenes 67, 68

Dichlorobenzhydrol 91
2,6-Dichlorobenzonitrile, properties and uses 144, 145
— synthesis 144, 145
Dichlorobenzyl alcohol 196
3,4-Dichlorobenzyl N-methylcarbamate, see Romate
Dichlorobromoacetic acid 311
2,5-Dichloro-4-bromophenol, synthesis 335
Dichlorobutane 35
Dichlorobutylene 260
Dichlorobutyne 41
2,4-Dichloro-6-(o-chlorophenylamino)-s-triazine, see Dyrene
Dichlorocyclohexadiene 46
Dichlorodiethyl ether 107
Dichlorodimethyl ether 107
1,8-Dichloro-3,6-dinitrocarbazole 398
2,3-Dichloro-1,4-dioxane 374
4,4'-Dichlorodiphenyl disulfide, see Mikazin
Dichlorodiphenylethane 75
Dichlorodiphenylethynylcarbinol 92
Dichlorodiphenylmethylcarbinol 281
1,2-Dichloroethane, explosive properties 38
— permissible concentration in air 38
— properties and uses 35, 36, 38, 276
— synthesis 38
2,5-Dichlorohydroquinone dimethyl ether 110
Dichloroisobutylenes 40
Dichloroisobutyric acid, properties and uses 119, 122
— acid, synthesis 122
Dichloromethoxybenzoic acid 176
Dichloromethylphenol 170
2,4-Dichloro-5-methylphenoxyacetic acid 158, 161, 172
1,3-Dichloro-2-methylpropene 40
2,3-Dichloronaphthoquine-1,4, see Dichlone
Dichloronitroaniline, see Dichloran
Dichloronitrobenzene 124, 125
2,5-Dichloro-3-nitrobenzoic acid, see Dinoben
Dichloronitrophenoxyacetic acids 164
5,2'-Dichloro-4'-nitrosalicylanilide, properties and uses 149
2,4-Dichlorophenol, dissociation constant 309
— properties and uses 273, 274, 332
— synthesis 335
Dichlorophenols 48, 49, 103, 166, 270, 334, 335
Dichlorophenoxyacetamides 165
3,4-Dichlorophenoxyacetic acid 161
Dichlorophenoxyacetic acids 157 ff., 172

Dichlorophenoxyaniline 109
Dichlorophenoxybutyric acid, see 2,4-DB
5,6-Dichloro-1-phenoxycarbonyl-2-trifluoromethylbenzimidazole, see Fenzaflor
2,4-Dichlorophenoxyethanol, properties and uses 108
— synthesis 180
Dichlorophenoxyfluoroacetic acids 159
Dichlorophenoxypropionic acid 173
2,4-Dichlorophenyl benzenesulfonate, see Genite
Dichlorophenyldimethylurea 234
Dichlorophenyl isocyanate 231, 232, 235
N-3,4-Dichlorophenyl-N'-methyl-N'-butylurea, see Neburon
N-3,4-Dichlorophenyl-N'-methyl-N'-methoxyurea, see Linuron
2,4-Dichlorophenyl-4'-nitrophenyl ether, see TOK E-25
2,6-Dichloro-4-phenylpyridine-dicarbonitrile-3,5, see Pyridinitril
1,2-Dichloropropane, permissible concentration in air 38
— properties and uses 35, 38, 39
— synthesis 38
1,3-Dichloropropene, permissible concentration in air 39
— properties and uses 35, 39
— synthesis 39
Dichloropropenes 35, 39
3,4-Dichloropropionanilide, see Propanide
Dichloropropionic acid, see Dalapon
Dichloropropylene 260
Dichlorosalicylanilide, properties and uses 149
— synthesis 148
3,4-Dichlorotetrahydrothiophene dioxide, properties, synthesis, and uses 398
2,6-Dichlorothiobenzamide, see Prefix
2,5-Dichlorothiophenol, synthesis 366
Dichlorotolan 76
4,5-Dichloro-2-trifluoromethyl-benzimidazole, see Chlorflurazole
Dichlorvos, persistence 310
— properties and uses 310, 311, 382, 383
— stabilization 311
— synthesis 310
Dicloran, see Dichloran
Dicofol, properties and uses 92
— synthesis 92
Dicoumarol, properties and uses 403
— synthesis 403
Dicryl, properties and uses 125
DID 139, 401
Dieldrin, permissible concentration in air 63

Dieldrin, properties and uses 42, 53, 61—63, 216, 396
— synthesis 63
— tolerances 63
O,O-Diethyl S-benzyl thiophosphate, see Kitazin
O,O-Diethyl S-carboethoxymethyl dithiophosphate, see Acethion
O,O-Diethyl S-carboethoxymethyl thiophosphate, see Acetophos
O,O-Diethyl S-(6-chlorobenzoxazolinyl-3-methyl) dithiophosphate, see Phosalone
O,O-Diethyl S-(6-chlorobenzoxazolinyl-3-methyl) thiophosphate 371
O,O-Diethyl O-[2-chloro-1-(2′,4′-dichlorophenyl)vinyl] phosphate, see Birlane
O,O-Diethyl O-(3-chloro-4-methylcoumarinyl-7) thiophosphate, see Coumaphos
O,O-Diethyl S-(4-chlorophenylthiomethyl) dithiophosphate, see Carbophenothion
Diethyl chlorothiophosphate, synthesis 326
O,O-Diethyl O-(2,2-dichloro-1-β-chloroethoxyvinyl) phosphate, see Phosthenon
O,O-Diethyl O-2,5-dichloro-4-iodophenyl thiophosphate, see Iodophos-ethyl
O,O-Diethyl S-(2,5-dichlorophenylthiomethyl) dithiophosphate, see Phenkapton
O,O-Diethyl O-2,4-dichlorophenyl thiophosphate, see Nemacide
O,O-Diethyl S-(2-diethylaminoethyl) thiophosphate, see Tetram
O,O-Diethyl S-(1,4-dioxenyl) dithiophosphate 373
O,O-Diethyl N-1,3-dithioanyl-2-imino phosphate, see Dithiolane
O,O-Diethyl S-(2-ethylmercaptoethyl) dithiophosphate, see Disulfoton
O,O-Diethyl 2-ethylmercaptoethyl thiophosphate, see Demeton
O,O-Diethyl S-(2-ethylsulfinyl) dithiophosphate 369
O,O-Diethyl O-2-ethylsulfonylethyl thiophosphate 344
O,O-Diethyl S-(ethylthiomethyl) dithiophosphate, see Phorate
Diethyl fluorophosphate 379
O,O-Diethyl (1-hydroxy-2,2,2-trichloroethyl) phosphonate 383
O,O-Diethyl S-(N-isopropylcarbamoylmethyl) dithiophosphate, see FAC 20
O,O-Diethyl O-(2-isopropyl-4-methylpyrimidyl-6) thiophosphate, see Diazinon

Diethyl mercury, synthesis and uses 287, 288
O,O-Diethyl S-(N-methyl-N-carboethoxycarbamoylmethyl) dithiophosphate, see Mecarbam
O,O-Diethyl O-(4-methylcoumarinyl-7) thiophosphate, see Potasan
O,O-Diethyl O-p-(methylsufinyl)phenyl thiophosphate, see Fensulfothion
O,O-Diethyl O-4-nitrophenyl phosphate, see Paraoxon
O,O-Diethyl O-4-nitrophenyl thiophosphate, see Parathion
Di-(4-ethylnonyl)amine 83
O,O-Diethyl phthalimidothiophosphate, properties and uses 353
O,O-Diethyl O-pyrazinyl thiophosphate, see Thionazin
O,O-Diethyl O-(3,4-tetramethylenecoumarinyl-7) thiophosphate, see Dition
Diethyltin dichloride 298
Diethyl-m-toluamide, see Deet
O,O-Diethyl O-2,4,5-trichlorophenyl thiophosphate 335
O,O-Diethyl O-3,5,6-trichloropyridyl thiophosphate, see Dursban
Difluoran 108
Difolatan, properties and uses 249
— synthesis 249
Difonate, properties and uses 385
2,3-Dihydro-5-carboxanilido-6-methyl-1,4-oxathiin, see Vitavax
2,3-Dihydro-5-carboxanilido-6-methyl-1,4-oxathiin-4,4-dioxide, see Plantvax
N-(5,6-Dihydrodicyclopentadienyl-5)-N′,N′-dimethylurea 226, 230
Dihydroheptachlor 42, 59
2,2′-Dihydroxy-5,5′-dichlorodiphenylmethane, properties, synthesis, and uses 107
4,4′-Dihydroxyphenyl sulfide 338
3,5-Diiodo-4-hydroxybenzonitrile, see Ioxynil
O,O-Diisopropyl S-(2-benzenesulfamidoethyl) dithiophosphate, see Compound R-4461
O,O-Diisopropyl S-benzyl thiophosphate, see Kitazin P
O,O-Diisopropyl S-(ethylsulfonylmethyl) dithiophosphate, see Aphidan
O,O-Diisopropyl O-4-nitrophenyl thiophosphate 304
N,N′-Diisopropylphosphorodiamidic fluoride, see Mipafox
Dilan, analytical method 80
— persistence 80

Dilan, photodecomposition 80
— properties and uses 80, 81
— synthesis 80
Diluents for dusts 14, 15
Dimedon 196
Dimefox, properties and uses 316
— syntheses 316
Dimelone, see Dimethyl carbate
Dimetan, analytical method 196
— properties and uses 196
— synthesis 196
Dimethirimol, properties and uses 423
Dimethoate isomerization 360
— metabolism 360, 361
— persistence 361
— properties and uses 355, 359—364
— stability 361, 362
— syntheses 361, 362
Dimethoate oxon, synthesis 360
Dimethrin, properties and uses 131 ff.
O,O-Dimethyl S-2-(acetylamido)ethyl dithiophosphate, see Amophos
Dimethylalkylhydrazonium salts 282
4-(N,N-Dimethylamino)-3,5-dimethylphenyl N'-methylcarbamate, see Zectran
2-Dimethylamino-4-hydroxy-5-n-butyl-6-methylpyrimidine, see Dimethirimol
4-Dimethylaminothiocyanobenzene, see Defoliant 2929 RP
4-Dimethylamino-3-tolyl N-methylcarbamate, see Matacil
Di-(3-methylamyl)amine 83
O,O-Dimethyl-(1-butyryl-2,2,2-trichlorophenyl) phosphonate, see Butonate
2-Dimethylcarbamoyl-3-methylpyrazolyl-5 dimethylcarbamate, see Dimetilan
Dimethyl carbate, properties and uses 139, 401
— carbate, synthesis 139
2,2-Dimethyl-6-carbobutoxy-2,3-dihydropyranone-4, see Indalone
O,O-Dimethyl S-(1-carboethoxybenzyl) dithiophosphate, see Cidial
O,O-Dimethyl S-carboethoxymethyl thiophosphate, see Methylacetophos
2,4-Dimethyl-5-carboxanilidothiazole, see G-696
O,O-Dimethyl S-(carboxymethyl) dithiophosphate 363
O,O-Dimethyl O-(2-chloro-2-N,N-diethylcarbamoyl-1-methylvinyl) phosphate, see Phosphamidon
O,O-Dimethyl 2-chloro-4-nitrophenyl thiophosphate, see Dicapthon

N,N-Dimethyl-N'-4-chlorophenyl-N'-trichloromethylthiosulfamide, synthesis and uses 263, 264
O,O-Dimethyl O-4-cyanophenyl thiophosphate, see Cyanox
O,O-Dimethyl S-(4,6-diamino-1,3,5-triazinyl-2-methyl) dithiophosphate, see Menazon
O,O-Dimethyl S-1,2-dicarboethoxyethyl dithiophosphate, see Malathion
O,O-Dimethyl S-1,2-dicarboethoxyethyl thiophosphate 355
O,O-Dimethyl O-2,5-dichloro-4-bromophenyl thiophosphate, see Bromophos
O,O-Dimethyl O-2,2-dichloro-1,2-dibromoethyl phosphate, see Naled
O,O-Dimethyl O-2,5-dichloro-4-iodophenyl thiophosphate, see Iodophos
O,O-Dimethyl O-(2,2-dichlorovinyl) phosphate, see Dichlorvos
2,2-Dimethyl-2,3-dihydrobenzofuranyl-7 N-methylcarbamate, see Furodan
O,O-Dimethyl S-(3,4-dihydro-4-keto-1,2,3-benzotriazinyl-3-methyl) dithiophosphate, see Azinphos methyl
5,5-Dimethyldihydroresorcinyl N,N-dimethylcarbamate, see Dimetan
O,O-Dimethyl O-(N,N-dimethylcarbamoyl-1-methylvinyl) phosphate, see Bidrin
Dimethyldiphenylacetamide, see Diphenamid
1,1'-Dimethyl-4,4'-dipyridylium chloride, see Paraquat
Dimethyldithiocarbamate salts 217, 218
O,O-Dimethyl S-(5-ethoxy-1,3,4-thiadiazolinyl-3-methyl) dithiophosphate 372
O,O-Dimethyl S-(N-ethylcarbamoylmethyl) dithiophosphate, see Fitios
O,O-Dimethyl S-ethylmercaptoethyl thiophosphate, see Metasystox-I
O,O-Dimethyl 2-ethylmercaptoethyl thiophosphate, see Methyl demeton
O,O-Dimethyl S-(2-ethylmercaptoethyl) dithiophosphate, see Thiometon
O,O-Dimethyl S-(2-ethylsulfinyl)ethyl dithiophosphate 369
O,O-Dimethyl S-(2-ethylsulfinyl)ethyl thiophosphate, see Oxydemetonmethyl
O,O-Dimethyl S-[2-(ethylsulfinyl)isopropyl] thiophosphate, see Metasystox-S
O,O-Dimethyl (1-hydroxy-2,2,2-trichloroethyl) phosphonate, see Trichlorfon
O,O-Dimethyl N-(isopropoxycarbamoyl) phosphate, see Avenin

O,O-Dimethyl S-(N-isopropylcarbamoyl-
	methyl) dithiophosphate 362
Dimethylmercury 291
O,O-Dimethyl S-(N-2-methoxyethyl-
	carbamoylmethyl) dithiophosphate,
	see Amidithion
O,O-Dimethyl S-(5-methoxypyronyl-2-
	methyl) thiophosphate, see Endothion
O,O-Dimethyl S-(5-methoxy-1,3,4-
	thiadiazolinyl-3-methyl) dithio-
	phosphate 372
O,O-Dimethyl S-2-(1-methylcarbamoyl-
	ethylmercapto)ethyl thiophosphate,
	see Vamidothion
O,O-Dimethyl S-(N-methylcarbamoyl-
	methyl) dithiophosphate,
	see Dimethoate
O,S-Dimethyl S-(N-methylcarbamoyl-
	methyl) dithiophosphate 360
O,O-Dimethyl S-(N-methylcarbamoyl-
	methyl) thiophosphate 360
O,O-Dimethyl O-(1-methyl-2-carbo-
	methoxyvinyl) phosphate,
	see Mevinphos
O,O-Dimethyl S-(N-methyl-N-formyl-
	carbamoylmethyl) dithiophosphate,
	see Formothion
O,O-Dimethyl 2-methylmercaptoethyl
	thiophosphate, see Tinox
O,O-Dimethyl O-(4-methylmercapto-3-
	methylphenyl) thiophosphate,
	see Fenthion
O,O-Dimethyl O-4-methylmercaptophenyl
	thiophosphate and oxon 337
N,N-Dimethyl-N'-3-methylphenyl-
	thiourea, see Methiuron
O,O-Dimethyl S-(morpholinocarbamoyl-
	methyl) dithiophosphate,
	see Morphothion
O,O-Dimethyl O-4-nitro-3-chlorophenyl
	thiophosphate, see Chlorthion
O,O-Dimethyl O-4-nitro-3-methylphenyl
	thiophosphate, see Fenitrothion
O,O-Dimethyl O-6-nitro-3-methylphenyl
	thiophosphate 330, 331
O,O-Dimethyl O-4-nitrophenyl thio-
	phosphate, see Methyl parathion
O,O-Dimethyl S-4-nitrophenyl thiophos-
	phate 328
O,S-Dimethyl O-4-nitrophenyl thiophos-
	phate 328
3,5-Dimethyl-4-nitrosopyrazoles, synthesis
	415
N,N-Dimethyl-N'-phenyl-N'-fluoro-
	dichloromethylthiosulfamide, see Eparen

3,4-Dimethylphenyl N-methylcarbamate,
	see MPCM
Dimethylphosphate of alpha-methylbenzyl-
	3-hydroxy-cis-crotonate,
	see Crotoxyphos
Dimethyl phosphite, synthesis 36
Dimethyl phthalate 139, 142, 151, 401
O,O-Dimethyl S-(N-phthalimidomethyl)
	dithiophosphate, see Imidan
Dimethyl tetrachloroterephthalate,
	see Dacthal
Dimethyl tetrachlorothioterephthalate,
	see Glenbar
3,5-Dimethyl-1,2,3,5-tetrahydro-1,3,5-
	thiadiazinethione-2, see Thiazone
N,N-Dimethyl-4-thiocyanoaniline 254, 259
Dimethylthiourea 430, 431
Dimethylthiuram disulfide, see Tridipam
O,O-Dimethyl O-2,4,5-trichlorophenyl
	thiophosphate, see Ronnel
Dimethylxanthogen disulfide 180
Dimetilan 188, 197
Dimite, properties and uses 91, 92
— stability 91
— synthesis 91
Dinex, properties and uses 101
— synthesis 101
Dinitroalkylphenols 93, 178
Dinitro-sec-amylphenol 100
Dinitroanisole 108
Dinitro-sec-butylphenol, see Dinoseb
Dinitrochlorobenzene 258, 259
Dinitro-o-cresol, see DNOC
Dinitrocyclohexylphenol, see Dinex
Dinitro-sec-octylphenyl crotonate,
	see Dinocap
2,4-Dinitrophenol, explosive properties
	97
— properties and uses 97, 389
— synthesis 97
Dinitrophenols 93, 177
2,4-Dinitrothiocyanobenzene, see DNTB
Dinoben, properties and uses 147, 148
— synthesis 148
Dinobuton, properties and uses 100
— synthesis 100
DINOC, see DNOC
Dinocap, properties and uses 97, 101, 102
— synthesis 101
Dinocton-o 100
Dinocton-p 101
Dinofen, see Dinibuton
Dinoseb, properties and uses, 98—101
— syntheses 99
Dinoseb methacrylate 100
Dinoterb-acetate 100

Di-*n*-octylamine 83
Dioctylbenzotriazolium chloride 427, 428
1,4-Dioxane-2,3-*S,S*'-bis(*O,O*-diethyldithio-
 phosphate, see Dioxathion
Dioxathion, isomers 373
— properties and uses 373, 374
— stability 373
— syntheses 373, 374
Dioxathion oxon 373
2-(1,3-Dioxolanyl-2)phenyl *N*-methyl-
 carbamate, see C 8353
1,1'-Di(pentamethylenecarbamoylmethyl)-
 4,4'-dipyridylium chloride, see PP-407
Diphenamid 144, 155
Diphenatrile 155
Diphenyl, see Biphenyl
Diphenylacetonitrile, see Diphenatrile
Diphenylamine as insecticide 83
Diphenyldiimide, see Azobenzene
Diphenylmercury 295
Diphenyl sulfones, properties and uses
 243—245
— sulfones, synthesis 245
Diphosgene 178
O,O-Dipropyl *O*-2,2-dichlorovinyl
 phosphate 311
Dipropyl isocinchomeronate, properties,
 synthesis, and uses 403
Dipterex, see Trichlorfon
Dipyridyls, synthesis 407
Dipyrrolidylthiuram disulfide, see DPDT
Diquat, properties and uses 409
— synthesis 409
Disodium ethylenebis(dithiocarbamate),
 see Nabam
Dispersed spraying 23
Disulfoton, metabolism 368
— properties and uses 367—369
— sulfonium compounds 368
— syntheses 369
Di-Syston, see Disulfoton
Dithanes C-31 and M-45 222
1,4-Dithiaanthraquinone-2,3-dinitrile,
 see Dithianon
Dithianon 425
Dithiocarbamates 216, 217
Dithiocarbamic acid esters, pesticidal
 properties 222 ff.
— acid salts, syntheses 211 ff.
Dithiocarbamic acids, pesticidal properties
 206 ff.
Dithiocarbonic acids, pesticidal properties
 177 ff.
— acids, synthesis 179 ff.
Dithiocyanoethane 253
Dithiocyanomethane 253

Dithiolane, properties and uses 318
— synthesis 318
Dithiophosphoric acid derivatives, pesticidal properties 354 ff.
— acid esters, syntheses 356 ff.
Dithiophosphoric acids, synthesis 357
Dithiosystox, see Disulfoton
Dition, properties and uses 341
— synthesis 341
Diuron, synthesis 231
DMSA 127
DN compounds, see specific compounds
DNOC, explosive properties 97
— metabolism 97, 98
— permissible concentration in air 98
— properties and uses 97, 98, 101
— synthesis 98
DNTB, explosive properties 259
— permissible concentration in air 259
— properties and uses 254, 259
— synthesis 259
Dodecylguanidine acetate, see Dodine
Dodecylguanidine tetrahydrophthalate 239
Dodecyl thiocyanate 253
Dodine properties and uses 228, 238, 239
— stability 238
Dorlone 39
Dormancy breaker 252
Dowco-109 321
Dow ET-14 and ET-57, see Ronnel
Dowfumes B-85 and EB-5 38
Dowicide-A 95
DPDT, synthesis 222
DPS, see Diphenyl sulfone
duPont fungicide 328, properties and
 uses 432
— fungicide 328, synthesis 432
— fungicide 1991, see Benlate
Dursban, properties and uses 321, 339, 340
— synthesis 339, 340
Dursban oxon 339
Dusts 12 ff.
Dylox, see Trichlorfon
Dyrene 433

E-236 424
E-605, see Parathion
E-838, see Potasan
EDB 409
Eggs 215
Ekatin, see Thiometon
Ekatin-F and -M, see Morphothion
EMMI, properties and uses 288, 289
— synthesis 289
Emulsifiers 22
Emulsive concentrates 12, 21 ff.

Enanthic aldehyde 111
Encapsulated insecticides 25
Endoif 200
Endosulfan, properties and uses 267, 268, 428
— stability 267
— synthesis 267, 268
Endosulfan-alcohol, synthesis 267
Endothall 200
Endothion, metabolism 343
— properties and uses 343
— synthesis 343
Endrin, properties and uses 42, 53, 63 ff., 396
— stabilization 64
— synthesis 64
— tolerance 64
Endrin ketone 64
ENT-21486, see Siglure
Enzymes and pesticides 7
Eparen, properties and uses 264
— synthesis 264
EPN, properties and uses 384
— synthesis 384
— tolerance 384
Eptam, see EPTC
EPTC, decomposition 209
— properties and uses 209
— synthesis 209
Eradex, properties and uses 182
— synthesis 182
Eradex-S-oxides 182
Erbon 108, 121, 122
Esters, pesticidal properties 122 ff.
Ester sulfonate, see Ovex
Ethanesulfonyl fluoride 276
Ethers, pesticidal properties 107 ff.
Ethion, properties and uses 364, 367
— synthesis 367
Ethoxynol, properties and uses 92
Ethrel, mode of action 384
— properties and uses 384, 385
— stability 384
O-Ethyl O,O-bis(4-nitrophenyl) thiophosphate 326
Ethyl bromide 35
O-Ethyl O-2,4-dichlorophenyl benzenethiophosphonate 384
Ethyl dimethyldithiocarbamate, see Compound No. 23
O-Ethyl S,S-diphenyl dithiophosphate, see Hinosan
S-Ethyl N,N-di-n-propylthiocarbamate, see EPTC
Ethylene, defoliant action 26

Ethylenebis(dithiocarbamates), pesticidal properties 218
Ethylenebis(dithiocarbamic) acid salts 217—219, 254
Ethylenebis(3-methyltetrahydro-1,3,5-thiadiazinethion-2-yl-5) 430
3,3'-Ethylenebis(tetrahydro-4,6-dimethyl-2H-1,3,5-thiadiazine-2-thione), see duPont fungicide 328
Ethylenebis(thiuram polysulfide), see Polyram
Ethylene chlorohydrin 90, 91
1,1'-Ethylene-2,2-dipyridylium bromide, see Diquat
Ethylene oxide 396
Ethylene sulfide 396
S-Ethyl hexahydro-1H-azepine-1-carbothioate, see Molinate
2-Ethylhexanediol-1,3, synthesis and uses 90, 401
2-Ethylhexylimide of bicyclo[2.2.1]heptenedicarboxylic acid, properties and uses 139
Ethyleneimine 396
N-Ethylmercuri-3,4,5,6,7,7-hexachloro-3,6-endomethylene-1,2,3,6-tetrahydrophthalimide, see EMMI
N-(Ethylmercuri)-p-toluenesulfonanilide, see Ceresan-M
Ethylmercury acetate 290
Ethylmercury chloride, permissible concentration in air 287
— chloride, synthesis and uses 50, 69, 285, 287, 290, 292, 295
Ethylmercury phosphate, permissible concentration in air 288
— phosphate, synthesis and uses 19, 287—289
Ethylmercury sulfate 286, 288
O-Ethyl O-(4-methylcoumarinyl-7) thiophosphoric acid 340
O-Ethyl S-methyl O-nitrophenyl thiophosphate 329
O-Ethyl O-(4-nitrophenyl) benzenethiophosphonate, see EPN
O-Ethyl O-(4-nitrophenyl) n-hexylthiophosphonate 381
O-Ethyl O-(4-nitrophenyl) isopropylthiophosphonate 381
O-Ethyl O-(4-nitrophenyl) methylthiophosphonate 381
O-Ethyl S-phenyl N-butylamidodithiophosphate, see Phosbutyl
O-Ethyl S-phenyl ethyldithiophosphonate see Difonate
Ethyltin trichloride 298

O-Ethyl O-2,4,5-trichlorophenyl ethylthio-
  phosphonate, see Trichloronate
Etrolene, see Ronnel
Eulan-AVA 275
Eulan-BL 275
Eulan-CN, synthesis 270
Eulan-N, synthesis 270
Eulan-SN 270
Evaporation inhibitors 22
— of drops 8 ff.

FAC 20, properties and uses 362
— synthesis 362
Falone, properties and uses 307, 308
— synthesis 308
Fenac, properties and uses 144, 153
— syntheses 153, 154
Fenchyl thiocyanoacetate 254
Fenidim, see Fenuron
Fenitrothion, explosive hazard 330
— properties and uses 330, 331, 337
— synthesis 330
Fenson, properties and uses 271, 274
— synthesis 274
Fensulfothion, properties and uses 337, 338
— syntheses 337, 338
Fenthion, persistence 336
— properties and uses 336, 337
— synthesis 336, 337
Fenthion sulfoxide and sulfone 336
Fenthiuram 104, 216
Fenthiuram-molybdate 104, 216
Fenuron, analytical method 230
— properties and uses 226, 230, 231
— synthesis 231
Fenzaflor, properties and uses 418
Ferbam, properties and uses 213, 215
— stability 215
— synthesis 215
Fermate, see Ferbam
Film-forming agents 17
Fitios, properties and uses 362
— synthesis 362
Flax 1, 98, 125, 147, 171, 175, 209—211,
  282, 438
Fluometuron, properties and uses 232, 233
— syntheses 233
Fluoride salts, pesticidal properties 444
Fluorine analogs of chlordane 59
— analogs of heptachlor 59
Fluoroacetamide 119
Fluoracetanilide 119
Fluoroacetic acid, mode of action 119
— acid, properties and uses 118, 119
— acid, synthesis 119
Fluoroalkylcarboxylic acids, toxicities 118

Fluorobromodinitrobenzene 81
Fluorodinitrobenzene and amino acids
  81, 82
Fluoroethyl esters of phenylphenylacetic
  acid, pesticidal properties 142
Fluoroethylformal 108
Fluoronitrobenzenes 81
Fluoroparacide, properties and uses 241
— synthesis 241
Fluoroparacide sulfoxide and sulfone 241
Fluorophenoxyacetic acid 172
4-Fluorophenyl-4'-chlorobenzyl sulfide,
  see Fluoroparacide
Fluorosulfacide, see Fluoroparacide
Fluorothiophenol, synthesis 242
5-Fluorouracil 439
5-Fluorouracilcarboxylic acid 439
Fluosilicates, pesticidal properties 444
FMA, see Ceresan
Folex, see Merphos
Folic acid derivatives, pesticidal properties
  439
Folidol-80, see Methyl parathion
Folithion, see Fenitrothion
Folpet 240, 248, 249
Folsan 82
Food control laboratories, U.S.S.R. 6
Formaldehyde, properties and uses
  111, 370—372, 430, 431, 443
— stabilization 111
Formamidines 129
Formative effect 427
Formothion, metabolism 363
— properties and uses 363, 364
— synthesis 364
Formulations of pesticides 9, 12 ff., 175 ff.
4a-Formyl-1,4,4a,5a,6,9,9a,9b-octahydro-
  dibenzofuran, properties, synthesis,
  and uses 397
1-Formyl-1,2,3,4-tetrahydroquinoline,
  see Kyuzol
Freons 20, 24, 36, 241, 256
Frescon 424
Fumazone 39
Fumigants (see also specific compounds)
  12, 25, 35, 177, 178, 181, 240, 276, 396,
  441
— definition 3
Fundal 129
Fungicides, definition 3
— systemic 4
Fungilon, properties and uses 417, 418
— synthesis 417
Fungus control 27, 30, 40, 68, 69, 81, 82,
  93 ff., 108, 111, 115 ff., 124, 130, 142,
  148 ff., 157 ff., 176, 182, 207, 213 ff.,

225, 227, 228, 238—240, 247 ff., 252 ff., 263 ff., 269, 276, 278, 283 ff., 297 ff., 321, 352—354, 356, 375, 380, 387, 390 ff., 396 ff., 400, 405, 409 ff., 414 ff., 426, 427, 430 ff., 441 ff., 445
Furan 396
Furethrin 131 ff.
Furfuralacetone 397
Furidazol, properties and uses 418
Furodan 192
2-(2-Furyl)benzimidazole, see Furidazol
Fusarex 82

G-4 107
G-696, properties and uses 419
— synthesis 419
Gardona analogs 315
— persistence 315
— properties and uses 315
— synthesis 315
Garlic 200
Garnitan, see Linuron
GC-7787, see Hexafluoroacetone
Genite, properties and uses 271, 273, 274
— synthesis 274
Germisan 293, 295
Gesaran, properties and uses 437, 438
— synthesis 438
Gibberellic acid 137
Gibberellins, action on plants 137, 138
— analogs 138
— properties and uses 130, 136 ff., 398
— stability 138
Gladiolas 226, 232, 292
Glenbar, properties and uses 153
Glucochloralose, properties and uses 117
Glyfluor 108
Glyodin, properties and uses 416, 417
— synthesis 416, 417
Grains (see also specific kinds) 27, 37, 68, 69, 98, 111, 128, 147, 149, 161, 169, 173, 176, 221, 222, 260, 263, 269, 290, 292, 294, 339, 394, 408, 411, 437, 438, 446
— pest-caused losses 5
Granosan 19, 287, 394
Granules 12, 15 ff.
Grapes and vines 35, 36, 40, 55, 138, 222, 247, 249, 259
— pest-caused losses 5
Grasses 125, 153, 206, 232, 418, 439
Green oil 29
Growth regulators 3, 5, 14, 112, 136, 154, 157 ff., 172 ff., 206, 207, 225, 259, 282, 384, 396, 398—400, 428

Guanidine and derivatives, pesticidal properties 225, 238 ff.
Gusathion, see Azinphos methyl
Gusathion-A, see Azinphos ethyl
Guthion, see Azinphos methyl
Gyplure, properties and synthesis 89

H-22 191
H-28 191
Haloacetic acids, pesticidal properties 118
Halobenzils, pesticidal properties 67
Halodinitrophenols, pesticidal properties 95
Halofumaric acids, pesticidal properties 124
Halogen compounds, pesticidal properties 443 ff.
Halogenated alicyclic compounds, pesticidal properties 42 ff.
Halogenated aliphatic compounds, see specific compounds
— aliphatic compounds, pesticidal properties 34 ff.
— aliphatic compounds, structure-activity relationships 35
Halogenated aromatic compounds, pesticidal properties 67 ff.
Halomethylphenoxyacetic acids as herbicides 158
Halonaphthalenes, pesticidal properties 67, 68
Halonitro compounds, biological action 78
Halophenols, pesticidal properties 102 ff.
Hanane, see Dimefox
HCCH, see BHC
Heavy oil 31
Hempa 318
Heptachlor, analytical methods 58
— dechlorination 58
— fluorine analogs 59
— permissible concentration in air 59
— persistence as epoxide 59
— properties and uses 42, 53, 56 ff., 59, 69, 104, 216
— synthesis 57
— tolerances 59
Heptachlor epoxide 59
Heptachlorocyclohexane 44, 47
Heptachloropropane 35
2-Heptadecylimidazoline acetate, see Glyodin
Herban, see Norea
Herbicides, definitions 3—5
— nonselective 4
— root or seed affecting 4, 5
— selective 4

Herbicides, wetting agents and mode of action 175 ff.
Herbisan 180
Heteroauxin, properties and uses 398, 399
— syntheses 399
Heterocyclic compounds, pesticidal properties 396 ff., 413 ff., 427 ff.
Hexabromobenzene 68
Hexachloroacetone, permissible concentration in air 112
— properties and uses 112
— synthesis 112
Hexachlorobenzene, permissible concentration in air 69
— properties and uses 46—50, 55, 67—69, 96, 106, 418
— synthesis 68
1,2,3,4,7,7-Hexachlorobicyclo[2.2.1]heptene-2,5,6-bis(methylene) sulfite, see Endosulfan
1,2,3,4,7,7-Hexachloro-5,6-bis(chloromethyl)-bicyclo[2.2.1]heptene-2, see Alodan
Hexachlorobutadiene, properties and uses 35, 40, 55
— synthesis 40
Hexachlorocyclohexenone 48
Hexachlorocyclopentadiene, properties and synthesis 53 ff.
Hexachlorodihydroxydiphenylmethane 48, 49
Hexachloroethane, properties and uses 38
— synthesis 38
Hexachloromethylcyclohexanes 43
Hexachlorophene, properties and uses 103
— synthesis 107
Hexachlorophenols, pesticidal properties 105
Hexachloropropene 35
Hexadecadienol and acetate 89
Hexadecyl thiocyanate 253
Hexafluoroacetone 112
Hexagamma, see Hexachlorobenzene
Hexakis(1-aziridinyl)phosphonitrile 318
Hexamethylenebenzamide, see Benzimine
Hexamethyl triamidophosphate 318
Hexylamine 83
Hexyl thiocyanate 253
Hinosan, properties and uses 375
— synthesis 375
HMPA 318
Hops 316
House fly reproductive rate 2
HS-14260 438
Hydantoins, properties and syntheses 233, 234

Hydram, see Molinate
Hydrazine derivatives, pesticidal properties 278 ff.
Hydrazobenzene 278
Hydrocarbons, pesticidal properties 26, 27
Hydrogen cyanide 37, 240, 252, 255, 284, 291, 399
Hydrogen phosphide, see Phosphine
Hydrogen sulfide 37, 214, 215, 284, 369, 377, 392
Hydrol, see APC
Hydroxybenzoic acids, pesticidal properties 142, 148 ff.
Hydroxybenzonitrile, synthesis 150
2-Hydroxy-4,6-bis(ethylamino)-s-triazine 435
Hydroxychlordene 58
4-Hydroxycoumarin, synthesis 402
2-Hydroxydiethyl sulfide 347
Hydroxydiphenyl, properties and uses 94—96
— synthesis 95
1-Hydroxyethyl-2-heptadecylimidazoline, synthesis and uses 417
3-Hydroxypyridazone, see Maleic hydrazide
8-Hydroxyquinoline, see Oxine
Hyvar, see Bromacil

Imidan, persistence 370
— properties and uses 370, 371
— synthesis 370, 371
Imidazolidinone-2, properties, synthesis, and uses 415
Imidazoline salts 415
Imides, pesticidal properties 124 ff.
Iminophosphate, see Dithiolane
Indalone, properties and uses 139, 401
— synthesis 401
3-Indolylacetaldehyde 112, 399
3-Indolylacetic acid, see Heteroauxin
3-Indolylacetonitrile 399
3-Indolyl-γ-butyric acid, properties, synthesis, and uses 159, 399, 400
Inferno, see Tetram
Insect-borne diseases 2
Insecticidal smokes 23
Insecticides, definitions 3
— in capsules 25
Insects, damaging species 1
— penetration by insecticides 3
Intrathion, see Thiometon
Inverted emulsions 22
Iodoaniline 83
Iodophos 335, 336
Iodophos-ethyl 335, 336
Ioxynil, properties and uses 144, 149, 150

Ioxynil, synthesis 150
Ioxynil caprylate 150
Ipazine 437
IPC, see Propham
Irish potato famine 1
Iron dimethyldithiocarbamate, see Ferbam
Isobenzan, properties and uses 397, 398
Isobornyl thiocyanoacetate 257
Isocyanatophosphoric acid 317
Isodrin, oxidation 64
— properties and uses 42, 53, 63
— synthesis 63
Isolan, properties and uses 188, 196, 197
— synthesis 197
Isopestox, see Mipafox
2-Isopropoxyphenyl N-methylcarbamate, see Baygon
Isopropyl N-acetoxy-N-phenylcarbamate, see Acylate
Isopropyl N-(3-chlorophenyl)carbamate, see Chloropropham
Isopropyl N-(4-fluorophenyl)carbamate 184
2-Isopropyl-4-methyl-6-hydroxypyrimidine, synthesis 338, 339
1-Isopropyl-2-methylpyrazolyl-5-dimethylcarbamate, see Isolan
Isopropyl N-phenylcarbamate, see Propham
3-Isopropylphenyl N-methylcarbamate, see UC-10854
Isothiocyanates, pesticidal properties 252 ff.
Isourea compounds 185
2-Isovalerialylindandione-1,3 114

K-69-79, see Avenin
Karathane, see Dinocap
Karsil 125
Keam 32
Kelthane, see Dicofol
Kepone 65
Kerosine 27 ff.
Ketoendrin 64
Ketones, pesticidal properties 112 ff.
2KF 146
Kitazin, properties and uses 352
— synthesis 352
Kitazin P 352
Korlan, see Ronnel
Krysid, see Antu
Kwiksan, see Ceresan
Kyuzol, properties and uses 411
— synthesis 411
Kyuzol-A, properties and uses 411, 412
— synthesis 411, 412

Langmuir formula 8
Lanoate, properties and uses 192

Lauric acid, pesticidal properties 118
Lawns 147, 152, 155, 294, 353
Lead arsenate 388
Lead compounds 286
Leather, protection 1, 81
Lebaycid, see Fenthion
Lenacil, properties and uses 421, 422
— synthesis 422
Lentils 210
Lethane 60, properties and uses 257
— 60, synthesis 257
Lethane 384, properties and uses 253, 256
— 384, synthesis 256, 257
Lethane 384, Special 257
Lettuce 76
Leutosan 295
Lichen control 115
Light oil 31
Limacides, definition 3
Lime sulfur, properties and uses 441
Lindane, penetration into insects 4
— properties and uses 25, 45, 59, 69, 76, 340
— separation from BHC 48 ff.
Linuron, decomposition 234
— properties and uses 234, 235
— syntheses 235
Lipsticks 142
Lithuram, see DPDT
Lovo 22
Low-volume spraying 10, 19, 24, 28

Magnesium chlorate 429, 445
Magnesium fluosilicate 444
Maize (see also Corn) 127, 439
Malaoxon, synthesis 358
Malathion, analytical method 358
— explosive hazard 357
— isomerization 357
— metabolism 355
— permissible concentration in air 357
— properties and uses 303, 304, 357—359
— syntheses 358, 359
— stability 358
— tolerance 358
Maleic hydrazide, properties and uses 419, 420
— hydrazide, synthesis 420
MALS, properties and uses 392
— synthesis 392
Maneb, decomposition 221
— properties and uses 211, 217, 221, 222
— stabilization 221
— synthesis 221, 222
Manganese dimethyldithiocarbamate, see Marbam

Manganese ethylenebis(dithiocarbamate), see Maneb
Maqbarl, properties and uses 191
Marbam 215
Marine growths, control 390
MAS, see Methylarsine sulfide
Matacil, properties and uses 191
MB 8882, properties and uses 205
2M-4C, see MCPA
2M-4CB, see MCPB
2M-4CP, see Mecoprop
MCPA, chlorocresol in 168
— esters 171
— properties and uses 120, 145, 147, 158, 159, 161, 162, 168—171, 438
— salts 168
— syntheses 168—170
MCPB, cyclization 174
— oxidation 173
— properties and uses 173, 174
— syntheses 174
MCPP, see Mecoprop
Mecarbam, metabolism 364
— properties and uses 364
— synthesis 364
Mecoprop, properties and uses 149, 172, 173
— salts 172, 173
— synthesis 173
Medinoterb-acetate as herbicide 100
Medium oil 31
Medlure 139, 140
Melons 423
MEMMI, properties and uses 292
— synthesis 292
Menazon, properties and uses 373
— synthesis 373
Meobal, see MPCM
Mercaptans, pesticidal properties 240 ff.
Mercaptobenzothiazole 215
Mercaptophos, see Demeton
Mercuran 50, 287
Mercuric chloride 283
Mercuric sulfide 285
Mercurihexane 69, 287
Mercury compounds, pesticidal properties 283 ff.
— compounds, reactions 284, 285
Mergamma 295
Merphos, properties and uses 308
— synthesis 308
Merthiolate 290
Mesulfan, properties and uses 249, 250
— stability 250
— synthesis 250
Mesurol, properties and uses 188, 194

Mesurol, synthesis 194, 195
Metaldehyde 111
Metaphos, see Methyl parathion
Metaphoxide 318
Metasystox, see Methyl demeton
Metasystox-I, properties and uses 348
— syntheses 348
Metasystox-R, see Oxydemetonmethyl
Metasystox-S, properties and uses 349
— synthesis 349
Metathion, see Fenitrothion
Metazol, properties and uses 292
— synthesis 292
Metepa 318
Methacrylic acid anilides 125
Methallyl bromide 35
Methallyl chloride, flammability 40
— chloride, properties and uses 35, 40
— chloride, synthesis 40
Methanesulfonic acid dichloramide, synthesis and uses 275
Methanesulfonyl fluoride, permissible concentration in air 276
— fluoride, properties and uses 276
— fluoride, synthesis 276
Methiotepa 318
Methiuron, properties and uses 236, 237
— synthesis 237
2-Methoxy-5-acetylbenzaldehyde, properties and uses 112
2-Methoxy-4,6-bis(ethylamino)-s-triazine 435
3-Methoxycarbonylaminophenyl N-3′-methylphenylcarbamate, see Betanal
Methoxychlor, dehydrochlorination 76
— properties and uses 50, 75 ff.
— synthesis 75
2-Methoxy-3,6-dichlorobenzoic acid, see Dicamba
2-Methoxy-4-ethylamino-6-isopropyl-amino-s-triazine, see Atratone
Methoxyethylmercuric acetate, synthesis 292
Methoxyethylmercuric chloride, see Aretan
Methoxyethylmercuric silicate, properties and uses 293
Methoxyethylmercury compounds 285—287
N-4-(4′-Methoxyphenoxy)phenyl-N',N'-dimethylurea, see C-3470
Methoxytrichlorobenzoic acid 48
Methylacetophos, methylating ability 351
— properties and uses 351
— synthesis 351
Methyl acrylate 123
Methyl N-(4-aminobenzenesulfonyl)-carbamate, see Asulam

Subject Index

Methylarsenic acid and salts 391—393
Methylarsine bis(dimethyldithiocarbamate), see Urbazid
Methylarsine bis(lauryl sulfide), see MALS
Methylarsine oxide 390—393
Methylarsine sulfide, properties and uses 391, 392
— sulfide, synthesis 392
N-Methyl-N'-(2-benzothiazolyl)urea 227
O-Methyl O,O-bis(4-nitrophenyl) thiophosphate 328
Methyl bromide, explosive properties 37
— bromide, methylating abilities 37
— bromide, permissible concentration in air 37
— bromide, properties and uses 35, 37, 128, 263
— bromide, synthesis 37
Methyl-1-(butylcarbamoyl)-2-benzimidazole carbamate, see Benlate
3-Methyl-5-carboxymethyltetrahydro-1,3,5-thiadiazinethione-2, see Thiadiazinethione
Methyl chloride, permissible concentration in air 36
— chloride, properties and uses 36, 37
— chloride, synthesis 36
O-Methyl O-2-chloro-4-*tert*-butylphenyl-N-methylamidophosphate, see Ruelene
Methyl crotonate 123
Methyl demeton, isomerization 346
— demeton, metabolism 346, 347
— demeton, oxidation products 347
— demeton, permissible concentration in air 346
— demeton, persistence 347
— demeton, properties and uses 346—350, 367, 368
— demeton, stability 346
— demeton, synergism 347
— demeton, synthesis 347
Methyldichloroarsine 391, 392
O-Methyl O-2,5-dichloro-4-bromophenyl phosphoric acid 335
Methyl N-(3,4-dichlorophenyl)carbamate, see Swep
O-Methyl O-2,4-dichlorophenyl N-isopropylamidothiophosphate, see Zytron
Methyl dimethyldithiocarbamate, see Cystogon
N-Methyldithiocarbamic acid 206
Methylene-bis(4-chlorophenyl sulfide) 241
3,3'-Methylene-bis(4-hydroxycoumarin), see Dicoumarol
Methylene bromide, toxicity 35
Methylene chloride, toxicity 35

Methylethyl chlorothiophosphate, synthesis 329
O-Methyl O-ethyl dithiophosphoric acid, synthesis 369
O-Methyl O-ethyl S-(2-ethylthioethyl) dithiophosphate, see Tetrathion
O-Methyl O-ethyl S-(N-methylcarbamoylmethyl) dithiophosphate 355
O-Methyl O-ethyl O-4-nitrophenyl thiophosphate, see Methylethylparathion
O-Methyl S-ethyl O-4-nitrophenyl thiophosphate 329
Methylethylparathion, permissible concentration in air 328
— properties and uses 328, 329
— synthesis 329
Methylethylthiophos, see Methylethylparathion
O-Methyl O-ethyl O-2,4,5-trichlorophenyl thiophosphate, see Trichlormetafos-3
Methyl formate 36, 123
Methyl isocyanate, properties and uses 186, 193—196, 198, 236
— isocyanate, synthesis 189
Methyl isothiocyanate, properties and uses 212, 213, 255, 260, 430, 431
— isothiocyanate, synthesis 260
3-Methylisoxazoledione-4,5-*o*-chlorophenylhydrazone 415
Methylmercapto-3,5-dimethylphenol, synthesis 194
4-Methylmercapto-3,5-dimethylphenyl N-methylcarbamate, see Mesurol
4-Methylmercapto-3-methylphenol, synthesis 337
Methylmercaptophos, see Methyl demeton
Methylmercuric chloride 292
Methylmercuric cyanide, see Chipcote
Methylmercuric hydroxide, synthesis 291
Methylmercuric sulfate 291
Methylmercuridicyanodiamide, see Panogen
N-(Methylmercuri)-3,4,5,6,7,7-hexachloro-3,6-endomethylene-1,2,3,6-tetrahydrophthalimide, see MEMMI
Methylmercury β-hydroxyquinolate, see Metazol
O-Methyl S-(5-methoxypyronyl-2-methyl) thiophosphate 343
N-Methyl-N'-methyl-N'-(2-benzothiazolyl)urea 227
Methyl-N-methylcarbamoylbenzenesulfonyl carbamate, see Compound MB-9555
2-Methyl-2-methylthiopropionaldoxime O,N-methylcarbamate, see Aldicarb
Methyl naphthylacetate 155

Methyl N-(4-nitrobenzenesulfonyl)-
  carbamate, see MB 8882
Methylnitrophos, see Fenitrothion
Methyl parathion, explosive hazard
  327, 328
— parathion, methylating ability 327
— parathion, properties and uses
  97, 303, 326-331, 336, 384
— parathion, synthesis 328
Methylphenoxyacetic acid 170
1-Methyl-2-phenyl-3-dodecylbenzimidazo-
  linium ferrocyanide, see Fungilon
3-Methylphenyl N-methylcarbamate,
  see MTMC
1-Methylpropyn-2-yl N-(m-chlorophenyl)-
  carbamate, see Bi-PC
6-Methyl-2,3-quinoxaline dithiocarbonate,
  see Chinomethionate
2-Methylthio-4,6-bis(ethylamino)-s-
  triazine 435
2-Methylthio-4,6-bis(isopropylamino)-s-
  triazine, see Prometryne
Methyl thiocyanate 260, 435
Methyl thiocyanoacetate 254
2-Methylthio-4-ethylamino-6-tert-
  butylamino-s-triazine, see HS-14260
2-Methylthio-4-isopropylamino-6-
  (3-methoxypropylamino)-s-triazine,
  see Gesaran
2-Methylthio-4-methylamino-6-isopropyl-
  amino-s-triazine, see Desmetryne
Methyl thiomethylketoxime N-methylcarbamate, see Lanoate
O-Methyl O-2,4,5-trichlorophenyl
  thiophosphoric acid 332
Methyl Trithion, properties and uses 366
— Trithion, synthesis 366
Methylumbelliferone, properties and uses
  403
— synthesis 341
Methylvaleric acid anilides 125
Metmercapturon, see Mesurol
Metramak, see Tetram
Meturin, properties and uses 235, 236
— synthesis 236
Mevinphos, persistence 312
— properties and uses 311, 312
— syntheses 311
Micronizers 441
Mikazin, properties and uses 241, 281
— synthesis 281
Milbex 281
Milk 2, 50, 74, 334
Mineral oils, see Petroleum oils
Minimum intervals, use of 6
Mipafox, properties and uses 316

Mipafox, synthesis 316
Mirex 65
Miscible oils 22
Mite control 55, 64, 91 ff., 129, 141, 176 ff.,
  182, 190 ff., 240 ff., 252, 266, 269,
  271 ff., 278 ff., 307 ff., 319 ff., 354,
  361 ff., 380, 384, 415, 418, 441, 442
Mitin-FF, synthesis 271
Mobam 192
MO Granules 109, 110
Molinate, properties and uses 210
— synthesis 210
Molluskicides, definition 3
Monalide 126
Monolinuron 234, 235
Monuron 184, 200, 204, 231, 236
Morestan, see Chinomethionate
Morphothion, properties and uses 363
— synthesis 363
Mothproofing, see Wool preservation
MPCM 190
MTMC 190
Mucochloric acid anilides 128
— acid, properties and syntheses
  397, 420, 421
Mulberries 376, 380
Murbetol 200
Murvesco, see Fenson
Muscatox, see Coumaphos
Mylone, see Thiazone

Nabam 217, 218
Naled, properties and uses 310, 311
— stability 311
Nankor, see Ronnel
Naphthalene 26, 27
Naphthoic acids 144
Naphthol 96
Naphthoquinones 115
Naphthoxyacetic acids 160, 161, 172
Naphthylacetaldehyde 112
Naphthylacetamide 155
Naphthylacetic acid, properties and uses
  154
— syntheses 154, 155
Naphthylacetonitrile, synthesis 154
Naphthyl carbamate, see Carbaryl
Naphthyldimethyl ether 108
1-Naphthyl N-methylcarbamate,
  see Carbaryl
Naphthylphthalamic acid, see Alanap
1-Naphthylthiourea, see Antu
Neburon, properties and uses 232
— synthesis 232
Nemacide, properties and uses 332
— synthesis 332

Nemacide oxon 332
Nemacure 40
Nemagon, properties and uses 35, 38, 39
— synthesis 39
Nematocides, definition 3
Nematode control 35, 38—40, 67, 82, 198, 206 ff., 213, 223, 332, 337, 340, 396, 398, 415, 430, 431
Neo-azazin 393
Neopynamin 131 ff.
Nicotine 396, 405, 406
Nirite, see DNTB
Nirosan 398
Nitralin, synthesis 84
Nitriles, pesticidal properties 128, 142
Nitrobenzenesulfonyl carbamate, see Compound MB-8882
Nitro compounds, pesticidal properties 78 ff., 81 ff.
Nitroethylene 78
Nitrofen 102
5-Nitrofurfural semicarbazone, properties, synthesis, and uses 397
4-Nitro-3-methyl phenol, synthesis 330
p-Nitrophenol, properties and uses 97, 324—329, 384
— synthesis 97
Nitrophenols, pesticidal properties 96 ff.
4-Nitrophenyl-2'-nitro-4'-trifluoromethylphenyl ether, see C-6989
4-Nitrophenyl-2,4,6-trichlorophenyl ether, see MO Granules
Nitrostyrenes, properties and uses 78
— synthesis 79
Nonachlor 56, 57
Nonyl alcohol 89
Norea, properties and uses 229, 230
— synthesis 229
Nornicotine 405, 406
NPD 379
Nuvan, see Dichlorvos

Oats 19, 68, 184, 202, 269
Octachlor 56—58
Octachlorocyclohexane 47
Octachlorocyclohexenone, see Octone
Octachlorodipropyl ether 108
1,3,4,5,6,7,8,8-Octachloro-3a,4,7,7a-tetrahydro-4,7-endomethylenenaphthalene, see Isobenzan
Octadecenediol acetate, see Gyplure
Octamethyl, see Schradan
Octamethylpyrophosphoramide, see Schradan
Octone, properties and uses 49, 112, 113
— synthesis 112

Octyl thiocyanate 253
Omadine preparations 405
OMPA, see Schradan
OMU, properties and uses 201, 226, 228, 229
— syntheses 229
Onions 40, 125, 152, 181, 200, 226, 232, 314, 384, 437
— pest-caused losses 5
OP-7, properties and uses 17, 18, 29, 176, 260
— synthesis 108
OP-10 17, 108
Optical glass, deterioration 2
Ordram, see Molinate
Organochlorine compounds, breakdown in petroleum fractions 30
Organolead compounds, see Lead
Organomercury compounds, see Mercury
Organophosphorus compounds, nomenclature 305 ff.
— compounds, pesticidal properties 303 ff.
Organotin compounds, pesticidal properties and uses 285, 297 ff.
— compounds, poisoning 301
— compounds, syntheses 299 ff.
Ortho 5353 191
Oryzon 92
Ovex, organoleptic effect 272
— permissible concentration in air 272
— properties and uses 269, 271—273, 275
— syntheses 272, 273
Ovotran, see Ovex
Oxime hydrazones 279
Oxine, properties and uses 409, 410
— syntheses 410
4-Oxo-3,4-dihydro-1,2,3-benzotriazine, synthesis 372
Oxydemetonmethyl, properties and uses 348, 349
— syntheses 349
— tolerance 349

Paints and lacquers, pesticidal 12, 20
Panogen, properties and uses 290—292
— synthesis 291
Paper, insecticidal 12, 23
Paraformaldehyde, uses 111
Paraldehyde, see Paraformaldehyde
Paraoxon 324, 325
Paraquat, properties and uses 36, 407—409
— synthesis 407
Parathion, analytical method 324
— dusts 14
— explosive properties 324
— metabolism 324 ff.
— permissible concentration in air 326

Parathion, properties and uses 97, 303, 304, 323—330, 332, 338, 379, 384
— synthesis 326
— tolerances 325
Paris green, properties and uses 388, 389
— green, syntheses 388
Parsley 125
Paste formulations 18—20, 22
Patoran 234, 235
PCBA, see Blastin
PCB compounds 44
PCPB, see Fenson
PCPCBS, see Ovex
Peaches 405
— pest-caused losses 1
Peanuts 232, 353
Pears 102, 173, 274, 289, 418
— pest-caused losses 5
Peas 109, 147, 155, 159, 210, 211, 229, 418
Peat carbolineum and tar 32, 33
— dry distillation 32
Pencils, insecticidal 12, 20
Pentac 55
Pentachloroacetone 112
Pentachloroacetophenone 315
Pentachloroanisole 106
Pentachlorobenzoic acid 147
Pentachlorobenzyl alcohol, see Blastin
Pentachlorocyclohexene 45
Pentachloroethane 35
Pentachloroethyl mercaptan, synthesis 249
Pentachloroisopropylbenzene 67
Pentachloromandelonitrile 92
Pentachloronitrobenzene, analytical method 82
— properties and uses 81, 82, 125, 201
— synthesis 82
Pentachloropentadienoic acid 122
Pentachlorophenol, analytical method 105
— properties and uses 48, 49, 104—107, 115, 178
— syntheses 106
Pentachlorophenoxyacetic acid 158
Pentachlorothiophenol 48, 69
3-sec-Pentylphenyl N-methylcarbamate, see Ortho 5353
Peppers 82, 112
Perchloro(methyl mercaptan), properties and uses 240, 245, 246, 248—251, 253, 264
— syntheses 246, 247
Perchloro(methyl mercaptans), pesticidal properties 245 ff.
Persistence of pesticides (see also specific compounds) 8 ff.
Perthane, properties and uses 76

Perthane, synthesis 76
— tolerance 76
Pest-caused losses in agriculture 1, 5 ff.
Pesticide-fertilizer combinations 16
Pesticide use, economic efficiency 5 ff.
Pesticides and enzyme systems 7
— and formulations, numbers of 6
— annual production 6
— application methods 9 ff.
— chemistry of (see also specific compounds) 1 ff.
— conversion products (see also specific compounds) 9
— definition and types 3
— formulations 12 ff.
— general requirements for 6 ff.
— ideal properties 9
— increased yields from 5 ff.
— in environment, U.S.S.R. 6
— mechanisms of action 7
— methods of use 9 ff.
— persistence 8 ff.
— production in the U.S.S.R. 6
— solvents (see also specific compounds) 20
— stability (see also specific compounds) 9
— toxicity of metabolites 7, 9
— volatility (see also specific compounds) 8
Pestox, see Dimefox
Pestox-III, see Schradan
Pestox-15, see Mipafox
Pests, types 1
— world losses from 3
Petrolatum 30
Petroleum oils, agricultural uses 27 ff.
— oils and frost resistance 29
— oils, mechanism of action 4, 30
— oils, oxidation 28, 30
— oils, photooxidation 28, 29
— oils, properties and uses 27 ff.
PH 40—21, properties and uses 418, 419
— synthesis 418, 419
Phaltan, see Folpet
Phenarsazine 390
Phenazon, see Chlorazine
Phenkapton, properties and uses 364, 366
— synthesis 366
Phenmedipham, see Betanal
Phenols, pesticidal properties 92 ff.
Phenothiazine, properties and uses 416, 424
— synthesis 424
Phenoxarsine 390
Phenoxarsine oxide 394
Phenoxyacetic acids 157 ff.
S-Phenoxyarsinyl O,O-diethyl dithiophosphate, see Thiarsine

Phenoxyethylamine, mode of action 84
Phenthiuram 50
Phenylacetic acids 143
1-Phenyl-4-amino-5-chloropyridazone-6, see Chlorazine
N-Phenyl-N',N'-dimethylurea, see Fenuron
N-Phenyl-N-hydroxy-N'-methylurea, see Meturin
Phenylmercuric acetate, see Ceresan
Phenylmercuric bromide 295, 298
Phenylmercuric chloride 295
N-(Phenylmercuri)-1,4,5,6,7,7-hexachlorobicyclo[2.2.1]heptene-5-dicarboximide 295
Phenylmercuripyrocatechin 295
Phenylmercuritriethanolammonium lactate, see Puratized
Phenylmercuriurea 295
N-Phenyl-N'-(2-methylcyclohexyl)urea 226
1-Phenyl-3-methylpyrazolyl-5 dimethylcarbamate, see Pyrolan
Phenyl sulfone, see Diphenyl sulfone
Phenylthiourea 238
Phorate, metabolism 364, 365
— persistence 365
— properties and uses 364—366
— syntheses 365, 366
Phorate sulfoxide and sulfone 364, 365
Phosalone, metabolism 371
— properties and uses 371
— synthesis 371
Phosbutyl, properties and uses 374, 375
— synthesis 375
Phosdrin, see Mevinphos
Phosgene 37, 54, 178, 182, 186—189, 193, 195, 198, 203, 208, 229
Phosphamide, see Dimethoate
Phosphamidon, metabolism 313
— properties and uses 313, 314
— synthesis 314
Phosphine 305—307, 446
Phosphinon, see Phosthenon
Phosphonic acid derivatives, pesticidal properties 380 ff.
— acid derivatives, syntheses 381
Phosphopyrone, see Endothion
Phosphoric acid derivatives, pesticidal properties 308 ff.
Phosphorous acid derivatives, pesticidal properties 307 ff.
Phosphorus compounds, pesticidal properties 445 ff.
Phosphorus pentasulfide, synthesis 357
Posthenon, properties and uses 311
— synthesis 311
Phostoxin 446

Photodecomposition 9, 15
Photosynthesis 186
Phthalan, see Folpet
Phthalophos, see Imidan
Phygon, see Dichlone
Picloram, properties and uses 403, 404
— synthesis 403, 404
PIMM 295
Pindone, properties and uses 114
— synthesis 114
Pineapples 422
Pinene, chlorinated 50
Piperazine 416
Piperonyl butoxide, properties and uses 108, 109, 413
— butoxide, synthesis 109
Piperonyl cyclonene, properties and uses 113, 114, 413
— cyclonene, synthesis 113
Pirofos, see Sulfotepp
Pival, see Pindone
Pivalin, see Pindone
2-Pivalylindandione-1,3, see Pindone
Plant sterilants 122
Plantvax, properties and uses 423, 424
— synthesis 424
PMAC, see Ceresan
Polarography 358
Polycarbacine, see Polyram
Polycarbazine 211
Polychlorobiphenyls, see PCB compounds
Polychlorocamphene 51
Polychlorocyclodienes 52 ff.
Polychloroethylenes 36
Polychloronaphthalenes 44
Polychloropinene, see Chlorothene
Polychloroterpenes 50 ff.
Polyethylenebis(thiuram sulfide and disulfide) 222
Polyram, properties and uses 220
— synthesis 220, 221
Potasan, properties and uses 340, 341
— synthesis 340, 341
Potassium chlorate 445
Potato famine, Irish 1
Potatoes 1, 3, 48, 82, 89, 91, 108, 155, 157, 186, 196, 209, 217, 221, 222, 226, 229, 232, 235, 236, 249, 252, 262, 276, 301, 302, 308, 369, 373, 419, 432
— pest-caused losses 1, 5
Powders 12 ff.
PP-407, properties, synthesis, and uses 408, 409
PP-745, properties, synthesis, and uses 409
Prefix, properties and uses 145
— synthesis 145

Preparation G-4-40 107
Prep-Defoliant 122
Probanil 200
Prolan 80
Prometryne, properties and uses 437
— syntheses 437
Promurite, properties and uses 237
— synthesis 237
Propachlor 126
Propanide 124, 125
Propargyl bromide 36, 41
Propazine, properties and uses 200, 436, 437
— synthesis 436
Propham and cancer 199
— properties and uses 184, 185, 199, 200, 202, 204
— stability 200
— synthesis 200
Propineb 222
S-n-Propyl N,N-di-n-propylthiocarbamate, see Vernam
Propylene oxide 396
S-Propyl N-ethyl-N-butylthiocarbamate, see Tillam
Propyl isome 413, 414
Propyl thiopyrophosphate, see NPD
Proxypham, analytical method 204
— persistence 204
— properties and uses 204
— synthesis 204
Puratized, properties and uses 295, 296
— synthesis 296
Pyramin, see Chlorazine
Pyrazon, see Chlorazine
Pyrethrins, air inactivation 132
— properties and uses 25, 63, 113, 130 ff., 256, 397, 413
Pyrethroids, synthesis and uses 108, 132 ff.
Pyridine-thiol-2 oxide-1, properties, synthesis, and uses 405
Pyridinitrile, properties and uses 405
— synthesis 405
Pyridylmercuric acetate and chloride 284
Pyrolan, properties and uses 188, 197
— synthesis 197
Pyrophosphoric acid derivatives, pesticidal properties 376 ff.
— acid derivatives, syntheses 377 ff.

Quaternary ammonium bases, analytical method 86
— ammonium bases as bactericides 85, 86
— ammonium bases, mode of action 87
— ammonium bases, pesticidal properties 83 ff.

Quaternary ammonium bases, properties and uses 86 ff.
— ammonium bases, synthesis 86
Quinces 366
Quinoline-8-sulfonic acid, synthesis 410
Quinolinol, see Oxine
8-Quinolinol benzoate or sulfate, see Quinosol
Quinomethionate, see Chinomethionate
Quinone oxime benzoylhydrazone, see Cerenox
Quinones, pesticidal properties 115 ff.
Quinosol, properties and uses 411
— synthesis 411
Quinoxaline-2,3-trithiocarbonate, see Eradex

R-1607, see Vernam
Randox and Randox-T 125
Ratindan, properties and uses 114, 115
— synthesis 114
Rats, destructive ability 1
Repellent MGK-326 403
Repellents 3, 117, 151, 152, 227, 237, 401, 403, 411
— definition 3
Residue laboratories, U.S.S.R. 6
Residues, safety in use 6
Resitox, see Coumaphos
Rhizoctol, see Methylarsine sulfide
Rhodan, properties and uses 254, 259, 260
— synthesis 260
Rice 92, 109, 113, 124, 137, 145, 153, 190, 191, 203, 206, 210, 290, 302, 315, 352, 375, 392, 393, 418
Rodent control 10, 12, 51, 108, 114, 115, 119, 237, 238, 263, 265, 388, 400—403, 416, 422, 445, 446
Rodenticides, definition 3
Rogor, see Dimethoate
Romate, properties and uses 196
— synthesis 196
Ronnel, properties and uses 320, 332—335, 337
— syntheses 333, 334
Root induction 398
Roses 102, 264, 278, 289, 427, 443
Rotenone 414
Round worms, see Nematode control
Rubber protection 97, 103
Ruelene, persistence 317
— properties and uses 317, 318
Rutgers-612 90, 401
Rye 438

Salicylanilide, properties and uses 142, 148
— synthesis 148
Salicylic acid esters 141
— acid in canning 148
Salithion 336
Sayfos, see Menazon
Schradan, metabolism 379, 380
— permissible concentration in air 379
— properties and uses 304, 376, 378—380
— stability 379
— synthesis 380
SD91—29 314
Seed disinfectants 4, 19, 50, 59, 63, 69, 78, 81, 82, 87, 95, 96, 104, 105, 111, 115—117, 149, 216, 247, 254, 260, 279, 281—285, 287—296, 301 ff., 371, 387, 388, 391, 394, 411, 415, 418, 423, 443, 446
Seedvax, properties and uses 419
— synthesis 419
Semesan preparations 293
Seredon, see Cerenox
Seredon Special 282
Sesamex 413
Sesamin 413, 414
Sesin 108
Sesoxane, see Sesamex
Sevin, see Carbaryl
Sex sterilants 3, 122, 318, 369, 415, 439
Sheep 387
— chemical shearing 446, 447
Siglure 139
Silica gel Aerogels 4, 446
Silk 1
Silvex, properties and uses 173
— synthesis 173
Simazine, persistence 435
— properties and uses 435, 436, 438
— synthesis 435
Simetone 435
Simetryne 435
Sinerfos 347
Slime control 112, 285, 287, 288, 296 ff., 301
Smoke pots 23
Smokes, insecticidal 23
Snail control 111, 149, 424
Soaps, insecticidal 12, 22
Sodium arsenate, synthesis 389
Sodium arsenite, properties and uses 387
— arsenite, synthesis 387
Sodium bisulfite 265
Sodium borate 445
Sodium chlorate, explosive hazard 445
Sodium chloride, pesticidal properties 443

Sodium *p*-dimethylaminobenzenediazo-sulfonate, see Dexon
Sodium ethylxanthate, synthesis and uses 179, 180
Sodium fluosilicate 444
Sodium isopropylxanthate, properties and uses 179, 180
— isopropylxanthate, synthesis 180
Sodium *N*-methyldithiocarbamate, see Carbothion
Sodium tetrathionate 264
Soil sterilization 78, 181, 206, 213, 230, 260, 262, 385, 428
Solan 125
Solbar 188, 442
Solid solutions of pesticides 20
Solutions as formulations 12, 19 ff.
Solvents, intermediate 24
— for formulations 20
— for pesticides (see also specific compounds) 20
Soy beans 3, 125, 148, 155, 203, 229, 234, 235
Spinach 76
Sprouting inhibitors 5, 48, 82, 89, 108, 155, 186
Stability of pesticides (see also specific compounds) 9
Stam F-34, see Propanide
Stannic and stannous chlorides 298
Stauffer R-4572, see Molinate
Stem thickeners 282
Sterilants, see under plant, sex, soil
Stomach insecticides, definition 3
Storage stability 9
Stored-product protection (see also specific products) 25, 35, 37, 38, 40, 65, 122, 123, 128, 136, 176, 240, 445, 446
Strawberries 125, 155, 216, 264, 308
— increased yield from pesticides 5
Strobane, properties and uses 50, 52
— synthesis 52
Sugarbeets 52, 59, 116, 120, 125, 126, 157, 184, 200, 201, 203, 206, 210, 226, 228, 229, 236, 279—282, 301, 302, 317, 366, 419, 420, 422, 423
Sugarcane 120, 276
Sulfides, pesticidal properties 240 ff.
Sulfites, acaricidal properties 265, 266
Sulfones, acaricidal properties 242, 243
Sulfonic acid derivatives, pesticidal properties 269 ff.
— acid amides, pesticidal properties 274 ff.
Sulfonyl fluorides 276
Sulfotepp 23, 379

Sulfur and compounds, pesticidal properties 6, 264, 265, 333, 357, 424, 441 ff.
Sulfur dioxide 37, 265, 441
Sulfuric acid derivatives, pesticidal properties 262 ff.
Sulfurous acid derivatives, pesticidal properties 262 ff., 265 ff.
Sulfuryl fluoride, properties and uses 263
— fluoride, synthesis 263
Sulphenone, properties and uses 243, 245
— synthesis 245
Sumithion, see Fenitrothion
Summer oils 27, 29
Sunflowers 157, 438
Surcopur, see Propanide
Swep, properties and uses 203, 204
— synthesis 204
Synergism 132, 136, 347, 413 ff.
Systam, see Schradan
Systemic herbicides, definition 4, 5
— insecticides, definition 3
Systox, see Demeton

2,4,5-T, properties and uses 48, 49, 103, 120, 158, 161, 171—173
— synthesis 171
2,4,5-T esters, properties and uses 171, 172
2,4,5-T salts, properties 171
Tabatrex, see Dibutyl succinate
Tag, see Ceresan
Taiga 152
Tanalite 389
2,4,5-TB, properties and uses 175
— synthesis 175
TDE and cancer 76
— properties and uses 76
— synthesis 76
Tedion, see Tetradifon
Telodrin, see Isobenzan
Telone 39
Temik, see Aldicarb
Tenoran, properties and uses 226, 232
— synthesis 232
Tepa 318
TEPP, properties and uses 304, 376—379
— smoke pots 23
— stability 378
Terbacil 421, 422
Terpenes, chlorination products (see also specific terpenes) 50 ff.
Terracur-p, see Fensulfothion
Terra-Systam, see Dimefox
Tetrabromocyclohexane 44
Tetrabromodichlorocyclohexanes 44
Tetrachloroacetophenone 314, 315

Tetrachlorobenzenes 48, 49, 68, 102, 103, 244
Tetrachlorobenzoic acids 147
Tetrachlorobenzoquinone, see Chloranil
Tetrachlorocyclohexadiene 45
Tetrachlorocyclohexene 46
2,4,5,4'-Tetrachlorodiphenyl sulfone, see Tetradifon
Tetrachloroethanes 35
Tetrachloroethylene 55
N-1,1,2,2-Tetrachloroethylthiotetrahydrophthalimide, see Difolatan
Tetrachloronitrobenzene 48, 81, 82
Tetrachlorophenols 105, 106
Tetrachlorophenoxyacetic acids 158 ff.
Tetrachloropyridine 340
Tetradecyl thiocyanate 253
Tetradichlone, see Tetradifon
Tetradifon, properties and uses 243, 244
— synthesis 244
Tetraethyl dithiopyrophosphate, see Sulfotepp
O,O,O,O-Tetraethyl S,S'-methylenebis(dithiophosphate), see Ethion
Tetraethyl monothiopyrophosphate 379
Tetraethyl pyrophosphate, see TEPP
Tetraethylthiuram disulfide 216
Tetraethyltin 298
Tetraethyltrithiopyrophosphate 373
N-(Tetrahydrodicyclopentadienyl)-N',N'-dimethylurea, see Norea
Tetrahydrophthalimide, synthesis 248
Tetrahydroquinoline, synthesis 412
Tetrahydrothiazine 428
3-(α-Tetralyl)-4-hydroxycoumarin, see Coumatetralyl
Tetram, isomers 352
— properties and uses 352, 353
— synthesis 352, 353
Tetramethylenedisulfotetramine, see Tetramine
Tetramethylphosphorodiamidic fluoride, see Dimefox
Tetramethyl pyrophosphate 376
O,O,O',O'-Tetramethyl O,O'-thiodi-p-phenylene phosphorothioate, see Abate
Tetramethylthiuram disulfide, see TMTD
Tetramethylthiuram monosulfide 216
Tetramine, properties and uses 263
— synthesis 263
1,3,6,8-Tetranitrocarbazole, see Nirosan
Tetra-n-propyl dithiopyrophosphate, see NPD
Tetrathion, properties and uses 369
— synthesis 369
Tetrazole 428

TH 052 H, see PH 40-21
Thallium sulfate 446, 447
Thanite, mode of action 255
— properties and uses 254, 257, 258
— synthesis 257, 258
Thiadiazinethione, analytical method 431
— properties and uses 430—432
— synthesis 432
Thiarsine, properties and uses 394
— synthesis 394
2-(Thiazolyl)benzimidazole 418
Thiazone, properties and uses 430, 431
— stability 430
— synthesis 431
Thimet, see Phorate
Thinol 424
Thiocarbamic acid esters, pesticidal properties 207 ff.
— acid esters, synthesis 207 ff.
Thiocarbamic acids, pesticidal properties 206 ff.
Thiocarbonic acids, pesticidal properties 177 ff.
— acids, synthesis 178 ff.
Thiocron, see Amidithion
Thiocyanates, mode of action 255
— pesticidal properties 252 ff.
Thiocyanoacetic acid 253
Thiocyanoacetone 254
4-Thiocyanoaniline, see Rhodan
Thiocyanobutyric acid 253
β-Thiocyanoethyl laurate, see Lethane 60
4-Thiocyanophenyl methanesulfonamide 275
Thiocyanopropionic acid 253
γ-Thiocyanopropyl-2,4,6-trimethylphenyl ether 254
Thiodan, see Endosulfan
Thiodiphenylamine, see Phenothiazine
Thiolo and thiono isomers 319 ff.
Thiolocarbamic acids, pesticidal properties 206 ff.
Thiolochlorocarbonates, synthesis 208
Thiometon, metabolism 367
— permissible concentration in air 368
— properties and uses 367—369
— stability 367
— syntheses 368
Thio-mevinphos 312
Thionazin, properties and uses 321, 340
— synthesis 340
Thioneb, synthesis and uses 222
Thionocarbonates, synthesis 179
Thiophone 396
Thiophos, see Parathion
Thiophosgene 179, 182, 245, 247

Thiophos-ME, see Methylethylparathion
Thiophosphates, syntheses 321 ff.
Thiophosphoric acid derivatives, pesticidal properties 319 ff.
Thiosulfonic acid esters 277
Thiotepa 318
Thiourea and derivatives, pesticidal properties 213, 225 ff., 236 ff., 260, 327, 437
Thiram, see TMTD
Thiuram-D, see TMTD
Thiuram disulfide, see TMTD
Tillam 209, 210
Tinox, isomerization 348
— properties and uses 347, 348
TMTD, explosive properties 215
— properties and uses 50, 104, 211, 215—217, 392
— synthesis 215
TO-2, see CMPT
Tobacco 89, 196, 406, 432
— alkaloids 405, 406
— arsenic in 387
TOK E-25, properties and uses 109
— E-25, synthesis 109
Tomato ripening 26
Tomatoes 19, 82, 112, 142, 148, 155, 161, 172
— increased yields from pesticides 5
Tooth damage by cryolite 444
Tordon, see Picloram
Toxaphene, analysis 51
— as rodenticide 51
— composition 51
— IR absorption 51
— properties and uses 51 ff., 58
— synthesis 51
— tolerances 51
Toxicity, acute vs. chronic 7
— dermal 7
— evaluation in E. and W. Germany 7
— evaluation in U.S.S.R. 7 ff.
— inhalation 7
— selective 8
2,4,5-TP, see Silvex
Trialkyltin acetates 298
Triallate, properties and uses 211
— synthesis 211
Triamylamine 83
s-Triazines, properties 432 ff.
— syntheses 434 ff.
— tautomerism 433
1,2,4-Triazolyl-3-urea 430
Tribenzylamine 83
Tribonate 100
Tribromophenoxyacetic acids 158

Tributyltin acetate, properties and uses 299, 301
— acetate, synthesis 301
Tributyltin chloride 297, 301
Tributyltin fluoride 297
Tributyltin hydroxide, properties and uses 301
— hydroxide, stability 301
— hydroxide, synthesis 301
S,S,S-Tributyl trithiophosphate, see Butifos
Tributyl trithiophosphite, see Merphos
Trichlorfon, methylating ability 382
— permissible concentration in air 382
— properties and uses 310, 381—383
— stability 382
— syntheses 382, 383
Trichlormetafos-3, properties and uses 334, 335, 337
— synthesis 334
Trichloroacetic acid, properties and uses 120, 121, 228, 231
— acid, syntheses 121
Trichloroacetonitrile, properties and uses 128
— stabilization 128
S-2,3,3-Trichloroallyl N,N-diisopropylthiocarbamate, see Triallate
3,5,6-Trichloro-4-aminopicolinic acid, see Picloram
Trichloroanisole 102, 103
1,2,4-Trichlorobenzene 335
Trichlorobenzenes 45, 46, 48, 49, 67, 68, 244
2,4,5-Trichlorobenzenesulfonyl chloride, synthesis 244
2,3,6-Trichlorobenzoic acid 145—147
Trichlorobenzoic acids 146, 147, 176
2,3,6-Trichlorobenzoic acid, synthesis 146
4,5,7-Trichloro-2,1,3-benzothiadiazole, see PH 40-21
Trichlorobenzyl chloride, see Randox-T
Trichlorocampene 50
Trichlorodibromopropane 35
Trichlorodinaphthylethane 75
Trichlorodinitrobenzene, see Brassisan
Trichlorodithienylethane 75
Trichloroethanes, toxicity 35
2,3,5-Trichloro-4-hydroxypyridine, see Daxtron
3,5,6-Trichloro-2-hydroxypyridine, synthesis 339, 340
Trichlorometaphos, see Ronnel
Trichloromethyl chloroalkyl sulfides 246
N-Trichloromethylthio-4-chloroanilide of methanesulfonic acid, see Mesulfan

Trichloromethyl thiocyanate, properties and uses 253
Trichloromethylthioethers, synthesis 251
N-Trichloromethylthiophthalimide, see Folpet
Trichloromethylthiosulfonates 250
N-Trichloromethylthiotetrahydrophthalimide, see Captan
Trichloronate 385
2,4,5-Trichlorophenol, analytical method 104
— dissociation constant 309
— properties and uses 48, 49, 102 ff., 106, 332—334
— synthesis 102
Trichlorophenols 103, 166, 334
2,3,4-Trichlorophenoxyacetic acid 172
2,4,6-Trichlorophenoxyacetic acid 164
Trichlorophenoxyacetic acids 157 ff.
Trichlorophenoxybutyric acid, see 2,4,5-TB
2,4,5-Trichlorophenoxyethyl ester of dichloropropionic acid, see Erbon
Trichlorophenoxypropionic acid, see Silvex
2,3,6-Trichlorophenylacetic acid, see Fenac
O-2,4,5-Trichlorophenyl dichlorophosphite, synthesis 333, 334
O-2,4,5-Trichlorophenyl dichlorothiophosphate, synthesis 333
O-(Trichlorophenyl) diethylthiophosphinate, see Agvitor
Trichlorophenyl ethanol 315
Tri(p-chlorophenyl)tin acetate 298
Trichloropropanes 39
Trichloropropenes 35
Trichloropropionic acid, properties and uses 122
— acid, synthesis 122
Trichloropropionitrile 128
Trichlorothiophenol 48
2,3,6-Trichlorotoluene, synthesis and uses 146, 153
Trichlorphon 351
Tricyclohexyltin acetate 297, 298
N-Tridecyl-2,6-dimethylmorpholine, see E-236
Tridipam 213
Trietazine 437
Triethyltin chloride and other salts 298, 299
Trifluoromethylaniline, synthesis 233
N-(3-Trifluoromethylphenyl)-N',N'-dimethylurea, see Fluometuron
Trifluorophenoxyacetic acids as herbicides 158
Trifluralin, properties and uses 83, 84
— synthesis 84

Subject Index

Trilon B 175
Trimedlure 140
Trimethylphenyl N-methylcarbamate 184
Trimethyl thiophosphate, synthesis 348
Trimeturon 233
Triphenylamine 83
Triphenyltin acetate, see Brestan
Triphenyltin hydroxide 299, 302
Triphenyltin oxide 299
Tris(1-aziridinyl) phosphine oxide 318
Tris(1-aziridinyl) phosphine sulfide 318
Trisben-200 146
Tris(2,4-dichlorophenoxyethyl) phosphite, see Falone
Tris(dimethyldithiocarbamoyl) arsine, see Azomat
Tris(ethyleneimino)-s-triazines 439
Tris(methylethyleneimino)-s-triazines 439
Trithion, see Carbophenothion
Tritox, see Trichloroacetonitrile
N-Tritylmorpholine, see Frescon
Trolene, see Ronnel
Tumacide, see MTMC
Turnips 281
Turpentine 28, 30, 50
Tuzet, see Urbazid

UC-8305, properties and uses 336
UC-10854, properties and uses 193
— synthesis 193
Undecachlorocyclohexane 46
Undecylenic acid as herbicide 118
Unsulfonatable residue 29
Urab 231
Urbasulf, see Methylarsine sulfide
Urbazid, properties and uses 392, 393
— synthesis 392, 393
Urea derivatives, pesticidal properties 225 ff.
Ureas, mechanism of action 225 ff., 236
Uspulun 293

Vamidothion, metabolism 350
— oxidation products 350
— properties and uses 349, 350
— synthesis 350
Vancide-F 215
Vancide-M 215
Vapam, see Carbothion
V-C 13 Nemacide, see Nemacide
Vegadex, see CDEC
Vernam 209
Vitavax, properties and uses 423
— synthesis 423
Volatility of pesticides (see also specific compound) 8

Volatility vs. vaporizability 8
Volman's salt 389
Voronit, see Furidazol

Warble fly, losses from in U.S.S.R. 2
WARF antiresistant 275
Warfarin 401, 402
Waxes, insecticidal 12, 20
Weed control 27, 68, 83, 84, 89, 91 ff., 96 ff., 105, 108—110, 112, 113, 119 ff., 124 ff., 142 ff., 152 ff., 157 ff., 172 ff., 178 ff., 183 ff., 198 ff., 206 ff., 223, 225 ff., 252, 262, 266, 269, 275, 276, 283, 293, 299, 307, 308, 353, 354, 375, 380, 387, 390 ff., 396, 400, 403, 404, 407 ff., 416, 418 ff., 426 ff., 430 ff., 445, 447
Weeds, consumption of nutrients per hectare 2, 3
— losses from 2, 3
— water requirements 2
Wepsin 427
Wettable powders 12, 16 ff.
— powders, composition 18
Wetting agents and herbicidal action 175 ff.
Wheat 127, 155, 175, 184, 202, 210, 211, 438
— disinfection 19
— increased yields from pest control 1, 5
Wine 247
Winter oils 28
WL 9385 and WL 19805 439
Wofatox, see Methyl parathion
Wood preservation 30, 32, 288, 301, 388, 389, 443, 444
Wool preservation 1, 2, 238, 269—271, 275, 295, 301

3,5-Xylenyl N-methylcarbamate, see Maqbarl

Yalan, see Molinate
Yellow oils 98
Yield increases from pesticide use 5 ff.

Zectran, metabolism 195
— properties and uses 188, 195, 196
— synthesis 195
Zerlate, see Ziram
Zinc 2-aminoethylenedithiocarbamate 220
Zinc arsenate 388, 389
Zinc arsenite 388

Zinc dichlorosalicylanilide 149
Zinc dimethyldithiocarbamate, see Ziram
Zinc ethylenebis(dithiocarbamate),
  see Zineb
Zinc phosphide 445, 446
Zinc 1,2-propylenebis(dithiocarbamate),
  see Propineb
Zinc salicylanilide 148
Zinc trichlorophenolate, synthesis and uses
  103, 104
Zineb, decomposition 219
— persistence 219

Zineb, properties and uses 211, 212, 217,
  219—222
— synthesis 219, 220
Zinophos, see Thionazin
Ziram, properties and uses 18, 211, 213,
  214, 216, 393
— stability 214
— synthesis 214
Zoocides, definition 3
ZPS, see Dibutyl adipate
Zytron, properties and uses 321, 353
— synthesis 353